Phytochemicals: Extraction, Optimization, Identification, Biological Activities, and Applications in the Food, Nutraceutical, and Pharmaceutical Industries

Phytochemicals: Extraction, Optimization, Identification, Biological Activities, and Applications in the Food, Nutraceutical, and Pharmaceutical Industries

Guest Editor

Ibrahim M. Abu-Reidah

MDPI

Basel • Beijing • Wuhan • Barcelona • Belgrade • Novi Sad • Cluj • Manchester

Guest Editor
Ibrahim M. Abu-Reidah
School of Science and the Environment
Memorial University of Newfoundland
St. John's, NL
Canada

Editorial Office
MDPI AG
Grosspeteranlage 5
4052 Basel, Switzerland

This is a reprint of the Special Issue, published open access by the journal *Processes* (ISSN 2227-9717), freely accessible at: https://www.mdpi.com/si/processes/7H8I8KB73J.

For citation purposes, cite each article independently as indicated on the article page online and as indicated below:

Lastname, A.A.; Lastname, B.B. Article Title. *Journal Name* **Year**, *Volume Number*, Page Range.

ISBN 978-3-7258-6145-3 (Hbk)
ISBN 978-3-7258-6146-0 (PDF)
https://doi.org/10.3390/books978-3-7258-6146-0

Contents

About the Editor

Ibrahim M. Abu-Reidah

Ibrahim M. Abu-Reidah is a researcher specializing in analytical chemistry, phytochemistry, and mass spectrometry-based metabolomics. His work centers on the extraction, characterization, and biological evaluation of plant-derived bioactive compounds and their applications in food, nutraceutical, and pharmaceutical industries. He has conducted extensive research on phenolics, phytosterols, and secondary metabolites, developing advanced analytical methods and contributing widely to high-impact scientific literature. His academic career includes serving as a Professor, mentoring graduate students, and undertaking visiting research appointments in leading international laboratories. He has received multiple competitive research grants, supervised MSc and PhD projects, and participated as an external examiner. As a Guest Editor for several journals, he curates Special Issues in analytical chemistry and phytochemistry, while also serving as an active peer reviewer and evaluator for numerous international journals and research initiatives.

Preface

This Reprint highlights recent progress in phytochemical science, covering extraction, optimization, identification, and biological activities. It brings together research and reviews that clarify the health benefits, mechanisms, and industrial applications of plant-derived compounds across food, nutraceutical, and pharmaceutical sectors.

Ibrahim M. Abu-Reidah
Guest Editor

processes

MDPI

Editorial

Special Issue on "Phytochemicals: Extraction, Optimization, Identification, Biological Activities, and Applications in the Food, Nutraceutical, and Pharmaceutical Industries"

Ibrahim M. Abu-Reidah

CREAIT, Core Science Facility, Memorial University of Newfoundland, St. John's, NL A1B 3X5, Canada; iabureidah@mun.ca

There is growing interest in using natural plant extracts in foods and beverages due to their ability to enhance the quality of food and provide therapeutic benefits [1,2]. Phytochemicals (organic compounds derived from plants) offer significant health-promoting properties beyond their fiber, vitamin, and mineral content [3]. These compounds, known for their antioxidant, antiviral, anticancer, and anti-inflammatory effects, play a vital role in disease prevention and health maintenance [4]. Owing to their diverse properties, they have found many applications in medicine, agriculture, and industry, particularly in the food, pharmaceutical, and nutraceutical sectors [5].

This Special Issue, "Phytochemicals: Extraction, Optimization, Identification, Biological Activities, and Applications in the Food, Nutraceutical, and Pharmaceutical Industries", focuses on recent advancements in the extraction of phytochemicals and explores their chemical compositions and biological activities. It highlights new analytical and bioanalytical techniques for studying these compounds and their mechanisms of action. It features twelve research papers and three reviews, covering topics such as novel extraction methods, phytochemical bioactivities, and the design of new functional foods and nutraceuticals. These studies underscore the increasing demand for natural bioactive compounds and eco-friendly processing methods, rendering this Special Issue a significant resource for professionals and researchers.

The included studies demonstrate a range of promising applications. For example, Thinh et al. [6] studied fucosylated chondroitin sulfate from *Bohadschia ocellata* and revealed its strong anticoagulant, anticancer, and PTP1B inhibitory activities, suggestive of its therapeutic potential. Moreover, Ramírez-Sucre et al. [7] developed an environmentally friendly method for extracting polyphenols from citrus peel using NADESs and ultrasound, emphasizing its potential for producing a high yield of functional foods and nutraceuticals. By combining green chemistry principles with innovative techniques, the study offers a sustainable approach to maximizing the recovery of bioactive compounds and reducing environmental impact in the food and nutraceutical industries.

A group of researchers optimized the extraction of essential oil from *Lippia graveolens* using ultrasound-assisted deep eutectic solvents, highlighting its potential as an eco-friendly antifungal agent for agriculture. Raucher et al. explored the antiproliferative effects of fruit juices on glioblastoma, finding cornelian cherry and wild blackberry juices more effective than standard chemotherapeutics, substantiating the therapeutic potential of dietary bioactives in cancer treatment. In addition, Carezzano et al. [8] evaluated essential oils and plant extracts as biopesticides against the soybean mosaic virus, offering sustainable alternatives to chemical pesticides. Their research supports environmentally friendly agricultural practices and highlights the potential use of botanical extracts in integrated pest management for safer, more sustainable crop protection [9].

Other studies in this Special Issue investigate the optimization of extraction methodologies and the pharmacological relevance of phytochemicals. Hihat et al. [10] explored the microwave-assisted extraction of total phenolic content from Coriandrum sativum leaves, presenting a model for optimizing antioxidant yield. This study demonstrates how process optimization can significantly enhance the efficiency of bioactive compound extraction, making it more viable for large-scale applications. Another significant study on *Achyranthes bidentata* aimed to enhance traditional salt-processing techniques to boost its bioactive properties and anti-osteoarthritis effects. This study combined traditional medicine with modern science, validating the use of ancient remedies through contemporary techniques and highlighting their relevance in current healthcare practices.

Jamous et al. [11] developed solid lipid nanoparticles for troxerutin to improve its bioavailability and controlled release, thus enhancing its therapeutic potential. This study demonstrated the vital role of nanotechnology in overcoming challenges in phytochemical delivery, ensuring stability and efficacy and expanding pharmaceutical applications. Other researchers have investigated the antimicrobial activity and bioactive components of *Chamaecyparis obtusa* essential oil. Key compounds, thujopsene and pinene, were identified through GC-MS, and commercial disinfectants containing the oil effectively reduced airborne microorganisms, highlighting its potential as a natural, eco-friendly indoor air disinfectant.

In a study conducted by Manjarrez-Quintero et al. [12], the extraction of essential oil from *Lippia graveolens* biomass was optimized using an ultrasound-assisted deep eutectic solvent method, enhancing its yield and bioactivity. This study underscored its potential as a natural antifungal agent for agricultural use, providing an eco-friendly alternative to synthetic pesticides. Additionally, Raucher and others [13] explored the antiproliferative effects of fruit juices against glioblastoma, revealing the potent effects of cornelian cherry and wild blackberry juice, which exhibited greater efficacy than standard chemotherapeutic agents in glioblastoma cell models. Their findings reinforce the therapeutic value of dietary bioactives and suggest a promising avenue for complementary cancer therapies.

Moreover, Belaiba et al. [14] studied *Ammoides pusilla*, identifying 20 key compounds, including perilic aldehyde and β-phellandrene. Extracts of the plant showed strong antioxidant, anti-diabetic, anti-inflammatory, and anticancer activities and demonstrated significant cytotoxicity against cancer cells, highlighting its therapeutic potential. This study underscores the potential of *A. pusilla* in facilitating drug discovery. The authors of another study highlighted the bioactive compounds present in *Citrus aurantium*, including flavonoids, essential oils, and vitamin C, with antioxidant, antimicrobial, and anti-inflammatory properties. They reviewed eco-friendly extraction methods, such as ultrasound-assisted techniques, to enhance the recovery of synephrine, emphasizing the plant's promising applications in food, pharmaceuticals, and cosmetics and promoting sustainable extraction technologies.

Grabska-Zielińska et al. [15] explored polymer films enhanced with olive leaf extract (OLE) for food packaging and other applications. OLE's antibacterial, antifungal, and antioxidant properties improved the films' functionality and structural integrity, emphasizing its potential application in sustainable food packaging, biomedicine, cosmetics, and future industrial research. Another study assessed the use of *Cnidoscolus quercifolius*, a versatile plant valued for its ecological and medicinal roles in Brazil's Caatinga biome. It exhibits antinociceptive, antioxidant, and anti-inflammatory properties, with applications in treating inflammation and infections. Its seed oil also shows promise for food consumption. The authors called for further research into its bioactive compounds and pharmacological mechanisms to support drug development.

Collectively, the papers in this Special Issue underscore the vast potential of phytochemicals across various applications, from medicine and agriculture to food science and biotechnology. The studies presented enhance our understanding of plant bioactives and their multifunctional roles and stress the need for environmentally friendly extraction and optimization processes. By integrating scientific advancements with practical applications, these studies pave the way for innovative solutions that address critical global challenges, from disease prevention to sustainable resource utilization.

We extend our thanks to the authors, reviewers, and editorial team for their important contributions to this Special Issue. This collection will serve as a significant resource for researchers, practitioners, and policymakers, inspiring further advancements and fostering innovation in phytochemistry and related fields. The continued exploration of phytochemicals will undoubtedly lead to new discoveries that benefit both human health and the environment, reinforcing the indispensable role of natural compounds in scientific progress.

Conflicts of Interest: The author declares no conflict of interest.

References

1. Chaachouay, N.; Zidane, L. Plant-Derived Natural Products: A Source for Drug Discovery and Development. *Drugs Drug Candidates* **2024**, *3*, 184–207. [CrossRef]
2. Tsoupras, A.; Moran, D.; Shiels, K.; Saha, S.K.; Abu-Reidah, I.M.; Thomas, R.H.; Redfern, S. Enrichment of Whole-Grain Breads with Food-Grade Extracted Apple Pomace Bioactives Enhanced Their Anti-Inflammatory, Antithrombotic and Anti-Oxidant Functional Properties. *Antioxidants* **2024**, *13*, 225. [CrossRef] [PubMed] [PubMed Central]
3. Amessis-Ouchemoukh, N.; Abu-Reidah, I.M.; Quirantes-Piné, R.; Madani, K.; Segura-Carretero, A. Phytochemical profiling, in vitro evaluation of total phenolic contents and antioxidant properties of *Marrubium vulgare* (horehound) leaves of plants growing in Algeria. *Ind. Crops Prod.* **2014**, *61*, 120–129. [CrossRef]
4. Abu-Reidah, I.M.; Taamalli, A. Promising Phytoconstituents in Antiangiogenesis Drug Development. *Nutraceuticals* **2024**, *4*, 450–468. [CrossRef]
5. Reguengo, L.M.; Salgaço, M.K.; Sivieri, K.; Maróstica Júnior, M.R. Agro-industrial by-products: Valuable sources of bioactive compounds. *Food Res. Int.* **2022**, *152*, 110871. [CrossRef]
6. Thinh, P.D.; Cao, H.T.T.; Trung, D.T.; Minh, D.K.; Cao, T.Q.; Van, T.T.T.; Zueva, A.O.; Ermakova, S.P.; Nguyen, T.-D. Fucosylated Chondroitin Sulfate from Bohadschia ocellata: Structure Analysis and Bioactivities. *Processes* **2024**, *12*, 2108. [CrossRef]
7. Ramírez-Sucre, M.O.; Avilés-Betanzos, K.A.; López-Martínez, A.; Rodríguez-Buenfil, I.M. Evaluation of Polyphenol Profile from Citrus Peel Obtained by Natural Deep Eutectic Solvent/Ultrasound Extraction. *Processes* **2024**, *12*, 2072. [CrossRef]
8. Carezzano, M.E.; Reyna, P.G.; Accotto, E.; Giordano, W.; Oliva, M.D.L.M.; Rodriguez Pardina, P.; Sabini, M.C. Plant-Derived Essential Oils and Aqueous Extract as Potential Ingredients for a Biopesticide: Phytotoxicity in Soybean and Activity against Soybean Mosaic Virus. *Processes* **2023**, *11*, 2265. [CrossRef]
9. Gupta, I.; Singh, R.; Muthusamy, S.; Sharma, M.; Grewal, K.; Singh, H.P.; Batish, D.R. Plant Essential Oils as Biopesticides: Applications, Mechanisms, Innovations, and Constraints. *Plants* **2023**, *12*, 2916. [CrossRef] [PubMed] [PubMed Central]
10. Hihat, S.; Touati, N.; Sellal, A.; Madani, K. Response Surface Methodology: An Optimal Design for Maximising the Efficiency of Microwave-Assisted Extraction of Total Phenolic Compounds from *Coriandrum sativum* Leaves. *Processes* **2024**, *12*, 1031. [CrossRef]
11. Jamous, Y.F.; Altwaijry, N.A.; Saleem, M.T.S.; Alrayes, A.F.; Albishi, S.M.; Almeshari, M.A. Formulation and Characterization of Solid Lipid Nanoparticles Loaded with Troxerutin. *Processes* **2023**, *11*, 3039. [CrossRef]
12. Mendoza-Osorno, A.E.; Avilés-Betanzos, K.A.; Uc-Varguez, A.; Carballo-Castañeda, R.; Moreno-Ulloa, A.; Ramírez-Sucre, M.O.; Rodríguez-Buenfil, I.M. Metabolomic Profiling (LC–MS^2) of Flowers and Bee Honey of Dzidzilche (*Gymnopodium floribundum* Rolfe) and Jabin (*Piscidia piscipula* L. Sarg.) from Yucatán, México. *Processes* **2023**, *11*, 3028. [CrossRef]
13. Raucher, D.; Rowsey, M.; Hinson, J.; Ćorković, I.; Lila, M.A.; Šimunović, J.; Kopjar, M. Bioactive Compounds, Antioxidant Activity, and Antiproliferative Potential on Glioblastoma Cells of Selected Stone Fruit Juices. *Processes* **2024**, *12*, 1310. [CrossRef]

14. Belaiba, M.; Saoudi, M.M.; Abedrabba, M.; Bouajila, J. *Ammoides pusilla* Aerial Part: GC-MS Profiling and Evaluation of In Vitro Antioxidant and Biological Activities. *Processes* **2024**, *12*, 1274. [CrossRef]
15. Grabska-Zielińska, S. Active Polymer Films with Olive Leaf Extract: Potential for Food Packaging, Biomedical, and Cosmetic Applications. *Processes* **2024**, *12*, 2329. [CrossRef]

processes MDPI

Article

Pham Duc Thinh, Hang Thi Thuy Cao, Dinh Thanh Trung, Duong Khanh Minh, Thao Quyen Cao, Tran Thi Thanh Van, et al.

Pham Duc Thinh [1,2,*], Hang Thi Thuy Cao [1], Dinh Thanh Trung [1], Duong Khanh Minh [2], Thao Quyen Cao [3], Tran Thi Thanh Van [1], Anastasia O. Zueva [4], Svetlana P. Ermakova [4] and Thanh-Danh Nguyen [5]

1. Nhatrang Institute of Technology Research and Application, Vietnam Academy of Science and Technology, 02 Hung Vuong, Nha Trang 650000, Vietnam; caohang.nitra@gmail.com (H.T.T.C.); dinhthanhtrung410@gmail.com (D.T.T.); tranthanhvan@nitra.vast.vn (T.T.T.V.)
2. Department of Chemistry, Graduate University of Science and Technology, Vietnam Academy of Science and Technology, 18 Hoang Quoc Viet, Ha Noi 11300, Vietnam; khanhminhchemistry@gmail.com
3. Institute of Agricultural Science and Technology, Kyungpook National University, Daegu 41566, Republic of Korea; quyen.cao.thao@gmail.com
4. G.B. Elyakov Pacific Institute of Bioorganic Chemistry, Far Eastern Branch of the Russian Academy of Sciences, 690022 Vladivostok, Russia; a.o.zueva@yandex.ru (A.O.Z.); swetlana_e@mail.ru (S.P.E.)
5. Institute of Chemical Technology, Vietnam Academy of Science and Technology, 1A, TL29, Thanh Loc Ward, District 12, Ho Chi Minh City 70000, Vietnam; danh5463bd@yahoo.com

* Correspondence: ducthinh.nitra@gmail.com

Abstract: Fucosylated chondroitin sulfate (FCS) was prepared from *Bohadschia ocellata* using protease hydrolysis. The structural characteristics of FCS were confirmed through chemical composition analysis using FTIR spectroscopy, ^{1}H NMR, and ^{13}C NMR. FCS from *B. ocellata* (FCS-Bo) exhibited an average molecular weight of approximately 122 kDa. The biological activities of FCS-Bo, including anticoagulant, anti-cancer, and Protein Tyrosine Phosphatase 1B (PTP1B) inhibition, were evaluated. FCS-Bo displayed potent anticoagulant properties, markedly extending activated partial thromboplastin time, prothrombin time, and thrombin time when compared to the heparin control. In anti-cancer bioactivity research, FCS-Bo efficiently inhibited colony formation in the colon cancer cell lines HCT-116, HT-29, and DLD-1, achieving inhibition rates of up to 65%. Additionally, FCS-Bo exhibited significant inhibition of PTP1B, with an IC_{50} as low as 0.0326 μg/mL, suggesting its potential for improving insulin sensitivity and managing conditions such as type 2 diabetes and obesity.

Keywords: *Bohadschia ocellata*; fucosylated chondroitin sulfate; anticoagulant activity; anti-cancer activity; protein tyrosine phosphatase 1B

1. Introduction

Sea cucumbers, invertebrates from the phylum Echinodermata and class Holothuroidea, are commonly found in benthic regions and deep-sea environments worldwide. The bioactive components of sea cucumbers have attracted significant interest due to their diverse biological activities, including anticancer, antiviral, antioxidant, and wound healing properties [1]. Recently, increasing attention has been directed toward the isolation, characterization, and study of the biological properties of sulfated polysaccharides derived from sea cucumbers [2–4]. These compounds predominantly exist in two structural forms: Fucosylated chondroitin sulfate (FCSs) and fucan sulfate (FSs).

Globally, the market for sea cucumber-derived products has expanded significantly, driven by both their medicinal properties and increasing consumer demand for nutraceuticals. The global sea cucumber market was valued at USD 540 million in 2020 and is expected to grow at a compound annual growth rate (CAGR) of 6.3% through 2027, reaching an estimated USD 850 million [5]. This growth is fueled by increased research and commercial interest in bioactive compounds such as FCS, which are being explored for their therapeutic potential in areas such as anticoagulation therapy and cancer treatment. With rising consumer awareness of natural health products and sustainable marine resources, the FCS-based pharmaceuticals could promote the industry for substantial future growth.

FCS is characterized by nearly equimolar proportions of *α-L*-fucose, *β-D*-glucuronic acid, and *N*-acetylgalactosamine residues, while FS is characterized by a backbone made up of fucose units linked through α (1 → 3) bonds or a combination of alternating α (1 → 3) and α (1 → 4) linkages [1]. FCS has attracted considerable interest due to its strong anticoagulant and antithrombotic properties. Their anticoagulant effect of FCS is intricately linked to its structural characteristics, particularly the sulfation patterns on both its fucose side chains and chondroitin backbone. These sulfation patterns, along with molecular weight (MW), play critical roles in modulating its biological activity. For instance, FCS from *Holothuria nobilis* demonstrates superior anticoagulant activity compared to that from *Acaudina molpadioidea* due to differences in sulfation patterns and sulfate content. Additionally, 2,4-O-disulfation of fucose residues has been identified as a key factor enhancing the anticoagulant efficacy of FCS, although variations in MW can also significantly influence its biological functions [6]. Moreover, FCSs derived from species such as *Stichopus herrmanni* [7], *Isostichopus badionotus* [8], and *Holothuria nobilis* [9] have shown strong inhibitory effects on thrombin and intrinsic factor Xase.

FCS extracted from sea cucumbers has also demonstrated anticancer activities. For example, low-molecular-weight FCS extracted from *Cucumaria frondosa* has demonstrated the ability to suppress both the growth and metastasis of Lewis lung carcinoma in mouse models [10]. FCS derived from *Isostichopus badionotus* [11] and *Hemioedema spectabilis* [12] has shown potent anticancer effects, such as inhibiting cancer cell migration and preventing cancer cells from adhering to platelet-coated surfaces.

In addition, FCS has attracted attention for its anti-diabetic effects. Many studies demonstrated that FCS can enhance glucose metabolism, improve insulin sensitivity, and attenuate hyperglycemia, such as FCS derived from *Stichopus japonicus* [13] and *Cucumaria frondosa* [14]. Thus, FCS exhibits significant therapeutic potential with its potent anticoagulant, anticancer, and antidiabetic activities, largely influenced by its structural characteristics. These findings highlight FCS as a promising candidate for further research and development in health and disease management.

Vietnam possesses approximately 70 species of sea cucumbers [15], but the commercial value of these species varies greatly, with some commanding prices from a few to several hundred U.S. dollars per kilogram. Despite growing commercial interest, research on the chemical composition and bioactive compounds of Vietnamese sea cucumbers is still scarce. The polysaccharides extracted from species such as *Stichopus variegatus*, *Bohadschia argus*, *Holothuria (Theelothuria) spinifera*, and *Holothuria (Stauropora) fuscocinerea* have been studied, and their structural characteristics have been reported in several previous studies [2,3,16]. Polysaccharides, including FCSs and FSs, from their species have been extracted and purified by anion exchange chromatography. These polysaccharides exhibit significant biological activities, particularly the anticoagulant properties of FCS and some FS. Structurally, FCS from *B. argus* and FCS from *H. spinifera* have a chondroitin backbone with sulfated fucosyl branches, though their fucosyl and GalNAc sulfation patterns differ. FS from *B. argus* is a linear polymer of *α-L*-fucopyranose 3-sulfate, while FS from *H. spinifera* contains a mixture of linear and branched sulfated fucan chains [2]. In *H. fuscocinerea*, FCS consists of alternating trisaccharide units with sulfated fucosyl branches, and FSs are linear and branched sulfate fucans, respectively [16]. The sulfated fucan SvF3 from *S. variegatus* is composed of *α-L*-fucopyranose residues with varying sulfation patterns [3]. Functionally, FCS from *B. argus* and *H. spinifera* show strong anticoagulant activity, while FS has minimal anticoagulant effects [16]. SvF3 also inhibited breast cancer cell migration and colony formation in vitro [3]. This highlights the untapped potential of Vietnam's sea cucumber resources for developing high-value bioactive compounds.

In this study, we extracted FCS from sea cucumber *Bohadschia ocellata* in Vietnamese sea. The FCS was investigated for structural characteristics and evaluated for its biological activities, including anticoagulant, anticancer, and antidiabetic activities. The results expanded our understanding of the structural diversity and biological activities of FCSs in Vietnamese sea cucumbers and provided potential applications in medicine and biotechnology.

2. Materials and Methods

2.1. Materials

Wild Sea cucumbers, *B. ocellata*, with lengths ranging from 20 to 30 cm and weights ranging from 300 to 500 g, were collected along the coastline of Van Gia, Khanh Hoa, Vietnam, in April 2020, as illustrated in Figure S1. All chemicals used for the extraction and chemical composition analysis of PS were analytical grade, purchased from Sigma (St. Louis, MO, USA) and Merck (Darmstadt, Germany). To evaluate cell cytotoxicity and soft agar assays, various materials were used, including phosphate buffered saline (PBS), *L*-glutamine, and penicillin-streptomycin solution, all sourced from Sigma-Aldrich (St. Louis, MO, USA). Other essential materials, such as Basal Medium Eagle, Dulbecco's Modified Eagle's Medium, Minimum Essential Medium, trypsin, and fetal bovine serum, were obtained from Thermo Fisher Scientific (Waltham, MA, USA), and 3-[4,5-dimethylthiazol-2-yl]-2,5-diphenyltetrazolium bromide was purchased from Promega Corporation (Madison, WI, USA).

2.2. Isolation of Polysaccharides

Polysaccharides from sea cucumber were extracted using an enzymatic hydrolysis method [3]. Polysaccharides in Sea cucumber cell walls were extracted in 0.1 M acetate buffer (pH 6) with a material-to-solvent ratio of 1:30 (g). The extraction buffer was supplemented with papain (2000 U/g) at 10% (*w*/*w*, based on dry weight), 5 mM EDTA, and 5 mM L-cysteine. The extraction process was carried out at 55 °C for 24 h. After extraction, the mixture was filtered and centrifuged at 6000 rpm, and the supernatant was boiled for 10–15 min to inactivate the enzyme. Subsequently, 10% Cetavlon (hexadecyltrimethylammonium bromide) solution was added to the extract to precipitate the polysaccharide at 4 °C for 24 h, followed by centrifugation at 10,000 rpm for 30 min to collect the precipitate. The Cetavlon-Polysaccharide complex was dissolved in a 3 M NaCl and 20% ethanol solution, and 98% ethanol was added to the solution until a final concentration of 80% was reached, further precipitating the sulfated polysaccharide (PS). After incubation at 4 °C for 24 h, the solution was centrifuged at 10,000 rpm for 30 min, and the precipitate was re-dissolved in distilled water. The solution was then dialyzed to remove salts and freeze-dried to obtain the PS.

The next step in the purification process involved separation using a Macro-Prep DEAE (Bio-Rad, Hercules, CA, USA) ion exchange column (2.5 × 15 cm). The column was rinsed with 50 mL of water, followed by elution with a gradient of NaCl solutions ranging from 0 to 2 M. The total gradient volume was 300 mL, with fractions collected at 3.75 mL intervals and subsequently analyzed using the phenol-sulfuric acid method.

2.3. General Methods

The total carbohydrate content was quantified by the phenol-sulfuric acid method [17]. A 50 µL aliquot of the sample, containing carbohydrates in the concentration range of 25–50 µg/mL, was transferred into a 10 mL test tube. Afterward, 50 µL of 5% phenol was added and thoroughly mixed. This was followed by the addition of 1 mL of concentrated H_2SO_4. The mixture was then heated in a water bath for 10–15 min, allowed to cool, and the absorbance was measured at 490 nm, using D-glucose as the reference standard.

The sulfate content was determined using the turbidimetric method with $BaCl_2$/gelatin after hydrolyzing the sample in 1N HCl at 100 °C for 6 h [18]. A 20 µL aliquot of the sample was transferred into a 96-well plate, followed by the addition of 150 μL of 4% trichloroacetic acid solution. Subsequently, 100 μL of 5% $BaCl_2$/gelatin solution was added, and the mixture was thoroughly mixed and allowed to stand for 10–15 min. The turbidity was then measured at 360 nm, with K_2SO_4 serving as the standard.

The uronic acid content was measured using the carbazole method with D-glucuronic acid as the standard [19]: The sample solution prepared at a concentration of 5 mg/mL was used for the quantification. First, 250 μL of the sample solution was placed in a glass tube, and 0.9% sodium borohydride in concentrated sulfuric acid was added. The mixture was heated at 100 °C for 10 min. Afterward, 50 μL of a 0.1% carbazole solution in absolute ethanol was added, and the mixture was incubated again at 100 °C for 15 min. Finally, the absorbance was recorded at 525 nm, and the uronic acid content was calculated based on a standard curve prepared from *D*-glucuronic acid.

Protein content was measured using bovine serum albumin as the standard [20]: Solution A was prepared by dissolving 0.1 M NaOH and 2% Na_2CO_3, while Solution B consisted of 0.5% $CuSO_4{\cdot}5H_2O$ dissolved in a 1% sodium-potassium tartrate solution. Solution C, a mixture of Solution A and Solution B in a 50:1 ratio, was freshly prepared just before use. The reaction began by adding 0.5 mL of the sample to a test tube, followed by 2.5 mL of Solution C. The mixture was thoroughly vortexed and allowed to stand for 20 min. Subsequently, 0.25 mL of 1N Folin reagent was added, and the tubes were vortexed again. The reaction was incubated for 60 min, after which the absorbance was measured at 750 nm. Protein concentration was calculated using a standard curve generated with an albumin standard (Thermo Scientific, Waltham, MA, USA).

Monosaccharide composition was analyzed as previously reported [6]. Polysaccharides were hydrolyzed into monomers using trifluoroacetic acid (TFA, 2N) prior to analysis by high-performance liquid chromatography (HPLC) on a Dionex CarboPac MA1 column with an electrochemical detector (ICS-6000, Thermo Fisher Scientific). The mobile phase consisted of 1 M NaOH, with a flow rate of 0.4 mL/min. The injection volume was 500 μL, and the run time was 30 min for standard solutions and 60 min for sample analysis. The monosaccharides, including *D*-galactose (Gal), *L*-fucose (Fuc), *N*-acetyl-galactosamine (GalNAc), Gluconic acid (Glu), and *N*-acetyl-glucosamine (GluNAc), were used to construct calibration curves. The monosaccharide composition of the polysaccharide fractions was expressed as the molar percentage based on the reference monosaccharide standards.

The infrared spectral analysis of FCS-Bo was performed on a Fourier transform infrared (FT-IR) spectrophotometer (Shimadzu Afinity-1S instrument equipped with a QATR-detector) in the wavenumber range of 4000–400 cm^{-1}. Data were recorded in LabSolutions IR software (Version 2.27).

For NMR analysis, the sample (10 mg) was dissolved in 99.9% D_2O and dried via vacuum evaporation at 40 °C. It was then re-dissolved in 99.96% D_2O and placed into an NMR tube. ^1H and ^{13}C NMR spectra were recorded using a Bruker Avance III HD 500 spectrometer at 333 K, with HOD suppression achieved through pre-saturation. COSY spectra were obtained at 333 K using standard pulse sequences.

The MWs of polysaccharides were determined by high-performance size-exclusion chromatography (HPSEC) using the HPLC instrument Shimadzu LC-20 Series (Shimadzu, Kyoto, Japan) equipped with an LC-20AD pump, degassing unit DGU-20A5R, autosampler SIL-20AHT, column oven CTO-20A, and refractive index de-tector RID-20A on the GPC column PSS SUPREMA combination ultrahigh (3 columns, dimensions of 8 mm × 300 mm, particle size of 10 μm, PSS, Mainz, Germany). Elution was performed with Lithium nitrate (0.1 M) at 40 °C with a flow rate of 1.0 mL/min. Different dextran standards (Sigma, Cibolo, TX, USA) of 1, 5, 12, 25, 50, 150, 270, and 670 kDa were used as reference standards.

2.4. Bioactivity

2.4.1. Determination of Coagulation Parameters

The anticoagulant activity is achieved through the following methods [21]: Blood samples were collected from albino rabbits or Wistar rats and immediately transferred into clean tubes containing sodium citrate (3.8%) to prevent clotting. The samples were centrifuged at 2000× g for 15 min to obtain platelet-poor plasma (PPP). Subsequently, 0.1 mL of PPP was added to tubes containing different concentrations of the test compounds or the control. The mixture was incubated at 37 °C for 3 min prior to the addition of pre-warmed thromboplastin reagent (Diagnostica Stago, Asnières sur Seine Cedex, France) for Prothrombin Time measurement. The clotting time was determined using a Sta-Evolution coagulation analyzer (Stago, Asnières sur Seine Cedex, France) according to the manufacturer's instructions.

For activated Partial Thromboplastin Time (aPTT) assays, 100 µL of the plasma sample (either test or control) was mixed with 100 µL of aPTT reagent containing ellagic acid and phospholipids. The mixture was incubated at 37 °C for 3 min, followed by the addition of 100 µL of pre-warmed 0.025 M $CaCl_2$ solution. The aPTT was automatically measured using the Sta-Evolution analyzer as the standard protocol provided by the manufacturer.

2.4.2. Cell Cytotoxicity and Soft Agar Assays

The cytotoxic activity of FCS_Bo and inhibition of colony formation of cancer cells were assessed using the methods described by Usoltseva et al. [22]. In this study, the cancer cell lines used were HT-29 (ATCC # HTB-38™) and HCT-116 (ATCC # CCL-247™). For the cytotoxicity assay, cells were seeded in 96-well plates, treated with varying concentrations of polysaccharides (100, 200, and 400 µg/mL), and incubated for 24 h. After adding MTS reagent, followed by a 4-h incubation at 37 °C, absorbance 490/630 was measured to assess cell viability. In the soft agar assay, cells were treated with polysaccharides (200 µg/mL) and cultured in soft agar (0.3% BME agar solution containing 10% FBS, 2 mM *L*-glutamine, and 25 µg/mL gentamicin) for 14 days at 37 °C with 5% CO_2 to evaluate colony formation using a microscope, and colonies were quantified with ImageJ 1.45 software. Statistical analysis was conducted using Student's t-test, with significance levels set at * $p < 0.05$, ** $p < 0.01$, and *** $p < 0.001$. This method allows for a standardized evaluation of the compound's effects on cell viability and provides a reliable framework for comparing its cytotoxic potential against these specific cancer cell lines.

2.4.3. Protein Tyrosine Phosphatase 1B (PTP1B) Assay

The enzyme activity of PTP1B was measured using 2 mM *p*-nitrophenyl phosphate (pNPP) as a substrate in a 50 mM citrate buffer (pH 6.0), which also contained 0.1 M NaCl, 1 mM EDTA, and 1 mM dithiothreitol (DTT). The reaction was conducted at 37 °C for 30 min, and then 10 N NaOH was added to stop the reaction. The absorbance of the resulting *p*-nitrophenol was measured at 405 nm to quantify the enzyme activity. The nonenzymatic hydrolysis of pNPP was accounted for by measuring absorbance in the absence of PTP1B. Ursolic acid was employed as a positive control [23], and IC_{50} values were calculated using nonlinear regression analysis with Microsoft Excel 365.

3. Results

3.1. Isolation and Fractionation of Sulfated Polysaccharide

Sulfated polysaccharide extracts were prepared from the body walls of *B. ocellata* sea cucumbers through conventional solubilization of the biomass using papain. The extract was treated with hexadecyltrimethylammonium bromide to precipitate the sulfated components, which were then transformed into water-soluble sodium salts by dissolving in 2 M NaCl and precipitation with ethanol to give rise to crude sulfated polysaccharides. The extraction yield of polysaccharides is 1.35%. As a result of fractionation of polysaccharide with a linear gradient of NaCl on an anion exchange carrier Macro-Prep DEAE, fractions of sulfated polysaccharides were isolated from the sea cucumber F1, F2, F3, and F4 (Figure S2) with yields of 0.82, 16.29, 4.93, and 4.24%, respectively (Table 1). The main monosaccharide residue of the F1–F4 fractions is obtained by anion exchange chromatography. The polysaccharide fractions of F1 and F2 possess monosaccharide components including Fuc, GalNAc, Gal, and Glu, while the F3 and F4 fractions—co Fuc, GalNAc, and Gal. Among them, fraction F2 has the highest extraction yield with the highest GalNAc content and an average molecular weight of 122 kDa (Figure S3).

Table 1. The extraction yield and chemical composition of polysaccharides from *B. ocellata*.

Samples	Yield (%) *	Total Carbohydrate (%) *	Monosaccharide Compositions (% mol)				Sulfate (% *w*/*w*) *	Uronic Acid (% *w*/*w*) *	Protein (% *w*/*w*) *
			Fuc	GalNAc	Gal	Glu			
F1	0.82	39.62 ± 2.55	1	0.29	0.92	0.39	10.99 ± 0.37	ND	ND
F2 (FCS-Bo)	16.29	36.24 ± 1.73	1	0.54	0.44	0.22	20.03 ± 0.65	12.47 ± 0.12	5.00 ± 0.2
F3	4.93	38.33 ± 2.53	1	0.06	0.16	-	21.06 ± 1.50	3.07 ± 0.12	1.82 ± 0.23
F4	4.24	53.80 ± 1.23	1	0.15	0.11	-	19.25 ± 2.45	2.47 ± 0.12	0.37 ± 0.01

Notes: *—by sample weight, ND—not identified.

The molar ratio of Glu to GalNAc in the F2 fraction is 1:0.54, indicating the presence of these two core building blocks of the chondroitin sulfate backbone. Furthermore, the molar ratio of Fuc to Glu to GalNAc is approximately 1:0.54:0.44, illustrating the fucosylation pattern of this compound. Therefore, fraction F2 can be preliminary identified as FCS-Bo. The structural characteristics of FCS-Bo were determined using FTIR spectroscopy and ^{1}H and ^{13}C NMR spectra (Figures 1–3).

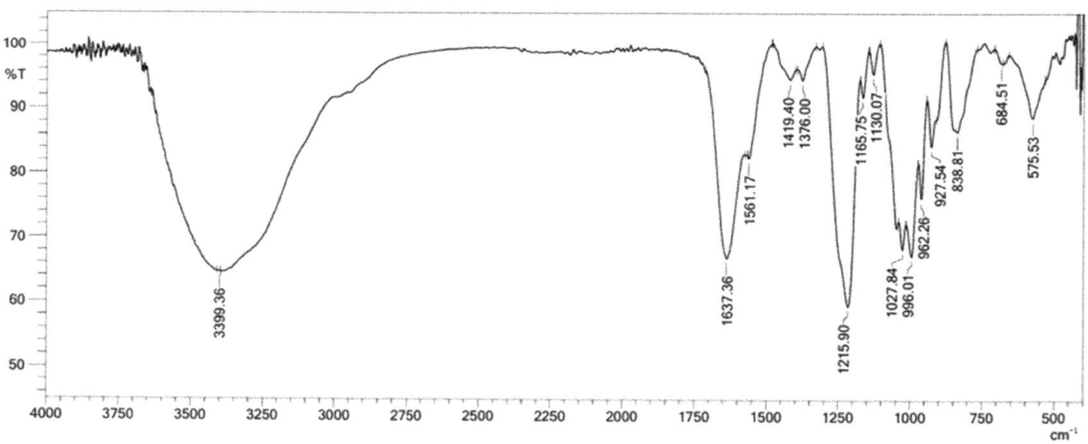

Figure 1. The IR spectra of fucosylated chondroitin sulfates FCS-Bo.

The FTIR spectrum of the FCS-Bo extract (Figure 1) exhibits characteristic absorption bands associated with a polysaccharide structure. The broad absorption band at 3390.36 cm^{-1} is attributed to O-H stretching vibrations, signifying the presence of hydroxyl groups within the polysaccharide. The strong absorption at 1637.36 cm^{-1} is attributed to the C=O stretching vibrations of uronic acid, a key component of chondroitin sulfate, while the band at 1561.17 cm^{-1} is associated with N-H bending vibrations, likely from the amide groups in GluNAc. The prominent band at 1215.90 cm^{-1} confirms the presence of sulfate groups through S=O stretching vibrations, a hallmark of sulfated polysaccharides. Additionally, the absorption bands at 1165.75 cm^{-1} and 1130.07 cm^{-1} are related to C-O-C glycosidic bond stretching, further substantiating the polysaccharide backbone. The absorption at 1027.04 cm^{-1}, attributed to C-O stretching, underscores the carbohydrate nature of FCS. The presence of fucose residues is suggested by the bands at 930.26 cm^{-1} and

888.81 cm^{-1}, which correspond to C-H deformation and β-glycosidic linkages, respectively. Lastly, the lower frequency bands at 684.51 cm^{-1} and 575.53 cm^{-1} may reflect the skeletal vibrations of the fucose residues and complex ring structures within the FCS-Bo. These results confirmed the presence of a saulfated and fucosylated nature of the chondroitin sulfate extract from sea cucumbers [24,25].

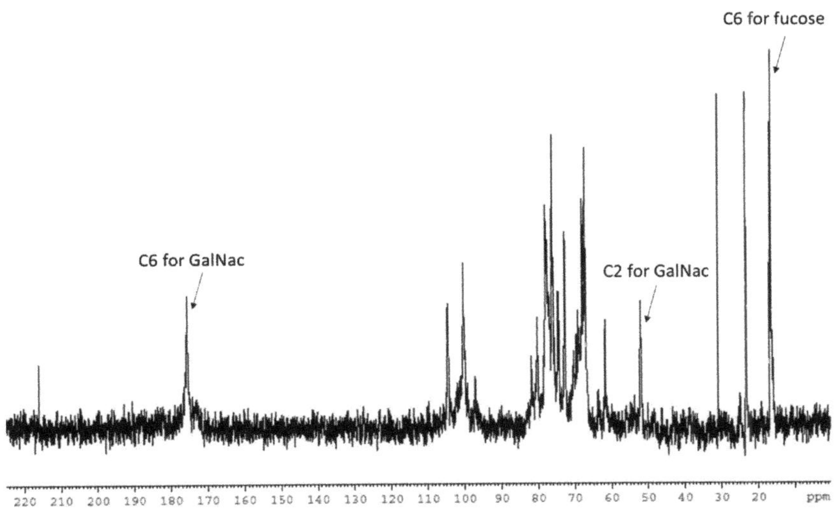

Figure 2. The ^{13}C NMR spectra of FCS-Bo fraction isolated from *B. ocellata*. The signals are assigned as follows: C6 (GalNAc): carbon-6 of *N*-acetylgalactosamine; C2 (GalNAc): carbon-2 of *N*-acetylgalactosamine; C6 (Fuc): carbon-6 of fucose.

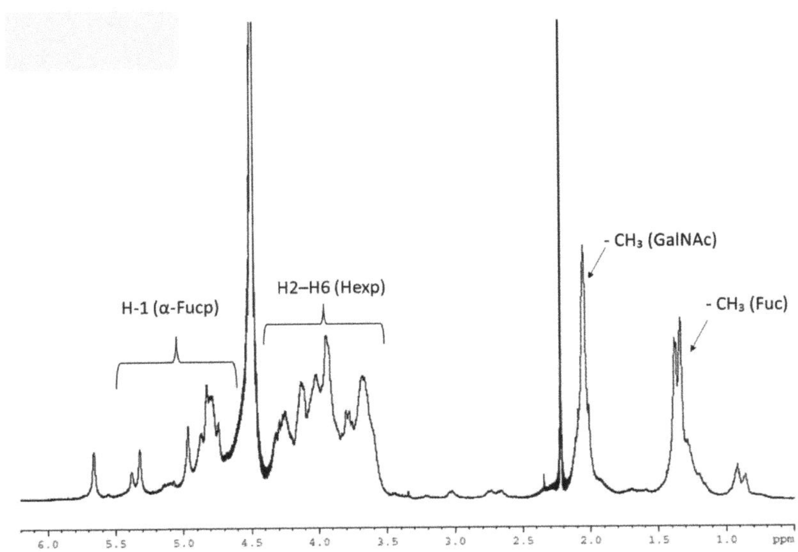

Figure 3. The ^{1}H NMR spectra of FCS-Bo fraction isolated from *B. ocellata*. The signals are assigned as follows: H-1 (α-Fucp): anomeric protons of fucose; H2–H6 (Hexp): protons on carbons C-2 to C-6 of the pyranose rings; CH_3 (GalNAc): methyl group from *N*-acetylgalactosamine; CH_3 (Fuc): methyl group from fucose (observed at ~1.5–1.2 ppm).

Structural characterizations of the polysaccharides were conducted utilizing both ^{1}H and ^{13}C NMR spectroscopic techniques [26] (Figures 2 and 3). The observed characteristic signals revealed the presence of particular monosaccharide units within the polysaccharide framework. The composition of FCS-Bo was confirmed to include fucose, galactosamine, and uronic acid as its primary monosaccharide constituents. This conclusion was based on the unique chemical shift values noted in the ^{13}C-NMR spectrum, specifically C-6 for fucose (δ 17.3 ppm) and *N*-acetylglucosamine (δ 176.3 ppm), along with C-2 for galactosamine (δ 52.7 ppm) (Figure 2).

In the ^{1}H NMR spectra, signals observed between 5.6 and 5.1 ppm corresponded to the anomeric protons, which are characteristic of the α-configuration in sulfated fucose residues within pyranose rings. The region spanning from 3.5 to 4.5 ppm was attributed to protons on carbons C-2 to C-6 of the hexosyl glycosidic ring. Additionally, distinct signals at 1.5–1.22 and 2.5–2.0 ppm were identified as the methyl protons of fucose and N-acetylgalactosamine, respectively. As illustrated in Figure 3, the small peaks at 5.16 and 5.28 ppm in the FCS-Bo spectrum were assigned to 4-O-sulfated fucose, which aligns with previously documented NMR data for FCS [27], reinforcing the consistency of the structural features.

The COSY spectrum of FCS-Bo reveals the correlations between proton signals, providing insight into the coupling patterns and the spatial arrangement of hydrogen atoms in the molecular structure (Figure 4). The cross-peaks observed in the COSY spectrum confirm the scalar coupling between the protons in the sugar ring, particularly those in the Fuc, Glu, and GalNAc residues. The strong cross-peaks between the anomeric protons (around 5.0–5.5 ppm) and their neighboring proton signals (in the range of 3.5–4.2 ppm) suggest typical glycosidic linkages within the polysaccharide chain. Based on the content of monosaccharides and sulfate and its characteristic spectra, we can assume that this polysaccharide is FCS.

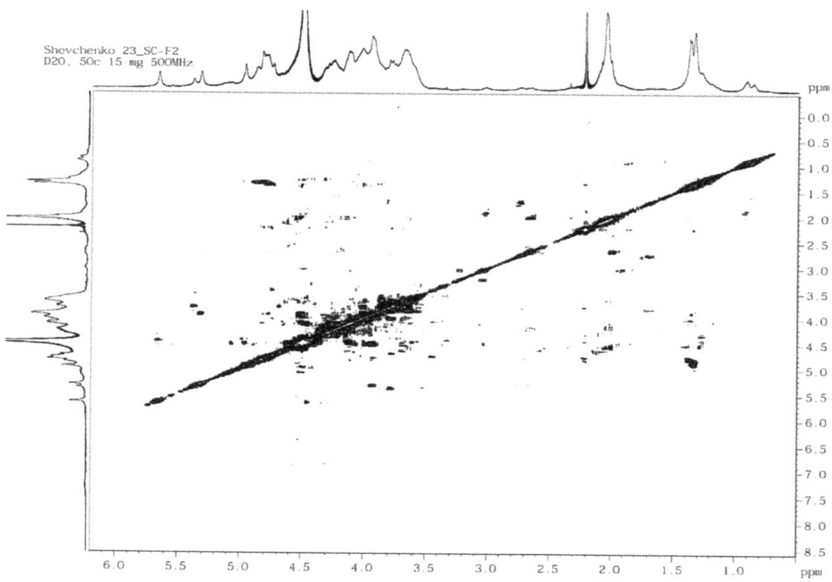

Figure 4. The COSY spectra of fucosylated chondroitin sulfates FCS-Bo isolated from *B. ocellata*.

3.2. Bioactivities of FCS-Bo

3.2.1. The Anticoagulant Activity

In this study, FCS-Bo was investigated for in vitro anticoagulant activity using aPTT, PT, and TT assays, and heparin with low molecular weight was used as a standard. As shown in Figure 5 (Table S2), the control group treated with DMSO exhibited normal clotting times in all assays, with an aPTT of 31.9 s, PT of 13.8 s, and TT of 17.4 s. In contrast, FCS-Bo at 100 µg/mL and 50 µg/mL significantly prolonged all coagulation parameters. Specifically, FCS-Bo at both concentrations extended the aPTT and PT beyond the upper detection limit (>120 s and >50 s, respectively). The TT was 63.4 s at 100 µg/mL of FCS-Bo, while it exceeded 240 s at 50 µg/mL. Heparin, which worked as a positive control at 100 µg/mL, also extended all parameters to the upper limits of detection. These findings suggest that FCS-Bo exhibits strong anticoagulant activity comparable to heparin standard, particularly in its ability to inhibit thrombin formation and prolong clotting times.

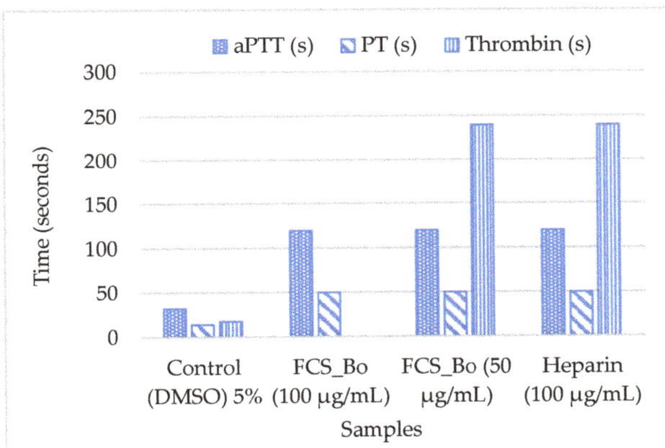

Figure 5. Effects of FCS_Bo and Heparin on Coagulation Time (aPTT, PT, and TT).

3.2.2. Cell Cytotoxicity Activity

This study of the biological activity of the substances at the first stage involves the assessment of their toxicity. Treatment of HCT-116, HT-29, and DLD-1 cells with the studied FCS-Bo at a concentration of up to 400 µg/mL did not lead to inhibition of their growth and was not accompanied by cell death. Consequently, the FCS-Bo at a concentration of up to 400 µg/mL is not toxic for HCT-116, HT-29, and DLD-1 cells.

Using the soft agar method, the effect of FCS-Bo on the spontaneous formation and growth of colonies of colon cancer cells HCT-116, HT-29, and DLD-1 was tested. For HCT-116 cells (Figure 6A), FCS-Bo at 400 µg/mL significantly inhibited colony formation by approximately 38% compared to the control ($p < 0.01$). A similar inhibitory effect was observed in HT-29 cells (Figure 6B), where colony formation was reduced by 43% at the same concentration ($p < 0.01$). In DLD-1 cells (Figure 6C), FCS-Bo exhibited a strong inhibitory effect, reducing colony formation by 65% at 400 µg/mL ($p < 0.05$). The microscopic images further corroborate the quantitative data, showing visibly fewer and smaller colonies in treated samples compared to the control. These findings suggest that FCS-Bo possesses potent anti-cancer activity, with the most substantial effects observed in the DLD-1 cells.

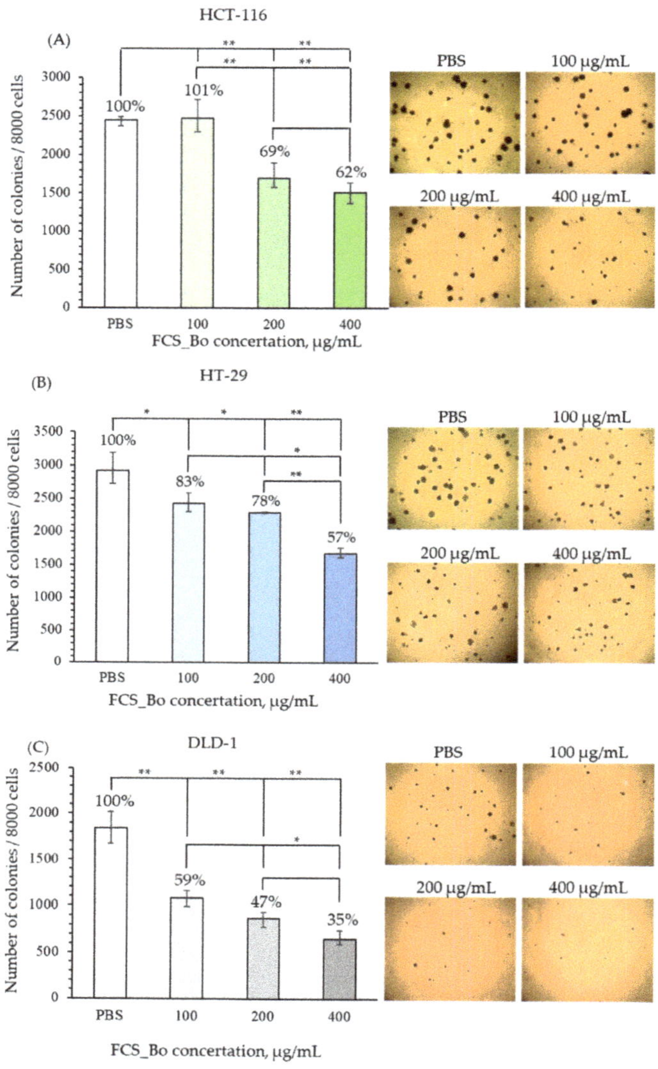

Figure 6. The inhibitory effects of FCS-Bo (100–400 µg/mL) on colony formation in human colorectal adenocarcinoma cell lines HCT-116 (**A**), HT-29 (**B**), and DLD-1 (**C**) are shown. Results are presented as the mean ± standard deviation (SD). Asterisks (*) denote a significant reduction in the number of colonies in polysaccharide-treated samples compared to the control (* $p < 0.05$, ** $p < 0.01$).

3.2.3. Protein Tyrosine Phosphatase 1B

PTP1B is a negative regulator of insulin signaling and is involved in the dephosphorylation of insulin receptors, which leads to a decrease in insulin sensitivity. Inhibition of PTP1B is therefore a promising strategy for enhancing insulin signaling, which can improve glucose uptake and metabolism. This makes PTP1B inhibitors potential therapeutic agents for managing conditions such as type 2 diabetes and obesity. By inhibiting PTP1B, it is possible to restore or improve insulin sensitivity, thereby assisting to control blood glucose levels and prevent the progression of metabolic disorders [28]. FCSs have been revealed to be potent in various bioactivity effects, including anti-inflammatory, antitumor, antiviral, and antithrombotic activities. However, scarce studies on the antidiabetic activity of FCS, FCS-Bo especially, have been discovered up to now [13].

The experimental results demonstrated that the FCS-Bo exhibits potent biological activities, underscoring its potential as a therapeutic agent (Figure 7). The IC_{50} value for the inhibition of PTP1B is 0.0326 μg/mL, which is significantly lower than that of the positive control, ursolic acid (IC_{50} = 1.467 μg/mL). This indicates that FCS-Bo is highly effective in inhibiting PTP1B, suggesting strong potential for enhancing insulin signaling and managing conditions of type 2 diabetes and obesity.

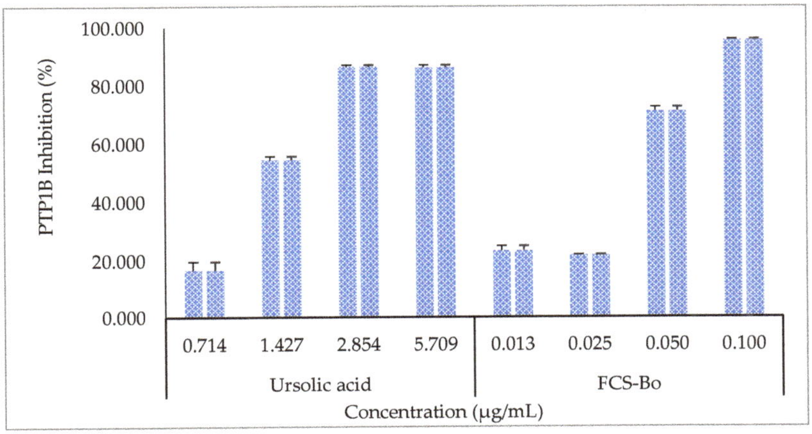

Figure 7. PTP1B inhibitory activity of ursolic acid and FCS-Bo.

4. Discussion

The bioactivities of FCS-Bo underscore its significant therapeutic potential across various biological systems. Notably, the potent anticoagulant activity of FCS-Bo is consistent with the well-established effects of sulfated polysaccharides from marine sources [29,30]. FCS-Bo effectively prolonged clotting times, including activated partial aPTT, PT, and TT, showing anticoagulant properties comparable to those of heparin, a standard therapeutic anticoagulant. The anticoagulant activity of FCS-Bo is significantly influenced by structural factors such as sulfation patterns, molecular weight, monosaccharide composition, and charge density. Consistent with these observations, FCS-Bo demonstrated potent anticoagulant properties, aligning with the structural determinants previously reported for other sea cucumber species. For example, the 2,4-O-disulfation in fucosyl residues, highlighted by Chen et al. [27] as critical for anticoagulant activity, may also be a key factor in the effectiveness of FCS-Bo. The ability of FCS-Bo to significantly prolong clotting times such as aPTT, PT, and TT suggests that its sulfation pattern effectively interferes with thrombin generation and intrinsic factor Xase (IF-Xase) activity. This is comparable to the strong anticoagulant activities observed in FCS from species such as *Holothuria polii* and *Bohadschia argus*, in which specific sulfation patterns played a major role in enhancing the bioactivity [2,31]. Furthermore, the molecular weight of FCS-Bo could be a critical factor in its anticoagulant efficacy. FCS-Bo with a molecular weight of 122 kDa demonstrated strong anticoagulant activity, as evidenced by its significant prolongation of clotting times. This activity is closely related to its molecular weight and structural features, particularly the sulfation patterns and the presence of fucose branches. The molecular weight of FCS-Bo places it within the range where the balance between anticoagulant efficacy and safety is optimized, as supported by previous studies on FCSs from other sea cucumber species. Comparing FCS-Bo to other species, it is notable that FCSs with similar or slightly different molecular weights exhibit varying degrees of anticoagulant activity. For example, the FCS extracted from *Isostichopus badionotus* has demonstrated that a reduction in molecular weight leads to a significant decrease in anticoagulant activities, including aPTT, TT, and thrombin inhibition [32]. This suggests that a higher molecular weight of FCS, such as the FCS-Bo, may be necessary for enhancing anticoagulant effects. Furthermore, studies have shown that depolymerized FCSs, with molecular weights in the range of 6–12 kDa, can sometimes exhibit enhanced anticoagulant properties by selectively inhibiting intrinsic factor Xase (IF-Xase) while reducing side effects such as platelet aggregation and spontaneous bleeding [33,34]. However, the relatively high molecular weight of FCS-Bo aligns with the strong anticoagulant activities observed, potentially offering a broader spectrum of anticoagulant effects, including thrombin inhibition and prolongation of clotting times, similar to the effects seen with other high molecular weight FCSs [35].

FCS-Bo exhibits a significant inhibitory effect on the growth and colony formation of human colon cancer cells, HCT-116, HT-29, and DLD-1, where it reduced colony formation by 38%, 43%, and 65%, respectively, at a concentration of 400 μg/mL. This inhibition suggests that FCS-Bo disrupts key signaling pathways critical for colony formation of cancer cells, aligning with the antitumor mechanisms observed in other FCS variants. The anticancer effects of FCSs have been linked to modulate several molecular pathways. For instance, low-molecular-weight FCS from *Cucumaria frondosa* has been shown to inhibit tumor growth and metastasis by increasing p53/p21 expression, inducing apoptosis through caspase-3 activation, and suppressing angiogenesis via VEGF inhibition [10]. Similar antitumor properties were reported for FCS from *Hemioedema spectabilis*, which interferes with cancer cell adhesion and tubulogenesis, key processes in tumor development and metastasis [12].

FCS from sea cucumbers demonstrates distinct bioactivities, largely due to its unique structural components, including sulfate groups, fucose branches, and a chondroitin sulfate backbone. These features influence its ability to interfere with various biological processes involved in tumor growth [36]. The degree and position of sulfation are critical determinants of FCS's bioactivity, as sulfation patterns dictate its interaction with proteins and enzymes involved in key processes such as cell proliferation and apoptosis [36]. Specifically, fucose branches are often attached to GlcA at the O-3 position and are sulfated at the 2 or 4 positions, depending on the sea cucumber species. This precise sulfation enhances FCS's ability to disrupt cancer cell signaling pathways. Furthermore, the diverse fucosylation patterns along the chondroitin sulfate backbone provide multiple functional groups, enabling FCS to target a variety of biological molecules. This structural complexity allows FCS to selectively exert cytotoxic effects on tumor cells while potentially minimizing harm to normal cells [36]. The presence of fucosylated residues is particularly important, as these compounds are known to significantly enhance biological activities, including anticancer effects [37]. Therefore, fucose is not merely a structural component but also serves as a functional modulator that contributes directly to FCS's antitumor properties.

Furthermore, the potent inhibition of PTP1B by FCS-Bo underscores its potential as a therapeutic agent for metabolic disorders, including type 2 diabetes and obesity. The significantly low IC_{50} value of FCS-Bo compared to the standard ursolic acid indicates a strong potential for enhancing insulin sensitivity and improving glucose metabolism. This is particularly relevant in the context of the increasing global burden of metabolic diseases, where safe and effective PTP1B inhibitors are in high demand. The dual activity of FCS-Bo as both an anticoagulant and a metabolic regulator further enhances its appeal as a multifunctional therapeutic agent. Further studies should investigate its bioavailability, pharmacokinetics, and efficacy in vivo to pave the way for potential clinical applications.

The data showed significant potential of FCS for applications in the pharmaceutical and biomedical fields. One key weakness of FCS is its structural variability depending on species, origin, harvesting season, and extraction methods, which can result in inconsistent biological effects [30]. Additionally, the high MW of FCS and the presence of complex sulfation patterns may reduce its bioavailability when administered orally or in vivo. Another limitation is that the specific sulfation positions and fucose branch points critical for its bioactivity can be difficult to standardize in large-scale production. To improve the application of FCS, one approach would be to optimize the extraction and purification processes to obtain more consistent structural forms of FCS across different species. Utilizing enzymatic hydrolysis or chemical modifications to reduce the molecular weight of FCS without losing its bioactivity can enhance its bioavailability. Moreover, refining the sulfation pattern through targeted chemical modification can enhance its specific biological functions, such as anticoagulant or anticancer activity.

Overall, the bioactivities observed in FCS-Bo highlight its potential as a versatile therapeutic agent with applications in anticoagulation, cancer treatment, and metabolic disorder management. Further exploration of its molecular mechanisms and clinical efficacy will be critical in harnessing its therapeutic potential.

5. Conclusions

This study provides a comprehensive examination of the fucosylated chondroitin sulfate extracted from the sea cucumber *B. ocellata*. The structural characterization using ^{1}H and ^{13}C NMR spectroscopy confirmed the presence of key monosaccharide components such as fucose, galactosamine, and uronic acid, solidifying the identification of FCS as a distinct biopolymer. Importantly, FCS-Bo demonstrated significant biological activities, particularly in its potent inhibition of PTP1B, with an IC_{50} value of 0.0326 μg/mL—substantially lower than the positive control, ursolic acid. This suggests strong potential for FCS-Bo in managing type 2 diabetes and obesity through the enhancement of insulin signaling. Additionally, FCS-Bo exhibited robust antioxidant properties, effectively scavenging free radicals and protecting cells from oxidative stress, as well as strong anticoagulant activity. These findings highlight the therapeutic potential of FCS-Bo across multiple domains, including metabolic regulation, oxidative stress mitigation, and anticoagulation, warranting further exploration into its clinical applications.

Supplementary Materials: The following supporting information can be downloaded at: https://www.mdpi.com/article/10.3390/pr12102108/s1, Figure S1. Sea cucumbers *Bohadschia ocellata* collected in the coastline of Van Gia, Khanh Hoa, Vietnam; Figure S2. Ion-exchange chromatography on Macro-Prep DEAE (Cl^- form, 8 × 2.5 cm) of polysaccharides from *Bohadschia ocellata*; Figure S3. Molecular weight of the fraction FCS_Bo isolated from *B. ocellata*; Table S1. Coagulation parameters (aPTT, PT, and TT) for FCS-Bo; Table S2. PTP1B inhibitory activity of ursolic acid and FCS_Bo at various concentrations.

Author Contributions: Conceptualization, P.D.T.; methodology, H.T.T.C.; software, S.P.E. and D.T.T.; validation, P.D.T. and H.T.T.C.; investigation, D.K.M., D.T.T., H.T.T.C., A.O.Z., T.Q.C. and S.P.E.; writing—original draft preparation, H.T.T.C. and T.-D.N.; writing—review and editing, P.D.T. and T.-D.N.; supervision, T.T.T.V., P.D.T. and S.P.E.; project administration, P.D.T.; funding acquisition, P.D.T. All authors have read and agreed to the published version of the manuscript.

Funding: This research was funded by the National Foundation for Science & Technology Development (NAFOSTED) (Project No. 106.02-2019.34).

Data Availability Statement: The original contributions presented in the study are included in the article/supplementary material, further inquiries can be directed to the corresponding author.

Acknowledgments: The NMR spectroscopy of polysaccharides was performed in the center for collective use of scientific equipment, "Far Eastern Center of Structural Studies".

Conflicts of Interest: The authors declare no conflicts of interest.

References

1. Pomin, V.H. Holothurian Fucosylated Chondroitin Sulfate. *Mar. Drugs* **2014**, *12*, 232–254. [CrossRef] [PubMed]
2. Ustyuzhanina, N.E.; Bilan, M.I.; Dmitrenok, A.S.; Tsvetkova, E.A.; Nikogosova, S.P.; Hang, C.T.T.; Thinh, P.D.; Trung, D.T.; Van, T.T.T.; Shashkov, A.S.; et al. Fucose-Rich Sulfated Polysaccharides from Two Vietnamese Sea Cucumbers *Bohadschia argus* and Holothuria (Theelothuria) Spinifera: Structures and Anticoagulant Activity. *Mar. Drugs* **2022**, *20*, 380. [CrossRef] [PubMed]
3. Thinh, P.D.; Ly, B.M.; Usoltseva, R.V.; Shevchenko, N.M.; Rasin, A.B.; Anastyuk, S.D.; Malyarenko, O.S.; Zvyagintseva, T.N.; San, P.T.; Ermakova, S.P. A Novel Sulfated Fucan from Vietnamese Sea Cucumber Stichopus Variegatus: Isolation, Structure and Anticancer Activity In Vitro. *Int. J. Biol. Macromol.* **2018**, *117*, 1101–1109. [CrossRef] [PubMed]

4. Mao, H.; Cai, Y.; Li, S.; Sun, H.; Lin, L.; Pan, Y.; Yang, W.; He, Z.; Chen, R.; Zhou, L.; et al. A New Fucosylated Glycosaminoglycan Containing Disaccharide Branches from *Acaudina molpadioides*: Unusual Structure and Anti-Intrinsic Tenase Activity. *Carbohydr. Polym.* **2020**, *245*, 116503. [CrossRef] [PubMed]
5. Sea Cucumber Market Size, Growth | Report, 2032. Available online: https://www.businessresearchinsights.com/market-reports/sea-cucumber-market-108483 (accessed on 23 September 2024).
6. Dong, X.; Pan, R.; Deng, X.; Chen, Y.; Zhao, G.; Wang, C. Separation, Purification, Anticoagulant Activity and Preliminary Structural Characterization of Two Sulfated Polysaccharides from Sea Cucumber *Acaudina molpadioidea* and *Holothuria nobilis*. *Process Biochem.* **2014**, *49*, 1352–1361. [CrossRef]
7. Li, X.; Li, S.; Liu, J.; Lin, L.; Sun, H.; Yang, W.; Cai, Y.; Gao, N.; Zhou, L.; Qin, H.; et al. A Regular Fucan Sulfate from *Stichopus herrmanni* and Its Peroxide Depolymerization: Structure and Anticoagulant Activity. *Carbohydr. Polym.* **2021**, *256*, 117513. [CrossRef]
8. Olivera-Castillo, L.; Grant, G.; Kantún-Moreno, N.; Barrera-Pérez, H.A.; Montero, J.; Olvera-Novoa, M.A.; Carrillo-Cocom, L.M.; Acevedo, J.J.; Puerto-Castillo, C.; Solís, V.M.; et al. A Glycosaminoglycan-Rich Fraction from Sea Cucumber *Isostichopus badionotus* Has Potent Anti-Inflammatory Properties In Vitro and In Vivo. *Nutrients* **2020**, *12*, 1698. [CrossRef]
9. Li, S.; Zhong, W.; Pan, Y.; Lin, L.; Cai, Y.; Mao, H.; Zhang, T.; Li, S.; Chen, R.; Zhou, L.; et al. Structural Characterization and Anticoagulant Analysis of the Novel Branched Fucosylated Glycosaminoglycan from Sea Cucumber *Holothuria nobilis*. *Carbohydr. Polym.* **2021**, *269*, 118290. [CrossRef]
10. Liu, X.; Liu, Y.; Hao, J.; Zhao, X.; Lang, Y.; Fan, F.; Cai, C.; Li, G.; Zhang, L.; Yu, G. In Vivo Anti-Cancer Mechanism of Low Molecular-Weight Fucosylated Chondroitin Sulfate (LFCS) from Sea Cucumber *Cucumaria frondosa*. *Molecules* **2016**, *21*, 625. [CrossRef]
11. Li, J.; Li, S.; Wu, L.; Yang, H.; Wei, C.; Ding, T.; Linhardt, R.J.; Zheng, X.; Ye, X.; Chen, S. Ultrasound-Assisted Fast Preparation of Low Molecular Weight Fucosylated Chondroitin Sulfate with Antitumor Activity. *Carbohydr. Polym.* **2019**, *209*, 82–91. [CrossRef]
12. Ustyuzhanina, N.E.; Bilan, M.I.; Dmitrenok, A.S.; Shashkov, A.S.; Ponce, N.M.A.; Stortz, C.A.; Nifantiev, N.E.; Usov, A.I. Fucosylated Chondroitin Sulfate from the Sea Cucumber Hemioedema Spectabilis: Structure and Influence on Cell Adhesion and Tubulogenesis. *Carbohydr. Polym.* **2020**, *234*, 115895. [CrossRef] [PubMed]
13. Gong, P.X.; Li, Q.Y.; Wu, Y.C.; Lu, W.Y.; Zeng, J.; Li, H.J. Structural Elucidation and Antidiabetic Activity of Fucosylated Chondroitin Sulfate from Sea Cucumber *Stichopus japonicas*. *Carbohydr. Polym.* **2021**, *262*, 117969. [CrossRef] [PubMed]
14. Hu, S.; Wang, J.; Xu, H.; Wang, Y.; Li, Z.; Xue, C. Fucosylated Chondroitin Sulphate from Sea Cucumber Inhibits High-Fat-Sucrose Diet-Induced Apoptosis in Mouse Pancreatic Islets via down-Regulating Mitochondrial Signaling Pathway. *J. Funct. Foods* **2014**, *7*, 517–526. [CrossRef]
15. Van Minh, C.; Quoc Long, P. Studying on the Bioactive Substances from Marine Organismsin Vietnam. 2009. Available online: http://tvhdh.vnio.org.vn:8080/xmlui/handle/123456789/18882 (accessed on 1 September 2024).
16. Bilan, M.I.; Dmitrenok, A.S.; Nikogosova, S.P.; Tsvetkova, E.A.; Ustyuzhanina, N.E.; Hang, C.T.T.; Thinh, P.D.; Trung, D.T.; Van, T.T.T.; Usov, A.I.; et al. The Structure of Sulfated Polysaccharides from the Sea Cucumber *Holothuria (Stauropora) fuscocinerea*. *Russ. J. Bioorganic Chem.* **2023**, *49*, 758–767. [CrossRef]
17. Dubois, M.; Gilles, K.A.; Hamilton, J.K.; Rebers, P.A.; Smith, F. Colorimetric Method for Determination of Sugars and Related Substances. *Anal. Chem.* **1956**, *28*, 350–356. [CrossRef]
18. Thuan, N.T.; Maria, M.D.M.; Tran, V.H.N.; Vo, T.T.D.; Holck, J.; Rasin, A.B.; Cao, H.T.T.; Van, T.T.T.; Meyer, A.S. Enzyme-Assisted Fucoidan Extraction from Brown Macroalgae *Fucus distichus* Subsp. *Evanescens* and *Saccharina latissima*. *Mar. Drugs* **2020**, *18*, 296. [CrossRef]
19. Bitter, T.; Muir, H.M. A Modified Uronic Acid Carbazole Reaction. *Anal. Biochem.* **1962**, *4*, 330–334. [CrossRef] [PubMed]
20. Lowry, O.H.; Rosebrough, N.J.; Farr, A.L.; Randall, R.J.; Lewis, A. Protein Measurement with the Folin. *J. Biol. Chem.* **1951**, *193*, 265–275. [CrossRef]
21. Chong, A.Y.; Blann, A.D.; Lip, G.Y.H. Assessment of Endothelial Damage and Dysfunction: Observations in Relation to Heart Failure. *QJM* **2003**, *96*, 253–267. [CrossRef] [PubMed]
22. Usoltseva, R.V.; Anastyuk, S.D.; Surits, V.V.; Shevchenko, N.M.; Thinh, P.D.; Zadorozhny, P.A.; Ermakova, S.P. Comparison of Structure and In Vitro Anticancer Activity of Native and Modified Fucoidans from *Sargassum feldmannii* and *S. duplicatum*. *Int. J. Biol. Macromol.* **2019**, *124*, 220–228. [CrossRef]
23. Le, T.T.; Ha, M.T.; Cao, T.Q.; Kim, J.A.; Choi, J.S.; Min, B.S. 1,5-Anhydro-d-Glucitol Derivative and Galloylated Flavonoids Isolated from the Leaves of *Acer ginnala* Maxim. as Dual Inhibitors of PTP1B and α-Glucosidase Enzymes: In Vitro and In Silico Studies. *Phytochemistry* **2023**, *213*, 113769. [CrossRef] [PubMed]
24. Ji, Y.; Yang, X.; Ji, Z.; Zhu, L.; Ma, N.; Chen, D.; Jia, X.; Tang, J.; Cao, Y. DFT-Calculated IR Spectrum Amide I, II, and III Band Contributions of N-Methylacetamide Fine Components. *ACS Omega* **2020**, *5*, 8572–8578. [CrossRef] [PubMed]
25. Mou, J.; Li, Q.; Qi, X.; Yang, J. Structural Comparison, Antioxidant and Anti-Inflammatory Properties of Fucosylated Chondroitin Sulfate of Three Edible Sea Cucumbers. *Carbohydr. Polym.* **2018**, *185*, 41–47. [CrossRef] [PubMed]
26. Pomin, V.H. NMR Structural Determination of Unique Invertebrate Glycosaminoglycans Endowed with Medical Properties. *Carbohydr. Res.* **2015**, *413*, 41–50. [CrossRef] [PubMed]
27. Chen, S.; Xue, C.; Yin, L.; Tang, Q.; Yu, G.; Chai, W. Comparison of Structures and Anticoagulant Activities of Fucosylated Chondroitin Sulfates from Different Sea Cucumbers. *Carbohydr. Polym.* **2011**, *83*, 688–696. [CrossRef]

28. Paul, A.; Sarkar, A.; Banerjee, T.; Maji, A.; Sarkar, S.; Paul, S.; Karmakar, S.; Ghosh, N.; Maity, T.K. Structural and Molecular Insights of Protein Tyrosine Phosphatase 1B (PTP1B) and Its Inhibitors as Anti-Diabetic Agents. *J. Mol. Struct.* **2023**, *1293*, 136258. [CrossRef]
29. Pangestuti, R.; Arifin, Z. Medicinal and Health Benefit Effects of Functional Sea Cucumbers. *J. Tradit. Complement. Med.* **2017**, *8*, 341–351. [CrossRef] [PubMed]
30. Myron, P.; Siddiquee, S.; Al Azad, S. Fucosylated Chondroitin Sulfate Diversity in Sea Cucumbers: A Review. *Carbohydr. Polym.* **2014**, *112*, 173–178. [CrossRef]
31. Li, C.; Niu, Q.; Li, S.; Zhang, X.; Liu, C.; Cai, C.; Li, G.; Yu, G. Fucoidan from Sea Cucumber *Holothuria polii*: Structural Elucidation and Stimulation of Hematopoietic Activity. *Int. J. Biol. Macromol.* **2020**, *154*, 1123–1131. [CrossRef]
32. Yan, L.; Li, J.; Wang, D.; Ding, T.; Hu, Y.; Ye, X.; Linhardt, R.J.; Chen, S. Molecular Size Is Important for the Safety and Selective Inhibition of Intrinsic Factor Xase for Fucosylated Chondroitin Sulfate. *Carbohydr. Polym.* **2017**, *178*, 180–189. [CrossRef] [PubMed]
33. Ustyuzhanina, N.E.; Bilan, M.I.; Anisimova, N.Y.; Dmitrenok, A.S.; Tsvetkova, E.A.; Kiselevskiy, M.V.; Nifantiev, N.E.; Usov, A.I. Depolymerization of a Fucosylated Chondroitin Sulfate from *Cucumaria japonica*: Structure and Activity of the Product. *Carbohydr. Polym.* **2022**, *281*, 119072. [CrossRef] [PubMed]
34. Guan, R.; Peng, Y.; Zhou, L.; Zheng, W.; Liu, X.; Wang, P.; Yuan, Q.; Gao, N.; Zhao, L.; Zhao, J. Precise Structure and Anticoagulant Activity of Fucosylated Glycosaminoglycan from *Apostichopus japonicus*: Analysis of Its Depolymerized Fragments. *Mar. Drugs* **2019**, *17*, 195. [CrossRef] [PubMed]
35. Luo, L.; Wu, M.; Xu, L.; Lian, W.; Xiang, J.; Lu, F.; Gao, N.; Xiao, C.; Wang, S.; Zhao, J. Comparison of Physicochemical Characteristics and Anticoagulant Activities of Polysaccharides from Three Sea Cucumbers. *Mar. Drugs* **2013**, *11*, 399–417. [CrossRef] [PubMed]
36. Xu, H.; Zhou, Q.; Liu, B.; Chen, F.; Wang, M. Holothurian Fucosylated Chondroitin Sulfates and Their Potential Benefits for Human Health: Structures and Biological Activities. *Carbohydr. Polym.* **2022**, *275*, 118691. [CrossRef] [PubMed]
37. Rai, D.B.; Solanki, R.; Medicherla, K.; Patel, S.; Pooja, D.; Kulhari, H. Fucosylated Dendrimer Mediated Enhancement of Solubility, Stability and Biological Activity of Genistein. *Food Biosci.* **2023**, *56*, 103200. [CrossRef]

processes

MDPI

Article

Evaluation of Polyphenol Profile from Citrus Peel Obtained by Natural Deep Eutectic Solvent/Ultrasound Extraction

Manuel Octavio Ramírez-Sucre †, Kevin Alejandro Avilés-Betanzos †, Anahí López-Martínez and Ingrid Mayanin Rodríguez-Buenfil *

Centro de Investigación y Asistencia en Tecnología y Diseño del Estado de Jalisco A.C., Subsede Sureste, Tablaje Catastral 31264, Km. 5.5 Carretera Sierra Papacal-Chuburná Puerto, Parque Científico Tecnológico de Yucatán, Mérida C.P. 97302, Yucatán, Mexico; oramirez@ciatej.mx (M.O.R.-S.); keaviles_al@ciatej.edu.mx (K.A.A.-B.); anahi0410vorstellen@gmail.com (A.L.-M.)

* Correspondence: irodriguez@ciatej.mx

† These authors contributed equally to this work.

Abstract: Citrus fruits are widely consumed worldwide; however, one of their primary uses is juice production, resulting in over 40 million tons of agro-industrial waste. Citrus peel is the main agro-industrial by-product in citrus production. In recent years, secondary metabolites of interest, mainly polyphenols such as hesperidin, have been identified in citrus peels. Currently, green alternatives like natural deep eutectic solvents (NADES) based on choline chloride and glucose (Glu), combined with ultrasound-assisted extraction, are studied to obtain polyphenol-rich extracts with potential health applications. This study aims to evaluate the effect of: (1) molar ratios (MR) of 1:0.5, 1:1 or 1:2 mol/mol of choline chloride (ChCl):glucose (Glu); (2) the percentage of added water (WA: 50, 60 or 70%) to NADES; and (3) different citrus peels of *Citrus aurantium* (bitter orange), *Citrus sinensis* (sweet orange), and *Citrus limon* (lemon) used for extraction, on polyphenol profiles, total polyphenol content (TPC), and antioxidant capacity (Ax) of the extracts. The extracts were analyzed using ultra-performance liquid chromatography (UPLC) and evaluated using the Folin–Ciocalteu method for TPC and DPPH assay for quantifying AC. A factorial experimental design 3^3 was implemented. The extract obtained with an MR of 1:1 (ChCl:Glu) from *Citrus aurantium* peel exhibited the highest concentration of hesperidin (2003.37 $\pm$ 10.91 mg/100 g dry mass), whereas an MR of 1:2 (ChCl:Glu) exhibited the highest concentration of neohesperidin (1045.94 $\pm$ 1.27 mg/100 g dry mass), both using 60% WA. This extract also showed the highest antioxidant capacity, achieving 100% inhibition. On the other hand, the highest concentration of total phenolic content (TPC) (96.23 $\pm$ 0.83 mg GAE/100 g dry mass) was obtained using *C. aurantium* peel with an MR of 1:0.5 (ChCl:Glu) and 60% WA. The extracts also presented high concentrations of rutin and catechin. These findings highlight the potential of revalorizing citrus peels, particularly *Citrus aurantium*, and their extracts obtained with NADES for possible health applications.

Keywords: citrus peel; eutectic solvent; NADES; ultrasound extraction; *Citrus aurantium*; polyphenol profile

1. Introduction

Citrus fruits, such as orange and lemon, are highly preferred and consumed worldwide. Their combined global production in 2023 reached 58.8 million tons, with Mexico ranking fourth. According to the Mexican Agri-Food and Fisheries Information Service (SIAP), this year, Mexico's citrus fruit production was 8.1 million tons, representing 13.77% of the global production, following Brazil, China, and the European Union (USDA, 2024) [1].

According to Kumar et al. [2], the primary agricultural waste from citrus fruit production is the peel (CPe), which accounts for 30% of the fruit's weight. Thus, in Mexico, this results in approximately 2.43 million tons of peel annually [3].

Although CPe is considered an agricultural waste, this plant matrix contains various components of interest, such as pectin (dietary fiber), carotenoids, essential oils, vitamin C (ascorbic acid), and polyphenols [4]; the latter have attracted particular interest due to their recognized antioxidant, anti-inflammatory, anticancer, antidiabetic, and other beneficial effects in humans [5]. High concentrations of these compounds have been reported in the peels of various citrus fruits. For example, Lagha-Benamrouche and Madani [6] determined the total polyphenol content (TPC) through maceration with methanol–water (4:1 *v*/*v*) in *C. sinensis* (25.60 ± 0.23 mg GAE/g dry mass) and *C. aurantium* (31.62 ± 0.88 mg GAE/g dry mass) peels. Similarly, Papoutsis et al. [7] registered the highest TPC (13.24 mg GAE/g dry mass) of *C. limon* peel by absolute methanol extraction.

While total polyphenols can be observed in citrus peel samples, the profile of individual polyphenols will not be the same for each type of citrus, even though they may share some characteristic compounds of the *Citrus* genus, such as hesperidin, naringenin, rutin, quercetin, and diosmin [8,9]. The characteristic phenolic compound of citrus fruits is hesperidin. This metabolite is classified as a flavonoid, and more specifically as a glycosylated flavanone, because its structure consists of the hesperetin molecule linked by a glycosidic bond to rutinose, a disaccharide formed by glucose and rhamnose [10].

Concentrations of this flavonoid have been reported in *C. sinensis* peel using various extraction methodologies, such as ultrasound (836 ± 29 mg/100 g dry mass), microwave (765 ± 12.13 mg/100 g dry mass), and ethanol (80%) maceration (551 ± 0.01 mg/100 g dry mass) [11]. For *C. limon* peel, extraction using the Soxhlet method, with absolute methanol at 85 °C for 12 h, has yielded concentrations of 331.50 ± 26.23 mg/100 g dry mass [9]. Finally, for *C. aurantium*, concentrations of 21.03 mg/100 g fresh mass have been reported. This metabolite has attracted significant interest in both the food and pharmaceutical industries, because of its various biological properties. It has notable cardioprotective and vasodilatory effects, which are harnessed in the creation of medications for circulatory system disorders [11]. Furthermore, it has been documented in relation to its capacity to enhance the body's antioxidant enzymes, such as catalase and superoxide dismutase, its neuroprotective capabilities [12], and its preventive effects against SARS-CoV-2 [13,14].

Various methodologies have been reported for conducting flavonoid extraction, including traditional methods such as maceration or steam distillation, where organic solvents like ethanol, methanol, and/or acetone have been used [15,16]. However, in the search for new green extraction technologies aimed at obtain high phenolic content extracts with characteristic individual polyphenols, ultrasound-assisted extraction (UAE) and natural deep eutectic solvents (NADES) have recently been implemented in agro-industrial by-products of citrus. These studies are recent and there is limited information available, but the initial findings suggest significant potential for their use in the cosmetic, pharmaceutical, or food industries without posing a risk to handlers and consumers [17–19]. In this context, there is growing interest in combining both green technologies, UAE and NADES, to improve the extraction of bioactive compounds from citrus peels.

UAE, through ultrasonic waves, generates cavitation in the solvent medium, which enhances mass transfer and disrupts plant cell walls, leading to higher extraction yields and shorter processing times. UAE's ability to improve solvent penetration and breakdown of plant structures makes it an ideal candidate for integration with other extraction methods to maximize efficiency [20], like NADES.

NADES have emerged as a promising alternative to conventional solvents, offering several favorable characteristics for the extraction of phenolic compounds. The effectiveness of NADES largely depends on the selection of their components, as these determine the solvent's polarity. Specifically, the choice of a hydrogen bond donor (common donors: glucose, fructose, glycerol, malic acid, and citric acid, among others) and its combination in a defined molar ratio with a hydrogen bond acceptor (such as choline chloride, betaine, urea, proline, etc.) can optimize the extraction of a particular compound of interest [21].

Furthermore, the addition of a certain percentage of water to NADES results in a mixture with greater polarity and affinity to phenolic compounds. This adjustment not only reduces the solvent's viscosity, thereby improving its diffusivity and mass transfer, but also promotes the formation of a hydrogen bond network between the solvent and the target metabolite. This network confers supramolecular stability on the mixture (through hydrogen bonds, van der Waals forces, and electrostatic interactions), preventing the degradation of bioactive compounds during the extraction process [22].

By integrating UAE with NADES, it is possible to harness the strengths of both technologies—UAE's capacity for cell disruption and enhanced mass transfer, combined with NADES' versatile properties and ability to stabilize bioactive compounds. This unified approach holds great potential for the efficient and sustainable extraction of bioactive compounds from citrus peels, contributing to the valorization of agricultural by-products and the advance of green chemistry, especially given the high production of citrus fruits and the substantial amount of peel generated as waste.

The innovative aspect of this study lies in evaluating different proportions of hydrogen bond donors, such as glucose, along with varying water content in NADES, to extract polyphenols from citrus peels, primarily the scarcely studied *C. aurantium*. There is very limited information available on the application of NADES and on the extracts obtained using these green solvents. Finally, the objective of this study is to evaluate the effect of the molar ratio and the percentage of added water in NADES on the polyphenol profile, highlighting the extraction combined with UAE from various citrus peels.

2. Materials and Methods

2.1. Citrus Peel

The peels of bitter orange (*Citrus aurantium*), sweet orange (*Citrus sinensis*), and lemon (*Citrus limon*) were used. Approximately 1 kg of each citrus was obtained at a local supermarket (August 2023) in Mérida, Yucatán, Mexico. The citrus peels were carefully removed using a knife, ensuring that both structures, the flavedo and the albedo, were preserved. This process was carried out without squeezing the fruits or recovering the juice.

2.2. Drying, Grinding, and Sieving of Citrus Peel

The peels were placed in a tray oven and dried following the method described by Chel-Guerrero et al. [23] with slight modifications, drying at 32 °C for 24 h to achieve a moisture content below 15%.

The dried peels were ground using a MasterChef® coffee grinder until a homogeneous powder was obtained. The powder was sieved to obtain a particle size of $\leq$500 µm. The sieved powder was stored at room temperature in resealable bags lined with aluminum foil until further use.

2.3. Preparation of Natural Deep Eutectic Solvent Based on Choline Chloride

For the preparation of natural deep eutectic solvents (NADES), the methodology described by Avilés-Betanzos et al. [24] was followed. Choline chloride (ChCl, 139.62 g/mol) and glucose (Glu, 180.16 g/mol) were mixed in different molar ratios (ChCl:Glu, 1:0.5, 1:1, or 1:2 mol/mol), with ChCl serving as the hydrogen bond acceptor (HBA) and Glu as the hydrogen bond donor (HBD). The mixtures were heated to a temperature of 90 °C in a water bath with continuous stirring (90 min) until a clear liquid phase was obtained. To each prepared NADES, a percentage of water (50, 60, or 70%) was added according to the final weight of the preparation.

2.4. Polyphenol Extraction from Citrus Peels Using NADES by Ultrasound Bath

To evaluate the effect molar ratio (MR), percentage of added water (WA) and of citrus peel type (CPe), on the NADES extraction of phenolic compounds, a 3^3 factorial experimental design was established.

The levels of these three factors were: for MR, 1:0.5 (−1), 1:1 (0) or 1:2 (1) of ChCl:Glu, mol/mol; for WA, 50% (−1), 60% (0), and 70% (1); and for CPe, *Citrus sinensis* (−1), *Citrus aurantium* (0), and *Citrus limon* (1), the nomenclature is shown in Table 1. The response variables were: (a) Polyphenol profile (mg/100 g dry mass); (b) Total polyphenol content (TPC, mg GAE/100 g dry mass); and (c) Antioxidant capacity (Ax, % DPPH Inhibition).

Table 1. Experimental design 3^3 for the evaluation of polyphenol extraction form citrus peel using different NADES.

#EXP	Encoded Values			Actual Values			Response Variable **
	X_1	X_2	X_3	MR	WA	CPe *	
1	−1	−1	−1	1:0.5	50	*C. aurantium*	Y_1
2	0	−1	−1	1:1	50	*C. aurantium*	Y_2
3	1	−1	−1	1:2	50	*C. aurantium*	Y_3
4	−1	0	−1	1:0.5	60	*C. aurantium*	Y_4
5	0	0	−1	1:1	60	*C. aurantium*	Y_5
6	1	0	−1	1:2	60	*C. aurantium*	Y_6
7	−1	1	−1	1:0.5	70	*C. aurantium*	Y_7
8	0	1	−1	1:1	70	*C. aurantium*	Y_8
9	1	1	−1	1:2	70	*C. aurantium*	Y_9
10	−1	−1	0	1:0.5	50	*C. sinensis*	Y_{10}
11	0	−1	0	1:1	50	*C. sinensis*	Y_{11}
12	1	−1	0	1:2	50	*C. sinensis*	Y_{12}
13	−1	0	0	1:0.5	60	*C. sinensis*	Y_{13}
14	0	0	0	1:1	60	*C. sinensis*	Y_{14}
15	1	0	0	1:2	60	*C. sinensis*	Y_{15}
16	−1	1	0	1:0.5	70	*C. sinensis*	Y_{16}
17	0	1	0	1:1	70	*C. sinensis*	Y_{17}
18	1	1	0	1:2	70	*C. sinensis*	Y_{18}
19	−1	−1	1	1:0.5	50	*C. limon*	Y_{19}
20	0	−1	1	1:1	50	*C. limon*	Y_{20}
21	1	−1	1	1:2	50	*C. limon*	Y_{21}
22	−1	0	1	1:0.5	60	*C. limon*	Y_{22}
23	0	0	1	1:1	60	*C. limon*	Y_{23}
24	1	0	1	1:2	60	*C. limon*	Y_{24}
25	−1	1	1	1:0.5	70	*C. limon*	Y_{25}
26	0	1	1	1:1	70	*C. limon*	Y_{26}
27	1	1	1	1:2	70	*C. limon*	Y_{27}

Note: EXP = Experiment number; MR = Molar Ratio of glucose per 1 mol of choline chloride; WA = Percentage of added water to NADES; CPe = Citrus Peels; TPC = Total Polyphenol Content; GAE = Gallic Acid Equivalent; DM = dry mass; * C. = *Citrus*; ** The response variables are: (a) Polyphenol profile (mg/100 g dry mass); (b) Total polyphenol content (TPC, mg GAE/100 g dry mass); and (c) Antioxidant capacity (Ax, % DPPH Inhibition).

The extraction of phenolic compounds was carried out following the procedure outlined by Avilés-Betanzos et al. [25], with some modifications. To a 1-g sample of citrus peel powder, 10 mL of NADES (MR and WA according to the experimental design) was added. The mixture was homogenized using a vortex mixer (Thermo Scientific®, Mexico City, Mexico, model Maxi Mix® II). The mixture was then sonicated in an ultrasonic bath for 30 min at 42 kHz (BRANSON®, model 351). Subsequently, the mixture was centrifuged (30 min, 4 °C, 4700 rpm) to recover the supernatant, which was then filtered through a

Nylon filter (0.22 μm). The extract was stored in chromatographic vials under refrigeration (<18 °C) until analysis.

2.5. Determination of Polyphenol Profile of Citrus Peel Extracts

The analysis and quantification of polyphenols in extracts from citrus peels were conducted using an advanced Acquity UPLC H-Class system equipped with a diode array detector and a high-performance column (Acquity UPLC HSS C18), as adapted from methodologies outlined by Avilés-Betanzos et al. [26]. To accurately measure the polyphenol concentrations, a calibration curve incorporating 18 distinct standards was employed, ranging from 5 to 75 μg/mL, including gallic acid, protocatechuic acids, chlorogenic acid, coumaric acid, cinnamic acid, catechin, rutin, kaempferol, quercetin, luteolin, vanillin, ferulic acid, diosmin, hesperidin, neohesperidin, naringenin, apigenin, and diosmetin. This setup ensured precise control over the analytical conditions, with the column maintained at 45 °C and samples introduced in 2-μL injections. Detection was optimized at 280 nm using a solvent system of water with 0.2% acetic (phase A) and acetonitrile with 0.1% acetic acid (phase B). Each injection had a duration of 15 min, where the gradient for elution was 99% A to 70% A and 1% B to 30% B (0 min to 10 min), followed by a steady state of 70% A and 30% B (10 min to 12 min), and finally elution from 70% A back to 99% A and 30% B back to 1% B (3 min). Quercetin and luteolina were quantified collectively, due to their overlapping peaks during analysis. Figure S2 shows the chromatogram with the lowest concentration (2 μgmL) of the calibration curve of selected individual polyphenols, including gallic acid, protocatechuic acid, catechin, chlorogenic acid, p-coumaric acid, rutin, hesperidin, quercetin + luteolin, and kaempferol.

2.6. Evaluation of Total Polyphenol Content in Citrus Peel Extracts

The polyphenolic content of the citrus peel extracts (CPex) was determined using the Folin–Ciocalteu method with some modifications, following the procedure described by Singleton et al. [26]. Initially, a dilution (1:10) of the CPex was made with distilled water, then 25 μL of extract diluted was mixed with 25 μL of distilled water, followed by the addition of 3 mL of water and 250 μL of Folin reagent (Sigma-Aldrich®, St. Louis, MO, USA). After 5 min, 750 μL of 20% sodium carbonate (Na_2CO_3, Sigma-Aldrich) and 950 μL of distilled water were added, and the mixture was allowed to stand for 30 min. The samples were then analyzed at 765 nm using a UV–Vis spectrophotometer (Thermo scientific®, Mexico City, Mexico, model Genesys 140). Prior to the sample analysis, a calibration curve was generated (Figure S1) with gallic acid from 5 μg/mL to 75 μg /mL (R^2 = 0.9965).

2.7. Antioxidant Capacity Assessment of Citrus Peel Extract

The assessment of antioxidant capacity (Ax) in CPex was performed using the DPPH assay, as outlined by Chel-Guerrero et al. [23]. Initially, 3.3 mg of DPPH was diluted to a total volume of 100 mL with methanol, and the solution subsequently was adjusted to an absorbance (Abs) of 0.700 ± 0.002 at 515 nm. The analysis was conducted using a Thermo Scientific® UV–vis spectrophotometer (Mexico City, Mexico, Genesys 140).

Following the standardization of the DPPH solution's Abs, a 100-μL sample of the CPex was added to 3.9 mL of the adjusted DPPH solution. This mixture was mixed and incubated for 30 min. The Abs was then measured at 515 nm. The antioxidant capacity was quantified as the percentage of inhibition, calculated according to Equation (1):

$$\%\ \text{DPPH Inhibition} = 100 - [(\text{Abs of CPex} \times 100)/(\text{Abs of adjusted DPPH solution})] \quad (1)$$

2.8. Statistical Analysis

The studies were carried out randomly. For each extract produced according to the experimental layout, measurements were taken in triplicate to ascertain both the total and specific polyphenols, along with the antioxidant capacity (Ax). The results are presented as mean values with standard deviations. The linear correlation analysis was conducted to examine the relationships between the total and specific polyphenols in by-products and their antioxidant capacities, as measured by DPPH assays. This involved calculating the Pearson correlation coefficient and conducting a principal component analysis (PCA). The statistical evaluations of the experimental design were performed using Statgraphics Centurion XVII.II-X64 (Statgraphics Technologies Inc.; Virgin, UT, USA), XLSTAT 2021.2.2 (Addison, Paris, France), and R software version 4.0.3 (The R Foundation for Statistical Computing, Vienna, Austria).

3. Results

3.1. Polyphenol Profile from Citrus Peel Extracts

Among the 18 polyphenols examined in each extract from the factorial design, only 14 individual polyphenols were identified. Cinnamic acid, coumaric acid, ferulic acid, and diosmin were not detected (Table S1). Additionally, Figure S3 shows the chromatograms of three extracts from different citrus peels, obtained under the same conditions, highlighting the variability in the profiles of individual polyphenols. For a better perspective on the polyphenol profile results from the experimental design 3^3 factorial, the data are presented in Figure 1, considering the polyphenol profile of (a) *C. aurantium*, (b) *C. sinensis*, and (c) *C. limon*. and only de majoritarian individual polyphenols (catechin, rutin, quercetin + luteolin, hesperidin, and neohesperidin).

The profile of major polyphenols in extracts from the *C. aurantium* peel is shown in Figure 1a. Extract of *C. aurantium* showed the highest TPC (>2000 mg GAE/100 g dry mass) from all the CPex analyzed. The two characteristic polyphenols, hesperidin and neohesperidin, were identified. Under the extraction conditions of experiment #5 (MR = 1:1, WA = 60%), the highest concentration of hesperidin (2003.37 ± 10.91 mg/100 g DM) was obtained while experiment #6 (MR = 1:2, WA = 60%) resulted in the highest concentration of neohesperidin (1045.94 ± 1.27 mg/100 g DM). Furthermore, the presence of all major polyphenols was observed in the extracts obtained using NADES with an MR of 1:2 mol and 70% of added water.

Hesperidin and rutin were the characteristic individual polyphenols in the peel of *C. sinensis* (Figure 1b). High concentrations of hesperidin were obtained under NADES conditions with an MR of 1:2 mol and 50% added water, while the extract with the highest concentration of rutin was achieved using a NADES with an MR of 1:1 mol and a WA of 70%. Under these same conditions, the extract contained all the major polyphenols (catechin, rutin, quercetin + luteolin, hesperidin, and neohesperidin).

In the extracts obtained from the peel of *C. limon*, rutin was identified as the characteristic and predominant polyphenol. Under two conditions of NADES ($p > 0.05$), a high concentration of rutin was observed: 541.60 ± 0.81 mg/100 g DM with an MR of 1:2 mol and AW of 50% (#21), and 526.42 ± 2.30 mg/100 g DM with an MR of 1:0.5 mol and AW of 60% (#22). Unlike the extracts from *C. aurantium* and *C. sinensis*, all the major polyphenols analyzed were identified and quantified in all extracts from the peel of *C. limon* (Figure 1c).

Amidst the minor polyphenols found in the various extracts of the design (Table S1), gallic acid was identified and quantified only in two extracts (#16, #22). The highest concentration (7.45 ± 0.02 mg/100 g DM) was present in the extract obtained from *C. sinensis* peel with an MR of 1:0.5 and AW of 70%, while the lowest concentration (4.85 ± 0.00 mg/100 g DM) was found in the extract from *C. limon* (MR = 1:0.5, WA = 60%). Chlorogenic acid is another compound that was identified and quantified in few extracts (#6, #9, #11, #12, #16) of the experimental design, within concentrations of 3.73 ± 0.27 up to 31.62 ± 0.38 mg/100 g DM in extracts #11 (MR = 1:1, WA = 50%, *C. sinensis*) and #16, respectively.

Other minor metabolites were found in the extracts, including protocatechuic acid, kaempferol, vanillin, naringenin, apigenin, and diosmetin (Table S1). The highest concentrations of protocatechuic acid (36.25 ± 0.24 mg/100 g DM) and diosmetin (47.54 ± 1.28 mg/100 g DM) were identified in the extracts obtained under the conditions of experiment #12 (MR = 1:2, WA = 50%, CPe = *C. sinensis*). Conversely, high concentrations of kaempferol

(27.77 ± 2.30 mg/100 g DM) and vanillin (15.33 ± 0.05 mg/100 g DM) were identified in experiment #5 (MR = 1:1, AW = 60%, CPe = *C. aurantium*).

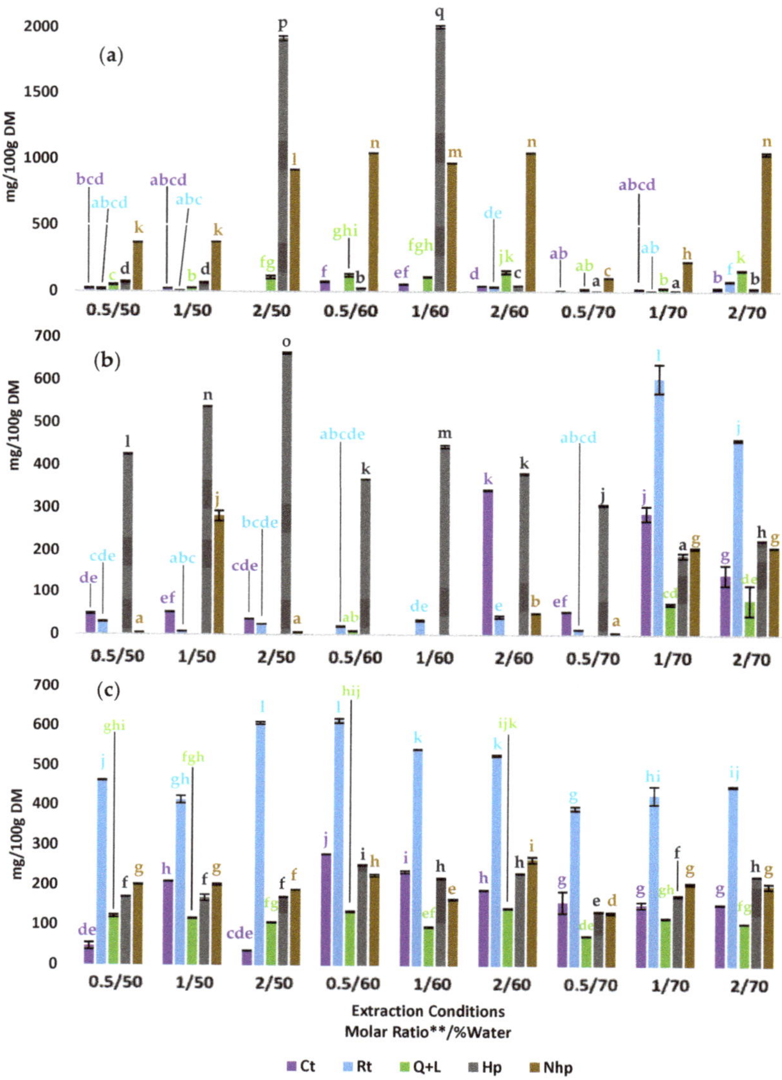

Figure 1. Major individual polyphenols in the peel of (**a**) *C. aurantium*, (**b**) *C. sinensis*, and (**c**) *C. limon*. Where Ct = catechin; Rt = rutin; Q + L = quercetin + luteolin; Hp = hesperidin; Nhp = Neohesperidin. Different letters with the same color indicate a statistically significant difference for a single individual polyphenol (LSD, $p < 0.05$). ** Molar ratio of glucose per 1 mol of Choline chloride. Values are means ± SD (n = 3).

The extract from experiment #20 (MR = 1:1, WA = 50%, CPe = *C. limon*) exhibited the highest concentration of naringenin (22.90 ± 0.05 mg/100 g DM), whereas the highest concentration of apigenin (10.50 ± 0.03 mg/100 g/DM) was observed in the extract obtained under the conditions of experiment #27 (MR = 1:2, WA = 70%, CPe = *C. limon*).

Table 2 shows the multifactorial ANOVA results on the polyphenol profile. These data indicate that concentrations of gallic acid (WA), protocatechuic acid (CPe), catechin (MR, CPe), rutin (MR, CPe), quercetin + luteolin (WA), kaempferol (WA), and diosmetin (WA, CPe) in citrus peel extracts are influenced ($p < 0.05$) by one or more primary factors, but not by their interactions (double or triple). Conversely, the concentrations of chlorogenic acid (WA*MR), hesperidin (WA*MR, MR*CPe, WA*MR*CPe), neohesperidin (WA*CPe), naringenin (WA*MR, WA*CPe), and apigenin (MR*CPe) are significantly affected ($p < 0.05$) by double or triple interactions.

Table 2. *p*-values of the main factors and their interactions on the polyphenol profile in citrus peel extracts.

Individual Polyphenol	Main Factors and Interactions						
	A	B	C	AB	AC	BC	ABC
Gallic acid	**0.0103**	0.1122	0.2965	0.0533	0.2024	1.0000	1.0000
Protocatechuic acid	0.5460	0.6303	**0.0003**	0.1288	0.5473	0.6643	0.3163
Chlorogenic acid	0.4602	0.5858	0.3388	**0.0188**	0.2426	0.5544	0.4694
Catechin	0.2820	**0.0428**	**<0.0001**	0.4860	0.7561	0.402	0.8375
Rutin	0.1186	**0.0495**	**<0.0001**	0.1930	0.9539	0.4421	0.5229
Quercetin + Luteolin	**0.0450**	0.4300	0.1299	0.1040	0.1298	0.6507	0.7122
Kaempferol	**0.0027**	0.4617	0.623	0.2498	0.1655	0.854	0.8114
Vanillin	0.3814	0.4521	**0.0124**	0.3908	0.998	0.5348	0.0877
Hesperidin	0.0893	**0.0224**	0.0544	**0.0458**	0.0762	**0.0462**	**0.0235**
Neohesperidin	**0.0297**	0.7984	**<0.0001**	0.3242	**0.0434**	0.7298	0.5838
Naringenin	0.3483	0.0651	**<0.0001**	**<0.0001**	**0.0078**	0.1065	0.1957
Apigenin	**0.0010**	0.0913	**<0.0001**	0.78	0.7784	**0.0331**	0.1322
Diosmetin	**0.0394**	0.2556	**0.0046**	0.3214	0.3621	0.9092	0.5612

Note: A = Added Water (%): B = Molar Ratio (mol/mol); C = Citric Peel; bold letters indicate that there is an effect ($p < 0.05$) from the main factors and/or their interactions on the concentration of individual polyphenols observed in the extracts.

3.2. TPC and Antioxidant Capacity on Citrus Peel Extracts

The lowest and highest concentrations of total polyphenols in the experimental design were found in *C. aurantium* extracts (Table S2). The lowest concentration (1.36 ± 0.24 mg GAE/100 g DM) of total polyphenol content (TPC) was achieved with a NADES having an MR of 1:1 mol and a WA of 50%. Using an MR of 1:0.5 mol with a higher WA percentage (60%) resulted in an extract with the highest concentration (96.23 ± 0.83 mg GAE/100 g DM) of phenolic compounds (Figure 2); these values were also the lowest and highest in the entire experimental design.

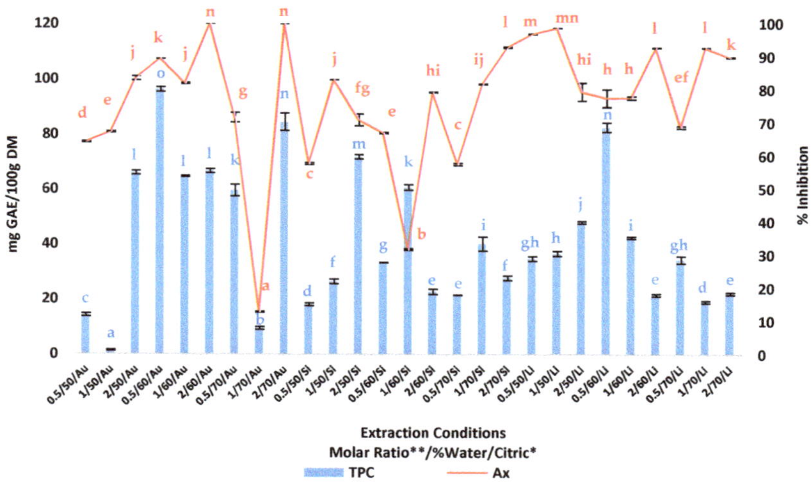

Figure 2. Total polyphenol content (TPC) and antioxidant capacity (Ax) of citrus peel extracts obtained using a choline chloride-based NADES with ultrasound. Different lowercase letters indicate statistically significant differences in TPC, while uppercase letters denote differences in antioxidant capacity. Different letters of the same color indicate a statistically significant difference (LSD, $p < 0.05$) for the total polyphenol content (blue) and antioxidant capacity (red); * Au = *Citrus aurantium;* Si = *Citrus sinensis;* Li = *Citrus Limon;* ** Molar ratio of glucose per 1 mol of Choline chloride. Values are means ± SD (n = 3).

On the other hand, only for *C. sinensis* extracts, the lowest concentration of TPC (18.03 ± 0.47 mg GAE/100 g DM) was obtained under the conditions of experiment #10 (MR = 1:0.5 mol, WAAW = 50%), while the highest concentration (71.73 ± 0.76 mg GAE/100 g DM) was observed in experiment #12 (MR = 1:2 mol, WA = 50%).

Figure 2 also displays the TPC concentrations for the *C. limon* extracts. Among the *C. limon* extracts obtained, the analysis revealed that the lowest (19.02 ± 0.43 mg GAE/100 g DM) concentration of phenolic compounds was achieved using a NADES with 1:1 (MR) and 70% added water (WA). In contrast, using 1:0.5 mol formulated with 60% WA resulted in an increased concentration (82.46 ± 1.74 mg GAE/100 g DM) of these metabolites of interest for *C. limon* extracts.

In a similar way, with TPC, the lowest and highest antioxidant capacity were found in *C. aurantium* peel extract. Under conditions of 11: (MR) and 70% of added water (WA), the lowest antioxidant capacity (12.82 ± 0.14% Inhibition) extract was obtained. For the highest antioxidant capacity extract, no difference ($p > 0.05$) was found between the conditions from experiments #6 (MR = 1:2 mol, WA = 60%, *C. aurantium*), #9 (MR = 1:2, WA = 70%, *C. aurantium*), both with 100 ± 0.00% de inhibition, and #20 (MR = 1:1 mol, WA = 50%, *C. limon*), with 98.58 ± 0.00% inhibition.

However, among extracts obtained from *C. sinensis* peel, the one with highest antioxidant capacity (83.05 ± 0.00% Inhibition) was obtained using an MR of 1:1 and 50% added water (Exp #11).

Figure 3a illustrates that any factor or interaction shows an effect ($p > 0.05$) on total polyphenol content in extracts obtained using choline chloride-based NADES. Conversely, the molar ratio was the only factor (main and interactions) to significantly affect ($p < 0.05$) the antioxidant capacity of extracts derived from a citrus by-product (Figure 3b).

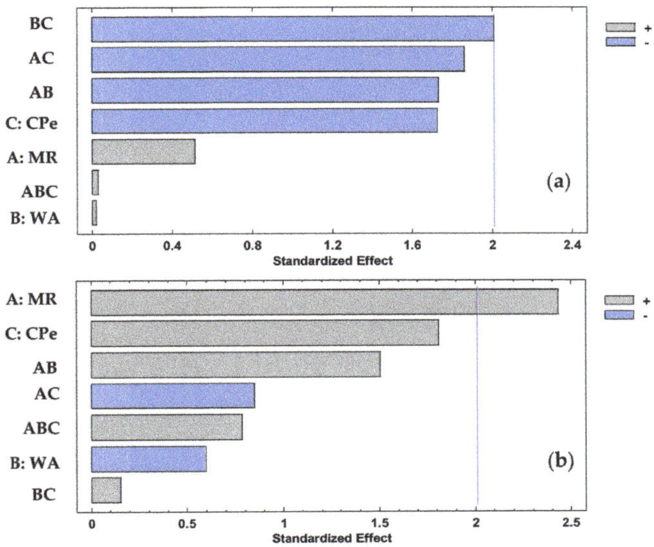

Figure 3. Pareto charts of (**a**) total polyphenol content (TPC) and (**b**) antioxidant capacity (Ax). Where MR = Molar Ratio; WA = Added Water; CPe = Citrus peel.

3.3. Linear and Pearson Correlation

According to the linear correlation analysis results (Table 3), the antioxidant capacities of extracts obtained from *C. limon* peel were well correlated with diosmetin (r = 0.7863) and vanillin (r = 0.8487). On the other hand, neohesperidin (r = 0.7486) and vanillin (r = 0.8072) showed a strong correlation with the antioxidant capacity of *C. sinensis* extract.

Table 3. Linear correlation of phenolic compounds with antioxidant capacity of extracts from different citrus peels.

Phenolic Compound	Antioxidant Capacity (DPPH)		
	C. aurantium	*C. sinensis*	*C. limon*
TPC	0.5564	0.1051	0.2280
Gallic acid	0.0000	−0.2372	−0.3056
Protocatechuic acid	0.0000	−0.0315	−0.0460
Catechin	0.3395	0.5492	−0.1918
Chlorogenic acid	−0.0394	0.1211	0.0000
Rutin	0.4420	0.5211	−0.3178
Q + L	**0.7706**	0.5747	0.6023
Kaempferol	**0.7110**	0.4312	0.3271
Vanillin	0.4496	**0.8072**	**0.8487**
Hesperidin	0.1803	−0.2467	0.0456
Neohesperidin	**0.7486**	**0.7129**	0.6004
Naringenin	0.3353	0.6091	0.6722
Apigenin	−0.3562	0.3604	−0.4075
Diosmetin	0.2755	−0.2951	**0.7863**

Additionally, the antioxidant capacity of extract from *C. aurantium* presented a good correlation with several phenolic compounds such as quercetin + luteolin (r = 0.7716), kaempferol (r = 0.7110), neohesperidin (r = 0.7484), and the total polyphenol content (r = 0.7514).

There was no linear correlation between phenolic compounds and antioxidant capacity when analyzing the complete experimental design 3^3 (Table S3).

The data obtained from all the response variables of the 3^3 experimental design were analyzed using the Pearson correlation (Figure 4).

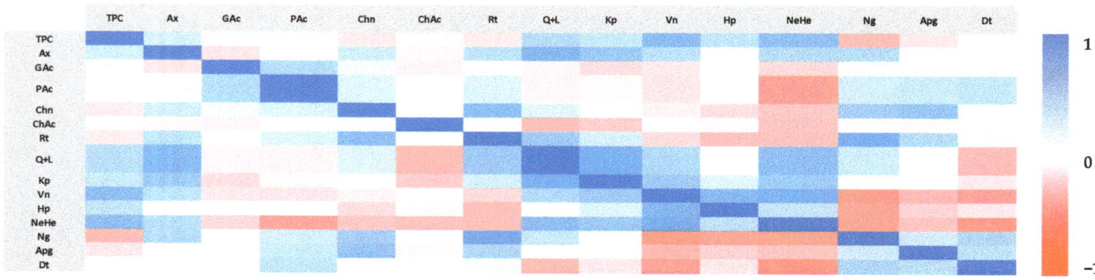

Figure 4. Pearson correlation heatmap of the 3^3 experimental design for the evaluation of polyphenol extraction from citrus peel using different NADES. TPC = Total polyphenol content; Ax = Antioxidant activity; GAc = Gallic acid; PAc = Protocatechuic acid; Chn = Catechin; ChAc = Chlorogenic acid; Rt = Rutin; Q + L = Quercetin + Luteolin; Kp = Kaempferol; Vn = Vanillin; Hp = Hesperidin; NeHe = Neohesperidin; Ng = Naringenin; Apg = Apigenin; Dt = Diosmetin.

A positive correlation was observed between naringenin and rutin (r = 0.7547), kaempferol and Q + L (r = 0.7193), and neohesperidin and vanillin (r = 0.7193).

In Figure S4, the analysis of TPC and antioxidant capacity by citrus peel extract is shown. For example, in the extracts of *C. aurantium* (Figure S4a), a positive correlation was observed between the TPC and the individual polyphenols kaempferol (r = 0.8448), Q + L (r = 0.7552), vanillin (r = 0.7738), and neohesperidin (r = 0.7846). Meanwhile, Ax showed a correlation with TPC (r = 0.7614), neohesperidin (r = 0.7486), Q + L (r = 0.7706), and kaempferol (r = 0.7110). Almost all the phenolic compounds that showed a positive correlation with TPC also exhibited a similar correlation with Ax.

C. sinensis (Figure S4b) extracts showed a positive correlation between TPC and diosmetin (r = 0.8497). On the other hand, Ax was positively correlated with vanillin (r = 0.8072) and neohesperidin (r = 0.7129). A positive correlation was found between TPC and gallic acid (r = 0.8631), as well as between Ax with both, vanillin (r = 0.8487) and diosmetin (r = 0.7863) in *C. limon* (Figure S4c) extracts.

3.4. Principal Analysis Component

The response variables (excluding gallic acid, protocatechuic acid, chlorogenic acid, apigenin, and diosmetin) and the experimental conditions were analyzed together using principal component analysis (PCA) with the aim of simplifying the number of variables and identifying patterns and/or data groupings. The first two components explain 72.65% of the data variability.

In Figure 5a, the PCA biplot of the data from the 3^3 experimental design is shown, where the lines indicate the response variables (TPC, Ax, and polyphenol profile), and the points represent the observations (experimental conditions). The K-means clustering of the experimental conditions and response variables is also represented in Figure 5b.

The results show that TPC (1) exhibits a small distance from vanillin (10) and neohesperidin (12), indicating correlation and a similar behavior of the variables. TPC is also near hesperidin (11) and kaempferol (9). Catechin (5), rutin (7), and naringenin (13) are located at an angle close to 90° relative to TPC, which can be interpreted as a very low or null correlation. On the other hand, Ax (2) is very close to quercetin + luteolin (8), as these polyphenols have shown higher concentrations in extracts with a high antioxidant capacity.

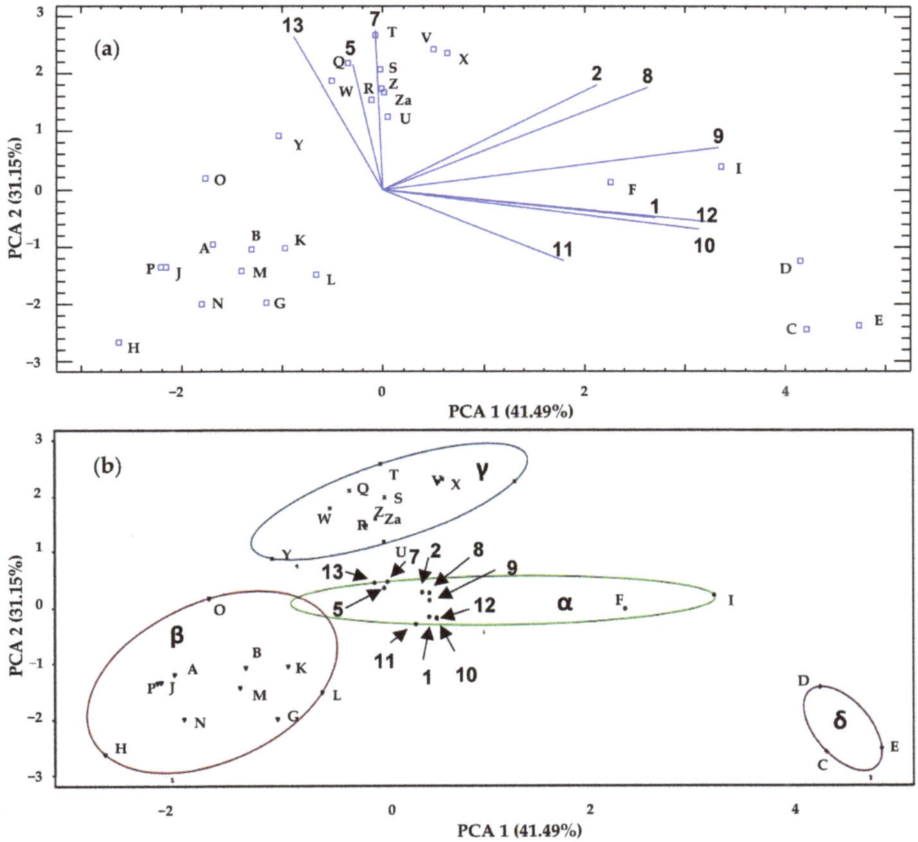

Figure 5. (**a**) Principal component analysis (PCA) and (**b**) *k*-means of the 3^3 experimental design for the evaluation of polyphenol extraction from citrus peel using different NADES. Where letters are conditions of experiments (Table 1): A = #1; B = #2; C = #3; D = #4; E = #5; F = #6; G = #7; H = #8; I = #9; J = #10; K = #11; L = #12; M = #13; N = #14; O = #15; p = #16; Q = #17; R = #18; S = #19; T = #20; U = #21; V = #22; W = #23; X = #24; Y = #25; Z = #26; Z.a = #27 and numbers are variables response that correspond to the order specified at Tables S1 and S2, 1 = Total polyphenol content; 2 = Antioxidant capacity; 5 = catechin; 7 = rutin; 8 = quercetin + luteolin; 9 = kaempferol; 10 = vanillin; 11 = hesperidin; 12 = neohesperidin; 13 = naringenin. Clusters are shown in Greek letters.

Additionally, four different clusters were identified. It was observed that the response variables TPC, Ax, and the individual polyphenols analyzed were primarily associated with *C. aurantium* and *C. sinensis*, as well as the extraction conditions situated in clusters α and β, which could potentially enhance the extraction of these bioactive compounds and their antioxidant capacity.

Finally, clusters γ and δ exhibit unique characteristics and conditions that are likely not optimal for obtaining extracts rich in polyphenols, antioxidant capacity, or high concentrations of individual polyphenols of interest.

4. Discussion

Each analyzed CPex showed differences in the polyphenol profile, which varied according to the molar concentration of the NADES and its added water percentage. The characteristic and predominant polyphenols in the extracts obtained in this study were hesperidin and neohesperidin, both identified in the peels of *C. aurantium* and *C. sinensis*. Conversely, catechin was observed in higher concentrations in the peel of *C. sinensis*, while rutin was the predominant phenolic compound in the peel extract of *C. limon*.

Hesperidin (Hp) is a flavonoid characteristic of citrus fruits, classified as a glycosylated flavanone derived from the structure of hesperitin. Hp has been previously reported in the peel of *C. aurantium* from Yucatán by Covarrubias et al. [27]. The extraction was performed using organic solvents, with the highest concentration of Hp obtained using distilled water (292.00 ± 0.00 mg/100 g dry mass), followed by 50% ethanol (107.00 ± 2.00 mg/100 g dry mass) and 96% ethanol (90.00 ± 0.00 mg/100 g dry mass); all the extractions were assisted by ultrasound. These concentrations were significantly lower than those reported in this study, observing a highest concentration of 2000.37 ± 10.91 mg/100 g dry mass under conditions of 1:1 mol/mol (ChCl:Glu) and 60% of added water from *C. aurantium* peel. This difference is primarily attributed to the use of NADES instead of traditional organic solvents.

According to Liu et al. [28], the extraction of Hp from the peel of *C. aurantium* shows better yields when using NADES (244.00 ± 24.00 mg/100 g dry mass), which can be based on choline chloride and feature hydrogen bond donors such as amines, alcohols, and even sugars like glucose, compared to organic solvents like methanol (216.00 ± 15.00 mg/100 g dry mass), 80% methanol (190.00 ± 17.00 mg/100 g dry mass), and water (210.00 ± 23.00 mg/100 g dry mass). This is due to various factors, such as low concentrations (40%) of water percentage that reduces the solvent's viscosity, allowing for proper mass transfer and the extraction of phenolic compounds like flavonoids. However, if the percentage reaches 80% of added water, it can disrupt the structure of the NADES, weakening the hydrogen bonds and decreasing the Hp extraction efficiency.

This same increase in added water would promote a highly polar environment, which, according to Pyrzynska [10] and Xu et al. [18], decreases the efficiency of Hp extraction due to its lower polarity behavior. Thus, the principle of "like dissolves like" would not apply. This explains the low concentrations of Hp found in the extracts of various citrus fruits, in this study, where a 70% water addition was used in NADES, compared to extracts with lower added water (50%, 60%).

Another factor in the extraction of hesperidin from citrus peels such as *C. sinensis*, as reported by Xu et al. [18], is the use of different ternary NADES (choline chloride-based, HBA) with various molar ratios and 25% added water. This study found that a neutral or slightly acidic pH (pH: 5.91–7.72) of the eutectic solvent resulted in higher extraction efficiency of Hp extraction. Under these conditions, the amorphous form of this flavanone is stabilized, improving its solubility [29]. In contrast, at a pH near 9, degradation of this metabolite is observed [18]. All factors involved in the extraction yield for Hp can be considered for the extraction of neohesperidin (Nhp) from citrus peels, as these metabolites are considered isomers due to their chemical structure. Both consist of hesperetin linked to a disaccharide; for Hp it is rutinose (Rhamnose-α1-6-glucose) and for Nhp it is neohesperidoside (Rhamnose-α1-2-glucose). However, Nhp might exhibit lower structural flexibility, potentially negatively affecting its extraction [30,31]. This behavior was observed in the extracts of *C. aurantium and C. sinensis*, where some extracts show similar concentrations of Hp, but was different in Nhp, verifying the data obtained from the PCA, where both metabolites presented a similar vector.

Another major polyphenol identified was rutin (quercetin-3-rutinoside), mainly in extracts from *C. limon* peels; this metabolite is a glycosylated flavonoid, meaning it is a quercetin molecule linked to a rutinose (glucose-α1-6-rhamnose) via a β-glycosidic bond at the 3rd carbon of the C-ring or pyran ring [32]. This coincides with the findings reported by Xi et al. [9], which indicate that the extract peel (methanol 80%:Dimethyl Sulfoxide [DMSO], 1:1 *v*/*v*, 12 h maceration) of *C. limon* contains a higher concentration of rutin (6.05 mg/100 g fresh peel) compared to other parts of the fruit (pulp > juice > seed). Nonetheless, the concentration is lower compared to hesperidin (331.5 mg/100 g fresh peel), which is opposite to the findings of this study by using NADES.

The concentrations of rutin found in the lemon extract can be attributed to the high solubility of this flavonoid in choline chloride-based NADES. Zang et al. [33] reported the extraction of rutin from *Sophora japonica* using various NADES preparations and added water combined with ultrasound. They observed that the increase in added water of a choline chloride-based NADES (choline chloride, 1:1 mol/mol) was from 10% to 20%, and the increase in extraction yield was from approximately 145 mg/g dry mass to 291 mg/g dry mass.

Regarding the extraction of total polyphenols, the highest concentration was observed in *C. aurantium* extracts (96.23 ± 0.83 mg/100 g DM), compared to *C. sinensis* and *C. limon*. This behavior was different from what has already been reported by Lagha-Benamrouche & Madani [6], where a maceration (80% methanol, 22 h) of *C. aurantium* and *C. sinensis* citrus peels (various varieties) was performed. The result was a TPC concentration in *C. aurantium* (bittersweet orange) peel extracts of 3162 ± 88.00 mg GAE/100 g DM, compared to *C. sinensis* with 2560 ± 23.00 mg GAE/100 g DM. The significant difference in these results is due to the extended extraction time when using maceration, which allows for greater diffusion of phenolic compounds into the solvent. However, the use of organic solvents and the required volume make it unsuitable for the extracts to be used for human consumption or at least to be considered GRAS (Generally Recognized as Safe) [34].

On the other hand, using NADES for the extraction of TPC from *C. aurantium* peel, it was observed that when using maceration of choline chloride-based NADES with a hydrogen donor such as 1,4-butanediol (1:3 mol/mol) and a water content of 49.9%, a concentration of 785.2 mg GAE/100 g DM was achieved [35]. This concentration does not align with that obtained in this study, with the use of glucose as the hydrogen donor at a ratio of 0.5 mol per 1 mol of choline chloride, 60% added water, and ultrasound-assisted extraction. These results are attributed to the affinity that choline chloride-based NADES exhibit with phenolic compounds, which is due to their ability to generate molecular stability. They present a slightly acidic pH, which, when combined with an appropriate percentage of water, can enhance diffusivity and solubility. Additionally, the 1,4-butanediol molecule has a linear configuration with two hydroxyl groups at the ends, unlike glucose, which has a cyclic structure. This allows 1,4-butanediol to generate NADES with greater stability among the hydrogen bond acceptor (HBA), water, and phenolic compounds [36]. Additionally, the use of ultrasound can further improve mass transfer by breaking down the plant matrix through the cavitation of ultrasonic waves and the differences in the concentration of TPC from these matrices reported in this same order, *C. aurantium* > *C. sinensis* > *C. limon* [18,28,37,38].

The antioxidant capacity observed in various citrus peel extracts was quite comparable, even though these extracts, such as those from *C. limon*, had significantly lower levels of total phenolic content (TPC) and individual polyphenols. This may be attributed to: (a) a more diverse polyphenol profile, creating a synergistic effect on DPPH radical inhibition [39]; and (b) the presence of other bioactive compounds such as vitamin C, limonoids, and coumarins, which are well-documented in citrus peels and possess an antioxidant capacity [40,41].

In contrast, extracts of *C. aurantium* and *C. sinensis* exhibited a less varied polyphenol profile but with high concentrations of hesperidin and neohesperidin, both reported to have a significant antioxidant capacity. Notably, neohesperidin showed a correlation with DPPH inhibition in both peels. According to Di Majo et al. [41], the glycosylation of hesperidin and neohesperidin could diminish their antioxidant capacity; however, it is comparable with molecules with a high antioxidant capacity like hesperitin.

Finally, all the extracts demonstrated a high capacity to inhibit the DPPH radical and exhibited elevated TPC concentrations. Most importantly, they showed high concentrations of polyphenols such as hesperidin, neohesperidin, rutin, and catechin, which have been reported for the prevention and treatment of various pathologies, including cardiovascular diseases, diabetes, hypercholesterolemia, and hypertension, among others [12,42,43].

5. Conclusions

The use of natural deep eutectic solvents (NADES) based on choline chloride and glucose, as a hydrogen bond donor, demonstrated high effectiveness in extracting phenolic compounds from the peels of various citrus fruits. These NADES showed a particular affinity for compounds such as hesperidin and neohesperidin in the peels of *C. aurantium* and *C. sinensis*, as well as rutin and catechin in the peels of *C. limon*, all of which are metabolites of interest for the pharmaceutical industry.

The addition of water to the eutectic solvents had a notable effect on the antioxidant capacity of the extracts. Furthermore, key factors, such as interactions among components (both double and triple), influenced the extraction of certain individual polyphenols, demonstrating that NADES can be modified to target the extraction of specific desired compounds.

Finally, extracts from *C. aurantium* and *C. sinensis* have great potential for use in the formulation of functional foods, nutraceuticals, or pharmaceuticals for the treatment of cardiovascular diseases. For this reason, it is recommended that extraction conditions be optimized to maximize their concentration.

Supplementary Materials: The following supporting information can be downloaded at: https://www.mdpi.com/article/10.3390/pr12102072/s1, Figure S1: Calibration curve of gallic acid for the determination of total polyphenols from citrus peel; Figure S2: Chromatogram of individual standards polyphenols mix. Numbering: 1 = gallic acid (RT = 0.6305 min, R^2 = 0.9940); 2 = protocatechuic acid (RT = 1.3252, R^2 = 0.9998 min); 3 = catechin (RT = 3.7071 min, R^2 = 0.9993); 4 = chlorogenic acid (RT = 3.8682 min, R^2 = 0.9993); 5 = coumaric acid (RT = 4.8438 min, R^2 = 0.9995); 6 = rutin (RT = 6.2894 min, R^2 = 0.9993); 7 = hesperidin (RT = 7.6707 min, R^2 = 0.9996); 8 = quercetin + luteolin (RT = 8.5638 min, R^2 = 0.9998); 9 = kaempferol (RT = 9.7681 min, R^2 = 0.9991); Figure S3: Chromatogram of selected individual polyphenols analyzed in citrus peel extracts of (a) *C. aurantium*, (b) *C. sinensis* and (c) *C. limon*. Numbering: 1 = gallic acid; 2 = protocatechuic acid; 3 = catechin; 4 = chlorogenic acid; 5 = coumaric acid; 6 = rutin; 7 = hesperidin; 8 = quercetin + luteolin; 9 = kaempferol; Figure S4: Pearson Correlation Heatmap of the 3^3 Experimental Design for the Evaluation of Polyphenol Extraction Using Different NADES from (a) *C. aurantium* (b) *C. sinensis* and (c) *C. limon*. TPC = Total polyphenol content; Ax = Antioxidant capacity; GAc = Gallic acid; PAc = Protocatechuic acid; Chn = Catechin; ChAc = Chlorogenic acid; Rt = Rutin; Q + L = Quercetin + Luteolin; Kp = Kaempferol; Vn = Vanillin; Hp = Hesperidin; NeHe = Neohesperidin; Ng = Naringenin; Apg = Apigenin; Dt = Diosmetin. Table S1: Citric peel extracts polyphenol profile obtained using choline chloride-based NADES by ultrasound-assisted extraction. Table S2: Citric peel extracts TPC and antioxidant capacity obtained using choline chloride NADES based by ultrasound-assisted extraction.

Author Contributions: Conceptualization, M.O.R.-S. and I.M.R.-B.; methodology, K.A.A.-B. and A.L.-M.; software, M.O.R.-S. and K.A.A.-B.; validation, I.M.R.-B. and M.O.R.-S.; formal analysis, K.A.A.-B.; investigation, K.A.A.-B. and I.M.R.-B.; resources, M.O.R.-S. and I.M.R.-B.; data curation, M.O.R.-S. and I.M.R.-B.; writing—original draft preparation, K.A.A.-B., M.O.R.-S. and I.M.R.-B.; writing—review and editing, M.O.R.-S., K.A.A.-B. and I.M.R.-B.; visualization, M.O.R.-S.; supervision, M.O.R.-S. and I.M.R.-B.; project administration, M.O.R.-S.; funding acquisition, M.O.R.-S. and I.M.R.-B. All authors have read and agreed to the published version of the manuscript.

Funding: Centro de Investigación y Asistencia en Tecnología y Diseño del Estado de Jalisco A.C.; which financed the project UEVISCOEUTECTIC No 2004500186, and the scholarship 661099 for Kevin Alejandro Avilés-Betanzos, financed by CONAHCYT.

Data Availability Statement: The original contributions presented in the study are included in the article/supplementary material, further inquiries can be directed to the corresponding author.

Conflicts of Interest: The authors declare no conflicts of interest.

References

1. United States Department of Agriculture (USDA): Foreign Agriculture Service. Citrus: World Markets and Trade. Available online: https://apps.fas.usda.gov/psdonline/circulars/citrus.pdf (accessed on 20 June 2024).
2. Kuma, V.; Kaur, R.; Aggarwal, P.; Singh, G. Underutilized citrus species: An insight of their nutraceutical potential and importance for the development of functional food. *Sci. Hortic.* **2024**, *296*, 110909. [CrossRef]
3. Servicio de Información Agroalimentaria y Pesquera (SIAP). Available online: https://nube.siap.gob.mx/cierreagricola/ (accessed on 20 June 2024).
4. Maqbool, Z.; Khalid, W.; Atiq, H.T.; Koraqi, H.; Javaid, Z.; Alhag, S.K.; Al-Shuraym, L.A.; Bader, D.M.D.; Almarzuq, M.; Afifi, M.; et al. Citrus Waste as Source of Bioactive Compounds: Extraction and Utilization in Health and Food Industry. *Molecules* **2023**, *28*, 1636. [CrossRef] [PubMed]
5. Rathod, N.B.; Elabed, N.; Punia, S.; Ozogul, F.; Kim, S.-K.; Rocha, J.M. Recent Developments in Polyphenol Applications on Human Health: A Review with Current Knowledge. *Plants* **2023**, *12*, 1217. [CrossRef]
6. Lagha-Benamrouche, S.; Madani, K. Phenolic contents and antioxidant activity of orange varieties (*Citrus sinensis* L. and *Citrus aurantium* L.) cultivated in Algeria: Peels and leaves. *Ind. Crops Prod.* **2013**, *50*, 723–730. [CrossRef]
7. Papoutsis, K.; Pristijono, P.; Golding, J.B.; Stathopoulos, C.E.; Scarlett, C.J.; Bowyer, M.C.; Vuong, Q.V. Impact of different solvents on the recovery of bioactive compounds and antioxidant properties from lemon (*Citrus limon* L.) pomace waste. *Food Sci. Biotechnol.* **2016**, *25*, 971–977. [CrossRef]
8. Pontifex, M.G.; Malik, M.; Connell, E.; Müller, M.; Vauzour, D. Citrus Polyphenols in Brain Health and Disease: Current Perspectives. *Front. Neurosci.* **2021**, *15*, 640–648. [CrossRef] [PubMed]
9. Xi, W.; Lu, J.; Qun, J.; Jiao, B. Characterization of phenolic profile and antioxidant capacity of different fruit part from lemon (*Citrus limon* Burm.) cultivars. *J. Food Sci. Tecnhol.* **2017**, *54*, 1108–1118. [CrossRef]
10. Pyrzynska, K. Hesperidin:A Review on Extraction Methods, Stability and Biological Activities. *Nutrients* **2022**, *14*, 2387. [CrossRef]
11. M'hiri, N.; Ioannou, I.; Mihoubi Boudhrioua, N.; Ghoul, M. Effect of different operating conditions on the extraction of phenolic compounds in orange peel. *Food Bioprod. Process.* **2015**, *96*, 161–170. [CrossRef]
12. Roy, J.; Azamthulla, M.; Mukkerjee, D. Hesperidin and Diosmin-A novel Drugs. *Int. J. Pharm. Res. Technol.* **2020**, *10*, 25–33.
13. Hajialyani, M.; Farzaei, M.H.; Echeverría, J.; Nabavi, S.M.; Uriarte, E.; Sobarzo-Sánchez, E. Hesperidin as a neuroprotective agent: A review of animal and clinical evidence. *Molecules* **2019**, *24*, 648. [CrossRef] [PubMed]
14. Cheng, F.J.; Huynh, T.K.; Yang, C.S.; Hu, D.W.; Shen, Y.C.; Tu, C.Y.; Wu, Y.C.; Tang, C.H.; Huang, W.C.; Chen, Y.; et al. Hesperidin is a potential inhibitor against SARS-CoV-2 infection. *Nutrients* **2021**, *13*, 2800. [CrossRef] [PubMed]
15. Safdar, M.N.; Kausar, T.; Jabbar, S.; Mumtaz, A.; Ahad, K.; Saddozai, A.A. Extraction and quantification of polyphenols from kinnow (*Citrus reticulate* L.) peel using ultrasound and maceration techniques. *J. Food Drug Anal.* **2017**, *25*, 488–500. [CrossRef]
16. Saini, A.; Panesar, P.S.; Bera, M.B. Comparative Study on the Extraction and Quantification of Polyphenols from Citrus Peels using Maceration and Ultrasonic Technique. *Curr. Res. Nutr. Food Sci. J.* **2019**, *7*, 678–685. [CrossRef]
17. Hilali, S.; Fabiano-Tixier, A.; Ruiz, K.; Hejjaj, A.; Nouh, F.A.; Idlimam, A.; Bily, A.; Mandi, L.; Chemat, F. Green Extraction of Essential Oils, Polyphenols, and Pectins from Orange Peel Employing Solar Energy: Toward a Zero-Waste Biorefinery. *ACS Sustain. Chem. Eng.* **2019**, *7*, 11815–11822. [CrossRef]
18. Xu, M.; Ran, L.; Chen, N.; Fan, X.; Ren, D.; Yi, L. Polarity-dependent extraction of flavonoids from citrus peel waste using a tailor-made deep eutectic solvent. *Food Chem.* **2019**, *297*, 124970. [CrossRef]
19. Elgharbawy, A.A.M.; Hayyan, A.; Hayyan, M.; Mirghani, M.E.S.; Salleh, H.M.; Rashid, S.N.; Ngoh, G.C.; Liew, S.Q.; Nor, M.R.M.; bin Mohd Yusoff, M.Y.Z.; et al. Natural Deep Eutectic Solvent-Assisted Pectin Extraction from Pomelo Peel Using Sonoreactor: Experimental Optimization Approach. *Processes* **2019**, *7*, 416. [CrossRef]
20. Molnar, M.; Gašo-Sokač, D.; Komar, M.; Jakovljević Kovač, M.; Bušić, V. Potential of Deep Eutectic Solvents in the Extraction of Organic Compounds from Food Industry By-Products and Agro-Industrial Waste. *Separations* **2024**, *11*, 35. [CrossRef]
21. Wu, K.; Ren, J.; Wang, Q.; Nuerjiang, M.; Xia, X.; Bian, C. Research Progress on the Preparation and Action Mechanism of Natural Deep Eutectic Solvents and Their Application in Food. *Foods* **2022**, *11*, 3528. [CrossRef]
22. Sailau, Z.; Almas, N.; Aldongarov, A.; Toshtay, K. Studying the Formation of Choline Chloride- and Glucose-Based Natural Deep Eutectic Solvent at the Molecular Level. *J. Mol. Model.* **2022**, *28*, 235. [CrossRef]
23. Chel-Guerrero, L.D.; Oney-Montalvo, J.E.; Rodríguez-Buenfil, I.M. Phytochemical characterization of by-products of habanero pepper grown in two different types of soils from Yucatán, Mexico. *Plants* **2021**, *10*, 779. [CrossRef] [PubMed]
24. Avilés-Betanzos, K.A.; Oney-Montalvo, J.E.; Cauich-Rodríguez, J.V.; González-Ávila, M.; Scampicchio, M.; Morozova, K.; Ramírez-Sucre, M.O.; Rodríguez-Buenfil, I.M. Antioxidant Capacity, Vitamin C and Polyphenol Profile Evaluation of a Capsicum chinense By-Product Extract Obtained by Ultrasound Using Eutectic Solvent. *Plants* **2022**, *11*, 2060. [CrossRef] [PubMed]
25. Avilés-Betanzos, K.A.; Cauich-Rodríguez, J.V.; González-Ávila, M.; Scampicchio, M.; Morozova, K.; Ramírez-Sucre, M.O.; Rodríguez-Buenfil, I.M. Natural Deep Eutectic Solvent Optimization to Obtain an Extract Rich in Polyphenols from *Capsicum chinense* Leaves Using an Ultrasonic Probe. *Processes* **2023**, *11*, 1729. [CrossRef]
26. Singleton, V.L.; Orthofer, R.; Lamuela-Raventós, R.M. Analysis of total phenols and other oxidation substrates and antioxidants by means of folin-ciocalteu reagent. In *Methods in Enzymology*; Academic Press: Cambridge, MA, USA, 1999; pp. 152–178.

27. Covarrubias C., A.; Jesús, P.V.; Hugo, E.A.; Teresa, A.T.; Ulises, G.C.; Neith, P. Antioxidant capacity and UPLC–PDA ESI–MS polyphenolic profile of *Citrus aurantium* extracts obtained by ultrasound assisted extraction. *J. Food Sci. Technol.* **2018**, *55* 5106–5114. [CrossRef]
28. Liu, Y.; Zhang, H.; Yu, H.; Guo, S.; Chen, D. Deep eutectic solvent as a green solvent for enhanced extraction of narirutin, naringin hesperidin and neohesperidin from Aurantii Fructus. *Phytochem. Anal.* **2019**, *30*, 156–163. [CrossRef]
29. Rosiak, N.; Wdowiak, K.; Tykarska, E.; Cielecka-Piontek, J. Amorphous Solid Dispersion of Hesperidin with Polymer Excipients for Enhanced Apparent Solubility as a More Effective Approach to the Treatment of Civilization Diseases. *Int. J. Mol. Sci.* **2022**, *23* 15198. [CrossRef]
30. Xu, F.; Liu, Y.; Zhang, Z.; Yang, C.; Tian, Y. Quasi-MSn identification of flavanone 7-glycoside isomers in Da Chengqi Tang by high performance liquid chromatography-tandem mass spectrometry. *Chin. Med.* **2009**, *4*, 15. [CrossRef]
31. Liu, F.; Han, S.; Ni, Y. Isolation and purification of four flavanones from peel of *Citrus changshanensis*. *J. Food Process. Preserv.* **2017** *41*, e13278. [CrossRef]
32. MacIkova, P.; Halouzka, V.; Hrbac, J.; Bartak, P.; Skopalova, J. Electrochemical behavior and determination of rutin on modified carbon paste electrodes. *Sci. World J.* **2012**, *1*, 394756. [CrossRef]
33. Zang, Y.Y.; Yang, X.; Chen, Z.G.; Wu, T. One-pot preparation of quercetin using natural deep eutectic solvents. *Process Biochem.* **2020**, *89*, 193–198. [CrossRef]
34. Kunaedi, A.; Indawati, I.; Karlina, N.; Fadhillah, Z.; Cantika, C.D.; Khulfiah, A.A. Study on the optimization of mulberry leaf extract by macerating ethanol and microwave assisted extraction method (MAE) with natural deep eutectic solvents (NADES). *J. Farm. Sains Dan Prakt.* **2023**, *9*, 11–19. [CrossRef]
35. Edrisi, S.; Bakhshi, H. Separation of polyphenolic compounds from *Citrus aurantium* L. peel by deep eutectic solvents and their recovery using a new DES-based aqueous two-phase system. *J. Mol. Liq.* **2024**, *402*, 124790. [CrossRef]
36. Mulia, K.; Muhammad, F.; Krisanti, E. Extraction of vitexin from binahong (*Anredera cordifolia* (Ten.) Steenis) leaves using betaine—1,4 butanediol natural deep eutectic solvent (NADES). *AIP Conf. Proc.* **2017**, *1823*, 020018. [CrossRef]
37. Ranjha, A.N.; Irfan, S.; Lorenzo, J.M.; Shafique, B.; Kanwal, R.; Pateiro, M.; Arshad, R.N.; Wang, L.; Nayik, G.A.; Roobab, U.; et al. Sonication, a potential technique for extraction of phytoconstituents: A systematic review. *Processes* **2021**, *9*, 1406. [CrossRef]
38. Olfa, T.; Gargouri, M.; Akrouti, A.; Brits, M.; Gargouri, M.; ben Ameur, R.; Pieters, L.; Foubert, K.; Magné, C.; Soussi, A.; et al. A comparative study of phytochemical investigation and antioxidative activities of six citrus peel species. *Flavour Fragr. J.* **2021**, *36*, 564–575. [CrossRef]
39. García, B.F.; Torres, A.; Macías, F.A. Synergy and other interactions between polymethoxyflavones from citrus byproducts. *Molecules* **2015**, *20*, 20079–20106. [CrossRef]
40. Nieto, G.; Fernández-López, J.; Pérez-Álvarez, J.A.; Peñalver, R.; Ros, G.; Viuda-Martos, M. Valorization of citrus co-products: Recovery of bioactive compounds and application in meat and meat products. *Plants* **2021**, *10*, 1069. [CrossRef]
41. di Majo, D.; Giammanco, M.; la Guardia, M.; Tripoli, E.; Giammanco, S.; Finotti, E. Flavanones in Citrus fruit: Structure-antioxidant activity relationships. *Food Res. Int.* **2005**, *38*, 1161–1166. [CrossRef]
42. Ganeshpurkar, A.; Saluja, A.K. The Pharmacological Potential of Rutin. *Saudi Pharm. J.* **2017**, *25*, 149–164. [CrossRef]
43. Ganeshpurkar, A.; Saluja, A. The pharmacological potential of catechin. *Indian J. Biochem. Biophys.* **2020**, *57*, 505–511. Available online: https://api.semanticscholar.org/CorpusID:226365739 (accessed on 15 June 2024).

processes

MDPI

Article

Optimized Ultrasonic Extraction of Essential Oil from the Biomass of *Lippia graveolens* Kunth Using Deep Eutectic Solvents and Their Effect on *Colletotrichum asianum*

Juan Pablo Manjarrez-Quintero [1], Octavio Valdez-Baro [1], Raymundo Saúl García-Estrada [1], Laura Aracely Contreras-Angulo [1], Pedro de Jesús Bastidas-Bastidas [1], J. Basilio Heredia [1], Luis Angel Cabanillas-Bojórquez [2,3,*] and Erick Paul Gutiérrez-Grijalva [4,*]

1 Centro de Investigación en Alimentación y Desarrollo, AC. Carretera a Eldorado Km 5.5, Col. Campo el Diez, Culiacán CP 80110, SI, Mexico; jmanjarrez224@estudiantes.ciad.mx (J.P.M.-Q.); ovaldez222@estudiantes.ciad.mx (O.V.-B.); rsgarcia@ciad.mx (R.S.G.-E.); lcontreras@ciad.mx (L.A.C.-A.); pbastidas@ciad.mx (P.d.J.B.-B.); jbheredia@ciad.mx (J.B.H.)

2 Posdoc CONAHCYT-Centro de Investigación en Alimentación y Desarrollo, AC. Carretera a Eldorado Km 5.5, Col. Campo el Diez, Culiacán CP 80110, SI, Mexico

3 Tecnológico Nacional de México–Instituto Tecnológico de Culiacán, Juan de Dios Bátiz 310 pte, Colonia Guadalupe, Culiacán CP 80220, SI, Mexico

4 Cátedras CONAHCYT-Centro de Investigación en Alimentación y Desarrollo, AC. Carretera a Eldorado Km 5.5, Col. Campo el Diez, Culiacán CP 80110, SI, Mexico

* Correspondence: luis.cabanillas@ciad.mx (L.A.C.-B.); erick.gutierrez@ciad.mx (E.P.G.-G.)

Abstract: Essential oils are emerging as alternatives to conventional pest control chemicals. *Lippia graveolens* Kunth (Mexican oregano) is a source of essential oils and during conventional extraction, the biomass generated is discarded as waste; however, reports show that this biomass is still a rich source of essential oils. Conventional essential oil extraction causes contamination and utilizes toxic solvents. Deep eutectic solvents (DESs) offer low toxicity, biodegradability, high selectivity, and yields comparable to organic solvents. This study obtained essential oil from *Lippia graveolens* biomass via hydrodistillation with ultrasound-assisted DES pretreatment. This research aimed to optimize the extraction of essential oil from *Lippia graveolens* biomass using ultrasound-assisted DESs and assess its in vitro and in vivo inhibitory effect on *C. asianum*. The response variables were extraction yield and total reducing capacity. Optimal conditions were determined using a central composite rotatable design, considering solid-to-liquid ratio (0.38 g/mL), ultrasonic amplitude (45.05%), and time (7.47 min). The optimized oil, with thymol (48%) as the predominant component, exhibited more volatile compounds than conventional hydrodistillation. Fungicidal assays highlighted its potential in controlling anthracnose in papaya fruits caused by *C. asianum*, making ultrasound-assisted DES pretreatment a promising alternative for obtaining essential oil from botanical byproducts.

Keywords: essential oils; deep eutectic solvents; ultrasound-assisted extraction; Mexican oregano; *Lippia graveolens*; antifungal activity

1. Introduction

The specie *Lippia graveolens* Kunth, also known as Mexican oregano, the most commercialized in Mexico, contains the compounds carvacrol and thymol, which are used industrially in multiple applications; oregano is sold as packaged leaves, in bottled oil or extract, in products for the perfume, and in the pharmaceutical, alcoholic beverage, natural preservative, cosmetic, and mechanical oil industries, among others [1]. In this context, the processing of plant raw materials generates solid organic residues, also known as biomass. These residues have been reported to be a source of bioactive compounds such as terpenes, alkaloids, flavonoids, polyphenols, and proteins, among others. These types of plant byproducts derived from industrial processes tend to be discarded, which impacts the environment as they emit large quantities of greenhouse gases such as CO_2. Considering this, it is important to minimize the environmental impact of these residues through processes that utilize their content [2].

There are multiple methods for obtaining secondary metabolites from plants. Among these, hydrodistillation is the most commonly used operation in the industry for the extraction of essential oils, which utilizes high temperatures that can alter the chemical composition and affect their quality; however, it achieves low extraction yields [3–5]. Also, an important content of solids is generated after hydrodistillation (biomass), which is usually discarded; however, reports show that biomass is a rich source of bioactive compounds that could be extracted by other technologies. Another widely used operation for obtaining essential oils is extraction with organic solvents; nonetheless, it presents disadvantages such as environmental harm and the tendency to contaminate extracts with toxic solvent residues. On the other hand, extraction with deep eutectic solvents has been proposed as an alternative to the conventional extraction processes of bioactive compounds like essential oils [6]. These solvents, composed of a mixture of hydrogen bond acceptors and donors, are characterized by being biodegradable, non-toxic, having high selectivity based on their composition, and being economical and easy to synthesize [7,8]. Also, ultrasound-assisted extraction has been coupled with deep eutectic solvents to increase the extraction yield of bioactive compounds, due to the reduced extraction times by increasing the contact surface between the solvent and the raw material [8].

Oregano essential oils (OEOs) have been extensively studied for their antimicrobial properties. Various studies indicate that these oils could be used in important crop plants to control and protect against phytopathogens. In this sense, Zhao et al. [9] found that essential oil of *Origanum vulgare* L., along with its components thymol and carvacrol, effectively inhibited *Botrytis cinerea* growth and spore germination, suggesting their potential as natural fungicides. Their study highlights the effectiveness of these natural compounds in combating one of the most common postharvest pathogens, providing an alternative to traditional chemical treatments. Similarly, Medina-Romero et al. [10] demonstrated that essential oil of *Lippia graveolens* Kunth inhibited the growth of various Fusarium species in vitro and on cherry tomato fruits, indicating its use as a biopesticide. This research underscores the versatility of *Lippia graveolens* Kunth essential oil in managing different fungal pathogens, enhancing the sustainability and safety of crop-protection practices. Additionally, Yılmaz et al. [11] identified essential oils of *Origanum vulgare* L., *Salvia officinalis* L., *Rosmarinus officinalis* L., *Eucalyptus* sp., and *Foeniculum vulgare* Mill. as effective inhibitors of *Colletotrichum gloeosporioides*, with *O. vulgare* L. being the most effective. Their findings revealed significant inhibitory effects in both in vitro and in vivo assays, showcasing the potential of these oils to act as natural protective agents against fungal infections. This supports the use of essential oils in postharvest operations due to their low toxicity, biodegradability, and multifunctionality, making them suitable for sustainable agricultural practices.

Due to the substantial losses in food crops caused by such pathogenic agents, it is crucial to address this issue by promoting research and the development of products that protect and enhance plant health [12]. Furthermore, essential oils could be extracted from oregano biomass; therefore, the objective of this research is to optimize the extraction of essential oil from *Lippia graveolens* Kunth biomass using ultrasound-assisted DESs (deep eutectic solvents) and to evaluate its inhibitory effect on *Colletotrichum asianum* using an in vitro and in vivo approach with papaya fruit as a model.

2. Materials and Methods

2.1. Plant Material

Oregano (*Lippia graveolens* Kunth) was acquired from Santa Gertrudis, Durango, México (coordinates: N 23°32′43.8″ W 104°22′20.8″). Oregano biomass from the hydrodistillation process was donated by the company Oreganic. *L. graveolens* Kunth samples were identified at the Herbarium of the School of Agriculture from the Universidad Autónoma de Sinaloa and given the catalogue number FA-UAS-017005. All aerial parts of the plants (leaves, flowers, and stems) were ground into a powder using an Ika Werke M20 grinder (Wilmington, NC, USA).

2.2. Chemical Reagents

Ethanol (Fermont CAS 64-17-5, Monterrey, NL, México), Folin–Ciocalteu (Sigma F9252), sodium carbonate (Na_2CO_3, Sigma 223530), gallic acid (Sigma G7384), choline chloride (Sigma C1879), and lactic acid (CTR 69775) were purchased from Sigma-Aldrich (St. Louis, MO, USA).

2.3. Preparation of Deep Eutectic Solvents

The preparation of deep eutectic solvents for the extraction of essential oils from *L. graveolens* biomass followed the methodology reported by Chen, Wu, Zhu, Wang, Su, and Yi [6] and Chen et al. [13]. Briefly, choline chloride was mixed as the hydrogen bond acceptor, and lactic acid was used as the hydrogen bond donor in a molar ratio of 1:2. A stirring and heating plate was used, maintaining the mixture at 80 °C until the solution was clarified. Thirty percent distilled water (*w*/*w*) was then added.

2.4. Ultrasonic Extraction

L. graveolens biomass powder was mixed with the choline chloride/lactic acid eutectic solvent in a 1:2 ratio and subjected to an ultrasonic extraction system (SONICS VCX 500, Newtown, CT, USA) following the methodology of Chen, Wu, Zhu, Wang, Su, and Yi [6]. The variables of time, ultrasonic amplitude, and solid/liquid ratio were studied according to the experimental design (Table 1). The sample container was placed in an ethylene glycol bath, maintaining a temperature range of 20 to 40 °C to prevent the degradation of the volatile compounds due to the ultrasonic power.

Table 1. Central composite rotatable design of response surface methodology for the optimization of ultrasonic extraction of Mexican oregano biomass.

	Process Variables			Response Variables	
Run	X_1: Solid/Liquid Ratio (g/mL)	X_2: Ultrasonic Amplitude (%)	X_3: Extraction Time (min)	Essential Oil Yield (%)	Total Reducing Capacity (mg GAE/g Biomass)
1	−1 (0.18)	−1 (36.22)	−1 (4.5)	0.9691	0.5949
2	−1 (0.18)	1 (83.78)	−1 (4.5)	0.6845	0.5783
3	−1 (0.18)	−1 (36.22)	1 (15.5)	0.6621	0.4489
4	−1 (0.18)	1 (83.78)	1 (15.5)	0.77	0.4656
5	1 (0.42)	−1 (36.22)	−1 (4.5)	1.2359	1.2170
6	1 (0.42)	1 (83.78)	−1 (4.5)	0.8689	1.002
7	1 (0.42)	−1 (36.22)	1 (15.5)	0.9030	1.0149
8	1 (0.42)	1 (83.78)	1 (15.5)	1.0488	1
9	0 (0.3)	−1.68 (20)	0 (10)	1.0310	1.1433
10	0 (0.3)	1.68 (100)	0 (10)	0.98	0.8152
11	0 (0.3)	0 (60)	−1.68 (0.75)	0.8575	1.085
12	0 (0.3)	0 (60)	1.68 (19.25)	0.8025	0.7625
13	−1.68 (0.1)	0 (60)	0 (10)	0.5106	0.5321
14	1.68 (0.5)	0 (60)	0 (10)	1.0181	1.1836
15	0 (0.3)	0 (60)	0 (10)	1.01	1.3382
16	0 (0.3)	0 (60)	0 (10)	1.03	1.2695
17	0 (0.3)	0 (60)	0 (10)	1.025	1.2650
18	0 (0.3)	0 (60)	0 (10)	1.02	1.4
19	0 (0.3)	0 (60)	0 (10)	1.02	1.4
20	0 (0.3)	0 (60)	0 (10)	0.9639	1.4913

GAE/g biomass: gallic acid equivalents per gram of biomass.

2.5. Optimization

The optimization of the ultrasonic extraction process was conducted using a rotatable central composite design of the response surface methodology with three factors (Table 1). The selected process variables were solid/liquid ratio (X_1, g/mL), ultrasonic amplitude (X_2, %) and extraction time (X_3, min), while the response variables were essential oil yield (%) and total reducing capacity (mg GAE/g biomass) [14]. The ranges for the process variable values were 0.1 to 0.5 g/mL for the solid/liquid ratio, 20 to 100% for ultrasonic amplitude, and 0.75 to 19.24 min for ultrasonic time. We used a design of 3 factors and high (1), low (−1), and axial (−1.68, 1.68) levels with a set of 20 runs. The optimal levels were obtained by contour plots of the graphs. The optimal treatment was carried out for quintupled. The optimization design and data analysis were performed using the statistical package Minitab 18 (Minitab Inc., State College, PA, USA).

2.6. Hydrodistillation

Essential oil extraction was carried out using a Clevenger-type apparatus. A measure of 80 mL of the extract from ultrasonic extraction was used, diluted with distilled water to a volume of 500 mL. The diluted extract was subjected to a temperature of 280 °C in a stirring plate (Cimarec, Thermo Scientific, Waltham, MA, USA) and stirred at 7000 rpm for 1.5 h. The oil yield was estimated based on dry weight (% *w*/*w*) (Equation (1)). Essential oil samples were stored in vials at 4 °C for chromatographic analysis [13,15].

$$\text{Essential oil yield}(\%) = \frac{\text{Essential oil weight (g)}}{\text{Total biomass weight (g)}} * 100 \tag{1}$$

2.7. Total Reducing Capacity

The total reducing capacity of the essential oils from *Lippia graveolens* Kunth was determined using the Folin–Ciocalteu method [16]. Ten microliters of oil were taken and placed in a 96-well microplate to which 230 µL of distilled water and 10 µL of 2 N Folin–Ciocalteu reagent were added. The samples were incubated for 3 min. Subsequently, 25 µL of 4 N Na_2CO_3 was added, and the samples were incubated for 2 h in the absence of light at room temperature. Absorbance was determined at a wavelength of 725 nm using a Synergy HT microplate reader (Bio-Tek Instruments, Inc., Winooski, VT, USA). A gallic acid curve was used, and the results were expressed in milligrams of gallic acid equivalents per gram of biomass (mg GAE/g biomass).

2.8. Gas Chromatography Analysis

The chromatographic analysis was performed using an Agilent 7890B gas chromatography system coupled with a 7000D GC/TD tandem mass spectrometry and an Agilent Technologies 7693 autosampler. High-purity helium was used as the carrier gas at a flow rate of 1 mL min^{-1}. The chromatographic column used was TG-WAXMS 30 m × 0.25 mm × 0.25 µm. The column inlet pressure was set at 9.37 psi. The column temperature was initially maintained at 40 °C for 1 min and then increased at a rate of 10 °C min^{-1} to 280 °C, where it was held for 6 min. A one-microliter volume of the optimized essential oil replicas was injected.

2.9. Mycelial Growth Inhibition

The mycelial growth assay was conducted using the methodology of Amini et al. [17] with modifications. The essential oil from the *Lippia graveolens* Kunth biomass from the optimal treatment was used. The assay was carried out on Petri dishes with potato dextrose agar (PDA) culture medium. Essential oils were dissolved in Tween 80 (0.5% *v*/*v*), and six concentrations (10, 25, 50, 75, 100, and 250 ppm) were evaluated. Culture media were prepared and sterilized, and the oils were added at the proposed concentrations before pouring into Petri dishes. After solidification of the PDA, a 5 mm mycelial disc of the *Colletotrichum asianum* fungus was added. The plates were incubated for 7 days at 27 °C or until the control plates (medium without fungicide) showed mycelial growth across their diameter. After incubation, the diameters of the mycelium on each plate were measured with a digital caliper (Gwong 150 mm). The inhibition diameter was calculated using Equation (1), where *Dc* and *Dt* are the radial growth (mm) of the pathogen in control plates and treated plates, respectively. Finally, the effective concentration at 50% (EC50) in parts per million for each treatment was calculated using Equation (2). This was performed through linear regression in Minitab 19 software. The assay was performed in triplicate.

$$\text{Micelial grownt inhibition}(\%) = \frac{Dc - Dt}{Dc} * 100 \quad (2)$$

2.10. Determination of Minimum Inhibitory Concentration and Minimum Fungicidal Concentration

The minimum inhibitory concentration of the essential oils from the optimal treatment was determined using the microdilution method in broth reported by Alhajj et al. [18] with modifications. A spore suspension of 1×10^5 conidia/mL was prepared from 7-day-old *Colletrotrichum asianum* growth; the phytopathology laboratory of CIAD Unidad Culiacán provided the isolated fungus. The assay was conducted in a 96-well microplate, testing concentrations of 600, 500, 475, 450, 425, 400, 375, 350, 325, 300, 275, 250, 225, and 200 ppm for the essential oils of *Lippia graveolens* Kunth. A measure of 180 microliters of potato dextrose broth (BD Difco), 20 µL of essential oil, and 50 µL of spore solution were added to each well for a total volume of 250 µL. The spore suspension without fungicide was used as a positive control, and the medium without extract served as a negative control. The plates were incubated at 27 °C for 7 days. The lowest concentration of treatments that showed no visible growth was considered the minimum inhibitory concentration (MIC). Wells showing growth were used to determine the minimum fungicidal concentration (MFC) by taking 10 µL and adding them to Petri dishes with potato dextrose agar (PDA). The assay was performed in triplicate.

2.11. In Vivo Fungicidal Activity

The in vivo antifungal activity was determined on papaya fruits (*Carica papaya* var. Maradol). Commercially mature fruits of comparable sizes, not subjected to fungicide application, and without mechanical damage, insect damage, or pathogen infection were selected. The fruit surface was disinfected by immersion in 2% sodium hypochlorite for 2 min, followed by a rinse with distilled water. Two methods, preventive (1) and corrective (2), were traced on the same fruit with two well-separated lines, each with four points. In the corrective method, the inoculum (spore suspension) was first applied at each point of line 1 on the upper surface of the fruit and allowed to dry. Subsequently, the fruit was immersed in the fungicide for 5 min; again, it was allowed to dry before inoculating the points of line 2, corresponding to the preventive method. The assay was conducted on unwounded fruits and repeated on wounded fruits. The wounds were made with a sterile needle at the four points of each line (1 and 2) on the upper surface of the fruit, and 5 µL of a 5×10^5 spores/mL suspension was applied. The fungicide concentrations were as follows: 0 (control), 300, 1000, and 2000 ppm for the optimized oregano essential oil, Timorex gold (Syngenta), and Tecto 60 (Syngenta). Treated fruits were stored in polyethylene bags for 14 days at 27 °C and 95% relative humidity. After 14 days, the fruits were examined for anthracnose lesions. In case of lesions, the lesion diameter was measured, and the effectiveness percentage was reported using Equation (3), where *Lc* is the lesion in the control fruit and *Lt* is the lesion in the treatment. The assay was performed in triplicate. For statistical analysis, a three-factor ANOVA (fungicide, concentration, method) was employed, and the response variable was the effectiveness percentage calculated using Equation (3) [19,20].

$$\text{Effective concentration}(\%) = \frac{Lc - Lt}{Lc} * 100 \quad (3)$$

2.12. Statistical Analysis

The data on essential oil yield (% w/w), total reducing capacity (mg GAE/g biomass), mycelial growth inhibition, and in vivo fungicidal activity were analyzed using the statistical package Minitab 19 (Minitab Inc., State College, PA, USA), which was employed for all data analyses. The assays were conducted in triplicate, and the results were reported as mean ± standard deviation. In the in vivo fungicidal activity assay, a three-factor ANOVA (fungicide, concentration, and extraction method) was employed, and the response variable was the effectiveness percentage. A Tukey test was performed to analyze mean differences, with a significance level set at $p < 0.05$.

3. Results

3.1. Predictive Models

Table 1 shows the combinations of process variables for ultrasonic-assisted extraction (amplitude, time, solid/liquid ratio) used during the essential oil extraction process of Mexican oregano (*Lippia graveolens* Kunth) biomass. These combinations display the obtained values of the response variables (essential oil yield and total reducing capacity).

Based on the experimental data included in Table 1, predictive models were constructed:

$$\hat{Y}_i = \beta_0 + \beta_1 X_1 + \beta_2 X_2 + \beta_3 X_3 + \beta_{12} X_1 X_2 + \beta_{13} X_1 X_3 + \beta_{23} X_2 X_3 + \beta_{11} {X_1}^2 + \beta_{22} {X_2}^2 + \beta_{33} {X_3}^2 \quad (4)$$

Equation (4) relates the selected process variables (X_1: solid/liquid ratio; X_2: ultrasonic amplitude; X_3: extraction time) to the response variables (essential oil yield and total reducing capacity).

3.1.1. Essential Oil Yield (EOY)

For the variable essential oil yield (EOY), a data range between 0.5106 and 1.2359% was found (Table 1). The analysis of variance (ANOVA) for the prediction model (Table 2) indicates that the overall, linear, and quadratic models are significant ($p < 0.001$), with a nonsignificant lack of fit ($p < 0.088$), a determination coefficient of $R^2 = 96.26\%$, and an adjusted determination coefficient (adjusted R^2) = 94.54%. Due to the significance of the model, a high contribution to variability (R^2), similar determination and adjusted coefficients, and null significance to the lack of fit, the model is shown to be adequate and reproducible. For the linear model, it presents the highest contribution to total variability (52.22%; $p < 0.000$), where the process variable with the highest contribution to variability is the solid/liquid ratio (45.01%; $p < 0.000$); this is followed by ultrasonic amplitude (3.23%; $p < 0.005$) and time (3.01%; $p < 0.006$), with the latter two contributing very little to the model. For the quadratic model, the contribution to variability (24.72%; $p < 0.000$) was lower than that of the linear model; here, the quadratic variable that contributes the most to variability is the solid/liquid ratio (17.79%; $p < 0.000$), followed by time (6.93%; $p < 0.000$). The interaction of variables contributes moderately to variability (19.32%; $p < 0.000$), where the combination of amplitude*time variables provides this value (Equation (5)).

$$\text{EOY}\ (\%) = 1.01 + 0.13X_1 - 0.035X_2 - 0.034X_3 + 0.11X_2X_3 - 0.081{X_1}^2 - 0.057{X_3}^2 \quad (5)$$

The prediction model for the response variable essential oil yield using coded and uncoded variables was represented by Equations (5) and (6), respectively, where the solid/liquid ratio variable was represented by SL, ultrasonic amplitude was represented by A, and extraction time was represented by T:

$$\text{EOY}(\%) = 0.646 + 4.538SL - 0.010A - 0.020T + 0.000865AT - 5.692A^2 - 0.00189T^2 \quad (6)$$

After conducting the analysis of variance and the prediction model for the essential oil yield variable, the contour and response surface plots were obtained (Figures 1 and 2). These plots illustrate the behavior of the process variables (amplitude, time, and solid/liquid ratio) on the response variable, i.e., the essential oil yield. Analyzing the contour and response surface plots, the process variable that most positively affects the response variable to maximize it is the solid/liquid ratio. The trend shows that as this variable increases, the essential oil yield is maximized. In contrast, for the variables time and ultrasonic amplitude, they have an opposite effect to the solid/liquid ratio, where a lower ultrasonic amplitude and time maximize the response variable.

Table 2. Analysis of variance for the essential oil yield variable.

Source	Degrees of Freedom	Contribution	SC Adjusted	MC Adjusted	F Value	*p* Value
Model	6	96.26%	0.510	0.085	55.83	0.000
Lineal	3	52.22%	0.276	0.092	60.57	0.000
S/L	1	45.01%	0.243	0.243	159.98	0.000
Amplitude	1	3.23%	0.017	0.017	11.25	0.005
Time	1	3.01%	0.015	0.015	10.48	0.006
Square	3	24.72%	0.131	0.065	43.01	0.000
$Time^2$	1	6.93%	0.047	0.047	31.38	0.000
S/L^2	1	17.79%	0.094	0.094	61.91	0.000
Interaction of two factors	1	19.32%	0.102	0.102	67.24	0.000
Amplitude × Time	1	19.32%	0.102	0.102	67.24	0.000
Error	13	3.74%	0.019	0.001		
Lack of fit	8	3.18%	0.016	0.002	3.59	0.088
Pure error	5	0.55%	0.002	0.0005		
Total	19	100%				

S = 0.0390341; R^2 = 96.26%; R^2 (adjusted) = 94.54%; R^2 (predicted) = 87.35%.

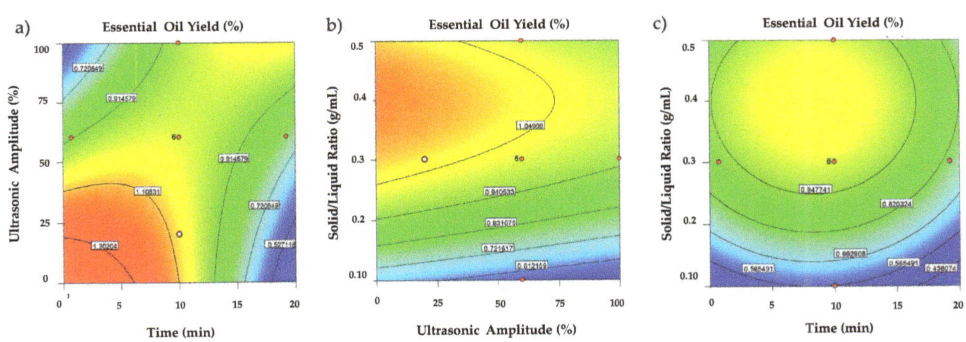

Figure 1. Contour plots for EOY with fixed S/L ratio (**a**) at 0.3 g/mL, fixed time (**b**) at 10 min, and fixed amplitude (**c**) at 60%.

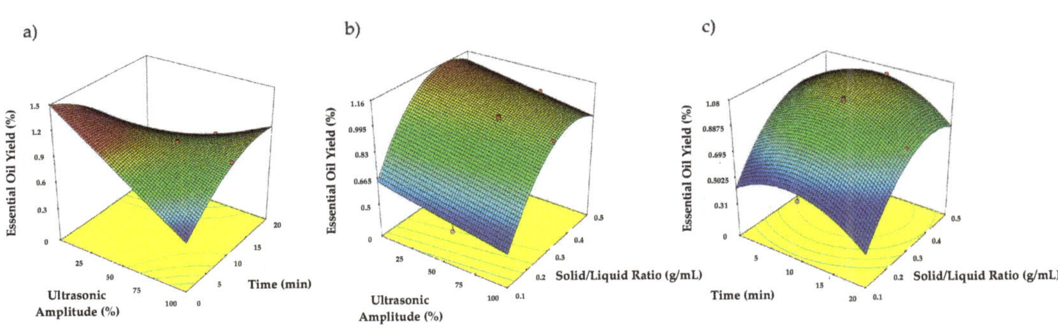

Figure 2. Response surface plots for EOY with fixed S/L ratio (**a**) at 0.3 g/mL, fixed time (**b**) at 10 min, and fixed amplitude (**c**) at 60%.

3.1.2. Total Reducing Capacity

The range of data obtained for the variable total reducing capacity by the Folin–Ciocalteu method is from 0.4489 to 1.4913 mg GAE/g biomass (Table 3). The prediction model, according to the analysis of variance, shows that the general, linear, and quadratic models are significant ($p < 0.000$), with an adjusted R^2 coefficient (91.34%) close to R^2 (94.08%) and a lack of fit p-value of 0.346. Since the model exhibits high significance, a high contribution to variability, and close values between the determination and adjusted coefficients, that the model is adequate and reproducible. The quadratic model is the largest contributor to variability with 53.11% and a p-value < 0.000. Regarding the process variables, the solid/liquid ratio proves to be the most significant contributor to variability (26.48%; $p < 0.000$), followed by time (16.44%; $p < 0.000$) and ultrasonic amplitude (10.19%; $p < 0.000$), which has the lowest contribution. The contribution to variability from the linear model (40.97%; $p < 0.000$) is less than that of the quadratic model; in this case, the variable with the greatest contribution to the linear model is the solid/liquid ratio (35.49%; $p < 0.000$), followed by time (3.41%; $p < 0.000$), and finally ultrasonic amplitude (2.06%; $p < 0.000$). Considering that the quadratic model contributes the most variability to the general model, it is possible to find the optimal conditions of the selected process variables that maximize the response variable total reducing capacity, highlighting the solid/liquid ratio and time variables as the most significant for the quadratic model. The prediction model for the response variable total reducing capacity using coded and uncoded variables was expressed by Equations (7) and (8), respectively:

$$\text{TRC} = 1.36 + 0.24X_1 - 0.057X_2 - 0.074X_3 - 0.20{X_1}^2 - 0.16{X_2}^2 - 0.18{X_3}^2 \quad (7)$$

$$\text{TRC} = -1.806 + 10.463SL + 0.030A + 0.103T - 14.111SL^2 - 0.000276A^2 - 0.005827T^2 \quad (8)$$

Table 3. Analysis of variance for the variable total reducing capacity (TRC).

Source	Degrees of Freedom	Contribution	Adjusted SS	Adjusted MS	F Value	p Value
Model	6	94.08%	2.039	0.339	34.41	0.000
Lineal	3	40.97%	0.888	0.296	29.97	0.000
S/L	1	35.49%	0.769	0.769	77.89	0.000
Amplitude	1	2.06%	0.044	0.044	4.53	0.053
Time	1	3.41%	0.073	0.073	7.49	0.017
Square	3	53.11%	1.151	0.383	38.84	0.000
Amplitude^2	1	10.19%	0.353	0.353	35.80	0.000
Time^2	1	16.44%	0.447	0.447	45.33	0.000
S/L^2	1	26.48%	0.574	0.574	58.10	0.000
Error	13	5.92%	0.128	0.009		
Lack of Fit	8	4.17%	0.090	0.011	1.48	0.346
Pure error	5	1.76%	0.038	0.007		
Total	19	100%				

S = 0.0994003; R^2 = 94.08%; Adjusted R^2 = 91.34%; Predicted R^2 = 80.17%.

To observe the behavior of the selected process variables (solid/liquid ratio, time, and ultrasonic amplitude) on the response variable total reducing capacity, contour and surface plots were generated (Figures 3 and 4). By observing these plots with the aim of maximizing the response variable and obtaining an optimum, it is possible to notice that the quadratic process variables exhibit similar behavior on the response in all arrangements. In this case, the quadratic model contributes more to the response variable than the linear model, resulting in curves instead of straight lines in the contour plots and domes or peaks in the surface plots. This indicates a nonlinear relationship between the independent variables and the response variable and indicates that the quadratic model is more suitable for describing the variability in the data. Considering that the quadratic variables with the highest contribution are the solid/liquid ratio and time, we focus our attention on the plots that include these variables. It can be seen in these plots that the optimal process conditions that maximize the response variable total reducing capacity are achieved, with these conditions being close to the central points of these variables. Regarding the rest of the plots, a very similar behavior is observed, where the region corresponding to the optimal conditions that maximize the response variable is close to the central values of the points.

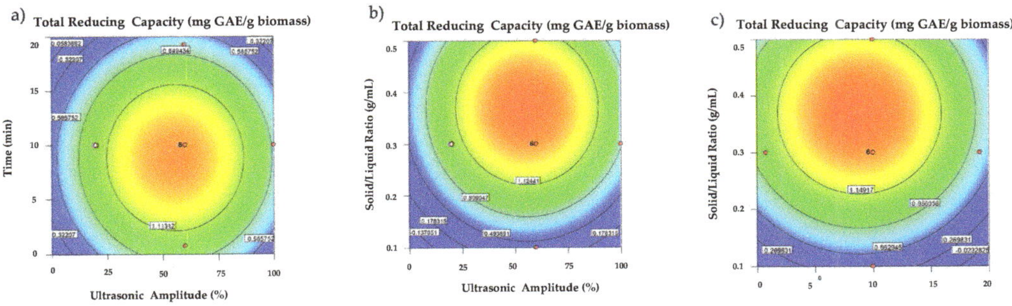

Figure 3. Contour plots for TRC with fixed S/L (**a**) at 0.3 g/mL, fixed time (**b**) at 10 min, and fixed ultrasonic amplitude (**c**) at 60%.

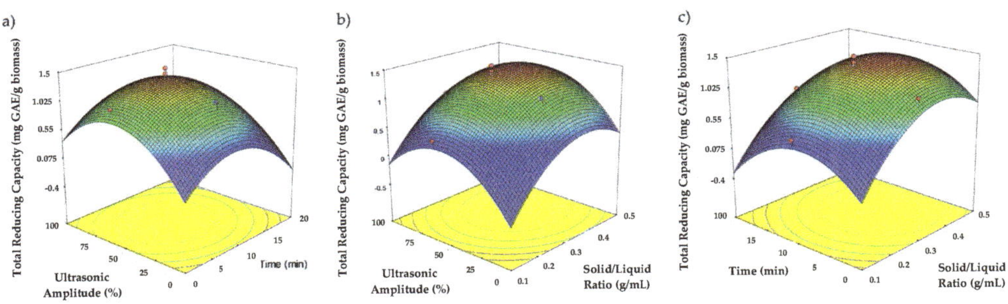

Figure 4. Response surface plots for TRC with fixed S/L ratio (**a**) at 0.3 g/mL, fixed time (**b**) at 10 min, and fixed amplitude (**c**) at 60%.

In a study by Michalaki et al. [21], the ultrasound-assisted extraction of phenolic compounds from *Origanum vulgare* ssp. hirtum was optimized. The authors reported that the total phenolic content determined by the Folin–Ciocalteu method increased with longer ultrasound extraction times, reaching a maximum of 362.1 ± 1.8 mg GAE/g DW after 40 min of extraction at 80 °C and 60% (v/v) ethanol. This finding is consistent with observations in this study, where total reducing capacity values also increased with longer extraction times, peaking at 1.4913 mg GAE/g DW around 10 min, 60% amplitude, and a S/L ratio of 0.3 g/mL, corresponding to the central values, before decreasing with further increases in time. The differences in results between these studies could be attributed to the solvents used, the nature of the raw materials, and the chemical profile of the extracts, as essential oils tend to have lower amounts of phenolic compounds.

3.1.3. Optimization of Ultrasonic-Assisted Extraction by DES of *Lippia graveolens* Biomass

To find the optimal process conditions to maximize the response variables (essential oil yield and total reducing capacity), the superposition of graphs from the contour plots was performed. The combination of process variables that maximizes the response variables was solid/liquid ratio = 0.38; ultrasonic amplitude = 45.05%; extraction time = 7.47 min (Figure 5). The predicted values for the response variables were essential oil yield = 1.127%

and total reducing capacity = 1.403 mg GAE/g biomass. Considering the optimal conditions and the predicted values obtained, five replicates (n = 5) were performed to confirm that the theoretical models are adequate and reproducible. The experimental values obtained under optimal conditions are included in Table 4: essential oil yield = 1.0932 ± 0.055%; total reducing capacity = 1.4788 ± 0.06 mg GAE/g biomass. Comparing the predicted values with the experimental values from the replicates, we can observe that they are close to each other, as these experimental values fall within the 95% confidence interval of the prediction model, indicating no significant difference. Therefore, the optimal ultrasound-assisted DES extraction conditions for essential oils to maximize essential oil yield and total reducing capacity are adequate and reproducible.

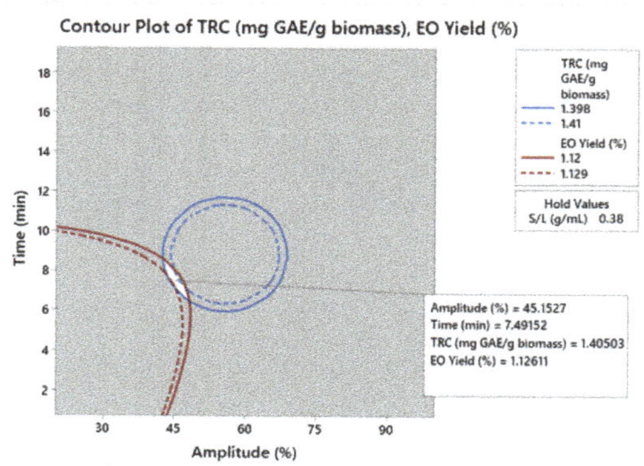

Figure 5. Graph showing the region with the best combination of process variables to maximize EOY and TRC.

Table 4. Confirmation report of optimized DES ultrasonic extraction.

Factor	Name	Optimal Level	Low Level	High Level	Code
A	Solid/Liquid Ratio (g/mL)	0.38	12	36	Actual
B	Amplitude (%)	45.15	12.5	37.5	Actual
C	Time (min)	7.49	32	42	Actual
Response	**Predicted Value**	**Measured Value Mean**	**Predicted SE**	**CI 95%**	**PI 95%**
Yield (%)	1.127	1.093	0.015	(1.09, 1.16)	(1.17, 1.63)
TRC (mg GAE/g biomass)	1.403	1.478	0.039	(1.318, 1.488)	(1.03, 1.21)

GAE/g biomass: gallic acid equivalents per gram of oregano biomass; SE: standard error; CI: confidence interval.

The optimal conditions obtained in this research (Table 4) to maximize the response variables (essential oil yield and total reducing capacity) differ from those reported by Biru et al. [22]. They optimized the ultrasound-assisted extraction process of essential oil from *Vernonia amygdalina*, obtaining optimal conditions of 17.263 min, sonication power of 150.677 W, and 6.811 mL/g, under which the essential oil yield was maximized at 4.185% g/g. The difference between the optimal conditions in this work and those of the mentioned study can be attributed to the different raw materials used, i.e., *Vernonia amygdalina* and *L. graveolens* biomass, respectively. It should be noted that the biomass in this study had already undergone a previous essential oil extraction process, so a lower extraction yield is expected. Additionally, the ultrasonic extraction of *Vernonia amygdalina* leaves did not include hydrodistillation as a separation step, which can degrade volatile compounds due to high temperatures and negatively impact the essential oil yield. Thus, thermolabile compounds were protected from degradation, favoring the oil yield.

In another study by Liu et al. [23], the extraction of essential oil from *Iberis amara* seeds was optimized using ultrasound-assisted hydrodistillation; they found that the optimal conditions for maximizing essential oil yield were 240 min, ultrasonic power of 45 W, and a solid/liquid ratio of 6.7 mL/g. The difference in values between this work and the study in question could be due to the nature of the raw materials used, as well as the different plant parts selected for extraction: seeds (*I. amara*) versus leaves, flowers, and fruit (*L. graveolens*).

Xu et al. [14] optimized the extraction of essential oil from turmeric (*Curcuma longa* L.) using microwave-assisted DES as a pretreatment for hydrodistillation. They found the following optimal conditions: NADES composed of choline chloride/oxalic acid in a 1:1 molar ratio, 60 g of plant material, pretreatment time of 5 min, pretreatment temperature of 84 °C, and hydrodistillation time of 76 min. Under these conditions, an essential oil yield of 0.85% was obtained. This differs from the results of this work; one of the causes could be the use of microwave technology, the different DES mixture used, and the nature of the plant material itself.

3.2. Gas Chromatography Analysis

All the essential oil samples from the *L. graveolens* biomass were analyzed using gas chromatography coupled to mass spectroscopy (GC-MS). Forty-six volatile compounds were identified in the optimized essential oil (EAU-DES-HD), while only twenty volatile compounds were detected in the oil extracted by conventional hydrodistillation (HD) (Table 5). Thymol was the most abundant essential oil in EAU-DES-HD samples, and its content was higher in the optimized than in the hydrodistillation samples (47% vs. 38%, respectively). Most of the remaining compounds were detected in amounts below 9%. The increased presence of compounds in the optimized oil through the use of DESs and ultrasound may be attributed to the capacity of eutectic solvents, particularly those with acidic properties, to dissolve the cellulose structures found in the plant cells of *L. graveolens*. This dissolution may lead to a greater release of secondary metabolites, specifically volatile compounds present in the intracellular medium.

Table 5. Profile of volatile compounds present in the biomass oil of *L. graveolens* optimized with ultrasonic–DES pretreatment and conventional hydrodistillation.

No	Compound	RT	Molecular Formula	Relative Content (%)		Classification
				UAE-DES-HD	HD	
1.	α-Fellandrene	3.46	$C_{10}H_{16}$	0.07 ± 0.04	ND	Monoterpene
2.	1,4-Cineole	3.50	$C_{10}H_{14}$	0.01 ± 0.004	ND	Monoterpene
3.	(+)-4-Carene	3.51	$C_{10}H_{16}$	0.22 ± 0.01	ND	Monoterpene
4.	p-Cymene	3.55	$C_{10}H_{14}$	1.66 ± 0.53	1.27 ± 0.45	Monoterpene
5.	Eucalyptol	3.59	$C_{10}H_{18}O$	0.68 ± 0.10	0.98 ± 0.36	Monoterpene
6.	γ-Terpinene	3.70	$C_{10}H_{16}$	0.16 ± 0.05	0.04 ± 0.02	Monoterpene
7.	α-Terpineyl propionate	3.78	$C_{13}H_{22}O_2$	0.16 ± 0.01	ND	Ester
8.	p-Propenyltoluene	3.85	$C_{10}H_{14}$	1.22 ± 0.17	ND	Aromatic hydrocarbon
9.	Linalool	3.87	$C_{10}H_{18}O$	ND	0.33 ± 0.11	Monoterpene
10.	Cis-3-Hexenyl butyrate	3.89	$C_{10}H_{18}O_2$	0.03 ± 0.007	ND	Ester
11.	Endo-borneol	4.25	$C_{10}H_{18}O$	0.54 ± 0.07	ND	Byciclic monoterpene
12.	Terpinen-4-ol	4.28	$C_{10}H_{18}O$	0.44 ± 0.05	ND	Monoterpene alcohol
13.	p-Tolyl acetate	4.36	$C_9H_{10}O_2$	0.24 ± 0.02	ND	Ester
14.	α-Terpineol	4.37	$C_{10}H_{18}O$	0.36 ± 0.23	ND	Monoterpene alcohol

Table 5. *Cont.*

No	Compound	RT	Molecular Formula	Relative Content (%)		Classification
				UAE-DES-HD	HD	
15.	Thymol methyl ether	4.50	$C_{11}H_{16}O$	0.55 ± 0.05	0.74 ± 0.23	Monoterpenoid
16.	Acetaldehyde O-methyloxime	4.82	$C_3H_5NO_2$	0.01 ± 0.01	ND	Oxime
17.	Thymol	4.82	$C_{10}H_{14}O$	47.04 ± 1.94	38.2 ± 0.01	Monoterpene
18.	Carvacrol	4.86	$C_{10}H_{14}O$	3.63 ± 0.97	4.86 ± 2.65	Monoterpene
19.	3-Methyl-4-isopropylphenol	4.86	$C_{10}H_{14}O$	3.64 ± 1.10	4.15 ± 1.84	Phenolic
20.	p-(Pentyloxy)acetophenone	4.90	$C_{13}H_{18}O_2$	0.03 ± 0.009	0.05 ± 0.02	Ketone
21.	Thymol acetate	5.09	$C_{12}H_{16}O_2$	0.40 ± 0.09	0.49 ± 0.18	Ester
22.	1,3-Indandione, 2-acetyl-	5.16	$C_{11}H_8O_3$	0.01 ± 0.0006	ND	Ketone
23.	1,1,5-Trimethyl-1,2-dihydronaphthalene	5.18	$C_{13}H_{16}$	0.06 ± 0.01	ND	Sesquiterpene
24.	(−)-Aristolene	5.22	$C_{15}H_{24}$	0.15 ± 0.06	ND	Sesquiterpenoid
25.	Isoledene	5.28	$C_{15}H_{24}$	0.41 ± 0.03	0.47 ± 0.18	Sesquiterpene
26.	β-selinene	5.61	$C_{15}H_{24}$	0.26 ± 0.05	ND	Sesquiterpene
27.	Aromadendrene	5.68	$C_{15}H_{24}$	3.08 ± 0.24	ND	Sesquiterpenoid
28.	1,4,7,-cicloundecatrieno, 1,5,9,9-tetrametil-, Z,Z,Z-	5.77	$C_{15}H_{24}$	6.15 ± 0.45	8.71 ± 3.08	Sesquiterpene
29.	Alloaromadendrene	5.82	$C_{15}H_{24}$	0.19 ± 0.01	ND	Sesquiterpenoid
30.	β-Panasinsene	5.85	$C_{15}H_{24}$	0.46 ± 0.03	0.54 ± 0.23	Sesquiterpene
31.	Cis-Calamenene	6.15	$C_{15}H_{22}$	ND	0.78 ± 0.01	Sesquiterpenoid
32.	Dehydro-aromadendrene	6.17	$C_6H_{12}O_2$	1.00 ± 0.05	ND	Sesquiterpenoid
33.	4′,6′-dihydroxy-2′,3′-dimethylacetophenone	6.23	$C_9H_{12}O_3$	0.76 ± 0.99	0.14 ± 0.03	Aromatic ketone
34.	Nerolidol 2	6.30	$C_{15}H_{26}O$	ND	0.88 ± 0.07	Sesquiterpene
35.	Ethanone, 1,1′,1″-(1,3,5-bencenethril)tris-	6.38	$C_{12}H_{12}O_3$	0.02 ± 0.003	ND	Aromatic ketone
36.	Cyclohexane, 1,2-dimethyl-3,5-bis(1-methylethenyl)-	6.40	$C_{14}H_{24}$	1.10 ± 0.23	ND	Cycloalkane
37.	1,3-Dimethyl-5-n-hexyladamantane	6.47	$C_{18}H_{32}$	0.03 ± 0.003	ND	Alkane
38.	Caryophyllenyl Alcohol	6.54	$C_{15}H_{26}O$	0.59 ± 0.11	ND	Alcohol
39.	(−)-spatulenol	6.57	$C_{15}H_{24}O$	0.92 ± 0.07	ND	Sesquiterpenoid
40.	(−)-globulol	6.62	$C_{15}H_{26}O$	1.75 ± 0.10	ND	Sesquiterpenoid
41.	Humuleno-1,2-epoxide	6.80	$C_{15}H_{24}O$	0.28 ± 0.02	1.14 ± 0.33	Epoxide
42.	(+)-rosifoliol	6.87	$C_{15}H_{26}O$	0.74 ± 0.44	0.92 ± 0.36	Sesquiterpene alcohol
43.	10-epi-γ -eudesmol	6.92	$C_{15}H_{26}O$	0.15 ± 0.01	ND	Sesquiterpene
44.	Humulenol-II	6.95	$C_{15}H_{24}O$	1.07 ± 0.03	2.38 ± 0.10	Sesquiterpene
45.	α-eudesmol	7.10	$C_{15}H_{26}O$	1.00 ± 0.02	ND	Sesquiterpenoid
46.	14-Hydroxycaryophyllene	7.19	$C_{15}H_{24}O$	0.65 ± 0.04	1.59 ± 0.67	Sesquiterpene
47.	Naphthalene, 1,6-dimethyl-4-(1-methylethyl)-	7.23	$C_{15}H_{18}$	0.33 ± 0.65	ND	Aromatic hydrocarbon
48.	p-anisole	9.62	$C_{10}H_{14}O$	0.07 ± 0.01	ND	Aromatic ether

RT: retention time; UAE-DES-HD: ultrasound-assisted extraction with deep eutectic solvents and hydrodistillation; HD: conventional hydrodistillation. Monoterpene: 3.11 ± 1.51; Ester: 0.86 ± 0.12; Aromatic hydrocarbon: 1.55 ± 0.65; Bicyclic monoterpene: 0.54 ± 0.07; Monoterpene alcohol: 0.80 ± 0.28; Monoterpenoid: 0.55 ± 0.05; Oxime: 0.01 ± 0.01; Phenolic: 3.64 ± 1.10; Ketone: 0.04 ± 0.02; Sesquiterpene: 11.68 ± 5.36; Sesquiterpenoid: 9.34 ± 4.01; Aromatic ketone: 0.78 ± 0.33; Cycloalkane: 1.10 ± 0.23; Alkane: 0.03 ± 0.003; Alcohol: 0.59 ± 0.11; Epoxide: 0.28 ± 0.02; Sesquiterpene alcohol: 0.74 ± 0.44; Aromatic ether: 0.07 ± 0.01.

Fan and Li [24] reported this pattern in a study on the extraction of essential oil from *Angelica sinensis* radix using microwave-assisted hydrodistillation with eutectic solvents. The oils extracted with DES and microwaves were analyzed and compared to those obtained by conventional hydrodistillation. The results indicated that the oils from the DES treatment (choline chloride and citric acid 1:3) contained a higher concentration of ligustilide, the major compound reported for *A. sinensis*.

Similarly to the findings in this study, Chen, Xu, Pang, Jin, Lv, Li, and Lee [13] extracted essential oil from cloves (*Syzygium aromaticum*) using microwave-assisted hydrodistillation with eutectic solvents. They found that the DES composed of choline chloride and lactic acid in a 1:2 ratio, the same used in our study, generated oils with a more diverse chemical profile than those generated through conventional hydrodistillation.

The ability of acidic DES to dissolve lignin was reported by Yu et al. [25]. They investigated the effect of DES composed of carboxylic acids as hydrogen bond donors (HBD) as a pretreatment to dissolve lignin from pine and poplar, with lactic acid being particularly effective. The authors suggest that the carboxyl groups (-COOH) in HBDs provide active protons, enhancing the strength of hydrogen bonds and facilitating the breaking of various chemical bonds in lignin. They observed that the β-O-4 bond content in lignin decreased after contact with acidic DES, highlighting these solvents as promising agents for treating plant biomass.

3.3. Determination of Minimum Inhibitory Concentration (MIC) and Minimum Fungicidal Concentration (MFC)

The minimum inhibitory concentration (MIC) of the optimized essential oil from *L. graveolens* biomass was 275 ppm, at which no visible growth was observed in the microplate. The minimum fungicidal concentration (MFC) was 300 ppm, confirmed by taking a sample from the well of the microplate at this concentration and applying it to a new PDA agar medium, where no growth was observed after 4 days. Similarly, Andrianjafinandrasana et al. [26] evaluated the effect of clove essential oil (*Syzygium aromaticum*) on the spore germination of *Colletotrichum asianum*. The authors reported a minimum inhibitory concentration (MIC) of 250 ppm. In this case, the values from that investigation and our work are close to each other. This closeness in MIC values could be attributed to the similarity in the lipophilic polarity of the major bioactive compounds present in the oils, which can penetrate the fungal cell and alter its function. For *S. aromaticum* essential oil, eugenol has been reported as the major compound, while for *L. graveolens* it is thymol, as confirmed in this work through chromatographic analysis. Both eugenol and carvacrol contain a phenol group in their structure, consisting of a benzene ring bonded to a hydroxyl group (-OH). This hydroxyl group can donate protons, giving the compound acidic properties. Phenolic groups can interact with the cell membranes of fungi through hydrogen bonds, potentially leading to membrane destabilization and loss of integrity. This interaction with cell membranes can alter essential functions of the fungi, such as nutrient and metabolite transport, ultimately leading to their death.

In another study by Muniz et al. [27], it was found that thyme essential oil (*Thymus vulgaris*) at 200 ppm completely inhibited the mycelial growth of *C. gloeosporioides*, considering this concentration as the MIC. It has been reported that essential oils have the capacity to damage the cytoplasmic membrane of microorganisms, affecting its structure and making it more permeable [28]. This means that the membrane becomes more permeable to substances that normally would not easily cross it, such as ions and large molecules. This can interfere with essential cellular processes and ultimately lead to the death or inhibition of the growth of the microorganism.

3.4. Mycelial Growth Inhibition

According to the results of the linear regression corresponding to the percentage of mycelial growth inhibition by the optimized essential oil from *L. graveolens* biomass, the effective concentration at 50% inhibition (EC_{50}) was 78.05 ppm. The highest concentration tested, 250 ppm, showed 100% inhibition. Therefore, the maximum concentration range was limited to 100 ppm (Figure 6).

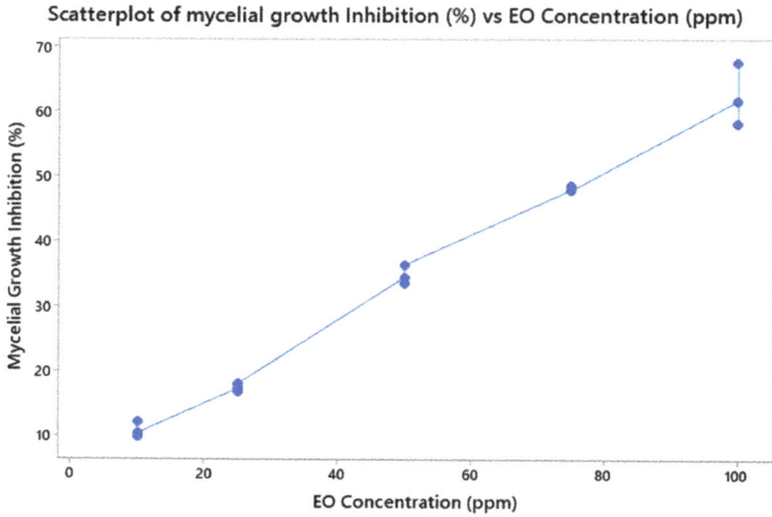

Figure 6. Graph showing the effective concentration at 50% (EC_{50}) of the optimized essential oil from *Lippia graveolens* biomass that inhibits the mycelial growth of *Colletotrichum asianum*. ($y = 0.5838x + 4.4342$; $R^2 = 98.79\%$; Adjusted $R^2 = 98.70\%$; EC50: 78.05 ppm).

The linear regression analysis presents a coefficient of determination $R^2 = 98.79\%$ and adjusted $R^2 = 98.70\%$, which are very close. In linear regression, this suggests that the model is capturing the variability in the data well and is not overfitting. This indicates that the model is robust and reliable in explaining and predicting the dependent variable.

Huang et al. [29] evaluated the effect of *Artemisia scoparia* essential oil on *Colletotrichum gloeosporioides* using an agar diffusion assay, where they obtained an EC_{50} of 9320 ppm. Additionally, the chemical profile of the oil was determined using gas chromatography–mass spectrometry. The major compounds reported were ethenyl-naphthalene (23.5%), 2,4-pentadienyl-benzene (11.8%), 1,2-dimethoxy-4-(2-propenyl)-benzene (10.0%), β-pinene (8.0%), and 1-methyl-4-(1-methylethyl)-1,4-cyclohexadiene (6.3%). The large difference between the EC_{50} values of the study in question and this work could be attributed to the chemical profile of the essential oils, as the major compound in *A. scoparia* was ethenyl-naphthalene, while in *L. graveolens* it was thymol. Previous reports in the literature indicate that thymol is a compound with high antimicrobial activity due to its ability to penetrate the fungal cell wall and disrupt its function, which could be the mechanism against this phytopathogenic species.

In addition, it is important to mention that, although they are pathogens of the same genus and share certain characteristics, they also present significant differences in terms of morphology, pathogenicity, genetics, and metabolism. These differences can affect their ability to infect different host plants and their responses to various treatments, including antifungal agents and essential oils.

On the other hand, Lima Oliveira et al. [30] applied *Cymbopogon citratus* (D.C. ex Nees) Stapf essential oil alone and in combination with chitosan on different species of *Colletotrichum*. When evaluating the essential oil alone at its highest concentration of 1250 ppm, they found a mycelial growth inhibition percentage of 37% on *C. asianum*. This greatly differs from the results obtained in this work, as the EC_{37} of *L. graveolens* was 55.78 ppm. Comparing both values, the oil of *L. graveolens* is 22.4 times more effective in inhibiting *C. asianum*. A possible reason for the difference between these values is the compounds present in the essential oils. According to the authors of the mentioned study, the major compounds in *C. citratus* oil were geranial (51.39%) and neral (29.29%). Although these compounds can affect the fungal cell wall, thymol (the major compound in *L. graveolens*) is generally considered to be more effective in terms of cell wall disruption and antifungal activity.

3.5. In Vivo Fungicidal Activity

The results of the fungicidal activity of the optimized essential oils and commercial fungicides are presented in Tables 6–9. The ANOVA for wounded fruits (Table 6) shows that the fungicide factor is the most contributive to total variability (48.68%, $p < 0.000$), followed by concentration (33.22%, $p < 0.000$), with the method factor contributing the least to variability (1.58%, $p < 0.000$). For the interaction's fungicide/concentration and fungicide/method, both have similar contributions to variability (6.23%, $p < 0.000$) and (6.38%, $p < 0.000$), respectively. The remaining interactions, concentration/method, and the three-way interactions have very low contributions to variability (0.46%, $p < 0.027$) and (1.25%, $p < 0.002$). The ANOVA shows a coefficient of determination R^2 = 97.92% and an adjusted coefficient of determination R^2 = 96.93%. The close values between these coefficients indicate that the model has a good capacity to explain the data.

Table 6. Analysis of variance for fungicide effectiveness on wounded fruits.

Source	Degrees of Freedom	Contribution	Adjusted SS	Adjusted MS	F Value	*p* Value
Fungicide	2	48.68%	12,808.6	6404.31	420.89	0.000
Concentration	2	33.22%	8767.6	4383.79	288.10	0.000
Method	1	1.58%	414.6	414.59	27.25	0.000
Fungicide × Concentration	4	6.23%	1639.9	409.98	26.94	0.000
Fungicide × Method	2	6.38%	1678.8	839.39	55.16	0.000
Concentration × Method	2	0.46%	122.2	61.09	4.01	0.027
Fungicide × Concentration × Method	4	1.25%	330.2	82.54	5.42	0.002
Error	36	2.08%	547.8	15.22		
Total	53	100%				

According to the results in Table 7, the most effective combination for preventing anthracnose in papaya fruits caused by *Colletotrichum asianum* in the presence of wounds consists of the corrective application of oregano essential oil at 2000 ppm using the corrective method. Our proved to be more effective than the chemical product Tecto 60 (thiabendazole). In this case, the highest mean, corresponding to the corrective method, does not share a letter with any other, indicating a significant difference in terms of fungicide effectiveness. This means that the group treated with the corrective method has a significantly higher effectiveness compared to the other groups. The least effective treatments correspond to Timorex (*Melaleuca alternifolia* essential oil) at concentrations of 1000 and 300 ppm, both applied with the corrective method (Figure 7). Consequently, it can be stated that these treatments are not suitable for correcting anthracnose.

Additionally, Sarkhosh, Schaffer, Vargas, Palmateer, Lopez, Soleymani, and Farzaneh [20] investigated the fungicidal effect of five essential oils from different plants on anthracnose caused by *C. gloeosporioides* in papaya fruits. The results of the assay on wounded and pathogen-inoculated papaya fruits showed that essential oils of *Satureja khuzistanica* and *Thymus* spp., at a concentration of 2000 ppm, caused a 59.26% and 58.40% reduction in lesion diameter and a 64.07% and 54.82% reduction in fruit decay, respectively. The application of these oils resulted in better preservation of fruit firmness. The results of our research showed greater effectiveness at the same concentration of 2000 ppm for the treatment with optimized oregano essential oil. The difference between the results of both studies could be due to the different chemical profiles of the essential oils used, as the nature of the compounds in the oils directly impacts their biological activity. Additionally, the species of phytopathogenic fungus must be considered, as one species may be more susceptible to the chemical compounds of the oil than the other.

Figure 7. Wounded papaya fruits 7 days and 14 days after inoculation.

Table 7. Tukey's method comparisons of mean fungicide effectiveness on wounded fruits.

Fungicide × Concentration × Method	Mean (% Effectiveness)	Groups
Oregano 2000 corrective	85.60	A
Tecto 2000 corrective	76.25	A B
Tecto 2000 preventive	74.30	A B
Tecto 1000 corrective	69.90	B
Tecto 1000 corrective	64.45	B C
Oregano 1000 corrective	56.41	C D
Oregano 2000 preventive	52.43	D E
Oregano 300 corrective	47.61	D E F
Oregano 1000 preventive	45.33	D E F G
Timorex 2000 preventive	40.93	E F G H
Tecto 300 corrective	36.22	F G H
Timorex 2000 corrective	35.20	G H
Oregano 300 preventive	32.69	H
Tecto 300 preventive	30.34	H I
Timorex 1000 preventive	20.61	I J
Timorex 300 preventive	20.56	I J
Timorex 1000 corrective	14.30	J
Timorex 300 corrective	10.03	J

Different letters indicate statistical significance ($p < 0.05$) according to Tukey's test. The results show the means of three replicates ($n = 3$).

The ANOVA for unwounded fruits (Table 8) shows that the concentration factor is the most contributive to total variability (38.82%, $p < 0.000$), followed by the method (11.73%, $p < 0.000$), with the fungicide factor contributing the least (2.53%, $p < 0.000$). For the interactions, the triple interaction was the most contributive (21.16%, $p < 0.000$), followed by the fungicide/method interaction (11.45%, $p < 0.000$) and the concentration/method interaction (8.45%, $p < 0.000$). The interaction with the least contribution to total variability was fungicide/concentration (3.98%, $p < 0.000$). The closeness between the coefficient of determination $R^2 = 98.12\%$ and the adjusted coefficient of determination $R^2 = 97.23\%$ indicates that the model can explain the variability of the data adequately and that unnecessary or redundant factors are not being included in the model. This gives us confidence in the model's predictive ability.

Table 8. Analysis of variance for fungicide effectiveness on unwounded fruits.

Source	Degrees of Freedom	Contribution	Adjusted SS	Adjusted MS	F Value	*p* Value
Fungicide	2	2.53%	698.3	349.17	24.23	0.000
Concentration	2	38.82%	10,718.8	5359.38	371.89	0.000
Method	1	11.73%	3238.9	3238.91	224.75	0.000
Fungicide × Concentration	4	3.98%	1099.7	274.93	19.08	0.000
Fungicide × Method	2	11.45%	3162.7	1581.34	109.73	0.000
Concentration × Method	2	8.45%	2332.4	1166.20	80.92	0.000
Fungicide × Concentration × Method	4	21.16%	5841.4	1460.36	101.34	0.000
Error	36	1.88%	518.8	14.41		
Total	53	100%				

S = 3.79, R^2 = 98.12%, R^2 (adjusted) = 97.23%, R^2 (predicted) = 95.77%.

In the case of the fungicidal activity assay on unwounded papaya fruits, according to Table 9, eight groups had the same mean effectiveness corresponding to 100%, including oregano essential oil with the corrective method at 2000, 1000, and 300 ppm. Within the same group A, Timorex Gold (*Melaleuca alternifolia* essential oil) at 2000 ppm with both methods and at 1000 ppm with the preventive method were also included, indicating that, at the highest concentration, this product is equally effective as a corrective and preventive treatment on papaya fruits; meanwhile, at a lower concentration of 1000 ppm, it is equally effective as a preventive treatment. The remaining means in group A correspond to the chemical control Tecto 60 (thiabendazole) at 2000 and 1000 ppm with the corrective method, confirming its high effectiveness as a corrective treatment. The least effective treatment in the assay corresponds to Timorex at 300 ppm with the preventive method, indicating that this treatment is not effective as a preventive fungicide (Figure 8).

Figure 8. Papaya fruits without wounds 7 days and 14 days after inoculation.

Uclaray et al. [31] studied the effect of encapsulated phenolic extracts of Cuban oregano (*Plectranthus amboinicus*) with plasmolyzed yeast on papaya fruits infected by *C. gloeosporioides*. When applied as a postharvest spray treatment for papaya, the microcapsules exhibited significant inhibitory action on the mycelial growth against *C. gloeosporioides* and greatly reduced anthracnose symptoms on the papaya fruits. At concentrations of 0.1, 1, and 10 mg/L of the non-encapsulated Cuban oregano phenolic extract, the severity percentage was around 50% for all three inoculated samples. Comparing these results with those of our research, corresponding to the lowest concentration of Mexican oregano by the preventive method, at 300 ppm, an effectiveness percentage of 48.28% was obtained. This indicates that the Cuban oregano phenolic extract is more effective in protecting the fruit from fungal attack, requiring a much lower concentration to achieve this effect. These differences could be attributed to the different fungal species tested (*C. gloeosporioides* and *C. asianum*), as the virulence of the species could be influencing the assay results. Additionally, differences in the chemical compositions of the phenolic extract and the essential oil play a crucial role in their respective fungicidal capacities. The phenolic extract of Cuban oregano may contain various phenolic compounds, including flavonoids and polyphenols, which could contribute to its fungicidal properties. In contrast, the essential oil of Mexican oregano is a complex mixture of volatile compounds, such as terpenes and phenylpropanoids, which may also exhibit fungicidal activities.

Table 9. Tukey's method comparisons of fungicide effectiveness means on unwounded fruits.

Fungicide × Concentration × Method	Mean (% Effectiveness)	Groups
Oregano 300 corrective	100	A
Oregano 1000 corrective	100	A
Tecto 1000 corrective	100	A
Timorex 1000 preventive	100	A
Tecto 2000 corrective	100	A
Timorex 2000 preventive	100	A
Oregano 2000 corrective	100	A
Timorex 2000 corrective	100	A
Tecto 2000 preventive	93.01	A B
Oregano 2000 preventive	88.12	B
Timorex 300 corrective	75.28	C
Oregano 1000 preventive	71.45	C D
Tecto 300 corrective	62.87	D E
Tecto 1000 preventive	57.81	E F
Tecto 300 preventive	48.42	F G
Oregano 300 preventive	48.28	F G
Timorex 1000 corrective	47.57	F G
Timorex 300 preventive	39.20	G

Different letters indicate statistical significance ($p < 0.05$) according to Tukey's test. The results show the means of three replicates ($n = 3$).

In another study by Barrera-Necha et al. [32], a fungicidal assay was conducted on papaya fruits inoculated with *C. gloeosporioides*, testing the oils of *C. zeylanicum* and *S. aromaticum* on papaya fruits during storage at room temperature and 14 °C. The lowest infection percentage of 3.3% was observed in papaya fruits treated with *S. aromaticum* at 47 ppm and 12.4% for *C. zeylanicum* at a concentration of 236 ppm at both evaluated temperatures. *S. aromaticum* oils could be an alternative for controlling *C. gloeosporioides* in papaya fruits. Comparing both studies, it can be noted that the results vary significantly. In this work, oregano essential oil and Timorex Gold showed high effectiveness in treating papaya fruits. However, in the study by Barrera-Necha, Bautista-Baños, Flores-Moctezuma, and Estudillo [32], *C. zeylanicum* and *S. aromaticum* oils were more effective. These differences could be attributed to variations in the concentrations used, storage conditions, and specific properties of the essential oils, including the presence of eugenol in clove oil, which could influence its antifungal activity. Both compounds, eugenol and thymol, could interact with fungi, causing anthracnose (*C. gloeosporioides* and *C. asianum*) similarly by affecting their cell membrane and vital metabolic processes. However, due to their structural differences and distinct mechanisms of action, each compound may exhibit varying affinities for specific pathways and cellular components of the fungus. This could explain the variations in effectiveness observed in previous studies and how each can have a distinct impact on the inhibition of growth and development of these Colletotrichum species. Given that they are different species, their responses to antimicrobial compounds, such as eugenol and thymol, may vary due to their distinct biochemical profiles and sensitivities to essential oils. Although both compounds may have general antimicrobial properties, the way they interact with the species may depend on the unique molecular and metabolic characteristics of each one. For example, eugenol and thymol may have different affinities for the specific metabolic pathways present in *C. gloeosporioides* and *C. asianum*, which could affect their viability and cellular function differently in each species. Additionally, the species may have differences in the composition of their cell membranes and other essential components, which could influence how the compounds interact and exert their effect.

4. Conclusions

The optimal process conditions for the ultrasound-assisted DES extraction of essential oil from Mexican oregano (*Lippia graveolens* Kunth) biomass as a pretreatment for hydrodistillation were 0.38 g/mL for solid/liquid ratio, 45.15% ultrasonic amplitude, and 7.49 min of ultrasonic time. The estimated values according to the response surface design were 1.127% for essential oil yield and 1.403 for TRC (mg GAE/g biomass). The experimental values obtained under optimal conditions were 1.0932 ± 0.055 and 1.4788 ± 0.06 for essential oil yield and TRC, respectively. The fact that these values are very close to the predicted ones indicates that the prediction models used are accurate and valid for predicting the results. A total of 47 compounds were found in the optimized essential oil and 20 compounds obtained by conventional hydrodistillation. In both cases, the major compound was thymol. The in vitro fungicidal activity of the optimized essential oils against *C. asianum* revealed that the essential oil of *L. graveolens* Kunth was effective in inhibiting the mycelial growth and spore germination of the fungus. The optimized essential oil of *L. graveolens* was effective on unwounded fruits at all concentrations using the corrective method. For wounded fruits, the highest concentration of oregano essential oil was the best treatment against the fungus *C. asianum*, deriving even better results than the positive controls corresponding to commercial products.

Therefore ultrasound-assisted DES extraction is a promising process to extract essential oils with antifungal potential against *C. asianum* from Mexican oregano biomass.

Author Contributions: Conceptualization, E.P.G.-G., J.B.H. and L.A.C.-B.; methodology, E.P.G.-G., L.A.C.-B., P.d.J.B.-B., J.P.M.-Q., L.A.C.-A. and O.V.-B.; formal analysis, J.P.M.-Q., L.A.C.-B., R.S.G.-E. and L.A.C.-A.; investigation, J.P.M.-Q.; writing—original draft preparation, J.P.M.-Q.; writing—review and editing, J.P.M.-Q., L.A.C.-B., J.B.H. and E.P.G.-G.; visualization, J.P.M.-Q. and L.A.C.-B.; supervision, L.A.C.-B. and E.P.G.-G.; project administration, L.A.C.-B. and E.P.G.-G. All authors have read and agreed to the published version of the manuscript.

Funding: This research received no external funding.

Data Availability Statement: The original contributions presented in the study are included in the article, further inquiries can be directed to the corresponding author/s.

Acknowledgments: J.P.M.-Q. thanks CONAHCYT for the scholarship granted for its graduate studies.

Conflicts of Interest: The authors declare no conflicts of interest.

References

1. SEMARNAT. *Technological Package for the Production of Oregano (Lippia Spp.) [Paquete Tecnológico para la Producción de Orégano (Lippia Spp.)]*; National Forestry Commission: Lerma de Villada, Mexico, 2008.
2. Bar-On, Y.M.; Phillips, R.; Milo, R. The biomass distribution on Earth. *Proc. Natl. Acad. Sci. USA* **2018**, *115*, 6506–6511. [CrossRef]
3. Soto-Armenta, L.C.; Sacramento-Rivero, J.C.; Acereto-Escoffié, P.O.; Peraza-González, E.E.; Reyes-Sosa, C.F.; Rocha-Uribe, J.A. Extraction Yield of Essential Oil from *Lippia graveolens* Leaves by Steam Distillation at Laboratory and Pilot Scales. *J. Essent. Oil-Bear. Plants* **2017**, *20*, 610–621. [CrossRef]
4. Ghavam, M. In vitro biological potential of the essential oil of some aromatic species used in Iranian traditional medicine. *Inflammopharmacology* **2022**, *30*, 855–874. [CrossRef] [PubMed]
5. Ghavam, M.; Markabi, F.S. Evaluation of Yield, Chemical Profile, and Antimicrobial Activity of *Teucrium polium* L. Essential Oil Used in Iranian Folk Medicine. *Appl. Biochem. Biotechnol.* **2024**, 1–17. [CrossRef] [PubMed]
6. Chen, Z.; Wu, K.; Zhu, W.; Wang, Y.; Su, C.; Yi, F. Chemical compositions and bioactivities of essential oil from perilla leaf (Perillae Folium) obtained by ultrasonic-assisted hydro-distillation with natural deep eutectic solvents. *Food Chem.* **2022**, *375*, 131834. [CrossRef]
7. Smith, E.L.; Abbott, A.P.; Ryder, K.S. Deep Eutectic Solvents (DESs) and Their Applications. *Chem. Rev.* **2014**, *114*, 11060–11082. [CrossRef] [PubMed]

8. Abbott, A.P.; Capper, G.; Davies, D.L.; Munro, H.L.; Rasheed, R.K.; Tambyrajah, V. Preparation of novel, moisture-stable Lewis-acidic ionic liquids containing quaternary ammonium salts with functional side chains. *ChemComm* **2001**, *19*, 2010–2011 [CrossRef]
9. Zhao, Y.; Yang, Y.-H.; Ye, M.; Wang, K.-B.; Fan, L.-M.; Su, F.-W. Chemical composition and antifungal activity of essential oil from *Origanum vulgare* against *Botrytis cinerea*. *Food Chem.* **2021**, *365*, 130506. [CrossRef]
10. Medina-Romero, Y.M.; Hernandez-Hernandez, A.B.; Rodriguez-Monroy, M.A.; Canales-Martínez, M.M. Essential oils of *Bursera morelensis* and *Lippia graveolens* for the development of a new biopesticides in postharvest control. *Sci. Rep.* **2021**, *11*, 20135 [CrossRef]
11. Yılmaz, A.; Ermiş, E.; Boyraz, N. Investigation of in vitro and in vivo anti-fungal activities of different plant essential oils against postharvest apple rot diseases-*Colletotrichum gloeosporioides*, *Botrytis cinerea* and *Penicillium expansum*. *J. Food. Saf. Food Qual.* **2016**, *67*, 122–131.
12. Bautista-Hernández, I.; Aguilar, C.N.; Martínez-Ávila, G.C.G.; Torres-León, C.; Ilina, A.; Flores-Gallegos, A.C.; Kumar Verma, D.; Chávez-González, M.L. Mexican Oregano (*Lippia graveolens* Kunth) as Source of Bioactive Compounds: A Review. *Molecules* **2021**, *26*, 5156. [CrossRef]
13. Chen, Y.; Xu, F.; Pang, M.; Jin, X.; Lv, H.; Li, Z.; Lee, M. Microwave-assisted hydrodistillation extraction based on microwave-assisted preparation of deep eutectic solvents coupled with GC-MS for analysis of essential oils from clove buds. *Sustain. Chem. Pharm.* **2022**, *27*, 100695. [CrossRef]
14. Xu, F.-X.; Zhang, J.-Y.; Jin, J.; Li, Z.-G.; She, Y.-B.; Lee, M.-R. Microwave-assisted Natural Deep Eutectic Solvents Pretreatment Followed by Hydrodistillation Coupled with GC-MS for Analysis of Essential Oil from Turmeric (*Curcuma longa* L.). *J. Oleo Sci.* **2021**, *10*, 1481–1494. [CrossRef] [PubMed]
15. Calvo-Irabién, L.M.; Parra-Tabla, V.; Acosta-Arriola, V.; Escalante-Erosa, F.; Díaz-Vera, L.; Dzib, G.R.; Peña-Rodríguez, L.M. Phytochemical Diversity of the Essential Oils of Mexican Oregano (*Lippia graveolens* Kunth) Populations along an Edapho-Climatic Gradient. *Chem. Biodivers.* **2014**, *11*, 1010–1021. [CrossRef] [PubMed]
16. Swain, T.; Hillis, W.E. The phenolic constituents of *Prunus domestica*: I—The quantitative analysis of phenolic constituents. *J. Sci. Food Agric.* **1959**, *10*, 63–68. [CrossRef]
17. Amini, J.; Farhang, V.; Javadi, T.; Nazemi, J. Antifungal Effect of Plant Essential Oils on Controlling Phytophthora Species. *Plant Pathol. J.* **2016**, *32*, 16–24. [CrossRef] [PubMed]
18. Alhajj, M.S.; Qasem, M.A.A.; Nabi, A.R.J.E.; Al-Mufarrej, S.I. In-Vitro Antibacterial and Antifungal Effects of High Levels of Chinese Star Anise. *Braz. J. Poult. Sci.* **2019**, *21*, eRBCA-2019. [CrossRef]
19. Cruz-Lachica, I.; Márquez-Zequera, I.; García-Estrada, R.S.; Carrillo-Fasio, J.A.; León-Félix, J.; Allende-Molar, R. Identificación de hongos mucorales causantes de la pudrición blanda en frutos de papaya (*Carica papaya* L.) en México. *Rev. Mex. Fitopatol.* **2017**, *35*, 397–417. [CrossRef]
20. Sarkhosh, A.; Schaffer, B.; Vargas, A.I.; Palmateer, A.J.; Lopez, P.; Soleymani, A.; Farzaneh, M. Antifungal activity of five plant-extracted essential oils against anthracnose in papaya fruit. *Biol. Agric. Hort.* **2018**, *34*, 18–26. [CrossRef]
21. Michalaki, A.; Karantonis, H.C.; Kritikou, A.S.; Thomaidis, N.S.; Dasenaki, M.E. Ultrasound-Assisted Extraction of Total Phenolic Compounds and Antioxidant Activity Evaluation from Oregano (*Origanum vulgare* ssp. hirtum) Using Response Surface Methodology and Identification of Specific Phenolic Compounds with HPLC-PDA and Q-TOF-MS/MS. *Molecules* **2023**, *28*, 2033. [CrossRef]
22. Biru, M.A.; Waday, Y.A.; Shumi, L.D. Optimization of Essential Oil Extraction from Bitter Leaf (*Vernonia amygdalina*) by Using an Ultrasonic Method and Response Surface Methodology. *Int. J. Chem. Eng.* **2022**, *1*, 4673031. [CrossRef]
23. Liu, X.; Ou, H.; Xiang, Z.; Gregersen, H. Optimization, chemical constituents and bioactivity of essential oil from *Iberis amara* seeds extracted by ultrasound-assisted hydro-distillation compared to conventional techniques. *J. Appl. Res. Med. Aromat. Plants* **2019**, *13*, 100204. [CrossRef]
24. Fan, Y.; Li, Q. An efficient extraction method for essential oil from angelica sinensis radix by natural deep eutectic solvents-assisted microwave hydrodistillation. *Sustain. Chem. Pharm.* **2022**, *29*, 100792. [CrossRef]
25. Yu, H.; Xue, Z.; Shi, R.; Zhou, F.; Mu, T. Lignin dissolution and lignocellulose pretreatment by carboxylic acid based deep eutectic solvents. *Ind. Crops Prod.* **2022**, *184*, 115049. [CrossRef]
26. Andrianjafinandrasana, S.N.; Chillet, M.; Ramonta, I.R.; Tsy, J.-M.L.P.; Minier, J.; Danthu, P. Biological activity *of Syzygium aromaticum* and *Ravensara aromatica* essential oils from Madagascar and their possible use against Postharvest mango Anthracnose. *Am. J. Plant Sci.* **2020**, *11*, 1682–1697. [CrossRef]
27. Muniz, A.C.C.; Marques, K.M.; Galati, V.C.; Pereira, F.D.; Mattiuz, B.H.; Glória, E.M.; Mattiuz, C.F.M. Evaluation of essential oil on the growth in vitro and in vivo of anthracnose in 'Hass' avocados. *Acta Hortic.* **2016**, *1120*, 207–213. [CrossRef]
28. Bakkali, F.; Averbeck, S.; Averbeck, D.; Idaomar, M. Biological effects of essential oils—A review. *Food Chem. Toxicol.* **2008**, *46*, 446–475. [CrossRef] [PubMed]
29. Huang, X.; Liu, T.; Zhou, C.; Huang, Y.; Liu, X.; Yuan, H. Antifungal Activity of Essential Oils from Three Artemisia Species against *Colletotrichum gloeosporioides* of Mango. *Antibiotics* **2021**, *10*, 1331. [CrossRef]
30. Lima Oliveira, P.D.; Anderson, W.; Paz, M.; Leite, E. Control of anthracnose caused by *Colletotrichum* species in guava, mango and papaya using synergistic combinations of chitosan and *Cymbopogon citratus* (D.C. ex Nees) Stapf. essential oil. *Int. J. Food Microbiol.* **2018**, *266*, 87–94. [CrossRef]

31. Uclaray, C.C.; Vidallon, M.L.P.; Almeda, R.A.; Cumagun, C.J.R.; Reyes, C.T.; Rodriguez, E.B. Encapsulation of wild oregano, *Plectranthus amboinicus* (Lour.) Spreng, phenolic extract in baker's yeast for the postharvest control of anthracnose in papaya. *J. Sci. Food Agric.* **2022**, *102*, 4657–4667. [CrossRef]
32. Barrera-Necha, L.L.; Bautista-Baños, S.; Flores-Moctezuma, H.E.; Estudillo, A.R. Efficacy of essential oils on the conidial germination, growth of *Colletotrichum gloeosporioides* (penz.) penz. and sacc and control of postharvest diseases in papaya (*Carica papaya* L.). *Plant Pathol. J.* **2008**, *7*, 174–178. [CrossRef]

processes

MDPI

Article

Bioactive Compounds, Antioxidant Activity, and Antiproliferative Potential on Glioblastoma Cells of Selected Stone Fruit Juices

Drazen Raucher [1,*], Mandy Rowsey [1], James Hinson [1], Ina Ćorković [2], Mary Ann Lila [3], Josip Šimunović [4] and Mirela Kopjar [2,*]

1 Department of Cell and Molecular Biology, University of Mississippi Medical Center, Jackson, MS 39216, USA; mrowsey@umc.edu (M.R.); jhinson@umc.edu (J.H.)

2 Faculty of Food Technology Osijek, Josip Juraj Strossmayer University of Osijek, F. Kuhača 18, 31000 Osijek, Croatia; ina.corkovic@ptfos.hr

3 Plants for Human Health Institute, Food Bioprocessing & Nutrition Sciences, North Carolina State University, North Carolina Research Campus, Kannapolis, NC 28081, USA; mlila@ncsu.edu

4 Department of Food, Bioprocessing and Nutrition Sciences, North Carolina State University, Raleigh, NC 27695, USA; simun@ncsu.edu

* Correspondence: draucher@umc.edu (D.R.); mirela.kopjar@ptfos.hr (M.K.)

Abstract: Glioblastoma presents one of the most formidable challenges in cancer treatment, remaining persistently incurable. There is a pressing need to explore less toxic alternatives, particularly natural remedies that could be applied in glioblastoma therapy. The aim of this research is to investigate the antiproliferative potential of selected stone fruit juices—tart cherry (*Prunus cerasus*), cornelian cherry (*Cornus mas*), and blackthorn (*Prunus spinosa*)—on U87-MG and GBM43 glioblastoma cells. Their effects were compared with temozolomide (TMZ), the current standard treatment. Additionally, the juices were assessed for their bioactive compounds and antioxidant potential. Unlike the other two juices, blackthorn juice did not exhibit an antiproliferative effect on U87-MG cells. However, all three juices, including blackthorn, demonstrated antiproliferative potential against TMZ-resistant GBM43 cells. Cornelian cherry exhibited an even stronger inhibitory effect than TMZ. This observation correlated with cornelian cherry being rich in iridoids, while tart cherry juice contained significant amounts of anthocyanins and proanthocyanidins. This research sheds light on the potential of cornelian cherry juice as a source of bioactive compounds with antiproliferative effects against glioblastoma cells, particularly TMZ-resistant GBM43 cells. Further research is warranted to explore the potential development of these compounds into therapeutic agents, either as single entities or in combination therapies for glioblastoma treatment.

Keywords: iridoids; anthocyanins; antiproliferative potential; tart cherry juice; cornelian cherry juice; blackthorn juice

1. Introduction

Over the past several decades, gliomas, the most frequent brain tumors in the adult population, have been the subject of numerous scientific studies. In recent years, increased interest has been dedicated to refining the diagnostic criteria for gliomas, particularly focusing on molecular biomarkers for categorization. This emphasis has led to the restructuring of the WHO Classification of Tumors of the Central Nervous System in 2021 [1]. According to this classification, which is based on molecular markers such as isocitrate dehydrogenase (IDH) expression and 1p/19q codeletion, adult-type diffuse gliomas are the most prevalent tumor types. These gliomas include subtypes such as astrocytoma (IDH-mutant astrocytoma), oligodendroglioma (IDH-mutant and 1p19q-codeleted), and glioblastoma (GBM) (IDH-wildtype) [1,2].

GBM stands out as one of the most challenging tumor types for treatment, and remains incurable. While angiogenesis, proliferation, and invasion are shared biological characteristics of all gliomas, GBM's invasiveness is particularly significant [3]. Despite extensive efforts in treatment, the survival rate for GBM remains low, presenting a significant obstacle. Following diagnosis, the prognosis for survival is typically limited to 13 months, with most patients succumbing within two years. While a 5-year survival rate has been achieved in clinical trials, it applies to only 4–5% of patients [4–9]. Despite the availability of various conventional treatments such as surgical resection, radiotherapy, and chemotherapy, the outcome for a majority of patients remains dismal as a result of high tumor recurrence rates [5,7,9,10].

The present standard chemotherapeutic drug for the treatment of patients with GBM is temozolomide (TMZ), known for its capability to penetrate the blood–brain barrier [11–14]. TMZ functions as an alkylating agent, inducing DNA mismatching through methylation [11]. It also has a significant part in halting cell progression at the G2/M phase, and triggering apoptosis in cancer cells [11,13]. Despite these mechanisms, the survival rate of patients remains very low due to the development of increased resistance to the drug [11]. Consequently, there is a strong interest in identifying and developing alternative strategies to achieve greater efficacy in reducing tumors and improving patient survival rates.

More recently, research focus has been directed to novel therapeutic interventions like the utilization of micronutrients [3], and the pursuit for active ingredients from natural sources which can be used for the prevention and treatment of malignant tumors is gaining considerable attention [15,16]. Over the years, plant-derived compounds have demonstrated a remarkable ability to inhibit the growth and development of cancer cells. These substances can be used to achieve inhibition of the proliferation of cells by various mechanisms such as the initiation of apoptosis, autophagy, and the blocking of cancer cells at different phases of the cell cycle [6]. In addition, many of these compounds have demonstrated the ability to deactivate the signaling pathways which are generally activated or conversely activate the signaling pathways which are generally deactivated in cancer cells [17–21]. One very important consideration is that most plant-derived compounds are considered safe for human consumption due to their minimal harmful effects. Consequently, it is believed that plant-derived anticancer drugs may cause mild or even no side effects on the overall health of the cancer patients [17,21,22]. In order to achieve the improvements of overall patient prognosis, besides the inhibition of the growth of cancer cells and the prevention of the development of recurrent tumors, these molecules need to possess the ability to pass through the blood–brain barrier [17]. In most cases, cancer suppression results from apoptosis or the programmed death of cells [23,24]. The proliferation of preneoplastic or neoplastic cells tends to be more significant than that of normal cells, making the induction of apoptosis or arrest of the cell cycle crucial for inhibiting the stimulation and progression of carcinogenesis, and ultimately for eliminating genetically damaged, preinitiated, or neoplastic cells from the organism [23,24]. Preinitiated cells refer to those that have undergone initial genetic changes or alterations that predispose them to malignancy but have not yet progressed to the stage of being fully initiated cancer cells. These cells represent an early stage of transformation and are potentially capable of developing into cancerous cells if not eliminated or halted through mechanisms such as apoptosis or cell cycle arrest. Generally, polyphenols are known for their anticancer effect, with anthocyanins known for apoptotic effects in human cancer cells [23–25], so their natural sources could potentially be used in cancer treatments. In a recent review of Ndongwe et al. [26], it was elaborated that iridoids, a large group of natural compounds, have the ability to generate conjugates with other drugs such as anticancer, antidiabetic, antileishmanial, and antimalarial drugs. These conjugations can result in synergistic effects and have the potential to increase drug efficiency. Additionally, the authors noted the importance of the full exploration of the role of iridoids in identifying less expensive and less toxic alternative/adjuvant cancer drugs [26].

The goal of this research was to assess the antiproliferative effects of specific stone fruit juices (tart cherry, cornelian cherry, and blackthorn) on two glioblastoma cell lines, U87-MG and GBM43. These two cell lines were selected since they are frequently used in cancer research on the glioblastoma multiforme. U87-MG cells are capable of generating tumors in experimental animals so they are frequently utilized as a model system for the investigation of glioblastoma. They are applied to the investigation of diverse aspects of the biology of glioblastoma, including growth of tumor, its invasiveness, and its response to selected therapies. Consequently, these cells are utilized in the testing of various drugs and preclinical studies [27–29]. The source and manipulation of the cells are key factors in determining the specific properties of GBM43 cells. They can easily share attributes that are mutual to glioblastoma cells, like accelerated proliferation and invasiveness. Mentioned cells are also utilized to acquire insights into glioblastoma biology, as well as for testing potential therapeutic interventions [30–32]. Additionally, a very important feature of these cells is that U87-MG glioblastoma cells are TMZ-sensitive, while GBM43 cells are TMZ-resistant [14]. Selected stone fruit juices' antiproliferative effects on glioblastoma cells were compared with those of TMZ, the present standard drug for glioblastoma treatment. Furthermore, the juices were analyzed for the contents of their bioactive compounds, total polyphenols and proanthocyanidins, as well as antioxidant potential.

2. Materials and Methods

2.1. Chemicals

Trolox, 4-dimethylaminocinnamaldehyde, 2,2-diphenyl-1-picrylhydrazyl (DPPH), 2,2′-azino-bis(3-ethylbenzothiazoline-6-sulfonic acid) diammonium salt (ABTS), gallic acid, chlorogenic acid, ellagic acid, rutin, (-)-epicatechin, and loganic acid were acquired from Sigma-Aldrich (St. Louis, MO, USA). Analytical standards of anthocyanins (cyanidin-3-glucoside, cyanidin-3-rutinoside), neochlorogenic acid, and hyperoside were acquired from Extrasynthese (Genay, France). From T.T.T. (Sveta Nedelja, Croatia), sodium carbonate was obtained, while potassium persulfate and Folin-Ciocalteu reagent were obtained from Kemika (Zagreb, Croatia). Orthophosphoric acid (HPLC-grade) was acquired from Fisher Scientific (Loughborough, UK), and methanol (HPLC-grade) from J.T. Baker (Deventer, The Netherlands). Neocuproine, cupric chloride, and 2,4,6-tri(2-pyridyl)-s-triazine (TPTZ) were acquired from Acros Organic (Geel, Belgium). Thiazolyl blue tetrazolium bromide (98%) (2-(3,5-diphenyltetrazol-2-ium-2-yl)-4,5-dimethyl-1,3-thiazole;bromide), Temozolomide (3-methyl-4-oxoimidazol[5,1-d][1,2,3,5]tetrazine-8-carboxamide), DMEM media (with L-glutamine, 4.5 g/L glucose and sodium pyruvate), solution of penicillin/streptomycin (10,000 U/mL penicillin, 10,000 µg/mL streptomycin in 0.85% NaCl), and 0.25% trypsin/0.1% EDTA were obtained from ThermoFisher Scientific (Waltham, MA, USA). Glioblastoma U87-MB and GBM43 cells were obtained from Mayo Clinic (Rochester, MN, USA) while fetal bovine serum was acquired from R&D Systems (Flowery Branch, GA, USA).

2.2. Preparation of Stone Fruit Juices

Stone fruits, namely tart cherry (*Prunus cerasus*), cornelian cherry (*Cornus mas*), and blackthorn (*Prunus spinosa*) fruits were grown near Varaždin (Croatia) at location 46°18′39.7″ N 16°32′67.7″ E. After the collection of these fruits (approximately 2 kg), they were first washed and pitted, and afterward pressed. The resulting juices were filtered through cheesecloth, followed by two min thermal treatment (300 mL of each juice) at 90 °C for inactivation of the naturally present enzymes and vegetative bacteria which could cause the degradation and spoilage of juices.

2.3. Evaluation of Total Polyphenols, Proanthocyanidins, and Monomeric Anthocyanins of Stone Fruit Juices

To determine the total polyphenols in the juices, the method of Singleton and Rossi [33] was applied. A total of 10 mL of Folin-Ciocalteu reagent (7.5%) was mixed with 0.2 mL of diluted juice and 1.8 mL demineralized water. A total of 8 mL of sodium carbonate solution (7.5%) was added to this mixture and it was kept for 120 min in a dark place before the reading of the absorbance at 765 nm took place. Results were presented as grams of gallic acid equivalents per L of juice (g GAE/L), so the calibration curve was generated with gallic acid. Readings were performed on a UV/Vis spectrophotometer (Cary 60 UV-Vis, Agilent Technologies, Santa Clara, CA, USA).

The concentration of proanthocyanidins was evaluated by the 4-(dimethylamino) cinnamaldehyde (DMAC) method [34]. Then, 1 mL of DMAC reagent was mixed with diluted juice and acidified ethanol. The mixture was kept for 30 min in a dark place before the reading of absorbance at 640 nm took place. The concentration of proanthocyanidins was presented as mg of procyanidin B2 equivalent per L of the juice (mg B2E/L), so the calibration curve was generated for procyanidin B2.

The pH differential method was applied to estimate the monomeric anthocyanins [35]. Two buffers (pH 1–0.025 M KCl and pH 4.5–0.4 M sodium acetate) were prepared to perform the analysis. A total of 2.8 mL of each buffer was mixed with 0.2 mL of the diluted juice. Mixtures were kept for 15 min in a dark place before the reading of absorbance at 515 nm and 700 nm took place. Concentrations of monomeric anthocyanins were presented as mg of cyanidin-3-glucoside per L of the juice (mg cyanidin-3-glucoside/L).

2.4. Evaluation of Antioxidant Activity of Stone Fruit Juices

To evaluate the antioxidant potential of the juices, ABTS, DPPH, FRAP, and CUPRAC assays were applied. The results of antioxidant activities were presented as μmol of Trolox equivalents per 100 mL of juice (μmol TE/100 mL), so calibration curves for all assays were generated using Trolox.

2.4.1. ABTS Assay

The ABTS assay was performed according to Arnao et al. [36]. A total of 3.2 mL of ABTS reagent was mixed with 0.2 mL of the diluted juice. The mixture was kept for 95 min in a dark place before the reading of absorbance at 734 nm took place.

2.4.2. DPPH Assay

For the DPPH method [37], 3 mL of DPPH solution was mixed with 0.2 mL of the diluted juice. The mixture was kept for 15 min in a dark place before the reading of absorbance at 517 nm took place.

2.4.3. FRAP Assay

The ferric-reducing ability was evaluated according to Benzie and Strain [38]. A total of 3 mL of FRAP reagent was mixed with 0.2 mL of the diluted juice. The mixture was kept for 30 min in a dark place before the reading of absorbance at 593 nm took place.

2.4.4. CUPRAC Assay

For the evaluation of cupric ion-reducing antioxidant capacity, the CUPRAC assay was utilized [39]. Copper chloride, neocuproine, and ammonium acetate buffer (pH 7) solution were mixed to a ratio of 1:1:1. To this mixture 0.2 mL of the diluted juice was added. The mixture was kept for 30 min in a dark place before the reading of absorbance at 450 nm took place.

2.5. Preparation of Stone Fruit Juices for High Performance Liquid Chromatography (HPLC)

Prior to HPLC analysis, to exclude impurities samples went through solid-phase extraction using commercial sorbent, Strata^{TM}-X 33 μm Polymeric Reversed Phase from Phenomenex (Torrance, CA, USA). Preconditioning with methanol (HPLC-grade) of cartridges was conducted after their insertion in a vacuum manifold which was operating at room temperature. Then, acetic acid solution (1% in water) was added, followed by the addition of the sample. The sample was dripping, and formed a compact ring in the cartridges. After the cartridges were dried, elution of bioactives was conducted with methanol [40,41]. The eluents were gathered and used for injection into the HPLC system.

2.6. Evaluation of Bioactive Compounds of Stone Fruit Juices Using Reversed Phase HPLC

The detection and quantification of individual bioactive compounds in juices was performed by HPLC (Agilent HPLC system 1260 Infinity II, Santa Clara, CA, USA). The whole system was composed of a quaternary pump, a vial sampler, DAD detector (which was recording spectra in the interval from 190 to 600 nm), and column (Poroshell 120 EC C-18, 4.6 × 100 mm, 2.7 μm). Buljeta et al. [42] have previously described the method. Orthophosphoric acid (0.1%) was used as a mobile phase A, and methanol was used as a mobile phase B. Both mobile phases were HPLC-grade. The injected volume of sample was set at 5 μL, and the flow rate was set at 1 mL/min. For the quantification of bioactive compounds, calibration curves for standards of anthocyanins (cyanidin-3-glucoside, cyanidin-3-rutinoside), hyperoside, rutin, gallic acid, ellagic acid, chlorogenic acid, neochlorogenic acid, (-)-epicatechin, and loganic acid were generated with a linearity of $R^2 > 0.99$. Anthocyanins were recorded at 520 nm, rutin and hyperoside at 360 nm, ellagic acid at 250 nm, gallic acid and (-)-epicatechin at 280 nm, neochlorogenic acid and chlorogenic acid at 320, and loganic acid at 245 nm. Cyanidin-3-glucosylrutinoside and derivate of pelargonidine-3-glucoside were expressed through cyanidin-3-glucoside, and cornuside through loganic acid. Concentrations of bioactives were presented as mg of bioactive compound per L of the juice (mg/L).

2.7. Evaluation of Antiproliferative Effects of Stone Fruit Juices on U87-MG and GBM43 Glioblastoma Cells

Two glioblastoma cell lines, U87-MG and GBM43, were used to assess the antiproliferative effects of selected stone fruit juices. Glioblastoma U87-MG cells were derived from a 44-year-old female patient, and GBM43 cells from a male patient. These cell lines were obtained from the American Type Culture Collection (ATCC) (U87-MG: ATCC® HTB-14™, GBM43: ATCC® CRL-3308™), ensuring their authenticity and reliability. The selected cells were cultured in 96 well plates. The cells were treated with stone fruit juices after 24 h. Fruit juices were applied in amounts of 1%, 2%, or 3% of the culture medium (DMEM—10% fetal bovine serum—1% penicillin/streptomycin). Positive and negative controls were also prepared. As the positive control for inducing cell death, cells treated with TMZ were used, while as the negative control untreated cells were used. The treatment period lasted for 3 days, after which the survival of glioblastoma cells was evaluated using the MTT (3-(4,5-dimethylthiazol-2-yl)-2,5-diphenyl-2H-tetrazolium bromide) method [43]. The method is based on the measurement of the metabolic activity of living cells. During reaction, MTT reagent is converted from yellow to purple due to the formation of formazan crystals by viable cells. Absorbance of the obtained purple formazan crystals was read and used for their quantification. An indication of greater cell viability is higher absorbance, reflecting the effectiveness of the treatment. The percentage of survival of glioblastoma cells for each treatment group was calculated according to following equation:

$$\text{Percentage of survival} = (A_{TS}/A_{UTS}) \times 100$$

where A_{TS} was absorbance of cells that were treated, and A_{UTS} the absorbance of untreated cells.

2.8. Statistical Analysis of the Obtained Results

For each assay, all samples were estimated in triplicate. The obtained results for bioactives and antioxidant activities of juices were compared using analysis of variance (ANOVA) and Fisher's least significant difference (LSD), with the significance defined as $p < 0.05$. Statistical analyses were conducted by the software program STATISTICA 13.1 (StatSoft Inc., Tulsa, OK, USA). The obtained results for antiproliferative effects of stone fruit juices on chosen glioblastoma cells were plotted and statistically analyzed using one-way analysis of variance (ANOVA) to evaluate the significant differences in survival percentages between groups, with the significance defined as $p < 0.05$, $p < 0.01$, $p < 0.001$. For this purpose, GraphPad Prism version 10.1.2. was applied.

3. Results

3.1. Bioactive Compounds and Antioxidant Potential of Stone Fruit Juices

Results of the determinations of total polyphenols, monomeric anthocyanins, proanthocyanidins, and antioxidant activities of selected stone fruit juices are presented in Table 1. Blackthorn and tart cherry juices had similar total polyphenol contents (approximately 2.6 g/L), which were higher than those of cornelian cherry juice (1.81 g/L). Tart cherry juice contained three times the amount of monomeric anthocyanins than blackthorn juice did (468.36 mg/L vs. 152.25 mg/L), while cornelian cherry juice had the lowest content of monomeric anthocyanins (15.51 mg/L). The highest content of proanthocyanidins was in tart cherry juice (1254.99 mg/L), whereas for the other evaluated juices this was 306.9 mg/L and 21.56 mg/L in blackthorn and cornelian cherry juices, respectively. Four methods were used to measure antioxidant activity, namely the DPPH, ABTS, FRAP, and CUPRAC methods. The highest antioxidant activity measured by the DPPH method was in cornelian cherry (9.08 µmol/100 mL), followed by blackthorn (8.40 µmol/100 mL) and tart cherry (6.51 µmol/100 mL) juice. However, the application of the ABTS method indicated that all juices had similar antioxidant activities (approximately 18 µmol/100 mL). Results of antioxidant activity obtained by the FRAP method indicated that blackthorn juice had the highest antioxidant potential, while cornelian cherry had the lowest (1.99 µmol/100 mL and 1.51 µmol/100 mL, respectively). The same trend was observed for the CUPRAC method (116.56 µmol/100 mL and 71.59 µmol/100 mL for blackthorn and cornelian cherry juices, respectively).

Table 1. Total polyphenols (TP), monomeric anthocyanins (ANT), proanthocyanidins (PAC), and antioxidant activity (µmol/100 mL) of stone fruit juices.

Parameters	Juice		
	Tart Cherry	Cornelian Cherry	Blackthorn
TP (g/L)	2.58 ± 0.05 [a]	1.81 ± 0.01 [b]	2.69 ± 0.02 [a]
ANT (mg/L)	468.36 ± 3.87 [a]	15.51 ± 0.51 [c]	152.25 ± 0.13 [b]
PAC (mg/L)	1254.99 ± 24.61 [a]	21.56 ± 1.13 [c]	306.90 ± 4.75 [b]
DPPH	6.51 ± 0.06 [c]	9.08 ± 0.15 [a]	8.40 ± 0.05 [b]
ABTS	17.65 ± 0.26 [a]	17.43 ± 0.32 [a]	18.13 ± 0.37 [a]
FRAP	1.72 ± 0.03 [b]	1.51 ± 0.00 [c]	1.99 ± 0.02 [a]
CUPRAC	91.56 ± 0.77 [b]	71.59 ± 0.08 [c]	116.56 ± 0.19 [a]

Values marked with various letters (a–c) in the same row are significantly different at $p < 0.05$.

HPLC analysis was utilized for the evaluation of bioactive compounds in chosen juices, and the results are presented in Table 2. In all juices, individual polyphenols were found, while iridoids were found only in cornelian cherry juice. The most abundant polyphenols in tart cherry juice were anthocyanins, cyanidin-3-rutinoside, and cyanidin-3-glucosylrutinoside (420.71 mg/L and 342.29 mg/L, respectively). Chlorogenic and neochlorogenic acids were also found in high concentrations (156.42 mg/L and 102.92 mg/L, respectively). Additionally, epicatechin (40.17 mg/L) and rutin (46.54 mg/L) were found. Blackthorn juice had the highest concentration of neochlorogenic acid (693.69 mg/L). It also contained anthocyanins (cyanidin-3-rutinoside and cyanidin-3-glucoside in concentration of 73.28 mg/L and 23.36 mg/L, respectively) and hyperoside (19.74 mg/L). Cornelian cherry juice contained cyanidin-3-glucoside (7.03 mg/L) and derivate of pelargonidin (7.96 mg/L). Also, gallic (23.93 mg/L), chlorogenic (10.87 mg/L), and ellagic (2.05 mg/L) acids were found. In addition to polyphenols, iridoids were identified in cornelian cherry juice, and they were the prevalent bioactives in this juice. Two iridoides were found: loganic acid, which was the dominant one (1285.82 mg/L), and cornuside (76.88 mg/L).

Table 2. Bioactive compounds (mg/L) found in stone fruit juices by HPLC analysis.

Bioactive Compounds	**Juice**		
	Tart Cherry	**Cornelian Cherry**	**Blackthorn**
Individual polyphenols			
Cyanidin-3-glucoside	ND	7.03 ± 0.08 [b]	23.36 ± 0.27 [a]
Cyanidin-3-rutinoside	420.71 ± 13.74 [a]	ND	73.28 ± 0.09 [b]
Cyanidin-3-glucosylrutinoside	342.29 ± 8.47	ND	ND
Pelargonidine *	ND	7.96 ± 0.03	ND
Hyperoside	ND	ND	19.74 ± 0.16
Rutin	46.54 ± 0.94	ND	ND
(-)-epicatechin	40.17 ± 1.60	ND	ND
Gallic acid	ND	23.93 ± 0.16	ND
Ellagic acid	ND	2.05 ± 0.09	ND
Chlorogenic acid	156.42 ± 3.35 [a]	10.87 ± 0.34 [b]	ND
Neochlorogenic acid	102.92 ± 2.04 [b]	ND	693.69 ± 12.66 [a]
Iridoides			
Loganic acid	ND	1285.82 ± 17.85	ND
Cornuside	ND	76.88 ± 3.52	ND

*—derivate; ND—not detected. Values marked with various letters (a, b) in the same row are significantly different at $p < 0.05$.

3.2. Antiproliferative Effects on U87-MG and GBM43 Cells of Stone Fruit Juices

The inhibition of the proliferation of glioblastoma U87-MG and GBM43 cell lines in response to stone fruit juices is presented in Figures 1 and 2. Their potential for inhibiting the proliferation of these cell lines was compared with the potential of TMZ, which is the standard chemotherapeutic drug for treating select tumor cell lines. The juices were applied at concentrations of 1%, 2%, and 3%. When TMZ was applied, the percentage of survival of glioblastoma U87-MG cells was 61.4%. Blackthorn juice did not exhibit an antiproliferative effect on these glioblastoma cells, while tart cherry and cornelian cherry juices showed slight effects. The survival rate ranged from 81% to 92% when tart cherry juice was applied, and from 71% to 82% when cornelian cherry juice was applied.

When TMZ was used for the inhibition of growth of GBM43 glioblastoma cells, their survival rate was 73.7%. All three juices demonstrated the potential for inhibition of GBM43 cells, however only cornelian cherry juice showed a good potency. Percentages of survival were 78.2%, 56.9%, and 41.8% when 1%, 2% or 3% of cornelian cherry juice was used for the inhibition of cells. The other two juices had lower potency for the inhibition of GMB43 cells than TMZ and cornelian cherry juice did. Survival rates ranged from 69% to 88% when tart cherry juice was applied, and from 76% to 87% when blackthorn juice was applied.

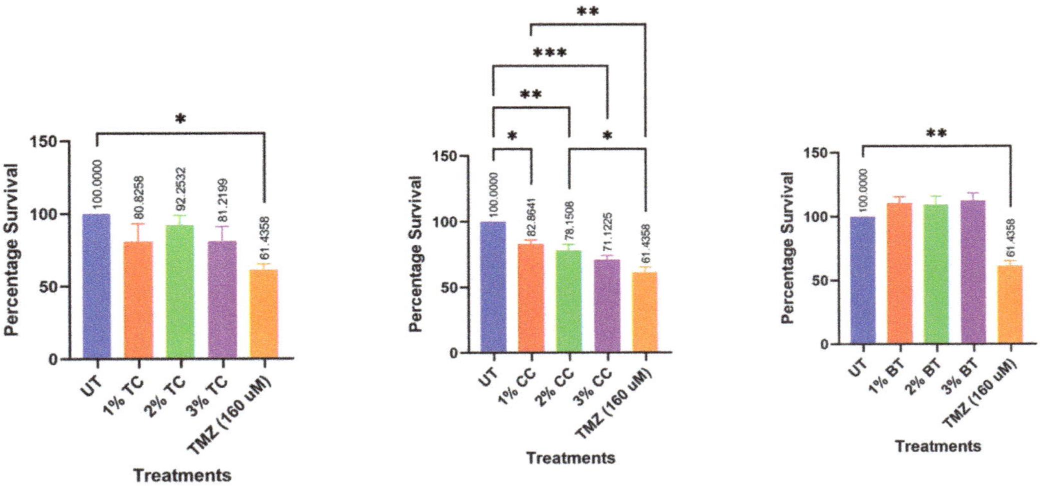

Figure 1. Antiproliferative effect of investigated stone fruit juices on U87-MG glioblastoma cells. (p values * < 0.05, ** < 0.01, *** < 0.001); UT—control; TMZ—Temozolomide; TC—tart cherry juice; CC—cornelian cherry juice; BT—blackthorn juice.

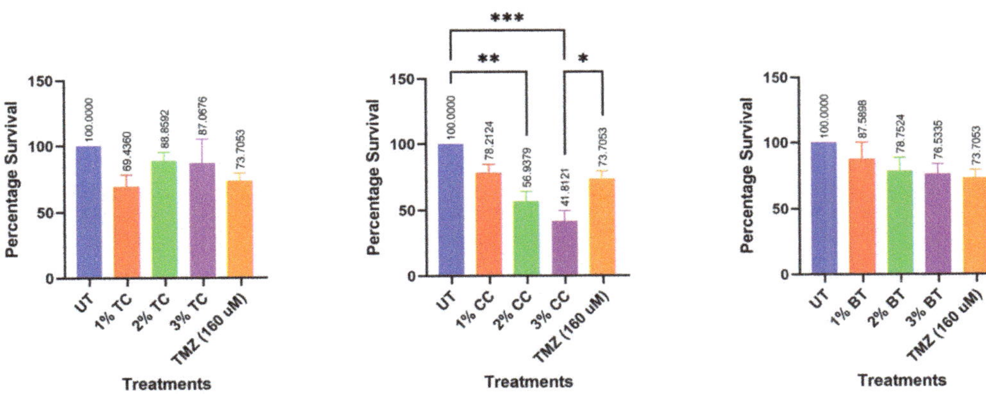

Figure 2. Antiproliferative effect of investigated stone fruit juices on GBM43 glioblastoma cells. (p values * < 0.05, ** < 0.01, *** < 0.001); UT—control; TMZ—Temozolomide; TC—tart cherry juice; CC —cornelian cherry juice; BT—blackthorn juice.

4. Discussion

The antiproliferative effects of selected stone fruit juices on U87-MG and GBM43 glioblastoma cells were compared with TMZ, the current standard chemotherapeutic drug of choice for the treatment of patients with glioblastoma. These two types of cell are commonly employed in glioblastoma research due to their ability to provide insights into the biology of glioblastoma, cancer cell growth, invasiveness, and suitability for testing potential therapeutic interventions [27–32]. As previously mentioned, there has been a growing interest in exploring active ingredients from natural sources for the prevention and treatment of malignant tumors [15,16]. Differences observed among the chosen stone fruit juices were in their profiles of bioactive compounds, their concentrations, and antioxidant activities. While the tart cherry and blackthorn juices had similar contents of total polyphenols, tart cherry juice had a considerably higher content of monomeric anthocyanins and proanthocyanidins. Tart cherry and cornelian cherry juices exhibited a slight potential for the inhibition of growth of glioblastoma U87-MG cells, however, in both cases it was lower than the one achieved with TMZ. Blackthorn juice was the only one among the tested juices to exhibit no potential for inhibition. All three juices showed a potential for inhibition of growth of glioblastoma GBM43 cells. Tart cherry and blackthorn juices had a similar or lower effect on the growth of GBM43 cells in comparison to TMZ. Only cornelian cherry juice, when applied in amounts of 2% and 3%, had a higher antiproliferative effect than TMZ. This positive influence of juices on the inhibition of glioblastoma cells can be ascribed to bioactive compounds which have been identified. In a previous study, we examined the impact of berry juices, namely raspberry, dwarf elderberry, and wild blackberry on the inhibition of growth of U87-MG and GMB46 glioblastoma cells [44], and it was demonstrated that the juices had the potential to inhibit both glioblastoma cells. It was determined that only wild blackberry juice exhibited higher potential to inhibit the growth of U87-MG cells compared to TMZ, while all berry juices demonstrated higher potency in inhibiting TMZ-resistant GBM43 cells [44]. This positive effect of berry juice is attributed to the polyphenols present in these juices, particularly anthocyanins. Although previous research indicated that berry juices with the highest concentration of anthocyanins had the highest potency in inhibiting the growth of both glioblastoma cells, the present study did not observe the same trend. Tart cherry juice had the highest anthocyanins concentration, but cornelian cherry had the highest potency in inhibition of growth of both cells, especially TMZ-resistant GMB43 cells. Cornelian cherry juice generally had the lowest content of polyphenols, thus its positive effect on glioblastoma cells could be ascribed to its possession of iridoids, especially loganic acid, bioactives that were not identified in the other two juices. Some studies showed the potential of iridoids as new anti-tumor drugs since they can prevent replication of DNA in cancer cells [21,26,45]. They specifically target neurotoxicity and oxidative stress, which is necessary in the treatment of disease through the improvement of antioxidant defenses and blocking cascades [26,46–49]. Significant biological activity of iridoids is assigned to the suppression of the expression of multiple important pro-inflammatory proteins, consequently achieving a variety of anti-inflammatory actions [26,50]. It was determined that anthocyanins and iridoids present in cornelian cherry fruits can modulate the redox system and pro-inflammatory cytokines [51]. Loganic acid inhibited the proliferation, invasiveness, and cellular migration of hepatocellular tumor cells by regulating the levels of protein of mesenchymal markers in CXCL12-treated cells. Additionally, it eliminated the MMP-9/2 gelatinolytic activity. It was suggested that loganic acid could be an anti-metastatic agent since it can suppress metastasis and epithelial mesenchymal transition processes throughout the reduction of expression of MnSOD in hepatocellular carcinoma cells [52]. Nano carriers of bioactive compounds of cornelian cherries also had the potential to inhibit the colorectal cancer cell line HT-29 by arresting the proliferation of cells in the G1 phase and causing its apoptosis [53]. Sweroside, an iridoid, also showed inhibitory effects on the growth of the U251 glioblastoma cells by causing the apoptotic death of cells. This was accompanied by upregulation of apoptotic proteins such as caspase 3 and 9, and Bax expressions. In addition, it induced arrest at the G0/G1 phase of the cell cycle and the JNK/p38 MAPK signal pathway [21].

The influence of tart cherry juice on glioblastoma cells could be attributed to high anthocyanins concentration, but also to high proanthocyanidin content. Anthocyanins have also been emphasized by other researchers as polyphenols important to the inhibition of the growth of different cancer cells. It was determined that they can be more efficient than other flavonoids for the inactivation of direct cell growth [23]. Petunidin, cyanidin, and delphinidin anthocyanidins could potentially cause the inhibition of glioblastoma cancer cells by affecting plasminogen activation and, through that, the inhibition of the migration of cancer cells. Their structure, more precisely the number of OH-groups on the B-ring, were highlighted as a cause for this action. Petunidin and cyanidin forms possess two OH-groups, and their inhibitory effect was 48%, while delphinidin had an inhibitory effect of 83% due to the possession of three OH-groups [9]. Regardless of their structure, when cyanidin and delphinidin were tested for their antiproliferative and apoptotic impact on MCF7 cell lines (human breast cancer), it was determined that cyanidin was more efficient [54]. Regarding structure, glycosylation can be a very important factor. Glycosides of aglycons could be more effective due to their ability to hinder glucose transport that results in the inhibition of energy metabolism, which, in turn, can cause mitochondrial damage and apoptosis of tumor cells [55]. Cyanidin-3-glucoside and cyanidin-3-glucosylrutinoside were found in high concentration in tart cherry juice, while in blackthorn juice cyanidin-3-glucoside and cyanidin-3-rutinoside were found, so those compounds were probably responsible for the positive influence of these juices on the proliferation of glioblastoma cells. Cyanidin-3-rutinoside and cyanidin-3-glucoside (fraction of mulberry anthocyanins) have been tested against human lung cancer cells, and it was found that they have an inhibitory influence on the migration as well as on the invasion of the highly metastatic A549 cells. The mechanism by which those anthocyanins inhibited lung cancer cells included the decrease in the expression of matrix matalloprotinase-2 (MMP-2) and urokinase-plasminogen activator (u-PA), and the increase in the expression of the tissue inhibitor of matrix matalloprotinase-2 (TIMP-2) and plasminogen activator inhibitor (PAI) [56]. Kang et al. [57] showed that anthocyanins and their aglycone, cyanidin, present in tart cherries, significantly diminished tumor development in the cecum of Apc^{Min} mice. In addition, these compounds caused the direct inhibition of the growth of HT 29 and HCT 116 cell lines (human colon cancer), with the aglycone cyanidin having much higher potential in comparison to the anthocyanin glycosides. Anthocyanin fractions of extracts of Mexican wild blackberries mostly possessed cyanidin-3-rutinoside and cyanidin-3-glucoside, and these extracts caused the initiation of apoptosis in C6 and RG2 cell lines. The inhibition of the C6 cell lines with these extracts was through the G0/G1 phase [4].

Procyanidins isolated from grape seeds have been one of the natural materials used to evaluate the effect of bioactives on glioblastoma cells. It was observed that they significantly inhibited the growth of glioblastoma by inducing G2/M arrest and decreasing mitochondrial membrane potential in U-87 cells. In addition, in these cells procyanidins caused the death of the non-apoptotic cell phenotype resembling paraptosis. They also caused the inhibition of U-87 cells by influencing a G-protein-coupled receptor, specifically the formyl peptide receptor, which is involved in the invasion and metastasis of tumor cells [58]. Cranberry proanthocyanidins and flavonoid-rich extract were tested against colon carcinoma (HT-29), glioblastoma multiforme (U87), and androgen-independent prostate carcinoma (DU145). It was found that the inhibition of these different cancer cells was achieved. The proanthocyanidin fraction was more effective than the flavonoid fraction at inducing the inhibition of glioblastoma cells in the G1 cell cycle in a time- and dose-dependent manner [59].

Karakas et al. [60] investigated the cytotoxicity of methanol extract of blackthorn fruit on glioblastoma cancer (LN229, T98G, U87) cell lines, and pancreatic cancer (PANC-1, ASPC-1) cell lines. In GBM cell lines, a decrease in cell viability was observed. However, by calculating IC_{50} values it was determined that the highest concentration was necessary for the inhibition of U87 cells. IC_{50} values were 5.245 mg/mL, 5.459 mg/mL, and 9.777 mg/mL of methanol extracts of blackthorn fruit for LN229, T98G, and U87 GBM cells, respectively. On the other hand, no effect was detected on either of the investigated pancreatic cancer cells [60].

It is important to consider that the presence of other phenols, as well as their concentrations and ratios with other phenolics, can significantly influence the availability of free-binding sites that could interact with glioblastoma cells. Interactions between compounds can result in achieving synergistic or antagonistic effects, as was also evident from the results of the determination of antioxidant activities.

Both the ABTS and DPPH methods are based on the reaction of H-atom donors with corresponding radicals, $ABTS^{\cdot+}$ and $DPPH^{\cdot}$. From our results, it is evident that higher antioxidant activity was achieved by the ABTS method, and there was no difference between samples in contrast to DPPH method. $ABTS^{\cdot+}$ is characterized by the reaction with any hydroxylated aromatics, irrespective of their actual antioxidative potential, which also includes OH-groups that do not contribute to the antioxidation [61]. DPPH˙ is characterized by higher selectivity in comparison to ABTS˙+ in the reactions with H-donors. As opposed to ABTS˙+, DPPH˙ does not react with flavonoids, which in B-ring have no OH-groups, or with aromatic acids, which have only one OH-group [61]. Even though the cornelian cherry juice contained the lowest concentration of polyphenols, it had the highest antioxidant activity estimated by the DPPH assay, probably due to the high concentration of loganic acid. It was observed that loganic acid had remarkable antioxidant activity in terms of DPPH scavenging [62], so its high concentration in cornelian cherry juice compensated antioxidant activity. The other two methods, FRAP and CUPRAC, are based on the reduction of metal ions by the action of antioxidants. The mechanism of the first one involves the reduction of the ferric ion (Fe^{3+})-ligand complex to ferrous (Fe^{2+}) complex [63], and the second one the reduction of cupric (Cu^{2+}) to cuprous ion (Cu^{+}) [64]. It is evident from the results of antioxidant activity determined by these methods that blackthorn juice had the highest antioxidant potential, even though the tart cherry juice contained considerably more anthocyanins and proanthocyanidins. This can be explained by the antagonistic phenomena that are the consequence of interactions between the phenolic compounds. It is well known that the number as well as position of OH-groups and OCH_3-groups on the phenolic rings predominantly affect the antioxidant potential of individual phenolic compounds. Nevertheless, the antioxidant potential of combined compounds, i.e., phenolic mixtures, is a complex outcome of different parameters, like intramolecular interactions, concentration of compounds, dissociation, ionization, matrix interference, etc. Consequently, the antioxidant potential of the complete mixture can be additive, synergistic, or antagonistic [65–67].

Generally, flavonoids (including anthocyanins) are mostly present in the form of metabolites in circulation since they are subjected to degradation reactions throughout the digestive system [68]. Consequently, anthocyanin forms, both intact and altered, can be detected in plasma. The most important altered forms are the corresponding phenolic acids and aldehydes, and various conjugates (for example methyl, sulfate, and glucuronyl conjugates) [69]. Nevertheless, studies have shown that flavonoids (including anthocyanins), as well as their altered forms (metabolites), may be detected in brain tissue due to their ability to pass across the blood–brain barrier [70], one of the obstacles in the efficient treatment of glioblastoma. Considering iridoids detected in cornelian cherry juice, it was determined that loganic acid was digested (in the experimental conditions) in contrast to cornuside. The fact that cornuside was present of in the colon fraction from gastrointestinal digestion in vitro makes this iridoid potentially bio-accessible [71]. Regarding their permeability through blood–brain barrier, it was determined that some iridoids have this ability. Valtrate, an iridoid component, has the ability to pass through the blood–brain barrier [72]. It was observed that valtrate had a potential antitumor activity against GBM cells in vitro and in vivo. Next to the inhibition of the proliferation, invasion, and migration of GBM cells, it also caused the induction of apoptosis in these cells [15]. Some other iridoids, such as gardenoside and agnuside, can also pass across the blood–brain barrier, which may be used for therapeutic purposes [73,74]. A variety of other iridoids can potentially also be used as therapeutic agents for intracerebral targeting because of their ability to pass across the blood–brain barrier. Even so, the key issue is the low content distribution within the brain, since these compounds go through rapid absorption as well as elimination in vivo, and are widely distributed to tissues and organs. Consequently, they have few biological benefits when they are orally administered [75].

Generally, delivering drugs to glioblastoma cells presents an extremely challenging objective for research and development. Considerable endeavor has been dedicated to developing effective delivery systems with the purpose of overcoming the heterogeneity of tumor cells (both molecular and cellular), their infiltrative nature, and the blood–brain barrier. Our results can contribute to the formulation of delivery systems, as some flavonoids and iridoids have demonstrated the ability to cross the blood–brain barrier. With this property and their potential to inhibit the growth of glioblastoma cells, these bioactives have the potential to be individually incorporated into delivery systems or combined with other natural bioactives and/or used in combination with TMZ to enhance its effects. This possibility should also be explored in future investigations.

5. Conclusions

The results of this study emphasized the diversity in composition of bioactive compounds and antioxidant capacities of chosen stone fruit juices (tart cherry, cornelian cherry, and blackthorn) and their potential for the inhibition of glioblastoma U87-MG and GBM43 tumor cells. While blackthorn juice did not inhibit growth of U87-MG cells, the other two juices did exhibit some potential to do so. TMZ-resistant GBM43 cells were affected by all three juices, but only the cornelian cherry juice showed higher potency than TMZ. For future studies, higher concentration levels of juice inclusion in the cell growth media should also be considered, as well as combinations of natural bioactives and their combinations with TMZ.

Supplementary Materials: The following supporting information can be downloaded at: https://www.mdpi.com/article/10.3390/pr12071310/s1, Figure S1: HPLC chromatogram at 520 nm of tart cherry juice (1—cyanidin-3-glucosylrutinoside; 2—cyanidin-3-rutinoside). Figure S2: HPLC chromatogram at 360 nm of tart cherry juice (3—rutin). Figure S3: HPLC chromatogram at 320 nm of tart cherry juice (4—neochlorogenic acid; 5—chlorogenic acid). Figure S4: HPLC chromatogram at 210 nm of tart cherry juice (6—(-)-epicatechin). Figure S5: HPLC chromatogram at 520 nm of cornelian cherry juice (7—cyanidin-3-glucoside; 8—pelargonidine derivate). Figure S6: HPLC chromatogram at 280 nm of cornelian cherry juice (9—gallic acid). Figure S7: HPLC chromatogram at 250 nm of cornelian cherry juice (10—ellagic acid). Figure S8: HPLC chromatogram at 320 nm of cornelian cherry juice (5—chlorogenic acid). Figure S9: HPLC chromatogram at 245 nm of cornelian cherry juice (11—loganic acid; 12—cornuside). Figure S10: HPLC chromatogram at 520 nm of blackthorn juice (7—cyanidin-3-glucoside; 2—cyanidin-3-rutinoside). Figure S11: HPLC chromatogram at 360 nm of blackthorn juice (13—hyperoside). Figure S12: HPLC chromatogram at 320 nm of blackthorn juice (4—neochlorogenic acid).

Author Contributions: Conceptualization, D.R., M.A.L., J.Š., and M.K.; methodology, D.R. and M.K.; formal analysis, M.R., J.H., I.Ć., D.R., and M.K.; investigation, D.R., M.A.L., and J.Š.; data curation, M.R., J.H., I.Ć., D.R., and M.K.; writing—original draft preparation, D.R. and M.K.; writing—review and editing, M.A.L. and J.Š.; supervision, D.R. and M.K.; project administration, M.K.; funding acquisition, M.K. All authors have read and agreed to the published version of the manuscript.

Funding: This study was funded and supported by Croatian Science Foundation under the grant number IP-2019-04-5749. Ina Ćorković acknowledges support from the Croatian Science Foundation program for Training New Doctoral Students (DOK-2020-01-4205).

Data Availability Statement: Data is contained within the article and Supplementary Materials.

Conflicts of Interest: The authors declare no conflicts of interest.

References

1. Louis, D.N.; Perry, A.; Wesseling, P.; Brat, D.J.; Cree, I.A.; Figarella-Branger, D.; Hawkins, C.; Ng, H.K.; Pfister, S.M.; Reifenberger, G.; et al. The 2021 WHO Classification of Tumors of the Central Nervous System: A Summary. *Neuro-Oncology* **2021**, *23*, 1231–1251. [CrossRef]
2. Muzyka, L.; Goff, N.K.; Choudhary, N.; Koltz, M.T. Systematic Review of Molecular Targeted Therapies for Adult-Type Diffuse Glioma: An Analysis of Clinical and Laboratory Studies. *Int. J. Mol. Sci.* **2023**, *24*, 10456. [CrossRef]
3. Thani, N.A.A.; Sallis, B.; Nuttall, R.; Schubert, F.R.; Ahsan, M.; Davies, D.; Purewal, S.; Cooper, A.; Rooprai, H.K. Induction of apoptosis and reduction of MMP gene expression in the U373 cell line by polyphenolics in *Aronia melanocarpa* and by curcumin. *Oncol. Rep.* **2012**, *28*, 1435–1442. [CrossRef]
4. Sánchez-Velázquez, O.A.; Cortés-Rodríguez, M.; Milán-Carrillo, J.; Montes-Ávila, J.; Robles-Bañuelos, B.; Santamaría del Ángel, A.; Cuevas-Rodríguez, E.O.; Rangel-López, E. Anti-oxidant and anti-proliferative effect of anthocyanin enriched fractions from two Mexican wild blackberries (*Rubus* spp.) on HepG2 and glioma cell lines. *J. Berry Res.* **2020**, *10*, 513–529. [CrossRef]
5. Thakkar, J.P.; Dolecek, T.A.; Horbinski, C.; Ostrom, Q.T.; Lightner, D.D.; Barnholtz-Sloan, J.S.; Villano, J.L. Epidemiologic and molecular prognostic review of glioblastoma. *Cancer Epidemiol. Biomarkers Prev.* **2014**, *23*, 1985–1996. [CrossRef]

6. Rooprai, H.K.; Christidou, M.; Murray, S.A.; Davies, D.; Selway, R.; Gullan, R.W.; Pilkington, G.J. Inhibition of Invasion by Polyphenols from Citrus Fruit and Berries in Human Malignant Glioma Cells In Vitro. *Anticancer Res.* **2021**, *41*, 619–633. [CrossRef]
7. Ramírez-Expósito, M.J.; Martínez-Martos, J.M. The Delicate Equilibrium between Oxidants and Antioxidants in Brain Glioma. *Curr. Neuropharmacol.* **2019**, *17*, 342–351. [CrossRef]
8. Stupp, R.; Mason, W.P.; van den Bent, M.J.; Weller, M.; Fisher, B.; Taphoorn, M.J.; Belanger, K.; Brandes, A.A.; Marosi, C.; Bogdahn, U.; et al. Radiotherapy plus concomitant and adjuvant temozolomide for glioblastoma. *N. Engl. J. Med.* **2005**, *352*, 987–996. [CrossRef]
9. Lamy, S.; Lafleur, S.; Bedard, V.; Moghrabi, A.; Barrete, S.; Gingras, D.; Béliveau, R. Anthocyanidins inhibit migration of glioblastoma cells: Structure-activity relationship and involvement of the plasmolytic system. *J. Cell Biochem.* **2007**, *100*, 100–111. [CrossRef]
10. Raucher, D. Tumor targeting peptides: Novel therapeutic strategies in glioblastoma. *Curr. Opin. Pharmcol.* **2019**, *47*, 14–19. [CrossRef]
11. Karachi, A.; Dastmalchi, F.; Mitchell, D.A.; Rahman, M. Temozolomide for immunomodulation in the treatment of glioblastoma. *Neuro Oncol.* **2018**, *20*, 1566–1572. [CrossRef] [PubMed]
12. Jiapaer, S.; Furuta, T.; Tanaka, S.; Kitabayashi, T.; Nakada, M. Potential Strategies Overcoming the Temozolomide Resistance for Glioblastoma. *Neurol. Med. Chir.* **2018**, *58*, 405–421. [CrossRef] [PubMed]
13. Tomar, M.S.; Kumar, A.; Srivastava, C.; Shrivastava, A. Elucidating the mechanisms of Temozolomide resistance in gliomas and the strategies to overcome the resistance. *Biochim. Biophys. Acta Rev. Cancer* **2021**, *1876*, 188616. [CrossRef] [PubMed]
14. Dhungel, L.; Rowsey, M.E.; Harris, C.; Raucher, D. Synergistic Effects of Temozolomide and Doxorubicin in the Treatment of Glioblastoma Multiforme: Enhancing Efficacy through Combination Therapy. *Molecules* **2024**, *29*, 840. [CrossRef] [PubMed]
15. Liu, X.; Hu, Y.; Xue, Z.; Zhang, X.; Liu, X.; Liu, G.; Wen, M.; Chen, A.; Huang, B.; Li, X.; et al. Valtrate, an iridoid compound in Valeriana, elicits anti-glioblastoma activity through inhibition of the PDGFRA/MEK/ERK signaling pathway. *J. Transl. Med.* **2023**, *21*, 147. [CrossRef] [PubMed]
16. Newman, D.J.; Cragg, G.M. Natural Products as Sources of New Drugs over the Nearly Four Decades from 01/1981 to 09/2019. *J. Nat. Prod.* **2020**, *83*, 770–803. [CrossRef]
17. Vengoji, R.; Macha, M.A.; Batra, S.K.; Shonka, N.A. Natural products: A hope for glioblastoma patients. *Oncotarget* **2018**, *9*, 22194–22219. [CrossRef]
18. Reddy, L.A.; Odhav, B.; Bhoola, K.D. Natural products for cancer prevention: A global perspective. *Pharmacol. Ther.* **2003**, *99*, 1–3. [CrossRef] [PubMed]
19. Prietsch, R.F.; Monte, L.D.; Da Silva, F.A.; Beira, F.T.; Del Pino, F.A.B.; Campos, V.F.; Collares, T.; Pinto, L.S.; Spanevello, R.M.; Gamaro, G.D.; et al. Genistein induces apoptosis and autophagy in human breast MCF-7 cells by modulating the expression of proapoptotic factors and oxidative stress enzymes. *Mol. Cell Biochem.* **2014**, *390*, 235–242. [CrossRef]
20. Liu, Y.; Jiang, Y.G. Podocalyxin promotes glioblastoma multiforme cell invasion and proliferation via β-catenin signaling. *PLoS ONE* **2014**, *28*, e111343. [CrossRef]
21. Ouyang, Z.; Xu, G. Antitumor Effects of Sweroside in Human Glioblastoma: Its Effects on Mitochondrial Mediated Apoptosis, Activation of Different Caspases, G0/G1 Cell Cycle Arrest and Targeting JNK/P38 MAPK Signal Pathways. *J. BUON* **2019**, *24*, 2141–2146. [PubMed]
22. Leung, H.W.; Lin, C.J.; Hour, M.J.; Yang, W.H.; Wang, M.Y.; Lee, H.Z. Kaempferol induces apoptosis in human lung non-small carcinoma cells accompanied by an induction of antioxidant enzymes. *Food Chem. Toxicol.* **2007**, *45*, 2005–2013. [CrossRef] [PubMed]
23. Kamei, H.; Kojima, T.; Hasegawa, M.; Koide, T.; Umeda, T.; Yukawa, T.; Terabe, K. Suppression of tumor cell growth by anthocyanins in vitro. *Cancer Investig.* **1995**, *13*, 590–594. [CrossRef] [PubMed]
24. Reddy, M.K.; Alexander-Lindo, R.L.; Nair, M.G. Relative inhibition of lipid peroxidation, cyclooxygenase enzymes, and human tumor cell proliferation by natural food colors. *J. Agric. Food Chem.* **2005**, *53*, 9268–9273. [CrossRef] [PubMed]
25. Michaud-Levesque, J.; Bousquet-Gagnon, N.; Beliveau, R. Quercetin abrogates IL-6/STAT3 signaling and inhibits glioblastoma cell line growth and migration. *Exp. Cell Res.* **2012**, *318*, 925–935. [CrossRef] [PubMed]
26. Ndongwe, T.; Witika, B.A.; Mncwangi, N.P.; Poka, M.S.; Skosana, P.P.; Demana, P.H.; Summers, B.; Siwe-Noundou, X. Iridoid Derivatives as Anticancer Agents: An Updated Review from 1970–2022. *Cancers* **2023**, *15*, 770. [CrossRef]
27. Allen, M.; Bjerke, M.; Edlund, H.; Nelander, S.; Westermark, B. Origin of the U87MG glioma cell line: Good news and bad news. *Sci. Transl. Med.* **2016**, *8*, 354re3. [CrossRef]
28. Pevna, V.; Wagnières, G.; Huntosova, V. Autophagy and Apoptosis Induced in U87 MG Glioblastoma Cells by Hypericin-Mediated Photodynamic Therapy Can be Photobiomodulated with 808 nm Light. *Biomedicines* **2021**, *9*, 1703. [CrossRef]
29. Mousavi, M.; Koosha, F.; Neshastehriz, A. Chemo-radiation therapy of U87-MG glioblastoma cells using SPIO@AuNP-Cisplatin-Alginate nanocomplex. *Heliyon* **2023**, *9*, e13847. [CrossRef]
30. Di Cintio, F.; Dal Bo, M.; Baboci, L.; De Mattia, E.; Polano, M.; Toffoli, G. The Molecular and Microenvironmental Landscape of Glioblastomas: Implications for the Novel Treatment Choices. *Front. Neurosci.* **2020**, *14*, 603647. [CrossRef]
31. Nguyen, H.-M.; Guz-Montgomery, K.; Lowe, D.B.; Saha, D. Pathogenetic Features and Current Management of Glioblastoma. *Cancers* **2021**, *13*, 856. [CrossRef] [PubMed]

32. Tsai, C.-Y.; Ko, H.-J.; Huang, C.-Y.F.; Lin, C.-Y.; Chiou, S.-J.; Su, Y.-F.; Lieu, A.-S.; Loh, J.-K.; Kwan, A.-L.; Chuang, T.-H.; et al. Ionizing Radiation Induces Resistant Glioblastoma Stem-Like Cells by Promoting Autophagy via the Wnt/β-Catenin Pathway. *Life* **2021**, *11*, 451. [CrossRef] [PubMed]
33. Singleton, V.L.; Rossi, J.A. Colorimetry of total phenolics with phosphomolybdic-phosphotonutric acid reagents. *Am. J. Enol. Vitic.* **1965**, *16*, 144–158. [CrossRef]
34. Prior, R.L.; Fan, E.; Ji, H.; Howell, A.; Nio, C.; Payne, M.J.; Reed, J. Multi-laboratory validation of a standard method for quantifying proanthocyanidins in cranberry powders. *J. Sci. Food Agric.* **2010**, *90*, 1473–1478. [CrossRef]
35. Giusti, M.M.; Wrolstad, R.E. Characterization and Measurement of Anthocyanins by UV-Visible Spectroscopy. In *Current Protocols in Food Analytical Chemistry Current Protocols*; John Wiley & Sons, Inc.: Hoboken, NJ, USA, 2001.
36. Arnao, M.B.; Cano, A.; Acosta, M. The hydrophilic and lipophilic contribution to total antioxidant activity. *Food Chem.* **2001**, *73*, 239–244. [CrossRef]
37. Brand-Williams, W.; Cuvelier, M.E.; Berset, C. Use of a free radical method to evaluate antioxidant activity. *LWT* **1995**, *28*, 25–30. [CrossRef]
38. Benzie, I.F.F.; Strain, J.J. The ferric reducing ability of plasma (FRAP) as a measure of "Antioxidant Power": The FRAP assay. *Anal. Biochem.* **1996**, *239*, 70–76. [CrossRef] [PubMed]
39. Apak, R.; Güçlü, K.; Ozyürek, M.; Karademir, S.E. Novel total antioxidant capacity index for dietary polyphenols and vitamins C and E, using their cupric ion reducing capability in the presence of neocuproine: CUPRAC method. *J. Sci. Food Agric.* **2004**, *52*, 7970–7981. [CrossRef]
40. Pambianchi, E.; Hagenberg, Z.; Pecorelli, A.; Grace, M.; Therrien, J.-P.; Lila, M.A.; Valacchi, G. Alaskan bog blueberry (Vaccinium uliginosum) extract as an innovative topical approach to prevent UV-induced skin damage. *Cosmetics* **2021**, *8*, 112. [CrossRef]
41. Može, Š.; Polak, T.; Gašperlin, L.; Koron, D.; Vanzo, A.; Poklar Ulrih, N.; Abram, V. Phenolics in Slovenian Bilberries (*Vaccinium myrtillus* L.) and Blueberries (*Vaccinium corymbosum* L.). *J. Agric. Food Chem.* **2011**, *59*, 6998–7004. [CrossRef]
42. Buljeta, I.; Pichler, A.; Šimunović, J.; Kopjar, M. Polyphenols and antioxidant activity of citrus fiber/blackberry juice complexes. *Molecules* **2021**, *26*, 4400. [CrossRef] [PubMed]
43. Mosmann, T. Rapid colorimetric assay for cellular proliferation and viability: Application to the proliferation and viability of murine macrophages. *J. Immunol. Methods* **1983**, *65*, 55–63. [CrossRef] [PubMed]
44. Kopjar, M.; Raucher, D.; Lila, M.A.; Šimunović, J. Anti-Glioblastoma Potential and Phenolic Profile of Berry Juices. *Processes* **2024**, *12*, 242. [CrossRef]
45. Gálvez, M.; Martín-Cordero, C.; Ayuso, M.J. Iridoids as DNA Topoisomers I Poisons. *J. Enzym. Inhib. Med. Chem.* **2005**, *20*, 389–392. [CrossRef] [PubMed]
46. Dinda, B.; Roy Chowdhury, D.; Mohanta, B.C. Naturally Occurring Iridoids, Secoiridoids and Their Bioactivity. An Updated Review, Part 3. *Chem. Pharm. Bull.* **2009**, *57*, 765–796. [CrossRef] [PubMed]
47. Lou, C.; Zhu, Z.; Xu, X.; Zhu, R.; Sheng, Y.; Zhao, H. Picroside II, an Iridoid Glycoside from Picrorhiza Kurroa, Suppresses Tumor Migration, Invasion, and Angiogenesis in Vitro and in Vivo. *Biomed. Pharmacother.* **2019**, *120*, 109494. [CrossRef] [PubMed]
48. Wang, C.; Xin, P.; Wang, Y.; Zhou, X.; Wei, D.; Deng, C.; Sun, S. Iridoids and Sfingolipids from Hedyotis Diffusa. *Fitoterapia* **2018**, *124*, 152–159. [CrossRef] [PubMed]
49. Villasenor, I. Bioactivities of Iridoids. *Antiinflamm Antiallergy Agents Med. Chem.* **2008**, *6*, 307–314. [CrossRef]
50. Viljoen, A.; Mncwangi, N.; Vermaak, I. Anti-Inflammatory Iridoids of Botanical Origin. *Curr. Med. Chem.* **2012**, *19*, 2104–2127. [CrossRef]
51. Sozański, T.; Kucharska, A.Z.; Szumny, A.; Magdalan, J.; Bielska, K.; Merwid-Ląd, A.; Woźniak, A.; Dzimira, S.; Piórecki, N.; Trocha, M. The protective effect of the *Cornus mas* fruits (cornelian cherry) on hypertriglyceridemia and atherosclerosis through PPARα activation in hypercholesterolemic rabbits. *Phytomed* **2014**, *21*, 1774–1784. [CrossRef]
52. Kim, N.Y.; Ha, I.J.; Um, J.-Y.; Kumar, A.P.; Sethi, G.; Ahn, K.S. Loganic acid regulates the transition between epithelial and mesenchymal-like phenotypes by alleviating MnSOD expression in hepatocellular carcinoma cells. *Life Sci.* **2023**, *317*, 121458. [CrossRef] [PubMed]
53. Radbeha, Z.; Asefi, N.; Hamishehkar, H.; Roufegarinejad, L.; Pezeshki, A. Novel carriers ensuring enhanced anti-cancer activity of *Cornus mas* (cornelian cherry) bioactive compounds. *Biomed. Pharmacother.* **2020**, *125*, 109906.
54. Tang, J.; Oroudjev, E.; Wilson, L.; Ayoub, G. Delphinidin and cyanidin exhibit antiproliferative and apoptotic effects in MCF7 human breast cancer cells. *Integr. Cancer Sci. Therap.* **2015**, *2*, 82–86.
55. Jing, N.; Song, J.; Liu, Z.; Wang, L.; Jiang, G. Glycosylation of anthocyanins enhances the apoptosis of colon cancer cells by handicapping energy metabolism. *BMC Complement. Med. Ther.* **2020**, *20*, 312. [CrossRef] [PubMed]
56. Chen, P.-N.; Chu, S.-C.; Chiou, H.-L.; Kuo, W.-H.; Chiang, C.-L.; Hsieh, Y.-S. Mulberry anthocyanins, cyanidin 3-rutinoside and cyanidin 3-glucoside, exhibited an inhibitory effect on the migration and invasion of a human lung cancer cell line. *Cancer Lett.* **2006**, *235*, 248–259. [CrossRef] [PubMed]
57. Kang, S.-Y.; Seeram, N.P.; Nair, M.G.; Bourquin, L.D. Tart cherry anthocyanins inhibit tumor development in Apc^{Min} mice and reduce proliferation of human colon cancer cells. *Cancer Lett.* **2003**, *194*, 13–19. [CrossRef] [PubMed]
58. Zhang, F.-J.; Yang, J.-Y.; Mou, Y.-H.; Sun, B.-S.; Ping, Y.-F.; Wang, J.-M.; Bian, X.-W.; Wu, C.-F. Inhibition of U-87 human glioblastoma cell proliferation and formyl peptide receptor function by oligomer procyanidins (F2) isolated from grape seeds. *Chem. Biol. Interact.* **2009**, *179*, 419–429. [CrossRef] [PubMed]

59. Ferguson, P.J.; Kurowska, E.M.; Freeman, D.J.; Chambers, A.F.; Koropatnick, J. In vivo inhibition of growth of human tumor lines by flavonoid fractions from cranberry extract. *Nutr. Cancer* **2006**, *56*, 86–94. [CrossRef] [PubMed]
60. Karakas, N.; Okur, M.E.; Ozturk, I.; Ayla, S.; Karadag, A.E.; Çiçek Polat, D. Antioxidant Activity of Blackthorn (*Prunus spinosa* L.) Fruit Extract and Cytotoxic Effects on Various Cancer Cell Lines. *Medeni. Med. J.* **2019**, *34*, 297–304.
61. Roginsky, V.; Lissi, E.A. Review of methods to determine chain-breaking antioxidant activity in food. *Food Chem.* **2005**, *92*, 235–254. [CrossRef]
62. Abirami, A.; Sinsinwar, S.; Rajalakshmi, P.; Brindha, P.; Rajesh, Y.B.R.D.; Vadivel, V. Antioxidant and cytoprotective properties of loganic acid isolated from seeds of Strychnos potatorum L. against heavy metal induced toxicity in PBMC model. *Drug Chem. Toxicol.* **2022**, *45*, 239–249. [CrossRef] [PubMed]
63. Shahidi, F.; Zhong, Y. Measurement of antioxidant activity. *J. Funct. Foods* **2015**, *18*, 757–781. [CrossRef]
64. Prior, R.L.; Wu, X.; Schaich, K. Standardized methods for the determination of antioxidant capacity and phenolics in foods and dietary supplements. *J. Agric. Food Chem.* **2005**, *53*, 4290–4302. [CrossRef] [PubMed]
65. Pinelo, M.; Manzocco, L.; Nunez, M.J.; Nicoli, M.C. Interaction among phenols in food fortification: Negative synergism on antioxidant capacity. *J. Agric. Food Chem.* **2004**, *52*, 1177–1180. [CrossRef] [PubMed]
66. Hang, D.T.N.; Hoa, N.T.; Bich, H.N.; Mechler, A.; Vo, Q.V. The hydroperoxyl radical scavenging activity of natural hydroxybenzoic acids in oil and aqueous environments: Insights into the mechanism and kinetics. *Phytochemistry* **2022**, *201*, 113281. [CrossRef]
67. Biela, M.; Kleinová, A.; Klein, E. Phenolic acids and their carboxylate anions: Thermodynamics of primary antioxidant action. *Phytochemistry* **2022**, *200*, 113254. [CrossRef] [PubMed]
68. Haizhou, W.; Oliveira, G.; Lila, M.A. Protein-binding approaches for improving bioaccessibility and bioavailability of anthocyanins. *Compr. Rev. Food Sci. Food Saf.* **2023**, *22*, 333–354.
69. Hribar, U.; Poklar Ulrih, N. The Metabolism of Anthocyanins. *Curr. Drug Metab.* **2014**, *15*, 3–13. [CrossRef] [PubMed]
70. Janle, E.M.; Lila, M.A.; Grannan, M.; Wood, L.; Higgins, A.; Yousef, G.G.; Rogers, R.B.; Kim, H.; Jackson, G.S.; Weaver, C. Method for evaluating the potential of 14C labeled plant polyphenols to cross the blood-brain barrier using accelerator mass spectrometry. *Nucl. Instrum. Methods Phys. Res.* **2010**, *268*, 1313–1316. [CrossRef]
71. Olędzka, A.; Cichocka, K.; Wolięski, K.; Melzig, M.F.; Czerwińska, M.E. Potentially Bio-Accessible Metabolites from an Extract of *Cornus mas* Fruit after Gastrointestinal Digestion In Vitro and Gut Microbiota Ex Vivo Treatment. *Nutrients* **2022**, *14*, 2287. [CrossRef]
72. Jugran, A.K.; Rawat, S.; Bhatt, I.D.; Rawal, R.S. Valeriana jatamansi: An herbaceous plant with multiple medicinal uses. *Phytother. Res.* **2019**, *33*, 482–503. [CrossRef] [PubMed]
73. Qu, K.; Zhao, L.; Luo, X.; Zhang, C.; Hou, P.; Bi, K.; Chen, X. An LC-MS method for simultaneous determination of five iridoids from Zhi-zi-chi Decoction in rat brain microdialysates and tissue homogenates: Towards an in depth study for its antidepressive activity. *J. Chromatogr. B* **2014**, *965*, 206–215. [CrossRef] [PubMed]
74. Ramakrishna, R.; Bhateria, M.; Singh, R.; Puttrevu, S.K.; Bhatta, R.S. Plasma pharmacokinetics, bioavailability and tissue distribution of agnuside following peroral and intravenous administration in mice using liquid chromatography tandem mass spectrometry. *J. Pharm. Biomed. Anal.* **2016**, *125*, 154–164. [CrossRef]
75. Wang, C.; Gong, X.; Bo, A.; Zhang, L.; Zhang, M.; Zang, E.; Zhang, C.; Li, M. Iridoids: Research Advances in Their Phytochemistry, Biological Activities, and Pharmacokinetics. *Molecules* **2020**, *25*, 287. [CrossRef] [PubMed]

processes

MDPI

Article

Ammoides pusilla Aerial Part: GC-MS Profiling and Evaluation of In Vitro Antioxidant and Biological Activities

Meriam Belaiba [1,2], Mohamed Marouane Saoudi [1], Manef Abedrabba [2] and Jalloul Bouajila [1,*]

1 Laboratoire de Génie Chimique, Université de Toulouse, CNRS, INP, UPS, F-31062 Toulouse, France; belaibameriam@gmail.com (M.B.); mohamed-marouane.saoudi@univ-tlse3.fr (M.M.S.)

2 Laboratoire des Matériaux Molécules et Applications, Université Tunis Carthage, IPEST, La Marsa 2070, Tunisia; abderrabbamanef@gmail.com

* Correspondence: jalloul.bouajila@univ-tlse3.fr; Tel./Fax: +33-562256885

Abstract: The study of *Ammoides pusilla*, a Tunisian medicinal plant, explored its chemical composition and biological activities, highlighting its under-exploited therapeutic potential. The essential oil, obtained by steam distillation, reveals twenty major compounds, including perilic aldehyde, β-phellandrene, and o-cymene. Two new natural constituents were identified in the cyclohexane extract and four in the dichloromethane extract. DPPH and ABTS tests showed that methanol extract exhibited the highest antioxidant activity, giving values of 78.9% and 65.5%, respectively, at 50 μg/mL. Its anti-diabetic activity (IC50 = 25.0 μg/mL) exceeds that of acarbose. The anti-SOD activity of methanol extract also showed promise, at 73.3% at 50 μg/mL. Essential oil and ethyl acetate extract showed notable inhibition of xanthine oxidase activity, reaching 69.0%. In addition, the essential oil demonstrated strong anti-AChE (63.23% at 50 μg/mL) and anti-inflammatory (IC50 = 31.0 μg/mL) activity. In terms of cytotoxicity, the methanol extract was effective against the HCT116 cell line (IC50 = 20.9 μg/mL), and all extracts showed activity against MCF7, OVCAR-3, and IGROV-1 cells, with IC50 values ranging from 4.0 to 25.0 μg/mL. This result underlines the potential of *Ammoides pusilla* extracts as important sources of bioactive compounds for therapeutic applications. Further research is needed to fully exploit these activities in drug development.

Keywords: *Ammoides pusilla*; GC-MS; HPLC; biological activities; cytotoxicity

1. Introduction

Metabolism and cellular respiration constantly produce free radicals in the human body; otherwise, this physiological production is already controlled by efficient internal "anti-oxidant" defense systems. In some cases, imbalances crop up either because of anti-oxidant deficiency or excessive release of free radicals. This phenomenon is called oxidative stress, which promotes many diseases, including cancer and other diseases [1]. The objective of the current study was to identify bioactive molecules having anti-oxidant effects from medicinal plants. In Algeria, the *Ammoides* genus (Apiaceae) comprises two distinct species. The first, *Ammoides atlantica* (Coss. et Dur.) Wolf, is found exclusively in Algeria, while the second, *Ammoides pusilla* (Brot.) Breistr., is widely distributed throughout the Mediterranean region. In traditional medicine, *Ammoides pusilla* (*A. pusilla*) is used as an infusion to treat a variety of ailments, including headaches, fever, flu, and diarrhea [2,3].

In the literature, we found works treating essential oils without any studies related to extracts. Using the disk diffusion method, the in vitro antibacterial activity of Algerian *A. pusilla* essential oil was evaluated against various bacterial strains. This essential oil was also analyzed using Gas Chromatography with Flame Ionization Detection (GC-FID) and Mass Spectrometry (GC-MS) [4,5]. The principal compounds present were thymol (53.2%), γ-terpinene (19.4%), and p-cymene (10.6%) [2]. Furthermore, the chemical composition of *Ptychotis verticillata* (*P. verticillate*) essential oil from Morocco was analyzed by GC-FID and GC-MS. Carvacrol (44.6%) and thymol (3.4%) are the most important compounds [6]. In a third study, at least 10 different compounds were detected in the essential oil extract of *P. verticillata*. The main compounds were thymol, gamma-terpinene, D-limonene, and m-cymene, accounting for 95.86% of the oil. They also demonstrated that *P. verticillata* exhibited antimicrobial activity in vitro against five Gram-positive bacteria [7]. A recent study investigated the chemical composition and antioxidant, antimicrobial, and antiproliferative activities of *A. pusilla* essential oil [8].

In this work, we focused only on the essential oil and organic extracts of *A. pusilla*. We were interested in: (a) chemical characterization of the essential oil using gas chromatography (GC-MS and GC-FID) and organic extracts (with and without derivatization by GC-MS), total phenolics, flavonoids, tannin, anthocyanins, and total reducing sugars; (b) antioxidant activity (DPPH and ABTS assays) of the essential oil and extracts; (c) evaluating biological activities (for the first time): anti-inflammatory, anti-SOD, anti-XOD, anti-diabetic, and cytotoxic effects with different cell lines HCT116: colorectal carcinoma; IGROV-1: ovarian cancer; MCF7: mammary gland/breast cancer; OVCAR-3: ovarian cancer.

2. Materials and Methods

2.1. Extraction

2.1.1. Organic Extract

The aerial parts of *A. pusilla* used in this report were collected during the flowering phase in Beja, a town located in the north of Tunisia (May, 2013). The plant has been identified by Professor Mohamed Bousaid from the Department of Biology (National Institute of Science, Application, and Technology) in Tunisia. The specimen reference under MB042013 was registered in the same department. The aerial part (stems, leaves, and flowers) of *A. pusilla*, dried, was ground to produce a fine powder, which was successively extracted with solvents of increasing polarity (Cyclohexane (Cyclo), Dichlorometane (DCM), Ethyl acetate (EtOAc), and finally Methanol (MeOH)). A total of 200 g of leaf powder was thus soaked in cyclohexane (2000 mL) for 4 h with frequent stirring at room temperature under pressure. Afterwards, all the mixture was filtered via Wattman paper (GF/A, 110 mm). Then, the solvent was evaporated by rotary vacuum evaporation at 35 °C (IKA, RV 10 auto V, Staufen Germany). The powder extracted with cyclohexane was then reused with the next solvent, dichloromethane, at the same conditions. The same process was followed for ethyl acetate and methanol. All extracts were then stored in sealed amber vials at 4 °C for further biological test activities and chemical composition analysis.

2.1.2. Essential Oil

The dry aerial parts of *A. pusilla* (200 g) were hydrodistilled for 3 h by means of a Clevenger-type apparatus. A yellowish oil was obtained, dried over a sodium sulfate anhydrous, then stored at +4 °C in amber vials for subsequent analysis. The optical densities were measured with a spectrophotometer (Multiskan Go, F1-01620, Thermo Fisher Scientific, Vantaa, Finland), enabling wavelength selection for 96-well plates and various types of cuvettes.

2.2. Chemical Composition of Essential Oil

GC-FID and GC-MS were used for quantitative and qualitative analysis of the essential oil [9]. For gas chromatographic analysis, we used a Varian Star (3400) chromatograph (Les Ulis, France) equipped with a DB-5MS fused silica capillary column (5% phenyl-methylpolysiloxane, 30 m × 0.25 mm, film thickness 0.25 μm). Initial chromatographic conditions included two gradients: the first increased the temperature from 60 °C to 260 °C at a rate of 5 °C/min, followed by a 15 min isothermal step at 260 °C. A second gradient, applied at 340 °C at a rate of 40 °C/min, completed the chromatographic program in 57 min.

A. pusilla essential oil was diluted in petroleum ether for analysis. A volume of 1 μL of the sample was injected using a 1:10 split mode. The carrier gas used was selected as helium (purity 99.999%) at a flow rate of 1 mL/min. It was kept at 200 °C. Using a mass spectrometer (Varian Saturn GC/MS/MS 4D), with an emission current of 10 µA and an electron multiplier voltage between 1400 and 1500 V. The trap temperature was 220 °C, and the transfer line temperature was 250 °C. The mass scan covered a range from 40 to 650 amu. In order to identify the compounds, their retention times (RTs) were compared with those of C5-C24n alkanes obtained on a non-polar DB-5MS column, and their mass spectra were compared with those in the NIST 08 database. Each compound was identified by comparing its retention time (RT) with that of C5-C24n alkanes obtained on a non-polar DB-5MS column and by comparing its mass spectra with those of the NIST 08 database. The percentage composition of the essential oil was determined by the GC-FID peak area normalization method, applying the same mass response factor to all compounds. Three injections of essential oil were used to establish the mean values for the analyses, which were performed in triplicate.

2.3. Organic Extract Profile by HPLC Analysis

The following methodology was used for obtaining the chromatogram of each standard compound as well as for all organic extracts of *A. pusilla*. A collection of phenolic standards was purchased from Sigma Aldrich (Saint-Quentin-Fallavier, France). A total of 1 mg of each standard compound (dilution 1/10) or 20 mg of each organic extract were dissolved in 1.5 mL of water (pH 2.6)/acetonitrile (90:10), sonicated, and passed through a nylon membrane filter (0.2 µm) before injection. Analysis of all solutions was performed using an Ultimate 3000 Dionex pump, autosampler thermosepration products, and UV-150 detector thermosepration products (Thermo Fisher Scientific, Waltham, MA, USA). A C18 column (5 µm, 4.6 mm × 250 mm) was used, and 50 µL was injected with a flow rate of 1.2 mL/min. A double solvent gradient elution was used: solvent A (acetic acid in water, pH 2.6) and solvent B (acetic acid in acetonitrile: water, (90:10), pH 2.6. The program of gradient was started with 10% B and increased to 30% B during 35 min. This was followed by a stage of 30% B for the following 5 min. A second gradient was applied between 30% B and 50% B for the following 5 min. Finally, the concentration of B increased up to 99.5% for the following 5 min. UV detection was stopped at 280 nm. The data analysis was performed using Chromeleon 6.8 software.

2.4. Examination of Organic Extracts' Volatile Components Both with and without Silylation

About 5 to 10 mg of each extract was added to 150 µL of BSTFA, adding 1% of TMCS, vial cap was closed, shaken for 30 s, then heated for 15 min at 40 °C. An additional 1 µL of the silylated mixture was directly analyzed by GC-MS. To avoid oxidation, measurements were taken in a sealed atmosphere and under nitrogen. The leaving group of the silylated compound must be weakly basic and capable of stabilizing a negative charge in the transition state. Samples without silylation were solubilized in their solvent of extraction at 3 mg/mL.

For the chromatographic conditions, we applied two gradients: the first was a temperature rise from 60 °C to 270 °C with a gradient of 15 °C/min and a 6 min isotherm at 270 °C. At 50 °C per minute, the second gradient was applied to 300 °C. The entire chromatographic program lasted 30 min. All other chromatographic conditions of GC-MS are identical to those of the essential oil. Analyses were established in triplicate, and results were determined based on the mean values of three injections of each sample. Identification of compounds was obtained by comparing spectral data obtained from the NIST 05 libraries with a percentage higher than 85% similarity.

2.5. Chemical Families

The quantification of total phenolics, flavonoids, total condensed tannins, and total anthocyanin contents was established according to the method cited in [10].

2.6. Reducing Sugar Determination with the 3,5-Dinitrosalicylic Acid (DNS) Assay

The DNS method is a colorimetric assay, using the reducing properties of glucose [11]. In hot and alkaline conditions, there is a reduction in dinitrosalycilic acid, which serves as an oxidant, and glucose, which is the reducing agent. The resulting compound is 3-amino-5-nitrosalicylic acid, a red compound. The amount of simple reducing sugars contained in each organic extract of *A. pusilla* is determined with a glucose standard solution (0–80 μg/mL). A total of 150 μL of each glucose solution or all organic extracts of *A. pusilla* (at 400 μg/mL) was mixed with 150 μL of DNS color reagent solution at 96 mM (5.31 M of 3,5-dinitrosalicylic acid, mixed with sodium potassium tartrate in an aqueous solution of NaOH 2 M). The reaction was conducted in Eppendorf tubes brought to ebullition in a water bath at 100 °C for 5 min. The cooling of the reaction mixture was affected by the addition of 750 μL of distilled water. The absorbance was determined at 540 nm.

2.7. Antioxidant Activity

2.7.1. Free Radical Scavenging Activity by 1-1-Diphenyl 2-Picryl Hydrazyl (DPPH•)

The hydrogen atom-or-electron release ability of the extract was measured by the variation from purple to white in the methanol solution of DPPH•. This method is based on the spectrophotometric assay using the stable radical, 2.2′-diphenylpicrylhydrazyl (DPPH•), as a reagent [12]. A total of 20 μL of each extract of *A. pusilla* (μg/mL) was mixed with 180 μL of 0.1 mM DPPH• in methanol. In plates, after an incubation of 30 min at room temperature in the dark, absorbance was measured at 517 nm using 200–1000 nm Multiskan Go microplate reader (Thermo Fisher Scientific, Vantaa, Finland). Methanol was used as a blank, and ascorbic acid (0.5–10 μg/mL) was used as the reference compound. The absorbance of solvent and DPPH• radical without extract was measured as a control. The radical-scavenging activities of extracts, expressed as percentage inhibition of DPPH•, were calculated according to the formula:

$$\text{Inhibition (\%)} = [(A_{517\text{control}} - A_{517\text{sample}})/A_{517\text{control}}] \times 100$$

2.7.2. ABTS (2,2′-Azinobis-3-ethylbenzothiazoline-6-sulfonate) Assay

The free radical scavenging capacity of antioxidants for the radical cation ABTS was evaluated using the following method [12]: a 7 mM ABTS solution was prepared at pH 7.4 (composed of 5 mM NaH_2PO_4, 5 mM Na_2HPO_4, and 154 mM NaCl), then potassium persulfate was added at a concentration of 2.5 mM. The mixture was kept at room temperature in the dark for 16 h before use, being prepared the day before the assay. After dilution in persulfate buffer, the absorbance of the mixture was adjusted to 0.70 ± 0.05 units at 734 nm. Next, 20 μL of each test sample was added to 180 μL of fresh ABTS solution, and absorbance was measured 6 min after initial mixing. As a reference standard, ascorbic acid was used. Free radical scavenging capacity was expressed as a percentage inhibition. Percentage inhibition was calculated using the same formula as for the DPPH assay.

2.8. Biological Activities

2.8.1. In Vitro α-Amylase Inhibitory Activity

The absorbance at 540 nm was used to evaluate α-amylase activity. The α-amylase inhibitory activity (%) was defined as the percent decrease in the maltose production rate over the control. Acarbose was used as a positive control. The α-amylase inhibition was expressed as a percentage of inhibition and calculated from the equation below, where A is the absorbance [12].

$$\text{Inhibition (\%)} = [(A_{540\text{control}} - A_{540\text{sample}})/A_{540\text{control}}] \times 100$$

2.8.2. In Vitro Xanthine Oxidase Inhibitory Activity

Xanthine oxidase (XOD or XO) is a form of xanthine oxidoreductase, an enzyme that generates reactive oxygen species. It catalyzes the oxidation of hypoxanthine to xanthine and can also catalyze the oxidation of xanthine to uric acid. The anti-Xanthine oxidase activity was assayed spectrophotometrically under aerobic conditions. The assay mixture consisted of essential oils tested at different concentrations (10, 50, and 100 mg/L). Whereas extracts were tested only at one concentration (50 mg/L), phosphate buffer (pH 7.5), and xanthine oxidase enzyme solution (0.01 units/mL in phosphate buffer, pH 7.5). After 15 min of pre-incubation at 25 °C, the reaction was started by adding the substrate solution (xanthine in the same buffer). At 25 °C, the mixture was incubated for 30 min. The absorbance was measured at 290 nm. The test was performed in triplicate. IC50 values were calculated from the mean values of the data, and for extracts, inhibition was determined at one concentration (50 mg/L) [12].

2.8.3. In Vitro Anti-SOD (Superoxide Dismutase) Activity

This activity was evaluated using the method of Belaiba, M et al. [12]. The auto-oxidation of pyrogallol by atmospheric oxygen could be inhibited by superoxide dismutase (SOD), an enzyme naturally present in the body. Moreover, variations in the rate of pyrogallol auto-oxidation in the presence of SOD have yet to be evaluated. The reaction mixture (200 μL) contains Tris buffer 50 mM. pH 7.5; diethylenetriaminepenta acetic acid 1 mM (DTPA); and superoxide dismutase (SOD). Extracts were tested at a concentration of 50 mg/L, followed by incubation for 37 min at 6 °C. After adding 30 mM of pyrogallocallol to initiate the reaction, the absorbance is measured for 4 min at 325 nm. The auto-oxidation of pyrogallol was tested in the absence of an enzyme. To perform this, pyrogallol was added to a Tris tampon (pH 8.5), and the same processes as an incubation of a positive control (without extracts) were repeated in the presence of 5% DMSO to measure the percentage of SOD inhibitory activity on pyrogallol auto-oxidation. The calculation of percent inhibition is as follows:

$$\%\text{Inhibition} = \text{Average } (A_{extract} - A_{controlextract})/(A_{Control} - A_{controlextract})) \times 100.$$

2.8.4. In Vitro Anti-Cholinesterase Activity

Alzheimer's disease is a chronic brain disease that alters the intellectual faculties irreversibly. This is due to a decreased level of the neurotransmitter acetylcholine (ACh). Acetylcholinesterase is an enzyme very important in the central nervous system. It catalyzes the cleavage of acetylcholine in the synaptic cleft after depolarization. Inhibitors of AChE, such as galanthamine, are frequently used in pharmacotherapy. Indeed, cells (neurons) in the nervous system release an enzyme called acetylcholinesterase (AChE). This enzyme decomposes, once released, acetylcholine (ACh) and choline acetate, resulting in the gradual reduction in the neurotransmitter acetylcholine (ACh). A simple method for evaluating the activity of AChE is the Ellman method [12]. The enzymatic activity was assessed by a modified colorimetric Ellman's method.

In this method, 50 μL of Tris-HCl buffer (pH 8), 25 μL of an extract buffer solution at different concentrations and 25 μL of an enzyme solution containing 2.8 U/mL AChE were used to assess the enzymatic activity of the extract. The reaction was then initiated via the addition of 125 μL of 3mM 5-5′-thiobis-2-nitrobenzoic acid (DTNB). After incubation of 15 min at 25 °C, 25 μL of a solution of 15mM ATCI (synthetic substrate for AChE) was added in a microplate of 96wells, and the final volume of each well was 225 μL. The absorbance of the mixture was measured at 412nm after 10min. A control mixture was prepared, using 75 μL of a solution similar to the sample mixture but with the respective solvent instead of extract. Each experiment was performed at least three times. Inhibition (%) was calculated in the following way:

$$(\%)\ \text{Inhibition} = 100 - (A_{sample}/A_{control}) \times 100.$$

where A_{sample} is the absorbance of the extract containing the reaction and $A_{control}$ is the absorbance of the reaction control. All tests were performed in triplicate. Extract concentrations providing 50% inhibition (IC_{50}) were obtained by plotting the inhibition percentage against extract solution concentrations.

2.8.5. In Vitro Anti-Inflammatory Activity

Lipoxygenase is known for catalyzing the oxidation of unsaturated fatty acids with 1.4-pentadiene groups. The activity was measured spectrophotometrically at 234 nm using the conjugated diene produced by the oxidation of linoleic acid by the 5-Lipoxygenease enzyme. Individually, 20 μL of essential oil concentrations and extracts were evaluated in sodium phosphate buffer (pH 7.4) with 5-LOX (500 U) and 60 μL of linoleic acid (3.5 mM). After a 10-min incubation at 25 °C, the absorbance at 234 nm was measured. The percentage of enzyme activity was plotted against the concentration of the essential oil. The IC_{50} value is the concentration of essential oil that causes 50% enzyme inhibition. While we determined only the percentage of inhibition for all extracts, Nordihydroguaiaretic acid (NDGA) (Sigma-Aldrich, Steinheim, Germany) was used as a positive control. The percentage inhibition of enzyme activity was calculated as shown below. All tests were carried out in triplicate [12].

$$\% \text{ Inhibition} = [(A_{Control} - A_{extract\ or\ essential\ oil})/A_{control}] \times 100.$$

2.8.6. Cytotoxic Activity with 3-[4,5-Dimethylthiazol-2yl]-2,5-diphenyl Tetrazolium Bromide (MTT) Assay

The cytotoxic activities of extracts and essential oils against cancer cell lines were evaluated by the MTT assay. The reagent used is the tetrazolium salt MTT (3-(4.5-dimethylthiazol-2-yl)-2.5-diphenyl tetrazolium). The tetrazolium ring is reduced by mitochondrial succinate dehydrogenase in active living cells to formazan. This forms a purple precipitate in the mitochondria. The amount of precipitate formed is proportional to the number of living cells (but also to the metabolic activity of each cell). A simple determination of optical density at 550 nm spectroscopy allows determining the relative amount of live cells.

The MTT colorimetric assay was carried out using 96-well plates. Cells were planted in a 96-well plate at a density of 10×10^3 cells/well (HCT116). Adherent cells (MCF7, IGROV-1 and OVAR) were delivered at a density of 12×10^3 cells/well and incubated overnight at 37 °C in a 5% CO_2 environment. Cells in the exponential growth phase were incubated at 37 °C for 72 h with each tested substance at 50 mg/L. After that, the medium was removed, and cells were treated with 50 μL of MTT solution (3 mg/mL in PBS) at 37 °C for 20 to 40 min.

To dissolve the mitochondria and therefore precipitate violet formazan, we added 80 μL of 100% DMSO. At the 540 nm wavelength, optical density was measured. All tests were established in triplicate. The anti-cancer effect of extracts was estimated in terms of growth inhibition percentage. To compare our results, we used an anti-cancer drug solution reference, tamoxifen (50 mM), at 3 different concentrations to determine the IC50 value. Tamoxifen is a selective estrogen receptor modulator used orally in breast cancer. It is currently the best treatment sold for this cancer [12].

2.9. Statistical Analysis of Data

SPSS (version 20.0) was used to compute significance, and Tukey's test was utilized to compare statistical differences among the solvents employed in the study. All measurements were carried out in quadruplicate using a one-way analysis of variance (ANOVA). To ascertain the relationship between the biological activities or antioxidants and the (TPC), the linear correlation coefficient (R2) was evaluated. Finally, a principal component analysis (PCA) was performed using XLSTAT (version 5.03) in order to verify how all the variables differed. The threshold for dependability was set at $p \leq 0.05$.

3. Results and Discussion

3.1. Extraction Yields

Extraction yields for extracts of the aerial part and essential oil of *A. pusilla* are shown in Table 1. For organic extracts, methanol had the highest yield (10.6%), followed by cyclohexane (5.4%), dichloromethane (0.7%), and ethyl acetate (0.1%). It is important to note that *A. pusilla* is an endemic plant, and there are no similar publications concerning its extracts. Nevertheless, this methodology is frequently employed by other research groups. For example, a recent study used organic solvents such as hexane, ethyl acetate, and methanol to extract bioactive compounds from eight plants in the same family (Apiaceae) [13]. The yield of essential oil was 4.3%, whereas in other studies, the essential oil yields were 0.81% and 2.3% [7]. These differences may be due to non-constant soil, climate change conditions, and maturity.

Table 1. Chemical composition and extraction yields of *A. Pusilla* extracts.

Extracts	Yields (%)	Phenolics (GAE) [a]	Flavonoids (QE) [a]	Tannins (CE) [a]	Anthocyanins (C3GE) [b]	Reducing Sugar (GE) [a]
Cyclohexane	5.4	89.3 ± 4.00	3.8 ± 0.10	6.5 ± 0.20	1.90 ± 0.02	2.74 ± 0.00
Dichloromethane	0.7	26.0 ± 2.90	2.2 ± 0.10	6.4 ± 0.40	0.90 ± 0.01	1.84 ± 0.00
Ethyl acetate	0.1	113.7 ± 2.40	47.6 ± 1.00	5.5 ± 0.50	0.10 ± 0.03	4.73 ± 0.00
Methanol	10.6	116.6 ± 2.70	91.4 ± 2.20	1.6 ± 0.20	0.50 ± 0.00	15.07 ± 0.00
Essential oil	4.3					

Note: GAE: gallic acid equivalents; QE: quercetin equivalents; CE: catechin equivalents; C3GE: cyanidin-3-glucoside equivalent; and GE: glucose equivalent. The letters [a,b] indicate [a]: g/kg dry mass; [b]: mg/kg dry mass.

3.2. Chemical Composition of Essential Oil

A total of 19 compounds were identified, representing 99.9% of the essential oil (Table 2). It consisted of 54.6% of oxygenated monoterpenes, 42.7% of monoterpene hydrocarbons, and 0.2% of sesquiterpenes. The main compounds in this broad range were peril aldehyde (47.5%), beta-phellandrene (27.4%), o-cymene (14.4%), and eugenol (5.5%). Ten compounds: peril aldehyde, tricyclene, 3-carene, kewda ether, cis-rose oxide, lavandulol, trans-carvyl acetate, alpha-copaene, and methyl eugenol were founded for the first time *A. pusilla* essential oil. Peril aldehyde was the predominant constituent in the essential oil of Perilla frutescens (72.07%).

Table 2. Main natural constituents of *A. pusilla* essential oil.

N°	Compound	RI	Percentage (%)
1	tricyclene	924	0.4
2	alpha-thujene	933	0.2
3	sabinene	972	0.7
4	beta-pinene	976	0.2
5	myrcene	988	0.6
6	3-carene	1014	0.5
7	o-cymene	1022	14.4
8	beta-phellandrene	1054	27.4
9	kewda ether	1080	0.2
10	trans-sabinene hydrate	1089	0.2
11	cis-rose oxide	1111	0.1
12	lavandulol	1162	0.7
13	terpinen-4-ol	1176	0.2
14	eugenol	1206	5.5
15	perilla aldehyde	1268	47.5
16	trans-carvyl acetate	1346	0.3
17	alpha-copaene	1377	0.2
18	methyl eugenol	1435	0.4
19	p-butylphenol	1793	0.1
	Total		99.9
	Monoterpene hydrocarbons (%)		42.7
	Oxygenated monoterpenes (%)		54.6
	Sesquiterpenes (%)		0.2
	Others (%)		2.4
	Total identified		99.9

In a previous study [2], a different composition from that of our sample was revealed, characterized by a predominance of thymol (53.2%), gamma-terpinene (19.4%), and p-cymene (10.6%) as the main constituents of the oil. In contrast, essential oil extracted from the aerial parts of *A. pusilla* in Morocco showed an abundance of phenolic compounds (48.0%), mainly carvacrol (44.6%) and thymol (3.4%) [6].

A recent study established the composition using the same method, and they found beta-phellandrene at a trace percentage (<0.05%), whereas in our study, it was considered the second major compound. Beta-pinene and alpha-thujene were found at the same percentage (0.2%). Sabinene and trans-sabinene hydrate were founded in trace percentages, whereas we founded them at 0.7 and 0.2%, respectively [8]. Such variations in chemical composition may be due to climatic conditions, geographical origin, and the plant's stage of development, all of which can have an impact on the chemical composition of its essential oil.

3.3. *Profile of the Molecular Composition of A. pusilla Extracts before and after Silylation*

The chemical composition of *A. pusilla* organic extracts with and without silylation has been established for the first time and has never been studied before. Analyses of cyclohexane extract (Table 3) demonstrated the presence of nine compounds, of which four were already identified in our essential oil (perilla aldehyde, o-cymene, 3-carene, and terpinen-4-ol). Furthermore, we found in this extract: thiocarbamic acid, N,N-dimethyl-S-1,3-diphenyl-2-butenyl ester, identified in the methanol extract of wild mushroom [14], 9-octadecenoic acid, (2-phenyl-1,3-dioxolan-4-yl)methyl ester, cis, which was already identified in *Lavandula coronopifolia* essential oil [15], and D-verbenone [16].

Table 3. Chemical composition of *A. pusilla* aerial part extracts before/after derivatization.

Compounds/Structure	Retention Time (min)	Samples				
		Essential Oil	Cyclohexane	Dichloromethane	Ethyl Acetate	Methanol
		Before derivatization				
o-Cymene 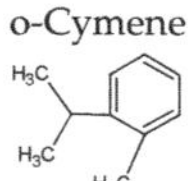	7.20	+	+			
Undecane 	7.72				+	
3-Carene 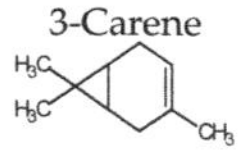	7.94	+	+			
Terpinen-4-ol 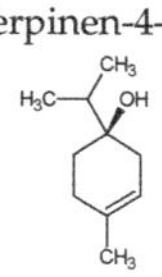	10.72	+	+			
Perilla aldehyde 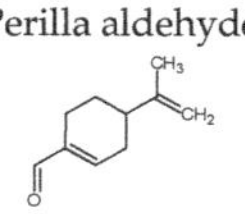	12.94	+++	++			

Table 3. *Cont.*

Compounds/Structure	Retention Time (min)	Samples				
		Essential Oil	Cyclohexane	Dichloromethane	Ethyl Acetate	Methanol
Thymol 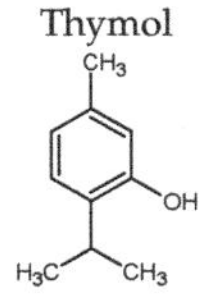	13.50					+
Cyclopropane, 1 methoxy 2,2-dimethyl-3-(3,3–dimethyl-1-propynyl) 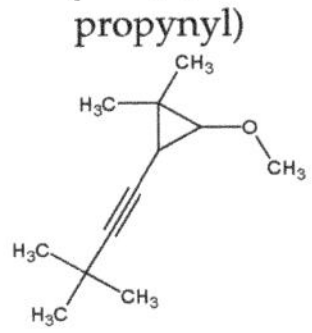	16.34		++			
Dodecanal 	16.84				+	
2-Allyl-1,4-dimethoxybenzene 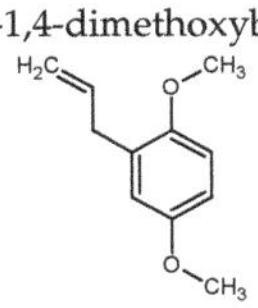	18.73		+			
2-Dodecenal (E) 	20.77				+	
D-Verbenone X30CDRA01	31.75		+			
Thiocarbamic acid, N,N-dimethyl, S-1,3-diphenyl-2-butenyl ester 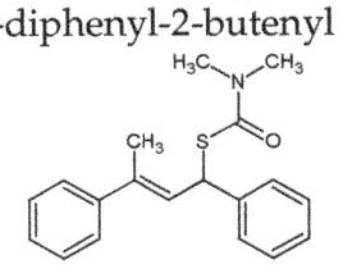	39.23		+			
9-Octadecenoic acid, (2-phenyl-1,3-dioxolan-4-yl)methyl ester, cis 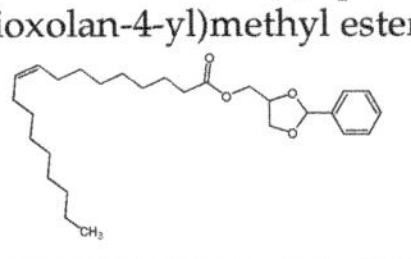	39.75		+			
After derivatization						
Glycerol 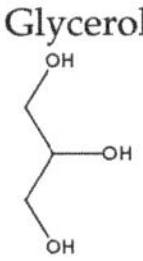	11.42			+	+	
Carvacrol 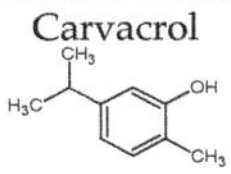	11.90				+	

Table 3. *Cont.*

Compounds/Structure	Retention Time (min)	Samples				
		Essential Oil	Cyclohexane	Dichloromethane	Ethyl Acetate	Methanol
2-methyl-alpha-D-Glucofuranoside 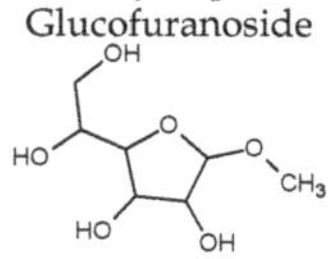	15.91			+		
2-Deoxy arabinohexonic acid 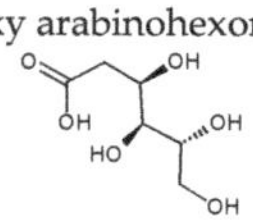	16.03			+		
2-Deoxy-ribopyranose 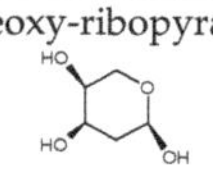	16.29			++		
2-methyl-beta-D-Mannopyranoside 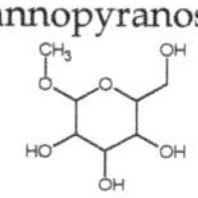	16.71			+		

Note: +++: High presence; ++: average presence; +: low presence.

In addition, two new compounds (Figure 1), never identified in natural substances, were presented in the cyclohexane extract: 2-allyl-1,4-dimethoxybenzene and cyclopropane-1-methoxy-2,2-dimethyl-3-(3,3–dimethyl-1-propynyl). 2-Dodecenal (E), dodecanal, and undecane were identified in the ethyl acetate extract. Dodecanal and 2-Dodecenal (E) were detected in the essential oil of *Coriandrum sativum,* belonging to the Apiaceae family [17].

Thymol was identified in methanol extract; it was also identified in the essential oil of *A. pusilla* [3]. It was found that thymol has not been identified in our essential oil. This may be due to both its low amount and its bioavailability; it was probably trapped in the cells of plants, and we could extract it with the fourth organic solvent.

After the silylation, we identified five compounds in the dichloromethane extract: four compounds were determined for the first time in the literature of natural products: 2-methyl-α-D-glucofuranoside, 2-Deoxy-arabinohexonic acid, 2-deoxy-ribopyranose, and 2-methyl-beta-D-mannopyranoside (Figure 2). We also noted the presence of glycerol in this derivate extract. In the ethyl acetate derivatives extract, we noted the presence of two compounds: 5-isopropyl-2-methylphenoxy (carvacrol) and glycerol. Carvacrol was founded as a major compound in the essential oil of *A. pusilla* from Morocco [5], at 44.6%. With this method of rapid analysis (without purification) by GC-MS, we identified an interesting number of original compounds.

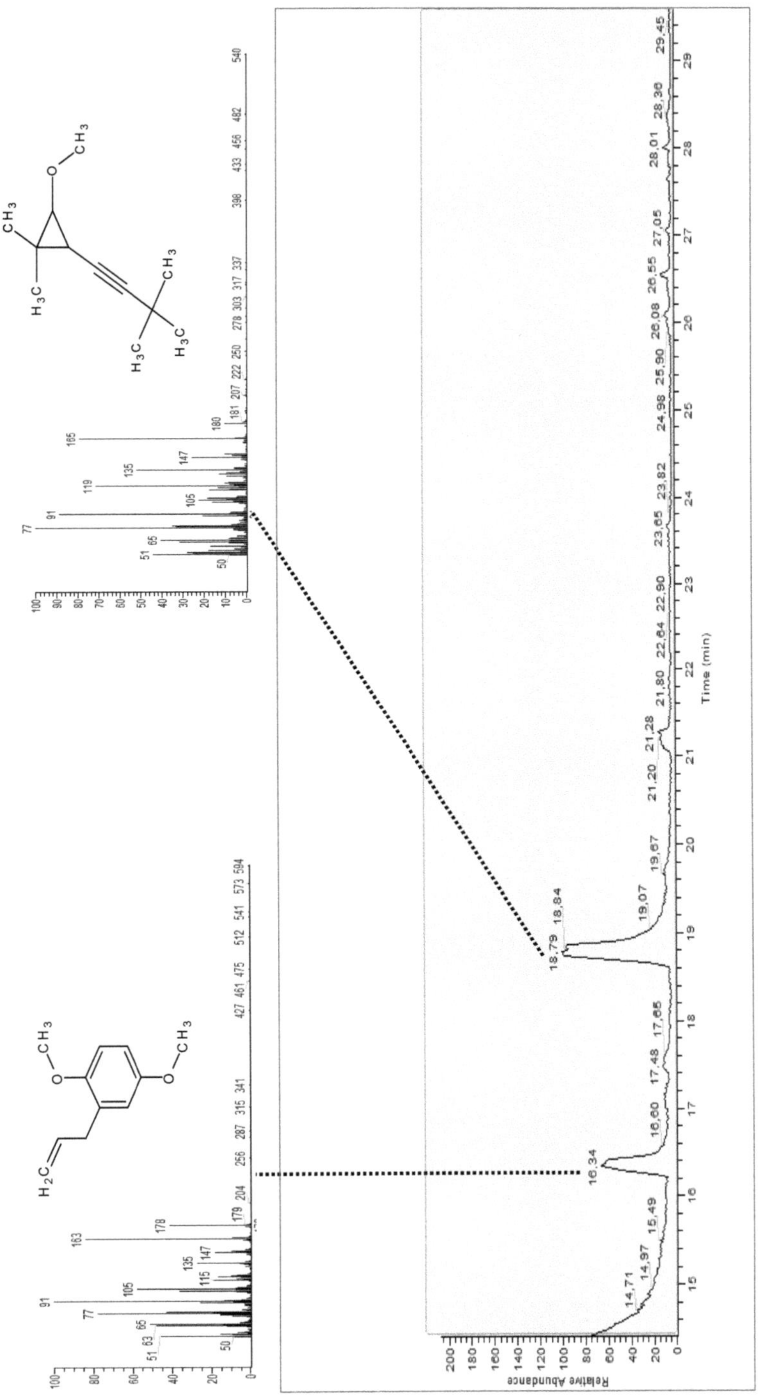

Figure 1. Molecules identified from cyclohexane extract of *A. pusilla* before derivatization using GC-MS.

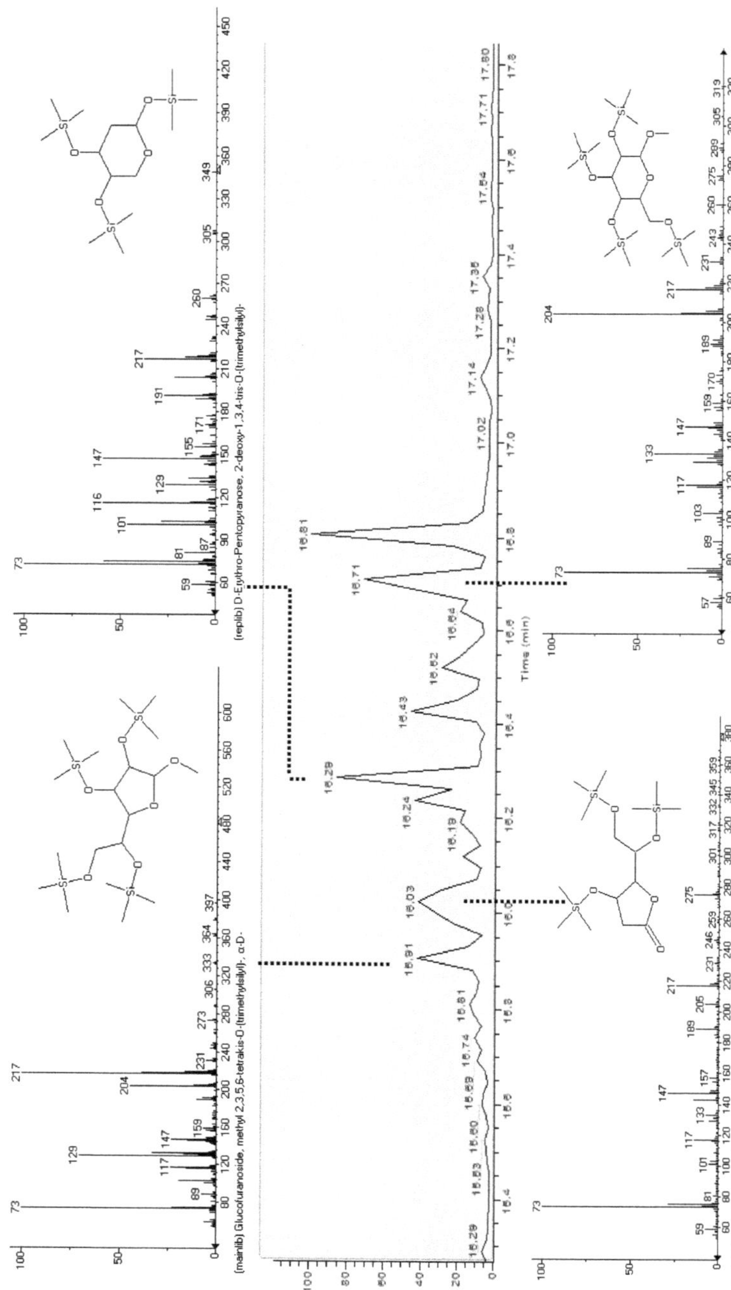

Figure 2. Molecules identified from dichloromethane extract of *A. pusilla* after derivatization using GC-MS.

3.4. Phenolics, Flavonoids, Tannins, and Anthocyanins

Table 1 shows the total phenolic content, flavonoids, tannins, and anthocyanins in extracts from the aerial part of *A. pusilla*. The highest amount of phenolic compounds was found in the methanol extract (116.6 ± 2.7 mg GAE/g dry weight), followed by the ethyl acetate extract (113.7 ± 2.4 mg GAE/g dry weight). For flavonoids, the methanol extract had the highest content (91.4 ± 2.2 mg EQ/g dry weight), while the ethyl acetate extract contained 47.6 ± 1.0 mg EQ/g dry weight. Total tannin content ranged from 1.6 ± 0.2 to 6.5 ± 0.2 mg/g dry weight catechin equivalent per gram dry weight. In terms of tannins, the cyclohexane extract had the highest concentration (6.5 ± 0.2 CE mg/g dry mass), followed by the dichloromethane extract (6.4 ± 0.4 CE g/kg dry mass). Ethyl acetate extract contained 5.5 ± 0.5 CE mg/g dry mass, while methanolic extract had the lowest concentration at 1.6 ± 0.2 CE mg/g dry mass. Anthocyanin was detected in lower quantities than in the other groups. The cyclohexane extract (1.9 ± 0.0 C3GE mg/kg dry mass) contained the highest concentration of anthocyanin. The chemical composition of several extracts revealed that the aerial part of *A. pusilla* is extremely rich in phenolic compounds. The total phenolics, tannins, flavonoids, and anthocyanin content of *A. pusilla* were determined for the first time.

3.5. Quantification of Sugar

The DNS assay gives us an idea of the total reducing sugar in the organic extracts. Results (Table 1) demonstrated that methanol extract was the richest extract in reducing sugar with 15.07 GE mg/g dry mass). Cyclohexane, dichloromethane, and ethyl acetate extracts reducing sugar contents to only 2.74, 1.84, and 4.73 GE mg/g dry mass, respectively. This assay has been established for the first time for *A. pusilla* aerial organics extracts.

3.6. HPLC Profile of Organic Extract and Comparison with Standards

HPLC analysis conducted on organic extracts revealed the presence of aromatic compounds (probably phenolics). These phytochemical compounds were compared to a standard solution injected under the same conditions. The identification of compounds was not established. This result showed that the phenolic structures present in extracts are little known. Chromatogram (280 nm) profile was mentioned in Figure 3. All the extracts were injected at 20 mg/mL.

Considering the total phenolic content in the cyclohexane extract, the two intense compounds (intensity towards 2500) and the non-polar eluted between 46 and 47 min were phenolic derivatives. Their structures must include a significant apolar group, which explains their late elution. The chromatogram of dichloromethane extract had a low intensity ($I < 500$). This was correlated with the total phenolics since it showed the lowest quantity. With regard to the ethyl acetate extract, the maximum intensity is 1400, and it has fairly polar (20 and 30 min) and non-polar compounds (40 and 50 min). Finally, the methanol extract gave a chromatogram that saturates in $I = 2500$ for the polar compound at 7.8 min. The other compounds between 20 and 50 min were also observed in the ethyl acetate extract. Thus, chromatogram intensity and total phenolic compound content in both extracts (ethyl acetate and methanol) were analogous.

3.7. Biological Properties

3.7.1. Anti-Oxidant Activity

By DPPH, methanol extract exhibited the best anti-radical power at 50 µg/mL with a percentage of 78.9 ± 0.2% (Figure 4), followed by ethyl acetate with 77.2 ± 0.5%, then cyclohexane with 56.8 ± 1.9%, and finally dichloromethane with 51.0 ± 4.8%. The essential oil tested at 50 µg/mL had 61.5 ± 1.2% inhibition for the DPPH assay.

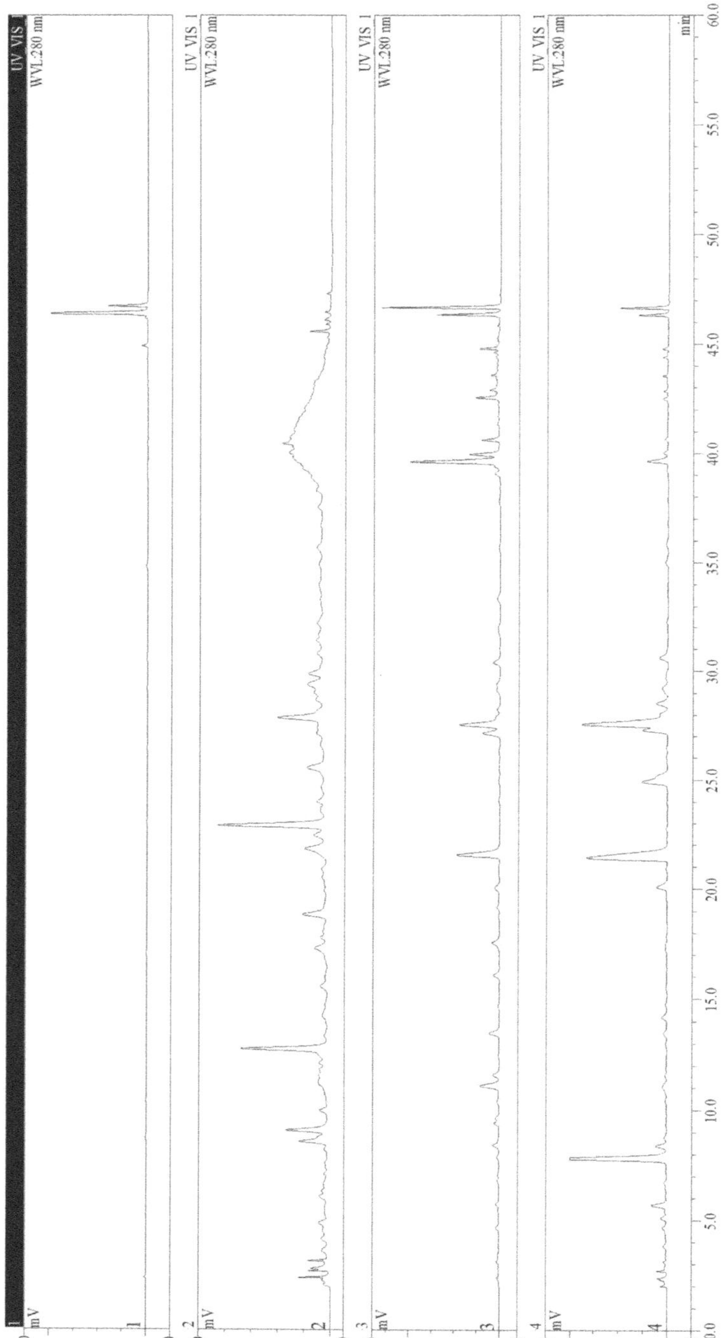

Figure 3. Chromatogram profiles of *A. pusilla* organic extracts (**1**: cyclohexane; **2**: dichloromethane; **3**: ethyl acetate; **4**: methanol).

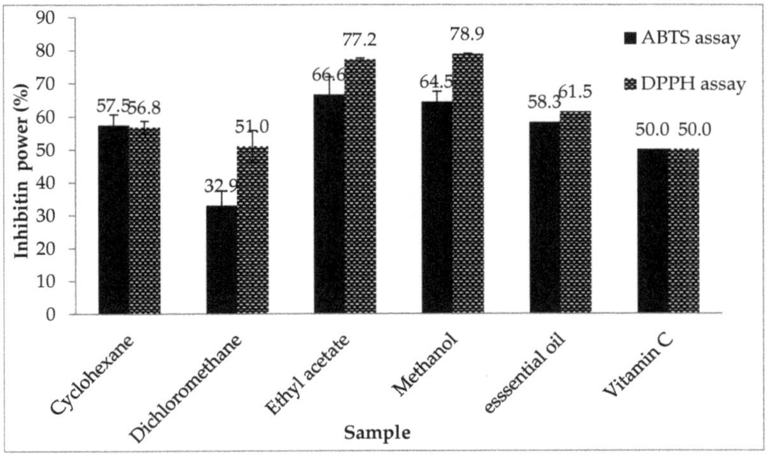

Figure 4. Chromatogram profiles of *A. pusilla* organic extract.

The anti-oxidant activity of extracts assayed using the ABTS assay is presented in Figure 4. At 50 µg/mL, the most effective extracts were methanol (78.9 ± 0.2%) and ethyl acetate (77.2 ± 0.5%). Dichloromethane and cyclohexane were less important at 50 µg/mL (51.0 ± 4.8 and 56.8 ± 1.9%, respectively). Essential oil tested at 50 µg/mL had 58.3 ± 2.3% inhibition for the ABTS assay.

As shown in Figure 5, significant correlation coefficients (R^2) between ABTS assay data and total phenolics were observed (R^2 = 0.99). In addition, the correlation coefficient R^2 between DPPH assay data and total phenolic contents was 0.78. Those results approve previous studies on the significant contribution of phenolics to anti-oxidant assays. Numerous studies have demonstrated that plants with a chemical composition similar to that of *A. pusilla* possess various biological activities, including anti-oxidant properties [18]. DPPH and ABTS test results show that *A. pusilla* essential oil and extracts (cyclohexane, dichloromethane, ethyl acetate, and methanol) have a significant capacity to neutralize free radicals, suggesting their potential as sources of natural anti-oxidants.

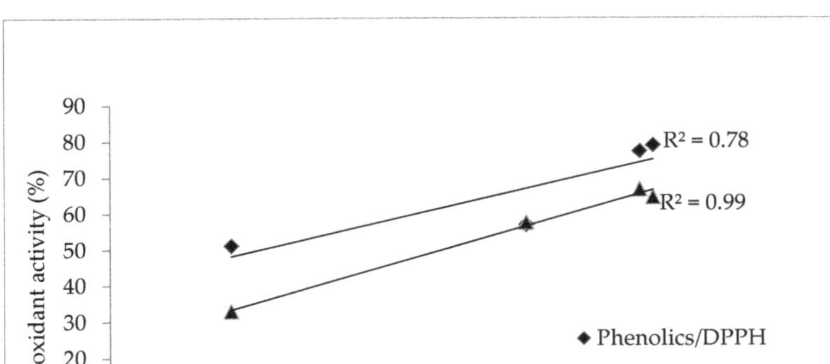

Figure 5. Correlation between anti-oxidant activity and phenolic contents in *A. pusilla* extracts.

3.7.2. Anti-Diabetic Activity

In the current study, we evaluated the α-amylase inhibition activity of the essential oil and organic extracts (Table 4). Methanol extract has an interesting anti-diabetic activity at 50 μg/mL; we obtained 94.2 ± 0.0% and an IC50 in order of 25.0 ± 0.1 μg/mL, more interesting than the positive control, acarbose (IC50 = 100 μg/mL). For cyclohexane and ethyl acetate extracts, they were less active, and we obtained only 17.2 ± 0.1 and 15.2 ± 0.2% inhibition at 50 μg/mL. IC50 were more than 50 mg/L for dichloromethane.

Table 4. Summary of biological activities of *A. pusilla* extracts and essential oil at 50 mg/L.

Activity	Anti-Diabetic		Anti-Alzheimer	Anti-Inflammatory	Anti-XOD	Anti-SOD
	(%)	(IC_{50})	(%)	(%)	(%)	(%)
Extracts	**α-Amylase**		**AchE**	**5-Lox**		
Cyclohexane	17.2 ± 0.1	>50	na	30.0 ± 0.1	na	5.0 ± 2.1
Dichloromethane	na	>50	na	34.3 ± 0.1	na	16.4 ± 0.9
Ethyl acetate	15.2 ± 0.2	>50	na	35.4 ± 0.1	69.0 ± 2.3	30.0 ± 0.5
Methanol	94.2 ± 0.0	25.0 ± 0.1	na	45.0 ± 0.0	54.8 ± 1.0	73.3 ± 0.9
Essential oil		na	63.2 ± 0.2	31.0 ± 0.0 *	69.0 ± 1.7	na
Acarbose *		100.0 ± 0.1				
Galantamin *			1.0 ± 0.2			
NDGA *				2.5 ± 0.4		
Allupurinol *					1.3 ± 0.1	

Note: * IC50 (μg/mL); na: not active.

This result proved that *A. pusilla* extracts have an interesting in vitro anti-diabetic activity, encouraging continued study of that specificity and identifying the compound responsible for that activity. Previous studies have demonstrated that. *A. pusilla* was traditionally used to treat diabetes and hypertension [19,20]. Methanol extract had the best anti-diabetic activity.

3.7.3. Effects of Samples on SOD and XOD Activities

We studied superoxide dismutase (SOD), a crucial anti-oxidant enzyme that scavenges the superoxide radical (O_2^-). Suggested to play a role in tumor suppression, SOD was mainly found in mitochondria, where most of the O_2^- was produced during cellular respiration. We revealed that increased SOD expression in ovarian cancer was a response to intrinsic oxidative stress caused by reactive oxygen species (ROS). As superoxide scavengers, SODs reduce oxidative stress and the stimulatory effect of ROS on cancer cell growth [21]. In our study, most of the tested extracts and essential oils of *A. pusilla* were slightly active on the SOD inhibition activity, except for the, methanol extract, which obtained 73.3 ± 0.9% at 50 μg/mL, as mentioned in Table 4. Among the extracts assayed for anti-XOD activity against 50 μg/mL, only ethyl acetate and methanol extracts showed any efficacy (69.0 ± 2.3 and 54.8 ± 1.0%, respectively). Likewise, the essential oil was very interesting as an XOD inhibitor, with a percentage of 69.0 ± 1.7%. This activity was established for the first time for this plant and has never been studied before. Methanol extract exhibited both anti-SOD and anti-XOD activities. Methanol extract was chosen due to its polarity and because it was anticipated that it would more effectively extract the polyphenols that possessed anti-oxidant activity. In the literature, a total of eighty-four different extracts from 27 medicinal plants and spices traditionally used against gout in Central and Eastern Europe were tested for XOD inhibition in vitro. Of the total, 25 extracts of 13 species showed inhibition in excess of 50%, while 16 extracts of 9 species showed comparable activity at 100 μg/mL. Moreover, both ethanolic extracts and methanolic extracts based on methylene chloride were the best performers [22].

3.7.4. Anti-AChE Activity

Galantamin is considered the most commonly used natural drug in Alzheimer therapy and acts as an enzyme inhibitor. This substance is isolated from the extract of snowdrops [23]. The anti-AChE activity of *A. pusilla* aerial part extracts and essential oils was evaluated. Only the essential oil exhibited anti-AChE activity with 63.23 ± 0.2% values at 50 µg/mL, and all the extracts were ineffective at this concentration (Table 4). To our knowledge, the anti-AChE activity of *A. pusilla* essential oil is established here for the first time.

3.7.5. Anti-Inflammatory Activity

5-Lox inhibition activity was evaluated for *A. pusilla* in Table 4. Essential oil was effective with an IC50 = 31.0 ± 0.0 µg/mL. This value of essential oil (compound mixture) was also comparable with the NDGA reference drug (2.5 ± 0.07 µg/mL). We outlined that most of the solvent extracts allowed us to obtain less than 50% inhibitory activity at a concentration of 50 µg/mL (IC50 > 50 µg/mL). Methanol extract showed the highest anti-inflammatory activity (45%). This suggests that the highest anti-inflammatory activity was achieved by the best anti-oxidant extracts, which contain the highest levels of total phenolics. In the literature, phenolics are considered good Lox inhibitors [24]. Anti-inflammatory in vitro activity was established for the first time in this study.

3.7.6. Cytotoxic Activity of Extracts

Four cancer cell lines, MCF-7, HCT116, IGROV-1, and OVCAR-3, were evaluated using the MTT test, which reliably detects cell proliferation. In this study, the cytotoxic activities of all the extracts were evaluated at 50 µg/mL, and the IC50 was calculated from the mean of the data (Table 5), established in triplicate.

Table 5. In vitro cytotoxic activity of *A. pusilla* extracts at 50 µg/mL.

Extracts	HCT116		MCF7		OVCAR-3		I-GROV1	
	%	IC_{50}	%	IC_{50}	%	IC_{50}	%	IC_{50}
Cyclohexane	na	>50	101.1 ± 1.6	20.0 ± 0.2	92.1 ± 2.4	4.0 ± 0.1	96.3 ± 1.3	15.0 ± 0.2
Dichloromethane	na	>50	99.1 ± 1.5	24.1 ± 0.4	91.3± 4.1	24.0 ± 0.1	91.6 ± 1.7	16.0 ± 0.1
Ethyl acetate	na	>50	94.2 ± 4.0	20.9 ± 0.1	93.3 ± 6.3	23.9 ± 0.1	93.3 ± 2.1	25.0 ± 0.1
Methanol	94.2 ± 2.3	20.9 ± 0.2	95.4 ± 3.6	19.53 ± 0.7	92.3 ± 6.3	24.8 ± 0.1	92.3 ± 2.1	20.8 ± 0.1
Tamoxifen		1.0 ± 0.2		1.0 ± 0.1		1.4 ± 0.3		2.2 ± 0.3

Note: na: not active.

For HCT116, only methanol extract was effective against this line cancer cell (IC50 = 20.9 ± 0.2 µg/mL) and was allowed to inhibit 94.2% of the total cells at 50 µg/mL. For MCF7, all extracts were active, and the IC50 varied from 19.53 to 24.1 µg/mL. Furthermore, all four organic extracts listed allowed inhibition of OVCAR-3, with IC50 values equal to 4.0 ± 0.1, 24.0 ± 0.1, 23.9 ± 0.1, and 24.8 ± 0.1 µg/mL, respectively. Also, for IGROV-1, all tested extracts were active to inhibit their growth, and their IC50 were 15.0 ± 0.2, 16.0 ± 0.1, 25.0 ± 0.1, and 20.8 ± 0.1 mg/L, respectively. We observed that *A. pusilla* aerial part extracts have different effects on each line of cancer. The study of anti-cancer *A. pusilla* property has never been cited before in the literature. It is possible that phenolic compounds, specifically flavonoid and tannin, in the aerial part of *A. pusilla* might be responsible for the cytotoxic activity of organic extracts against HCT116, MCF7, OVCAR-3, and IGROV-1. The better IC50 value observed in cyclohexane extract with 4.0 ± 0.1 µg/mL indicates that the anti-cancer agent has a stronger cytotoxic effect against the ovarian cell line (OVCAR-3).

3.8. Principal Components Analysis (PCA)

Total phenolics, flavonoids, total condensed tannins, total anthocyanin contents, reducing sugar, and biological activity contents of *A. pusilla* extracts were also analyzed according to PCA (Table 6). All axes of measurement were excluded from the results obtained.

Table 6. Variable factor contribution to the principal components analysis (%).

	F1	F2	F3
Phenolics	12.249	0.000	0.059
Flavonoids	8.685	3.787	9.496
Tannins	6.790	8.807	2.476
Anthocyanins	4.150	5.458	35.995
Reducing sugar	2.899	15.799	0.983
Cytotoxicity/HCT116	1.026	17.233	8.720
Cytotoxicity/MCF7	12.272	0.142	0.315
Cytotoxicity/OVAR	12.167	0.049	1.699
Cytotoxicity/IGROV	12.353	0.000	0.594
Antidiabetic	1.026	17.233	8.720
Anti-alzheimer	12.263	0.002	0.682
Anti-inflamatory	8.176	6.784	1.218
Anti-XOD	3.231	8.678	29.017
Anti-SOD	2.714	16.027	0.027

As indicated in Figure 6, the overall variation was 91.51%. PC1 and PC2 axes were responsible for 57.40% and 34.11% of the variability, respectively.

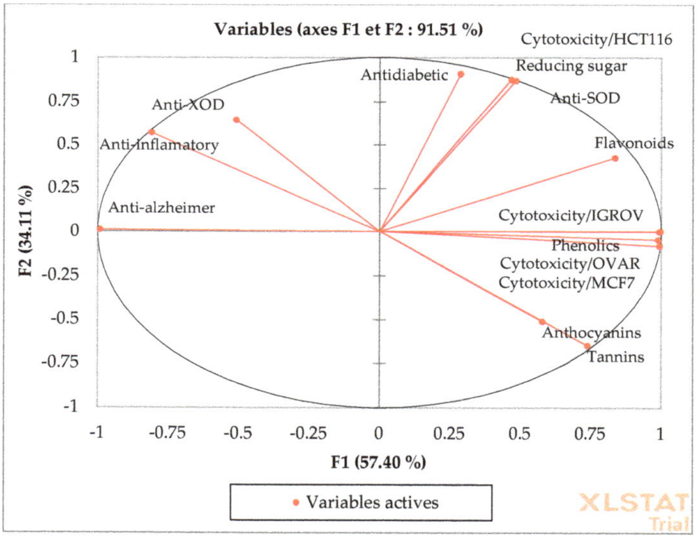

Figure 6. Principal component analysis "loading plot" of total phenolics, flavonoids, total condensed tannins, total anthocyanins contents, reducing sugar, and biological activity assays of *A. pusilla* extracts.

The loadings in the principal component analysis (PCA) diagram represent both the correlation of the principal components against the original variables and the corresponding correlations between various activities: total phenolics, flavonoids, total condensed tannins, total anthocyanins, and reducing sugar.

Referring to the data in Table 7, many correlations were identified. The first principal component (PC1) is strongly correlated with total phenolics, total condensed tannins, and total anthocyanins, as well as with cytotoxic activity against IGROV cells (r = 0.996), MCF7 cells (r = 0.993), and OVAR cells (r = 0.989). Furthermore, the second principal component (PC2) correlated well with cytotoxic activity against HCT116 (load = 0.907), anti-diabetic activity (load = 0.907), and anti-SOD activity (load = 0.875). For the third principal component (PC3), a correlation was observed with anti-Alzheimer's (load = 0.086) and anti-inflammatory (load = 0.115) activities, although these loads were moderate.

Table 7. Correlations among variables and factors.

	F1	F2	F3
Phenolics	0.992	−0.002	−0.025
Flavonoids	0.835	0.425	−0.321
Tannins	0.739	−0.649	−0.164
Anthocyanins	0.577	−0.511	0.625
Reducing sugar	0.483	0.869	0.103
Cytotoxicity/HCT116	0.287	0.907	0.307
Cytotoxicity/MCF7	0.993	−0.082	−0.058
Cytotoxicity/OVAR	0.989	−0.048	−0.136
Cytotoxicity/IGROV	0.996	0.001	−0.080
Anti-diabetic	0.287	0.907	0.307
Anti-alzheimer	−0.993	0.010	0.086
Anti-inflamatory	−0.811	0.569	0.115
Anti-XOD	−0.510	0.644	−0.561
Anti-SOD	0.467	0.875	−0.017

In Figure 7, the Biplot shows the position of extracts according to their content of total phenols, flavonoids, total condensed tannins, total anthocyanins, reducing sugars, and biological activities, as known from their specific chemical profiles. The results show that the methanolic extract exhibits strong anti-DPPH activity, attributable to its richness in TPCs [11], underscoring a significant correlation between the availability of phenolic compounds and anti-oxidant activity.

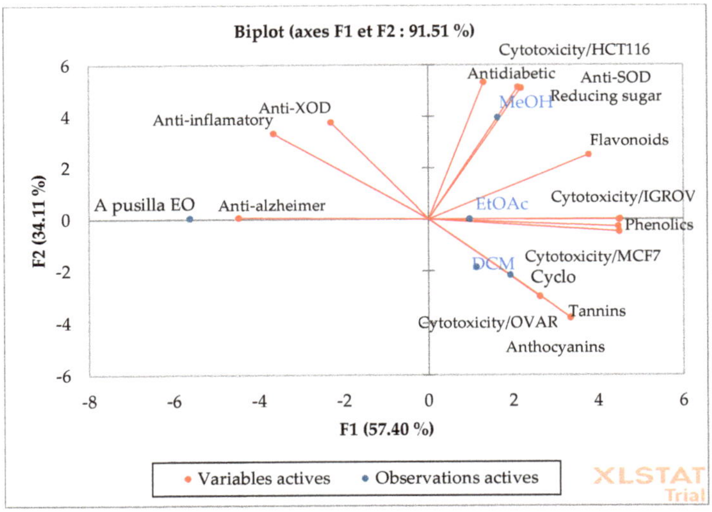

Figure 7. Biplot of principal component analysis results for the anti-oxidant and biological activities of different extracts of *A. pusilla*. (Cyclo: Cyclohexane; DCM: Dichloromethane; EtOAc: Ethyl acetate; MeOH: Methanol).

Furthermore, CYHA, EtOAc, and DCM extracts showed cytotoxic activity against IGROV, MCF7, and OVAR cells, suggesting the presence of other constituents such as fatty acids, known for their cytotoxic activity in the literature. A significant correlation was also observed between *A. pusilla* essential oil (EO) and anti-Alzheimer's activity.

4. Conclusions

In the present study, we investigated the chemical composition, anti-oxidant, anti-inflammatory, anti-SOD, anti-XOD, anti-diabetic, and cytotoxic effects (HCT116, MCF7, IGROV-1, and OVCAR-3 cell lines) of essential oil and organic extracts of *A. pusilla* aerial part. The chemical composition of organic extracts demonstrated the presence of thiocarbamic acid, N, N-dimethyl, S-1,3-diphenyl-2-butenyl ester, 9-Octadecenoic acid, (2-phenyl-1,3-dioxolan-4-yl) methyl ester, and cis. D-verbenone, identified in the essential oils of thyme and rosemary and having interesting anti-oxidant and anti-microbial activities, and was also identified here for the first time in *A. pusilla* and in the *Ammoides* genre. An interesting anti-diabetic activity was observed in methanol extract at 50 μg/mL; we obtained 94.2 ± 0.0% and an IC50 = 25.0 ± 0.1 μg/mL, four times more active than acarbose (IC50 = 100.0 ± 0.1 μg/mL). The study of the anti-cancer activity of *A. pusilla* was cited for the first time. The better IC50 value was observed in cyclohexane extract with 4 ± 0.1 μg/mL against the ovarian cell line (OVCAR-3). It will be interesting to discover the molecules responsible for all good activities, with applications in the pharmaceutical industry and further in vivo experiments. For the molecules responsible for these activities, it will be interesting to optimize their extractions using a process/green solvent pair (supercritical CO_2 without or with solvent, Accelerated Solvent Extraction (ASE), water, ethanol, etc.). These subsequent steps will focus on developing environmentally friendly extraction technologies for possible therapeutic uses.

Author Contributions: M.B. completed the practical tasks and drafted the manuscript. M.M.S., M.A. and J.B. contributed to the manuscript's correction, work orientation, and project coordination. All authors have read and agreed to the published version of the manuscript.

Funding: This research received no external funding.

Data Availability Statement: Data are contained within the article.

Conflicts of Interest: The authors declare no conflicts of interest.

References

1. Andrés, C.M.C.; Pérez de la Lastra, J.M.; Juan, C.A.; Plou, F.J.; Pérez-Lebeña, E. Polyphenols as Antioxidant/Pro-Oxidant Compounds and Donors of Reducing Species: Relationship with Human Antioxidant Metabolism. *Processes* **2023**, *11*, 2771. [CrossRef]
2. Ouasti, M.; Fatima, B.; Ouafae, M.; Bouchra, L.; Elachouri, M. Ammoides Pusilla (Brot.) Breistr. Apiaceae. In *Ethnobotany of Northern Africa and Levant*; Springer: Cham, Switzerland, 2023; pp. 1–7. [CrossRef]
3. Senouci, H.; Benyelles, N.G.; Dib, M.E.; Costa, J.; Muselli, A. Ammoides Verticillata Essential Oil as Biocontrol Agent of Selected Fungi and Pest of Olive Tree. *Recent Pat. Food Nutr. Agric.* **2020**, *11*, 182–188. [CrossRef] [PubMed]
4. Benyoucef, F.; Dib, M.E.A.; Tabti, B.; Zoheir, A.; Costa, J.; Muselli, A. Synergistic Effects of Essential Oils of Ammoides Verticillata and Satureja Candidissima Against Many Pathogenic Microorganisms. *Antiinfect. Agents* **2020**, *18*, 72–78. [CrossRef]
5. Bendjabeur, S.; Bensouici, C.; Hazzit, M. Investigation of Chemical Composition, Anticholinesterase, Antioxidant, Antihemolytic and Antibacterial Activities of Essential Oil and Ethanol Extract from Aerial Parts of Algerian Ammoides Verticillata (Brot.) Breistr. *J. Essent. Oil Res.* **2024**, *36*, 185–199. [CrossRef]
6. Kant, R.; Kumar, A. Review on essential oil extraction from aromatic and medicinal plants: Techniques, performance and economic analysis. *Sustain. Chem. Pharm.* **2022**, *30*, 100829. [CrossRef]
7. Ziani, B.E.C.; Barros, L.; Boumehira, A.Z.; Bachari, K.; Heleno, S.A.; Alves, M.J.; Ferreira, I.C.F.R. Profiling polyphenol composition by HPLC-DAD-ESI/MSn and the antibacterial activity of infusion preparations obtained from four medicinal plants. *Food Funct.* **2018**, *9*, 149. [CrossRef] [PubMed]
8. Elbouny, H.; Ouahzizi, B.; Bouhlali, E.D.T.; Sellam, K.; Alem, C. Pharmacological, Biological and Phytochemical Aspects of Thymus Munbyanus Boiss. & Reut.: A Review. *Plant Sci. Today* **2022**, *9*, 399–404. [CrossRef]
9. Peng, X.; Liu, N.; Wang, M.; Liang, B.; Feng, C.; Zhang, R.; Wang, X.; Hu, X.; Gu, H.; Xing, D. Recent Advances of Kinetic Model in the Separation of Essential Oils by Microwave-Assisted Hydrodistillation. *Ind. Crops Prod.* **2022**, *187*, 115418. [CrossRef]

10. Kisiriko, M.; Anastasiadi, M.; Terry, L.A.; Yasri, A.; Beale, M.H.; Ward, J.L. Phenolics from Medicinal and Aromatic Plants: Characterisation and Potential as Biostimulants and Bioprotectants. *Molecules* **2021**, *26*, 6343. [CrossRef]
11. Ayadi, J.; Debouba, M.; Rahmani, R.; Bouajila, J. The Phytochemical Screening and Biological Properties of *Brassica napus* L. var. napobrassica (Rutabaga) Seeds. *Molecules* **2023**, *28*, 6250. [CrossRef]
12. Belaiba, M.; Aldulaijan, S.; Messaoudi, S.; Abedrabba, M.; Dhouib, A.; Bouajila, J. Evaluation of Biological Activities of Twenty Flavones and In Silico Docking Study. *Molecules* **2023**, *28*, 2419. [CrossRef] [PubMed]
13. Tripathi, S.; Morya, V.; Prakash, V. Profiling of Antifungal Activities from the Leaf Extract of Selected Apiaceae Family Plants against *Aspergillus fumigates*. *Am. J. Plant Sci.* **2023**, *14*, 1–14. [CrossRef]
14. Sharma, N.; Kaur, M.; Gupta Phutela, U.; Bhatia, S. Characterization of Halotolerant Microalga Isolated from Waterlogged Habitats: Deciphering the Biochemical Profiling and Unraveling the Molecular Identity. *J. Basic Microbiol.* **2024**, *64*, 2300496. [CrossRef] [PubMed]
15. Abdoul-Latif, F.M.; Elmi, A.; Merito, A.; Nour, M.; Risler, A.; Ainane, A.; Bignon, J.; Ainane, T. Essential Oils of Tagetes Minuta and Lavandula Coronopifolia from Djibouti: Chemical Composition, Antibacterial Activity and Cytotoxic Activity against Various Human Cancer Cell Lines. *Int. J. Plant Biol.* **2022**, *13*, 315–329. [CrossRef]
16. Ayub, M.A.; Hanif, M.A.; Blanchfield, J.; Zubair, M.; Abid, M.A.; Saleh, M.T. Chemical Composition and Antimicrobial Activity of Boswellia Serrata Oleo-Gum-Resin Essential Oil Extracted by Superheated Steam. *Nat. Prod. Res.* **2023**, *37*, 2451–2456. [CrossRef]
17. Amiripour, A.; Jahromi, M.G.; Soori, M.K.; Torkashvand, A. mohammadi Changes in Essential Oil Composition and Fatty Acid Profile of Coriander (*Coriandrum sativum* L.) Leaves under Salinity and Foliar-Applied Silicon. *Ind. Crops Prod.* **2021**, *168*, 113599. [CrossRef]
18. Taibi, M.; Elbouzidi, A.; Ou-Yahia, D.; Dalli, M.; Bellaouchi, R.; Tikent, A.; Roubi, M.; Gseyra, N.; Asehraou, A.; Hano, C.; et al. Assessment of the Antioxidant and Antimicrobial Potential of Ptychotis Verticillata Duby Essential Oil from Eastern Morocco: An In Vitro and In Silico Analysis. *Antibiotics* **2023**, *12*, 655. [CrossRef]
19. Bouyahya, A.; El Omari, N.; Elmenyiy, N.; Guaouguaou, F.-E.; Balahbib, A.; Belmehdi, O.; Salhi, N.; Imtara, H.; Mrabti, H.N.; El-Shazly, M.; et al. Moroccan Antidiabetic Medicinal Plants: Ethnobotanical Studies, Phytochemical Bioactive Compounds, Preclinical Investigations, Toxicological Validations and Clinical Evidences; Challenges, Guidance and Perspectives for Future Management of Diabetes Worldwide. *Trends Food Sci. Technol.* **2021**, *115*, 147–254. [CrossRef]
20. Idm'hand, E.; Msanda, F.; Cherifi, K. Ethnopharmacological Review of Medicinal Plants Used to Manage Diabetes in Morocco. *Clin. Phytosci.* **2020**, *6*, 18. [CrossRef]
21. Alateyah, N.; Gupta, I.; Rusyniak, R.S.; Ouhtit, A. SOD2, a Potential Transcriptional Target Underpinning CD44-Promoted Breast Cancer Progression. *Molecules* **2022**, *27*, 811. [CrossRef]
22. Cherbal, A.; Bouabdallah, M.; Benhalla, M.; Hireche, S.; Desdous, R. Phytochemical Screening, Phenolic Content, and Anti-Inflammatory Effect of Foeniculum Vulgare Seed Extract. *Prev. Nutr. Food Sci.* **2023**, *28*, 141–148. [CrossRef] [PubMed]
23. Brah, A.S.; Armah, F.A.; Obuah, C.; Akwetey, S.A.; Adokoh, C.K. Toxicity and Therapeutic Applications of Citrus Essential Oils (CEOs): A Review. *Int. J. Food Prop.* **2023**, *26*, 301–326. [CrossRef]
24. Lončarić, M.; Strelec, I.; Moslavac, T.; Šubarić, D.; Pavić, V.; Molnar, M. Lipoxygenase Inhibition by Plant Extracts. *Biomolecules* **2021**, *11*, 152. [CrossRef] [PubMed]

processes

MDPI

Article

Response Surface Methodology: An Optimal Design for Maximising the Efficiency of Microwave-Assisted Extraction of Total Phenolic Compounds from *Coriandrum sativum* Leaves

Soraya Hihat [1,2], Noureddine Touati [2,3,*], Abdelhakim Sellal [3,4] and Khodir Madani [1,5]

[1] Laboratoire de Biomathématiques, Biophysique, Biochimie, et Scientométrie, Faculté des Sciences de la Nature et de la Vie, Université de Bejaia, Bejaia 06000, Algeria; soraya.hihat@univ-bba.dz (S.H.); khodir.madani@univ-bejaia.dz (K.M.)

[2] Département des Sciences Alimentaires, Faculté des Sciences de la Nature et de la Vie et des Sciences de la Terre et de l'Univers, Université Mohamed el Bachir el Ibrahimi, Bordj Bou Arreridj 34030, Algeria

[3] Laboratoire Santé et Environnement, Faculté des Sciences de la Nature et de la Vie et des Sciences de la Terre et de l'Univers, Université Mohamed el Bachir el Ibrahimi, Bordj Bou Arreridj 34030, Algeria; sellalhak@yahoo.fr

[4] Département de Biochimie, Faculté des Sciences de la Nature et de la Vie, Université Ferhat Abas Sétif 1, Sétif 19000, Algeria

[5] Centre National de Recherche en Technologies Agroalimentaires, Route de Targa-Ouzemour, Bejaia 06000, Algeria

* Correspondence: n.touati@univ-bba.dz; Tel.: +213-791-568-864

Abstract: The optimization of total phenolic compounds (TPC) extraction yield and maximization of total antioxidant capacity (TAC) from coriander leaves were investigated using response surface methodology. The extraction of TPC was carried out using microwave-assisted extraction. A Box-Behnken design was used to study the effects of the three independent variables, solvent concentration (ethanol/water 20–80%), microwave power (100–500 watt) and irradiation time (30–150 s) on the response. A second-order polynomial model was used to predict the reaction. The regression analysis showed that about 99% of the variations could be explained by the models. The predicted values were 50.97 GAE/g dw and 5.75 mg GAE/g dw for TPC and TAC, respectively. The reaction surface analysis showed that the optimum extraction parameters that maximized the extraction of antioxidants yield were 52.62% ethanol, 452.12 watt and 150 s. Under optimal conditions, the experimental values for TPC and TAC were 49.63 ± 0.93 mg GAE/g dw and 5.55 ± 0.07 mg GAE/g dw, respectively. The experimental values are in agreement with the predicted values, indicating the suitability of the model used and the success of the response surface methodology in optimizing the extraction conditions.

Keywords: *Coriandrum sativum*; microwave-assisted extraction; response surface methodology; total antioxidant capacity; total phenolic compounds

1. Introduction

Polyphenols are bioactive compounds found abundantly in various plant-based foods that have antioxidant properties [1,2]. They are known for their potential health benefits, including reducing inflammation and protecting against certain chronic diseases such cardiovascular diseases, cancer, diabetes, and neurodegenerative disorders [3–6]. Polyphenols also play a crucial role in food preservation due to their antioxidant and antimicrobial activities. When used in food preservation, polyphenols help to prevent oxidation and spoilage by inhibiting the growth of microorganisms. This can extend the shelf life of food products and maintain their quality for longer periods of time [7].

Coriander, also known as cilantro, is a versatile herb that has a distinctive taste and aroma. It is widely used in different cuisines around the world and can be used in various forms, including leaves and seeds [1]. Cilantro leaves are often used as a garnish or ingredient in dishes like salsa, guacamole, and curries, while the seeds can be ground into a powder or used whole in dishes like pickles, sausages, and curries. In addition to its culinary uses, coriander has numerous health benefits that make it a valuable addition to any diet. Coriander leaves are a good source of antioxidants, specifically polyphenols, that can protect the body against damage caused by free radicals. According to a study conducted by Ashika et al. [8], coriander leaves contain vanillic, p-coumaric, cisferulic, and trans-ferulic acids, which are responsible for the antioxidant properties of the plant. Additionally, Scandar et al. [9] found that the leaves and stems of coriander are the most nutritious and beneficial parts of the plant. Nhut et al. [10] found that coriander leaves contain rutin, a type of flavonoid that has antioxidant, anti-inflammatory, and anti-cancer properties. The study also found that the total flavonoids content, the compound considered as one of the secondary plant metabolites, demonstrates the effectiveness of biological roles such as inhibiting plasma platelet aggregation, histamine release, and antiviral activity.

Extraction is the first and most important step in the recovery and purification of bioactive compounds from plant materials. Among the various novel and sophisticated extraction techniques that have been introduced and studied to improve efficiency, microwave-assisted extraction (MAE) has garnered significant attention in recent years for its efficient extraction of bioactive compounds from various plant sources [11,12]. MAE offers advantages such as reduced extraction times, higher yields, and lower environmental impact compared to traditional methods like Soxhlet extraction and maceration [13]. The success of MAE relies on factors like solvent choice, exposure time, temperature, and equipment specifications [14]. Additionally, the unique heating mechanism of microwaves allows for quicker extraction processes and better preservation of thermolabile compounds compared to conventional heating methods [15,16]. Furthermore, MAE has been recognized as an environmentally friendly technique that minimizes energy consumption and solvent usage, making it ideal for extracting bioactive compounds from agro-industrial waste [17].

Response surface methodology is a statistical and mathematical technique for modelling and optimizing process parameters [18]. Response surface methodology is highly useful in various fields such as engineering design optimization, chemical and biological process optimization, and natural product extraction [19]. It provides a systematic approach to relate the input variables to the responses and is a cost-effective and efficient method for optimizing experimental processes. The importance of response surface methodology lies in its ability to approximate the relationship between input and output variables, which supports the optimization of complex systems and reveals synergistic effects of process parameters. The methodology includes the design of experiments, the analysis of results and the creation of models to improve the results in different applications. The planes used in the response surface methodology study are quadratic planes such as the Box-Behnken planes or the central composite planes (Box-Wilson). Response modelling is performed using regression techniques that allow a response to be linked to a set of factors [20].

The aim of this study is to enhance the extraction efficiency of antioxidants from coriander leaves using microwave-assisted extraction. This will be achieved by modeling and optimizing the extraction conditions, which include solvent concentration, microwave power, and irradiation time. Response surface methodology will be utilized to determine the optimal conditions for extracting total phenolic compounds and total antioxidant capacity.

2. Materials and Methods

2.1. Chemical Reagents

DPPH reagent (2,2-diphenyl-1-picrylhydrazyl) and Sodium carbonate (Na_2CO_3) were purchased from Sigma–Aldrich (Darmstadt, Germany), Folin–Ciocalteu phenol reagent ($H_3PW_{12}O_{40}$) and gallic acid from Biochem, Chemopharma (Montreal, QC, Canada). Ethanol (Sigma–Aldrich, ≥99.8% (GS)) of analytical grade of purity were used. All chemicals and solvents used were of analytical grade.

2.2. Sample Preparation

In this study, coriander plant (*Coriander sativum*) was procured from a local market located in the Bejaia region of Northern Algeria. Leaves were separated from the stem and washed with tap water followed by distilled water, and left to dry for approximately 48 h at room temperature in a ventilated and dark room. The dried sample was then ground to obtain a fine powder with a diameter of less than 250 μm using a grinder (A11 basic grinder from Ika, Staufen in Germany). The powder was stored at −20 °C before analysis. The moisture content was determined by constant weight at 105 °C and was found to be 5.0 ± 0.5%. The water activity (Aw) was measured using a Hygro Palm Aw instrument (Rotronic AG Bassersdorf, Switzerland) and was found to be 0.18 ± 0.2 at a temperature of 20.6 °C.

2.3. Experimental Work

To optimize the microwave-assisted extraction procedure, a series of single-factor experiments were conducted to determine the effects of individual process parameters. This approach was chosen to minimize the total experimental work required while still providing valuable insights for improving the procedure. Through this method, the impact of each parameter on the process could be assessed and the most significant factors could be identified. This allowed for a more targeted and efficient optimization process (Table 1). The variable was kept constant (50%, 500 watt and 150 s for ethanol concentration, microwave power and irradiation time, respectively) when it was not studied.

Table 1. Results of single-factor experiments for microwave-assisted extraction from coriander leaves.

Solvent Concentration			Microwave Power			Irradiation Time		
%	TPC	TAC	watt	TPC	TAC	s	TPC	TAC
10	37.85 ± 0.24 ^{c}	2.92 ± 0.11 ^{c}	100	36.43 ± 0.65 ^{bc}	3.25 ± 0.13 ^{bc}	30	38.66 ± 0.14 ^{c}	3.06 ± 0.13 ^{c}
20	39.35 ± 0.34 ^{ab}	3.11 ± 0.09 ^{ab}	300	40.97 ± 0.93 ^{a}	3.89 ± 0.08 ^{a}	90	43.08 ± 0.27 ^{a}	4.03 ± 0.07 ^{a}
50	42.91 ± 0.48 ^{a}	3.91 ± 0.21 ^{a}	500	38.46 ± 0.29 ^{b}	2.48 ± 0.01 ^{b}	150	40.98 ± 0.75 ^{b}	3.59 ± 0.05 ^{b}
80	40.80 ± 0.82 ^{b}	3.50 ± 0.19 ^{b}	700	37.89 ± 0.83 ^{bc}	3.33 ± 0.15 ^{ab}	210	39.71 ± 0.53 ^{bc}	3.26 ± 0.14 ^{ab}
100	38.65 ± 0.59 ^{c}	2.89 ± 0.04 ^{c}	900	35.96 ± 0.47 ^{c}	2.98 ± 0.09 ^{bc}	270	38.88 ± 0.71 ^{c}	2.92 ± 0.04 ^{c}

Results are reported as mean ± SD. Same letters in the same column refer to mean not statistically different according to ANOVA test. TPC: total phenolic compounds (mg GAE/g dw); TAC: total antioxidant capacity (mg GAE/g dw); GAE: gallic acid equivalent; dw: dry weight of leaves.

In both the single-factor trials and the consecutive response surface methodology optimizations, the focus was on evaluating the extraction yield of total phenolic compounds and the maximization of total antioxidant capacity.

Following that, an optimization of the processes was carried out utilizing a response surface methodology approach based on a Box-Behnken design to refine the conditions (Table 2).

Table 2. Independent variables affecting the microwave-assisted extraction.

Independent Variables	Factor Levels		
	−1	0	+1
x1: Solvent concentration (%)	20	50	80
x2: Microwave power (watt)	100	300	500
x3: Irradiation time (s)	30	90	150

The experimental design applied was a three-level three-factor Box-Behnken design, and the required number of experiments (*N*) was determined using the formula outlined in Equation (1):

$$N = 2k(k-1) + C_0 \quad (1)$$

where k is the number of factors and C_0 is the number of central points (3).

To analyze the data, a regression analysis was conducted to fit a second-order polynomial equation (quadratic model) based on the general equation (Equation (2)) in order to predict the optimal conditions for the extraction process.

$$y = a_0 + \sum_{i=1}^{3} a_i x_i + \sum_{i=1}^{3} a_{ii} x_i^2 + \sum_{i=1}^{3} \sum_{j=1}^{3} a_{ij} x_i x_{j\ (i \neq j)} \quad (2)$$

where y represents the response function; a_0 is a constant coefficient; a_i, a_{ii} and a_{ij} are the coefficients of the linear, quadratic and interactive terms, respectively, and x_i and x_j represent the coded independent variables.

The factor levels were coded as −1 (low), 0 (central point or middle) and 1 (high). The variables were coded according to the following equation (Equation (3)):

$$x_i = \frac{X_i - X_0}{\Delta X} \quad (3)$$

where x_i represents a new variable (dimensionless) encoded from the original variable X_i. X_0 represents the reference value or initial value, and ΔX represents the increment or discretization step.

After conducting the analysis of variance, the regression coefficients for the individual linear, quadratic, and interaction terms were determined. To illustrate the impacts of independent variables and their interactions, three-dimensional surface plots were created from the fitted polynomial equation using the regression coefficients.

To confirm the reliability of the model, further extraction experiments were conducted at the predicted optimal conditions determined by the response surface methodology. The experimental results obtained were then compared to the values forecasted by the regression model.

2.4. *Microwave-Assisted Extraction Process*

For the optimization of the microwave-assisted extraction process, the following parameters influencing the extraction process were selected: solvent concentration, microwave power and irradiation time. A quantity (1 g) of coriander leaf powder was placed in a 250 mL bottom flask containing ethanol-water quantities. The suspension was extracted at different concentrations of solvent, microwave power and irradiation time. The extracts were separated by centrifugation at 3000 rpm (NF 200, Nüve, Turkey) for 10 min and stored at 4 °C until use. Fifteen tests were performed before the optimum was determined. An extraction was then performed under the optimum conditions determined using response surface methodology. The total phenolic compounds and total antioxidant capacity were considered in the response surface methodology optimization and model validation tests.

2.5. *Response Parameters*

2.5.1. Determination of the Total phenolic Compounds

The total phenolic compounds (TPC) of the plant samples was determined using the Folin-Ciocalteu reagent as reported by Singleton and Rossi [21]. A volume of 0.1 mL of the extract was mixed with 0.8 mL Folin-Ciocalteu reagent (10% *v*/*v*) and 0.4 mL sodium carbonate (7.5% *w*/*v*). The absorbance was measured at 720 nm (UV/Vis spectrophotometer, Biotech Engineering Management Co., Ltd., Nicosia, Cyprus) after 60 min of incubation at room temperature against a blank (made as reported for the sample but with 0.1 mL of sample solvent). The results were expressed as milligrams of gallic acid equivalent per gram of dry weight (mg GAE/g dw) by referring to a calibration curve.

2.5.2. Evaluation of the Total Antioxidant Capacity

The total antioxidant capacity (TAC) of the plant samples was evaluated using the DPPH radical method as reported by Brand-William [22]. A volume of 0.2 mL of the extract was mixed with 1 mL of methanolic DPPH solution (60 µM). The absorbance was measured at 515 nm after 30 min of incubation at room temperature against a blank (made as reported for the sample but with 0.2 mL of sample solvent). The results were expressed as milligrams of gallic acid equivalent per gram of dry weight (mg GAE/g dw) by referring to a calibration curve.

2.6. Statistical Analysis

All extraction trial and subsequent analyses were performed in triplicate, and the results are reported as means ± standard deviation (SD). The impact of each factor on the TPC yield and TAC in the single-factor experiment for microwave-assisted extraction was evaluated through ANOVA and Tukey's post hoc test at a 95% confidence level to determine their statistical significance. The data collected from the Box-Behnken design experiments for microwave-assisted extraction were subjected to ANOVA to assess the significance and appropriateness of the model for the response variable. A significance level of $p < 0.05$, $p < 0.01$ and $p < 0.001$ was considered as indicating significant, highly and very highly significant results, respectively. The JMP (Version 17.0, SAS, Cary, NC, USA) software was utilized to analyze all the experimental results and create the Box-Behnken design.

3. Results and Discussion

3.1. Effect of Single Factors

The influence of different extraction parameters on TPC yield and TAC was investigated using a method that varied one parameter at a time. The factors examined were ethanol concentration, microwave power, and irradiation time.

Choosing the right extraction solvent is crucial for determining the quantity and quality of the extracted phenolic compounds. Acetone, ethanol, and methanol are widely used solvents for extracting phenolic compounds from plants [23]; nevertheless, given the toxicity concerns associated with methanol and acetone, which are not suitable for food applications, ethanol solvent was chosen instead. Ethanol offers several advantages over other solvents, including higher extraction efficiency, environmental friendliness, and lower cost. The efficiency of extraction is significantly influenced by the concentration of ethanol, as reported in the literature [24]. Therefore, the impact of varying ethanol concentration was evaluated to optimize the extraction efficiency. Based on Table 1, the value of TPC and TAC ranged from 37.85 ± 0.24 to 42.91 ± 0.48 mg GAE/g dw and 2.92 ± 0.11 to 3.91 ± 0.21 mg GAE/g dw, respectively. As can be seen, the TPC and TAC increased significantly from the concentration of 10 to 50%, and then decreased to achieve a value of 38.65 ± 0.59 mg GAE/g dw and 2.89 ± 0.04 mg GAE/g dw, respectively. A similar tendency was reported for the extraction of total polyphenols from other plant material [25–27]. The effect of ethanol concentration on the extraction of antioxidants can be attributed to polarity changes. As the ethanol concentration in the solvent increases, its polarity decreases, which enhances the extraction of less polar components [28]. Moreover, the increase in ethanol concentration also promotes the breakdown of cell membranes, which in turn enhances the solvent's ability to penetrate the solid matrix during the extraction process [29,30]. At higher ethanol concentrations, the resulting polarity becomes unsuitable for extracting antioxidants from coriander leaves, making it less effective for this purpose. Given these findings, the concentration range of 20–80% was chosen for the response surface methodology trials.

The selection of microwave power is crucial in determining the efficiency and yield of the phenolic compounds extraction process. Higher microwave power, which leads to the increase in the temperature, can accelerate the extraction by disrupting hydrogen bonds, enhancing solvent penetration into the matrix, and facilitating the release of target compounds [31]; however, it is important to note that there is a limit to this effect and beyond a certain point, increasing the microwave power may not lead to any further improvements in the extraction yield as higher microwave power levels can induce degradation of phenolic compounds due to elevated temperatures [32]. As shown in Table 1, the value of TPC yield and TAC ranged from 35.96 ± 0.47 to 40.97 ± 0.93 mg GAE/g dw and 2.98 ± 0.09 to 3.89 ± 0.08 mg GAE/g dw, respectively. The variation of microwave power over the range of 100–900 watt caused the increment in TPC yield and TAC which was achieved at 300 watt followed by a significant decrease. Our result was lower than those reported in the literature [33]. Based on these results, the microwave power range 100–500 watt was selected for the response surface methodology trials.

According to the Table 1, the value of TPC and TAC ranged from 38.66 ± 0.14 to 43.08 ± 0.27 mg GAE/g dw and 2.92 ± 0.04 to 4.03 ± 0.07 mg GAE/g dw, respectively. As can be seen, the TPC and TAC increased significantly from the irradiation time of 30 to 90 s, and then decreased to achieve the value of 38.88 ± 0.71 mg GAE/g dw and 2.92 ± 0.04 mg GAE/g dw, respectively, after an extended extraction time. Our result was higher than that reported in the literature [12]. Based on these results, the irradiation time range 30–150 s was selected for the response surface methodology trials.

3.2. Optimization by Response Surface Methodology

3.2.1. Construction of the Experimental Plan

The optimization of the antioxidants extraction from coriander leaves by response surface methodology is operated using the Box-Behnken model based on the maximization of TPC extraction yield and TAC as response variables. Three parameters (independent variables), solvent concentration (ethanol 20–80%), microwave power (100–500 watt) and irradiation time (30–150 s) were investigated. The ranges (the lower and upper ends) of each independent variable were determined based on the result of the single factor effect from preliminary work (Section 2.2). The factors levels, observed and predicted values of TPC and TAC were summarized in Table 3.

Table 3. Box-Behnken design matrix, experimental and predicted values of total phenolic compounds (TPC) and total antioxidant capacity (TAC).

Run	Variable Levels			TPC (mg GAE/g dw)		TAC (mg GAE/g dw)	
	x_1	x_2	x_3	Observed Value	Predicted Value	Observed Value	Predicted Value
1	20 (−)	100 (−)	90 (0)	24.34	26.72	2.15	3.08
2	20 (−)	500 (+)	90 (0)	33.33	35.96	3.30	4.38
3	80 (+)	100 (−)	90 (0)	38.84	40.22	3.86	4.79
4	80 (+)	500 (+)	90 (0)	39.41	41.04	3.57	4.64
5	50 (0)	100 (−)	30 (−)	40.79	42.54	3.72	4.87
6	50 (0)	100 (−)	150 (+)	41.28	43.78	4.15	5.15
7	50 (0)	500 (+)	30 (−)	45.18	46.68	4.42	5.42
8	50 (0)	500 (+)	150 (+)	47.45	49.70	4.90	5.75
9	20 (−)	300 (0)	30 (−)	32.91	34.79	3.08	4.00
10	80 (+)	300 (0)	30 (−)	41.73	44.61	3.89	4.82
11	20 (−)	300 (0)	150 (+)	36.32	37.45	3.06	4.13
12	80 (+)	300 (0)	150 (+)	44.08	46.21	4.22	5.30
13	50 (0)	300 (0)	90 (0)	47.81	49.48	4.27	5.20
14	50 (0)	300 (0)	90 (0)	46.81	49.48	4.22	5.20
15	50 (0)	300 (0)	90 (0)	47.81	49.48	4.12	5.20

x_1, solvent concentration; x_2, microwave power; x_3, irradiation time; GAE: gallic acid equivalent; dw: dry weight of leaves.

The amount of TPC and the maximization of TAC from *Coriandrum sativum* leaf extract using the MAE method ranged from 26.34 to 49.70 mg GAE/g dw and 3.08 to 5.75 mg GAE/g dw, respectively. The highest yield in TPC and TAC was observed under extraction conditions of 50% (*v*/*v*) solvent concentration, 500 watt microwave power, and an irradiation time of 150 s. Hihat et al. [1], who investigated the effect of oven and microwave drying on the total polyphenols and antioxidant capacity of coriander leaves, reported values of 48.44 mg GAE/g dw and 82.21%, respectively.

3.2.2. Analysis of the Model

To assess the significance of the model, an ANOVA analysis was conducted. Table 4 shows the ANOVA results for the effects of solvent concentration, microwave power and irradiation time on TPC and TAC, relative to the dry weight of coriander leaves.

Table 4. Analysis of model variance and lack of fit for total phenolic compounds (TPC) and total antioxidant capacity (TAC) of coriander leaves.

Source	DF	Sum of Squares	Mean Square	F Ratio	Prob. > F
TPC (mg GAE/g dw)					
Model	9	619.553	68.839	87.880	<0.0001 *
Error	5	3.916	0.783		
Corrected total	14	623.470			
Lack of fit	3	3.250	1.083	3.250	0.2441
Pure error	2	0.666	0.333		
Total error	5	3.916			
R^2	0.994				
R^2 adjusted	0.982				
TAC (mg GAE/g dw)					
Model	9	6.401	0.711	35.498	0.0005 *
Error	5	0.100	0.020		
Corrected total	14	6.502			
Lack of fit	3	0.088	0.029	5.058	0.1695
Pure error	2	0.011	0.005		
Total error	5	0.100			
R^2	0.984				
R^2 adjusted	0.956				

* Statistically significant values ($p < 0.05$).

The coefficients of determination (R^2) for the TPC and TAC models are 0.994 and 0.984, respectively, indicating that only a very small percentage (0.06 and 0.16%, respectively) of the total variation remains unexplained by these models. In addition, the adjusted coefficients of determination (R^2 adj) for the TPC and TAC models are 0.982 and 0.956, respectively, indicating close agreement between the experimental and predicted values. The lack of fit test is used to determine if the model is appropriate for describing the experimental data or if another model should be selected. The lack of fit test values for TPC and TAC are 0.2441 and 0.1695, respectively, which are higher than 0.05 and not significant compared to the pure error. This suggests that the current model is sufficient to fit the experimental data.

The regression coefficients for the intercept, the linear, quadratic and interaction terms of the models were calculated using the least squares method and presented in Table 5. The ANOVA of regression coefficient showed a linear response based on the p-value of solvent concentrations (x_1), which were highly significant ($p < 0.0001$), followed by microwave power (x_2) and irradiation time (x_3) for both TPC and TAC. In the interaction between variables, only solvent concentration-microwave power (x_1x_2) had a significant effect on TPC ($p < 0.0051$) and TAC ($p < 0.0038$). While in the quadratic model, only solvent concentrations (x_1^2) and microwave power (x_3^2) were significant for TPC ($p < 0.0001$ and $p < 0.0002$, respectively), and solvent concentration (x_1^2) and irradiation time (x_3^2) for TAC ($p < 0.0001$ and $p < 0.0314$, respectively). The mathematical equations correlating the TPC and TAC (Equation (4) and Equation (5), respectively) with process variables are given below in terms of coded factors excluding non-significant terms.

$$TPC = 49.476 + 4.645x_1 + 2.515x_2 + 1.065x_3 - 2.105x_1x_2 - 9.205x_1^2 - 4.290x_2^2 \quad (4)$$

$$TAC = 5.203 + 0.493x_1 + 0.288x_2 + 0.152x_3 - 0.360x_1x_2 - 0.859x_1^2 + 0.218x_3^2 \quad (5)$$

Table 5. Regression coefficient, standard error, and Student's t-test results of response surface of TPC and TAC referred to dry weight of coriander leaves.

Parameter	Estimate	Standard Error	t Ratio	Prob > |t|
TPC (mg GAE/g dw)				
Intercept	49.476	0.510	96.83	<0.0001 *
Solvent concentration (x_1)	4.645	0.312	14.84	<0.0001 *
Microwave power (x_2)	2.515	0.312	8.04	0.0005 *
Irradiation time (x_3)	1.065	0.312	3.40	0.0192 *
x_1*x_2	−2.105	0.442	−4.76	0.0051 *
x_1*x_3	−0.265	0.442	−0.60	0.5754
x_2*x_3	0.445	0.442	1.01	0.3608
x_1*x_1	−9.205	0.460	−19.99	<0.0001 *
x_2*x_2	−4.290	0.460	−9.32	0.0002 *
x_3*x_3	0.489	0.460	1.06	0.3368
TAC (mg GAE/g dw)				
Intercept	5.203	0.081	63.67	<0.0001 *
Solvent concentration (x_1)	0.493	0.050	9.87	0.0002 *
Microwave power (x_2)	0.288	0.050	5.77	0.0022 *
Irradiation time (x_3)	0.152	0.050	3.05	0.0285 *
x_1*x_2	−0.360	0.070	−5.09	0.0038 *
x_1*x_3	0.087	0.070	1.24	0.2713
x_2*x_3	0.012	0.070	0.18	0.8667
x_1*x_1	−0.859	0.073	−11.66	<0.0001 *
x_2*x_2	−0.124	0.073	−1.69	0.1527
x_3*x_3	0.218	0.073	2.96	0.0314 *

* Values statistically significant at $p < 0.05$.

3.2.3. Analysis of Plots Describing Factors Effect

The most effective method for illustrating the impact of an independent variable on the yield of TPC extraction and TAC is to generate response surface plots using a model. This involves manipulating two variables within the experimental range being studied, while holding the remaining variable at its central level (0 level) [34]. Figure 1 demonstrates the influence of solvent concentration, microwave power, and irradiation time on the yield of TPC extraction and TAC from coriander leaves.

As can be seen in Figure 1a, the increase in solvent concentration (x_1) and microwave power (x_2) resulted in an augmented yield of TPC and TAC. However, higher solvent concentration (x_1) and microwave power (x_2) had a negative effect on the extraction, probably due to the decrease in polarity of the solvent [35] and/or the degradation of antioxidants [36]. Moreover, analysis of the surface plots showed that there were optimal levels of concentration of TPC and TAC, which is due to the quadratic effect of the solvent concentration (x_1^2) as corroborated in Table 5. Additionally, the impact of solvent concentration was observed to be statistically significant in interaction with microwave power (x_1x_2) (p value 0.0051 and 0.0038 for TPC and TAC, respectively). This finding was in agreement with the study which reported the significance of ethanol percentage on the yield extraction of total polyphenols from *Pistacia lentiscus* [33]. Polarity plays a crucial role in extracting antioxidants from various plant and marine sources. Several studies have shown that the polarity of the solvent affects the antioxidant activity of the extracts. By increasing the ethanol concentration in the solvent, the cell membrane breaks down, allowing the solvent to more easily penetrate the solid matrix [29]. Moreover, increasing the ethanol concentration in the solvent reduces its polarity, which enhances the extraction of less polar components [28]. However, using a very high ethanol concentration is not suitable for extracting TPC from plant material.

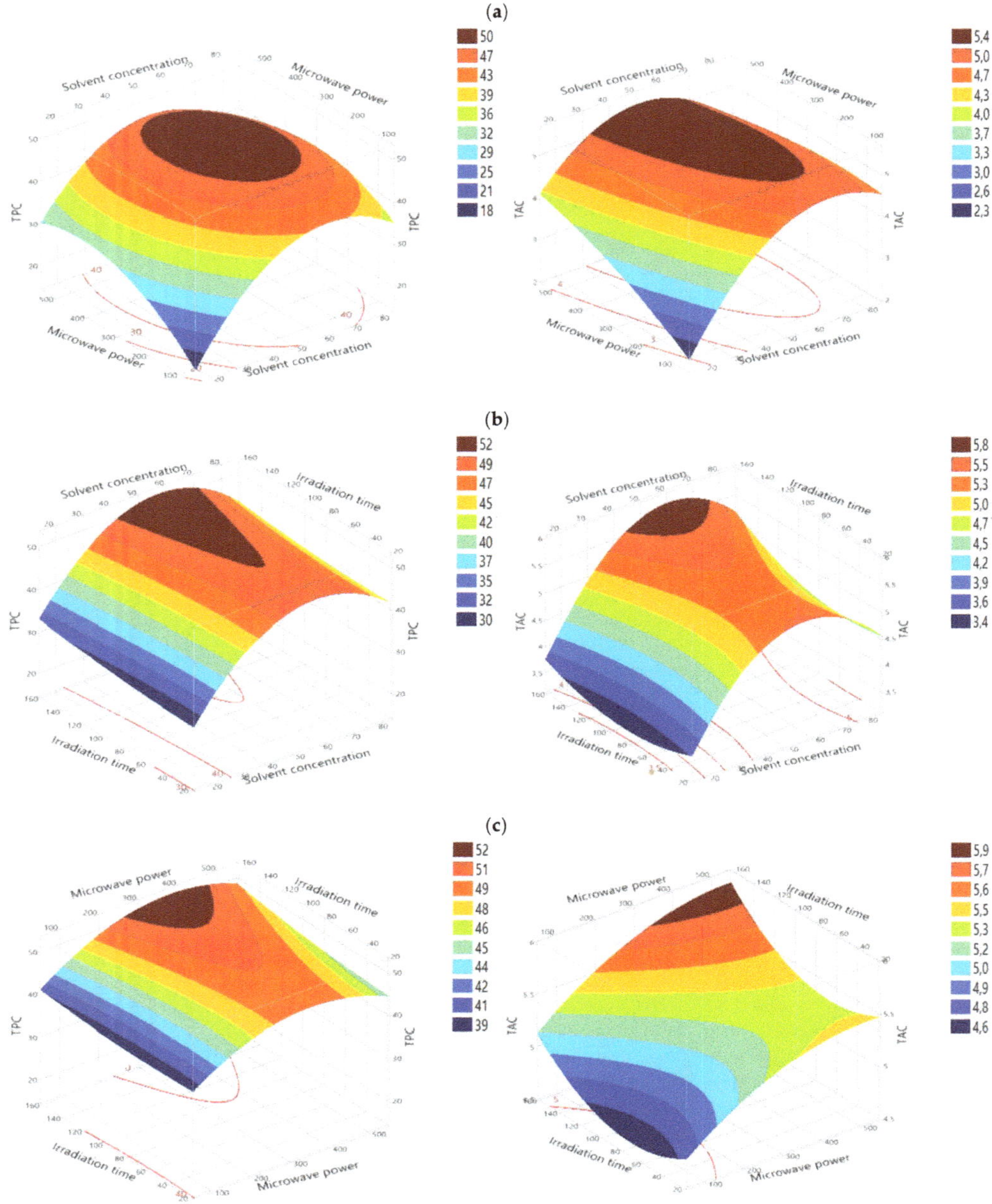

Figure 1. Effects of solvent concentration and microwave power (**a**), solvent concentration and irradiation time (**b**), and microwave power and irradiation time (**c**) on TPC (mg GAE/g dw) and TAC (mg GAE/g dw).

According to Figure 1b, the yield of TPC and TAC is affected by the increase in solvent concentration (x_1) and irradiation time (x_3). It was observed that both factors independently influenced the extraction of TPC and TAC. Additionally, analyzing the surface plots revealed that there were optimal concentration levels for TPC and TAC, which is attributed to the quadratic effect of irradiation time (x_3^2) as supported by Table 5. The impact of solvent concentration in interaction with irradiation time (x_1x_3) was found to be not significant (p value 0.5754 and 0.2713 for TPC and TAC, respectively). An elongated extraction time allows for prolonged contact between the plant material and the solvent, facilitating the transfer of phenolic compounds into the solvent. This can lead to increased extraction yields as more compounds have the opportunity to be released from the plant matrix [37].

Based on the Figure 1c, increasing microwave power (x_2) and irradiation time (x_3) resulted in higher yields of TPC and TAC. In other words, both microwave power (x_2) and irradiation time (x_3) simultaneously influenced TPC and TAC. However, it was observed that microwave power had a greater impact on TPC extraction than irradiation time. It was reported that the higher absorption of microwave energy led to increased temperatures within the sample, causing cell rupture and facilitating the release of antioxidant compounds [38]; however, excessive power may also degrade sensitive compounds [39]. Further, many studies have reported a decline in the levels of recovered TPC when using microwave-assisted extraction for long periods. Zhang et al. [40] observed that the yields of phenolic compounds of A. blazei increased significantly from 1 to 5 min during MAE, followed by a slight decrease.

3.2.4. Validation of the Model

To assess the predictive power of the model, the optimal conditions were determined based on maximum desirability. The optimal conditions for achieving the highest TPC and maximum TAC were found to be 52.62% ethanol concentration, 452.12 watt microwave power, and 150 s irradiation time. Under these conditions, the experimental values for TPC and TAC were 49.63 ± 0.93 mg GAE/g dw and 5.55 ± 0.07 mg GAE/g dw, respectively. These experimental results were in good agreement with the predicted values for TPC and TAC, which were 50.97 and 5.75 mg GAE/g dw, respectively. Zeković et al. [12] investigated coriander seeds and reported optimal microwave-assisted extraction conditions with an ethanol concentration of 63%, an extraction time of 19 min, and an irradiation power of 570 watt to simultaneously maximize total polyphenols yield and increase antioxidant activity. Their predicted values for TPC and antioxidant activity (IC_{50}) were 311.23 mg GAE/100 g and 0.0315 mg/mL, respectively.

4. Conclusions

The response surface methodology was employed to explore the individual and interactive effects of three variables, namely solvent concentration, microwave power and irradiation time, with the objective of optimizing the microwave-assisted extraction of TPC and the maximization of TAC from *Coriander sativum*. The high correlation of the mathematical model indicates that a quadratic polynomial model can be used for modelling the solid–liquid extraction of TPC and TAC. From the response surface plots, all three investigated factors significantly affected the TPC extraction yield and the maximization of TAC. The experimental values (49.63 ± 0.93 mg GAE/g dw and 5.55 ± 0.07 mg GAE/g dw for TPC and TAC, respectively) agreed with the predicted values (50.97 GAE/g dw and 5.75 mg GAE/g dw for TPC and TAC, respectively) and clearly showed the suitability of the developed quadratic models. These results confirm the predictability of the model for TPC from coriander leaves and TAC under the experimental conditions used (52.62% aqueous ethanol as solvent, 452.12 watt as microwave power and 150 s as irradiation time). This optimized process, which is simple, fast, efficient and non-denaturing, can be used for the extraction of substances of interest in both research and industrial settings.

Author Contributions: Conceptualization, S.H. and N.T.; Methodology, S.H. and N.T.; Software, S.H. and N.T.; Validation, S.H. and N.T.; Formal analysis, S.H. and N.T.; Investigation, S.H. and N.T.; Writing—original draft, S.H., N.T. and A.S.; Visualization, S.H. and N.T.; Supervision, K.M.; Project administration, K.M. All authors have read and agreed to the published version of the manuscript.

Funding: This work was financed by the Algerian Ministry of Higher Education and Scientific Research.

Data Availability Statement: Data are contained within the article.

Conflicts of Interest: The authors declare no conflict of interest.

References

1. Hihat, S.; Remini, H.; Madani, K. Effect of oven and microwave drying on phenolic compounds and antioxidant capacity of coriander leaves. *Int. Food Res. J.* **2017**, *24*, 503–509.
2. Touati, N.; Barba, F.J.; Louaileche, H.; Frigola, A.; Esteve, M.J. Effect of storage time and temperature on the quality of fruit nectars: Determination of nutritional loss indicators. *J. Food Qual.* **2016**, *39*, 209–217. [CrossRef]
3. Marinova, D.; Ribarova, F.; Atanassova, M. Total phenolics and total flavonoids in bulgarian fruits and vegetables. *J. Univ. Chem. Technol. Metal.* **2005**, *40*, 255–260.
4. Kang, M.H.; Lee, J.S.; Kim, H.Y.; Kwon, S.; Choi, Y.S.; Chung, H.R.; Kwak, T.K.; Cho, Y.H. Selecting items of a food behavior checklist for development of Nutrition Quotient (NQ) for children. *Korean J. Nutr.* **2012**, *45*, 372–389. [CrossRef]
5. Rubio, L.; Motilva, M.J.; Romero, M.P. Recent advances in biologically active compounds in herbs and spices: A review of the most effective antioxidant and anti-inflammatory active principle. *Food Sci. Nut.* **2013**, *53*, 943–953. [CrossRef] [PubMed]
6. Forzato, C.; Vida, V.; Berti, F. Biosensors and Sensing Systems for Rapid Analysis of Phenolic Compounds from Plants: A Comprehensive Review. *Biosensors* **2020**, *10*, 105. [CrossRef] [PubMed]
7. Sakhraoui, A.; Touati, N.; Hihat, S. Effect of Time and Temperature Storage on the Quality of unpasteurized Prickly Pear Juice Enriched with Hydro-soluble Opuntia ficus indica seeds Extract. *Turk. J. Agric. Food Sci. Technol.* **2023**, *11*, 1817–1824. [CrossRef]
8. Ashika, B.D.; Chitrali, L.R.; Naresh, S.; Sunil, K.S.; Akki, S.; Balasubramanian, S. Phytochemical studies on the methanolic extract of coriander sativum leaves-an in vitro approach. *Eur. J. Biomed. Pharm. Sci.* **2018**, *5*, 494–500.
9. Scandar, S.; Zadra, C.; Marcotullio, M.C. Coriander (*Coriandrum sativum*) polyphenols and their nutraceutical value against obesity and metabolic syndrome. *Molecules* **2023**, *28*, 4187. [CrossRef]
10. Nhut, P.T.; Quyen, N.T.N.; Truc, T.T.; Minh, L.V.; An, T.N.T.; Anh, N.H.T. Preliminary study on phytochemical, phenolic content, flavonoids and antioxidant activity of Coriandrum Sativum l. originating in Vietnam. *IOP Conf. Ser. Mater. Sci. Eng.* **2020**, *991*, 012022. [CrossRef]
11. Putnik, P.; Lorenzo, J.M.; Barba, F.J.; Roohinejad, S.; Jambrak, A.R.; Granato, D.; Montesano, D.; Kovačević, D.B. Novel Food Processing and Extraction Technologies of High-Added Value Compounds from Plant Materials. *Foods* **2018**, *7*, 106. [CrossRef]
12. Zeković, Z.; Vladić, J.; Vidović, S.; Adamović, D.; Pavlić, B. Optimization of microwave-assisted extraction (MAE) of coriander phenolic antioxidants–response surface methodology approach. *J. Sci. Food Agric.* **2016**, *13*, 4613–4622. [CrossRef]
13. López-Salazar, H.; Camacho-Díaz, B.H.; Ocampo, M.L.A.; Jiménez-Aparicio, A.R. Microwave-assisted extraction of functional compounds from plants: A Review. *Bioresources* **2023**, *18*, 6614–6638. [CrossRef]
14. Das, A.; Basak, S.; Chakrabartty, S.; Dhibar, M. Microwave: An ecologically innovative, green extraction technology. *Curr. Anal. Chem.* **2022**, *18*, 858–866. [CrossRef]
15. Dean, J.R. *Extraction Techniques for Environmental Analysis: Microwave-Assisted Extraction*; John Wiley & Sons Ltd.: Hoboken, NJ, USA, 2022; pp. 205–217.
16. Verma, A.; Alsayadi, G.M.H.; Choudhary, Y.; Sandal, P.; Kurmi, B.D. Microwave-assisted processes and their applications. *Pharmaspire* **2022**, *14*, 59–64. [CrossRef]
17. Arellano, C.A.; Corpuz, A.; Nguyen, L.T. Microwave-Assisted Extraction for the Valorization of Agro-Industrial Waste. In *Valorization of Agro-Industrial Byproducts*, 1st ed.; Anal, A.K., Panesar, S., Eds.; ImprintCRC Press: Boca Raton, FL, USA, 2022; pp. 31–48. [CrossRef]
18. Bouallegue, A.; Casilloc, A.; Chaari, F.; Cimini, D.; Corsaro, M.M.; Bachoual, R.; Ellouz-Chaabouni, S. Statistical optimization of levan: Influence of the parameter on levan structure and angiotensin I-converting enzyme inhibitory. *Int. J. Biol. Macromol.* **2020**, *158*, 945–952. [CrossRef]
19. Vesa, I.; Spraggon, M.; Fattah, I.M.R. Response surface methodology (RSM) for optimizing engine performance and emissions fueled with biofuel: Review of RSM for sustainability energy transition. *Results Eng.* **2023**, *18*, 101213. [CrossRef]
20. Myers, R.H.; Montgomery, D.C.; Anderson-Cook, C.M. *Response Surface Methodology: Process and Product Optimization Using Designed Experiments*, 3rd ed.; John Wiley & Sons: Hoboken, NJ, USA, 2009; ISBN 978-1-118-91601-8.
21. Singleton, V.L.; Rossi, J.A.J.R. Colorimetry of total phenolics with phosphomolybdic phosphotungstic acid reagents. *Am. J. Enol. Vitic.* **1965**, *16*, 144–158. [CrossRef]
22. Brand-Williams, W.; Cuvelier, M.E.; Berset, C. Use of a free radical method to evaluate antioxidant activity. *Food Sci. Technol.* **1995**, *28*, 25–30. [CrossRef]
23. Boeing, J.S.; Barizão, É.O.; Silva, B.S.; Montanher, P.F.; Almeida, V.D.C.; Visentainer, J.V. Evaluation of solvent effect on the extraction of phenolic compounds and antioxidant capacities from the berries: Application of principal component analysis. *Chem. Cent. J.* **2014**, *8*, 48. [CrossRef]
24. Jiménez-Moreno, N.; Volpe, F.; Moler, J.A.; Esparza, I.; Ancín-Azpilicueta, C. Impact of Extraction Conditions on the Phenolic Composition and Antioxidant Capacity of Grape Stem Extracts. *Antioxidants* **2019**, *8*, 597. [CrossRef] [PubMed]

25. Spigno, G.; Tramelli, L.; De Faveri, D.M. Effects of extraction time, temperature and solvent on concentration and antioxidant activity of grape marc phenolics. *J. Food Eng.* **2007**, *81*, 200–208. [CrossRef]
26. Li, H.; Deng, Z.; Wu, T.; Liu, R.; Loewen, S.; Tsao, R. Microwave-assisted extraction of phenolics with maximal antioxidant activities in tomatoes. *Food Chem.* **2012**, *130*, 928–936. [CrossRef]
27. Pan, X.; Niu, G.; Liu, H. Microwave-assisted extraction of tea polyphenols and tea caffeine from green tea leaves. *Chem. Eng. Process.* **2003**, *42*, 129–133. [CrossRef]
28. Cheok, C.Y.; Chin, N.L.; Yusof, Y.A.; Talib, R.A.; Law, C.L. Optimization of total phenolic content extracted from *Garcinia mangostana Linn.* hull using response surface methodology versus artificial neural network. *Ind. Crops Prod.* **2012**, *40*, 247–253. [CrossRef]
29. Vatai, T.; Škerget, M.; Knez, Ž. Extraction of phenolic compounds from elder berry and different grape marc varieties using organic solvents and/or supercritical carbon dioxide. *J. Food Eng.* **2009**, *90*, 246–254. [CrossRef]
30. Zhang, S.; Chen, R.; Wu, H.; Wang, C. Ginsenoside extraction from *Panax quinquefolium* L. (American ginseng) root by using ultrahigh pressure. *J. Pharm. Biomed. Anal.* **2006**, *41*, 57–63. [CrossRef] [PubMed]
31. Zeng, S.; Wang, B.; Lv, W.; Wu, Y. Effects of microwave power and hot air temperature on the physicochemical properties of dried ginger (*Zingiber officinale*) using microwave hot-air rolling drying. *Food Chem.* **2023**, *404*, 134741. [CrossRef]
32. Zahoor, I.; Khan, M.A. Microwave assisted convective drying of bitter gourd: Drying kinetics and effect on ascorbic acid, total phenolics and antioxidant activity. *J. Food Meas. Charact.* **2019**, *13*, 2481–2490. [CrossRef]
33. Dahmoune, F.; Spigno, G.; Moussi, K.; Remini, H.; Cherbal, A.; Madani, K. *Pistacia lentiscus* leaves as a source of phenolic compounds: Microwave-assisted extraction optimized and compared with ultrasound-assisted and conventional solvent extraction. *Ind. Crops. Prod.* **2014**, *61*, 31–40. [CrossRef]
34. Hayat, K.; Hussain, S.; Abbas, S.; Farooq, U.; Ding, B.; Xia, S. Optimized microwave-assisted extraction of phenolic acids from citrus mandarin peels and evaluation of antioxidant activity *in vitro*. *Separ. Puri. Technol.* **2009**, *70*, 63–70. [CrossRef]
35. Zhang, Z.S.; Li, D.; Wang, L.J.; Ozkan, N.; Chen, X.D.; Mao, Z.H.; Yang, H.Z. Optimization of ethanol-water extraction of lignans from flaxseed. *Sep. Purif. Technol.* **2007**, *57*, 17–24. [CrossRef]
36. Cheng, Y.; Xue, F.; Yang, Y. Hot water extraction of antioxidants from tea leaves—Optimization of brewing conditions for preparing antioxidant-rich tea drinks. *Molecules* **2023**, *28*, 3030. [CrossRef]
37. Jha, A.K.; Sit, N. Extraction of bioactive compounds from plant materials using combination of various novel methods: A review. *Trends Food Sci. Technol.* **2022**, *119*, 579–591. [CrossRef]
38. García-Baños, B.; Reinosa, J.J.; Peñaranda-Foix, F.L. Temperature assessment of microwave-enhanced heating processes. *Sci. Rep.* **2019**, *9*, 10809. [CrossRef] [PubMed]
39. Bachir Bey, M.; Meziant, L.; Benchikh, Y.; Louaileche, H. Deployment of response surface methodology to optimize recovery of dried dark fig (*Ficus carica* L., var. Azenjar) total phenolic compounds and antioxidant activity. *Int. Food Res. J.* **2014**, *21*, 1477–1482. [CrossRef]
40. Zhang, Z.; Lv, G.; Pan, H.; Fan, L. Optimisation of the microwave-assisted extraction process for six phenolic compounds in *Agaricus blazei* murrill. *Int. J. Food Sci. Technol.* **2012**, *47*, 24–31. [CrossRef]

processes

MDPI

Article

Optimizing the Salt-Processing Parameters of *Achyranthes bidentata* and Their Correlation with Anti-Osteoarthritis Effect

Jieqiang Zhu [1], Lisha Shen [1], Guofang Shen [2] and Yi Tao [1,*]

[1] College of Pharmaceutical Science, Zhejiang University of Technology, Hangzhou 310014, China; zhujieqiang@zjut.edu.cn (J.Z.); 221122070262@zjut.edu.cn (L.S.)
[2] Hangzhou Food and Drug Inspection and Research Institute, Hangzhou 310022, China; shengf@126.com
* Correspondence: taoyi1985@zjut.edu.cn

Abstract: *Achyranthes bidentata* is always salt-processed before being prescribed for treating osteoarthritis. Yet the salt-processing parameters have not been optimized, and the specific bioactive constituents responsible for the osteoarthritis effect of salt-processed *A. bidentata* have not been fully elucidated. In this study, a Box–Behnken experimental design was chosen for the optimization of the salt-processing parameters of *A. bidentata*, including stir-frying time, concentration of brine, and soak time. Meanwhile, HPLC–Q-TOF-MS was utilized to analyze the chemical profiles of various batches of raw and salt-processed *A. bidentata*. The anti-inflammatory potential of nine batches of both raw and salt-processed *A. bidentata* was assessed via a cyclooxygenase-2 (COX-2) inhibitory assay. A gray correlation analysis was conducted to correlate the peak areas of the compounds in raw and salt-processed *A. bidentata* with their COX-2 inhibitory effects. Finally, the optimal salt-processing conditions are as follows: soak time: 29 min; concentration of brine: 1.8%; stir-frying time: 4.4 min. Twenty-nine compounds were identified. Eight compounds were found to have a strong positive correlation with anti-inflammatory activity, as confirmed by the COX-2 inhibitory assay. Notably, this is the first report of the COX-2 inhibitory effects of sanleng acid, stachysterone D, dihydroactinidiolide, *N-cis*-feruloyl-3-methoxytyramine, 9,12,13-trihydroxy-10-octadecenoic acid, azelaic acid, and dehydroecdysone.

Keywords: *Achyranthes bidentata*; salt-processing; cyclooxygenase-2; anti-inflammatory

1. Introduction

Traditional methods for processing crude herbs distinctly differentiate traditional Chinese medicine from Western medicine. These methods encompass sauteing with rice wine [1,2] or brine, steaming with water or rice wine [3], frying with sand [4,5] or oil, and braising with rice wine or licorice liquids. The primary objective of processing is to eliminate or diminish toxicity and side effects, alter nature and actions, and enhance therapeutic effects. Simultaneously, to guarantee safety and efficacy, it is essential to rigorously control and regulate the quality of processed products.

Achyranthes bidentata, as one of the "four major Huai medicines," is mainly cultivated in Henan province. The chemical constituents of *A. bidentata* include oleanolic acid glycosides, saponins, ecdysterone, ketosteroids, and flavonoids [6]. Although modern pharmacological studies have shown that *A.bidentata* has a plethora of bio-activities, including anti-osteoarthritis [7], improving memory [8], regulating blood sugar level [9], preventing apoptosis [10], promoting angiogenesis, UV protection [11], anti-convulsion [12], anti-osteoporosis [13], improving xerophthalmia [14], and alleviating acute kidney injury [15], this herb is commonly used in the clinical treatment of osteoarthritis. After being salt-processed, the effects will be dramatically enhanced. However, the salt-processing parameters for *Achyranthes bidentata* have not been optimized yet.

Quality by design (QbD) is a concept in the pharmaceutical and biopharmaceutical industries endorsed by regulatory agencies such as the U.S. Food and Drug Administration

(FDA) [16]. The key principles of QbD involve the systematic identification of critical product attributes and process parameters and understanding how these factors influence product quality. Experimental design methods were always used. The salt-processing method for *Achyranthes bidentata* includes several process parameters, such as stir-frying time, concentration of brine, and soak time. How these process parameters affect the quality of salt-processed *Achyranthes bidentata* needs to be understood.

Salt-processed *Achyranthes bidentata* is prescribed for treating osteoarthritis. The anti-inflammatory effect is the critical product attribute of salt-processed *Achyranthes bidentata.* Cyclooxygenase-2 (COX-2), also known as prostaglandin oxidase synthase (PTGS-2), is a bifunctional enzyme with both cyclooxygenase and catalase activities. COX-2 inhibitors have played a significant anti-inflammatory role in vivo and in vitro [17]. *A. bidentata* can reduce the expression of COX-2 in the synovial tissue of joints through the arachidonic acid pathway and exert therapeutic effects on osteoarthritis [7]. The results suggested that the anti-osteoarthritis effect of *A. bidentate* is associated with COX-2; however, the compounds of *A. bidentate* that are attributed to the anti-inflammatory effects are poorly understood.

The aim of this work is to optimize the salt-processing parameters of *A. bidentate* by using a Box–Behnken experimental design, apply HPLC–Q-TOF-MS analysis in conjunction with orthogonal partial least squares discriminant analysis to identify the differential markers between raw and salt-processed *A. bidentate*, and obtain the quality markers for the anti-inflammatory activity of salt-processed *A. bidentate* by using COX-2 inhibition assay and gray correlation analysis.

2. Materials and Methods

2.1. Instruments and Chemicals

Agilent 1290 high-performance liquid chromatography (Agilent Technology Co., Santa Clara, CA, USA); PD-1D-50 freeze dryer (Beijing Boyikang Experimental Instrument Co., Beijing, China); UPLC-Triple TOF 5600+ high-resolution mass spectrometry system (Waters Corporation/AB Sciex Corporation, Framingham, MA, USA); Tecan sunrise multifunctional microplate reader (Diken Trading Co., Shanghai, China).

Acetonitrile was obtained from Tedia Company, Inc. (Fairfield, OH, USA); Methanol from Sinopharm Chemical Reagent Co., Shanghai, China; *n*-butanol from Sinopharm Chemical Reagent Co.; PBS buffer from Shanghai Yuanye Biotechnology Co., Shanghai, China; COX-2 inhibitor screening kit from Biyuntian Biotechnology Co., Shanghai, China; Nine batches of *A. bidentata* were collected from Wen County, Henan Province, and authenticated by Professor Wang Ping of Zhejiang University of Technology. Voucher specimens were deposited in the herbarium of the College of Pharmaceutical Science, Moganshan campus of Zhejiang University of Technology. Standard substances with purities > 98%, including ecdysterone, 25*R*-inokosterone, and 25*S*-inokosterone, were obtained from the China Institute for Food and Drug Control; Sanleng acid, 9,12,13-trihydroxy-10-octadecenoic acid, and azelaic acid were obtained from Chengdu Must Biological Technology Co., Chengdu, China (purities > 98%); Dihydroactinidiolide, *N-cis*-feruloyl-3-methoxytyramine, *N-trans*-feruloyltyramine, dehydroecdysone, and stachysterone D were obtained from Sichuan Weikeqi Biological Technology Co., Chengdu, China (purities > 98%). Maleic acid and fumaric acid were obtained from Sinopharm Chemical Reagent Co., Shanghai, China (purities > 98%).

2.2. Optimization of Salt-Processing Procedure for A. bidentata

Dried roots of *A. bidentata* were immersed in brine for a period. After that, the roots were transferred to the wok with gentle heat and stir-fried for a while. Finally, the salt-processed *A. bidentata* was allowed to cool.

Three parameters, including soak time, concentration of brine, and stir-frying time, were investigated in the present study to optimize the processing conditions. The software Design Expert (Trial Version 7.0.3, Stat-Ease Inc., Minneapolis, MN, USA) was employed for experimental design, data analysis, and model building. Box–Behnken designs with three variables were used to determine the response pattern and then to establish a model. Three variables with three levels of each variable were used for the optimization of magnetic solid-phase extraction: soak time (X1), concentration of brine (X2), and stir-frying time (X3). The 2020 edition of Chinese Pharmacopeia takes ecdysterone as a quality indicator of *Achyranthes bidentata*. In the optimization of processing conditions, the dependent variable (Y) was the concentration of ecdysterone.

2.3. HPLC–Q-TOF-MS Analysis

The conditions for HPLC chromatography were as follows: The chromatographic column was Agilent Kromasil C_{18} (4.6 mm × 250 mm, 5.0 μm). The injection volume was set to 20 μL. The mobile phase consisted of acetonitrile (A) and 0.1% formic acid water (B). The gradient elution program was as below: 0 min–3 min, 16–20% A; 3 min–10 min, 20–23% A; 10 min–20 min, 23–26% A; 20 min–35 min, 26–40% A; 35 min–45 min, 40–50% A; 45 min–50 min, 50–60% A; 50 min–51 min, 60% to 15% A; 51 min–60 min, 15% A. The flow rate was set to 1 $mL \cdot min^{-1}$. The column temperature was adjusted to 40 °C. The detection wavelength was 250 nm.

The samples were analyzed with Waters Synapt G2 Q-TOF (Milford, MA, USA) mass spectrometry, which was equipped with an electron spray ionization (ESI) source. The internal calibration for mass accuracy is sodium formate. The first step is an initial correction with sodium formate. After passing the test, the standard curve will be made. After passing the test, the real-time internal standard correction will be made. The MS conditions were as follows: the mass acquisition was performed under negative ionization mode; the scanning range was 100 Da to 1000 Da; the scanning time was 0.2 s; the capillary voltage was 3000 V; the cone voltage was 30 V; the ion source temperature was 120 °C; the desolvent-free flow rate was 800 $L \cdot h^{-1}$; and the desolvation temperature was 350 °C. The mass data were analyzed using Masslynxtv 4.2 software.

2.4. COX-2 Inhibitory Assay

The inhibitory assay of COX-2 was measured according to the method previously described with slight modifications [18]. A total of 75 μL aliquot of COX-2 buffer was added to a 96-well plate, and then 5 μL of COX-2 cofactor solution, 5 μL of COX-2 solution, and 5 μL of test solution of *A. bidentata* sample or single compound were added subsequently. The 96-well plate was incubated at 37 °C for 10 min. Then, a 5 μL aliquot of the COX-2 fluorescence probe was added. After that, a 5 μL aliquot of COX-2 substrate solution was added and mixed at 37 °C for 5 min constantly. The 96-well plate was put into the microplate reader. The excitation wavelength was set to 560 nm, and the emission wavelength was set to 590 nm. Celecoxib was selected as the positive control. A total of 5.0 mL of aqueous extract of raw and salt-processed *A. bidentata* was added to 5.0 mL phosphate-balanced saline (pH = 7.4) and thoroughly mixed. The mixture was centrifuged, and the supernatant was sent for a COX-2 inhibitory assay. A gradient concentration of standard substance solutions including sanleng acid, stachysterone D, dihydroactinidiolide, *N-cis*-feruloyl-3-methoxytyramine, *N-trans*-feruloyltyramine, 9,12,13-trihydroxy-10-octadecenoic acid, azelaic acid, and dehydroecdysone were prepared with phosphate-balanced saline (pH = 7.4). The inhibition rate of the sample solutions was calculated.

2.5. Gray Correlation Analysis

The peak area of each compound in nine batches of raw and salt-processed *A. bidentata* and the COX-2 inhibition rate were input into the gray modeling software (Version 7) to conduct the correlation analysis. The correlation coefficient r between each compound and COX-2 inhibitory activity was obtained. The correlation coefficient $r > 0.85$ can be considered a very strong positive correlation [19].

2.6. Chemometric Analysis

The peak area of each compound in the nine batches of raw and salt-processed *A. bidentata* was subjected to the software SIMCA-P (version 13) for pattern recognition. An unsupervised pattern recognition method, i.e., principal component analysis, was first used for modeling. After that, a supervised method, i.e., orthogonal partial least squares discriminant analysis, was used to identify the differential compounds between raw and salt-processed *A. bidentata*. Variable importance plots of PLS-DA models were used to find significantly altered compounds (VIP value $\geq$ 1.0) [20].

3. Results and Discussion

3.1. Optimization of the Salt-Processing Procedure for A. bidentata

The Box–Behnken response surface method is a multifactor nonlinear experimental optimization method. This method is continuous over a range of experimental conditions and can be analyzed for any of these test-point conditions. Therefore, the Box–Behnken response surface method was chosen to optimize the salt-processing procedure [21]. The experimental framework for the Box–Behnken design (BBD) of salt-processing parameters, encoded numerically, included 17 randomized permutations. The optimization of the salt-processing protocol was gauged by the concentration of ecdysterone as the dependent variable. The specific experimental configurations and their outcomes are enumerated in Table 1.

Table 1. Box–Behnken design for optimization of the salt-processing procedure of *A. bidentata*.

Run	Independent Variable			Conc. of Ecdysterone (μg/mL)
	X1 (Soak Time, min)	X2 (Concentration of Brine, %)	X3 (Stir-Frying Time, min)	
1	30	2	5	485.571
2	30	1	7	305.3
3	10	2	7	395.933
4	30	3	7	395.356
5	50	2	7	391.509
6	50	3	5	389.812
7	30	2	5	488.776
8	10	3	5	421.638
9	30	2	5	484.89
10	10	2	3	410.132
11	30	2	5	486.46
12	30	1	3	446.888
13	10	1	5	436.884
14	50	2	3	457.694
15	30	2	5	480.762
16	50	1	5	438.679
17	30	3	3	463.877

An Analysis of Variance (ANOVA) was utilized to evaluate the statistical significance and adequacy of the model, with findings delineated in Table 2. Significance testing of each factor was conducted using F-tests and p-values. The model's F-value of 3.89 suggests statistical significance. Factors with p-values below the 0.05 threshold were considered significant. The ANOVA results indicated that the linear term for stir-frying time, as well as the quadratic terms for brine concentration and stir-frying time, had a significant impact on the concentration of ecdysterone.

Table 2. Estimated regression coefficients for the quadratic polynomial model.

Factor	Regression Coefficients	Standard Error	DF	F Value	Prob > F
B_0	485.29	13.60	1	3.89	0.0435
Linear					
B_1	1.64	10.75	1	0.0232	0.8831
B_2	5.37	10.75	1	0.2493	0.6329
B_3	−36.31	10.75	1	11.41	0.0118
Interaction					
β_{12}	−8.41	15.20	1	0.3058	0.5975
β_{13}	−13.0	15.20	1	0.7311	0.4208
β_{23}	18.27	15.20	1	1.44	0.2685
Quadratic					
β_{11}	−26.29	14.81	1	3.15	0.1193
β_{22}	−37.25	14.81	1	6.32	0.0401
β_{33}	−45.19	14.81	1	9.30	0.0186
Lack of fit			3	250.38	<0.0001
Pure error			4		
R^2	0.8334		Adjusted R^2	0.6192	

As is shown in Table 2, the value of the determination coefficient R^2 of 0.8334 indicated a good agreement between the observed and predicted values by the model. Response surface plots were employed for the visualization of the effect of two factors on the response. The interacting effect of soak time and concentration of brine is displayed in Figure 1A. A higher concentration of ecdysterone should be expected in the middle of the soak time, ranging from 10 to 50 min. This is because the brine regulates the osmotic pressure of the plant cells, which affects the texture of the hyssop after cooking. The interaction of stir-frying time with soak time is demonstrated in Figure 1B. The investigated system seems to perform better at the middle value of stir-frying time, ranging from 3 min to 7 min. High temperatures may destroy the chemical constituents. The effect of stir-frying time in relation to the concentration of brine is shown in Figure 1C. With the increase in brine concentration from 1% to 2.6%, the concentration of ecdysterone remains unchanged. The selected values of salt-processing conditions are presented in Table 3.

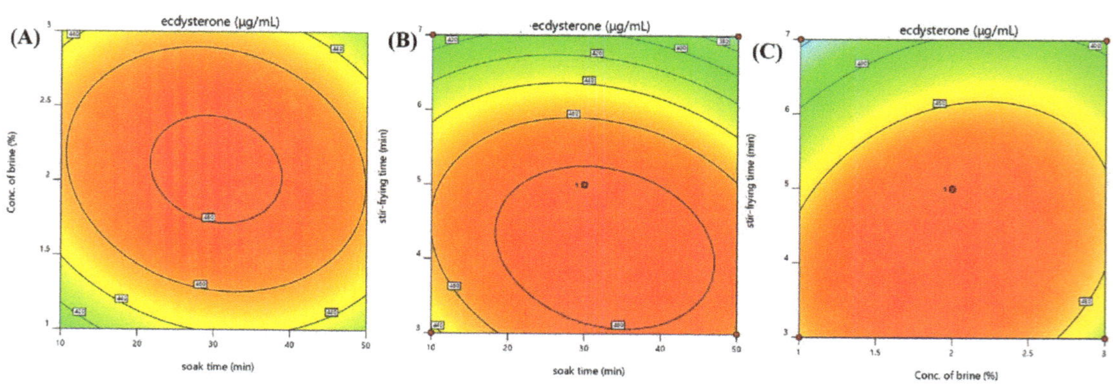

Figure 1. Response surface plots of the three factors. (**A**) interaction between concentration of brine and soak time; (**B**) interaction between stir-frying time and soak time; (**C**) interaction between stir-frying time and concentration of brine.

Table 3. Predicted and experimental response values under optimal conditions.

Factor	Value
Soak time (min)	29
Concentration of brine (%)	1.8
Stir-frying time (min)	4.4
Predicted values (μg/mL)	490.1
Experimental values (μg/mL)	491.79 ± 0.83

The real and predicted mean concentrations of ecdysterone under repeatability conditions are shown in Table 3. The experimental values of the concentration of ecdysterone were calculated at 491.79 μg/mL. No significant difference ($p < 0.05$) was observed between the theoretical and experimental responses. The optimal conditions are as follows: soak time: 29 min; concentration of brine: 1.8%; stir-frying time: 4.4 min.

3.2. Characterization of Chemical Constituents of Raw and Salt-Processed A. bidentata

The HPLC–Q-TOF-MS method was employed to analyze the chemical compounds of raw and salt-processed materials under negative ion mode, and the total ion chromatograms (TIC) are shown in Figure 2. Twenty-nine compounds were identified by comparing them with standard compounds and related works of literature. The identification of chemical compounds and the variation ratio of the compounds in *Achyranthes bidentata* after and before salt processing are displayed in Table 4. Of note, the peak area ratios of compounds 2, 4, 7, 12, 13, and 22 were all above 10 after salt processing. Their structures are shown in Figure S1.

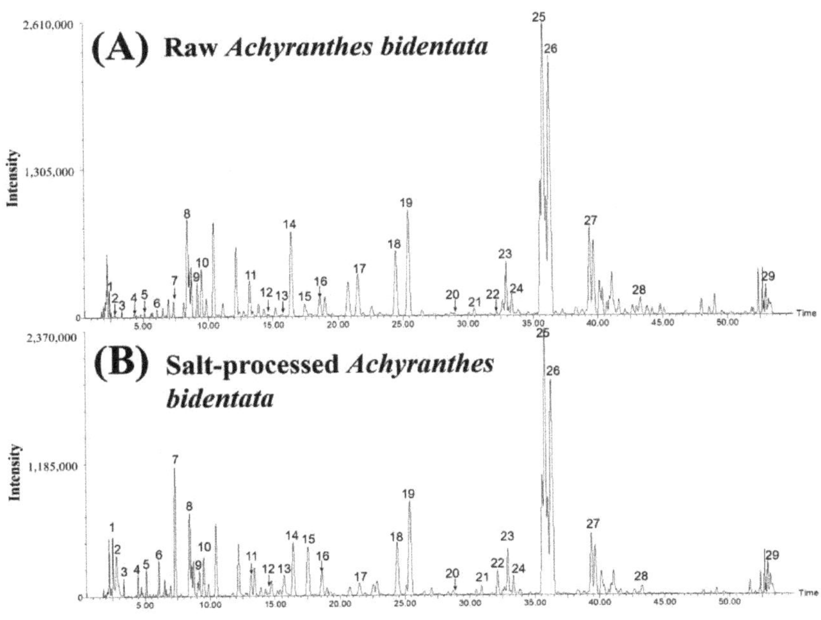

Figure 2. HPLC–Q-TOF-MS total ion chromatograms of raw and salt-processed *Achyranthes bidentata*. (**A**) raw *Achyranthes bidentata;* (**B**) salt-processed *Achyranthes bidentata*.

Table 4. Identification of chemical compounds in *Achyranthes bidentata*.

No.	t_R (min)	Detected Mass (m/z)	MS^2	Molecular Formula	Mass Error (ppm)	Compound (Name)	Identification	Peak Area Ratio
1	2.50	115.0036	100.0234, 99.0085, 99.9332	$C_4H_4O_4$	4.3	Maleic acid	standards	1.91
2	2.81	115.0032	111.1115, 104.7421, 87.0104, 77.8366, 73.9177	$C_4H_4O_4$	0.9	Unknown	—	26.59
3	3.34	115.0037	111.0400, 108.0446, 105.0223, 97.0370, 89.0271, 87.0083, 59.0153	$C_4H_4O_4$	5.2	Fumaric acid	standards	2.04
4	4.47	204.0664	186.0639, 159.0353, 131.0420, 108.0486, 69.8676	$C_{11}H_{11}NO_3$	1.5	Indolyllactic acid	[22]	12.55
5	5.1	235.0757	220.0741, 187.0871, 141.6590, 122.0250, 94.0303, 93.0204, 77.4244	$C_{16}H_{12}O_2$	−0.9	Methylflavone	[23]	6.97
6	6.07	541.3029	495.2991, 477.2920, 299.1640, 249.1525, 129.0598, 99.0488	$C_{27}H_{44}O_8$	3.0	Polypodine B	[24]	4.93
7	7.32	210.0773	177.0686, 154.0703, 124.0408, 122.0292, 94.0312, 93.0255, 66.0354	$C_{10}H_{13}NO_4$	3.3	Methoxyty-rosine	[25]	10.63
8	8.43	525.3055	479.3034, 319.1958, 301.1819, 210.0828, 159.1058	$C_{27}H_{44}O_7$	−1.7	Ecdysterone	standards	0.80
9	9.18	525.3040	479.3073, 443.1053, 377.1133, 346.1055, 319.1931, 301.1936, 282.1422, 261.1450, 217.1015, 186.0648, 159.1079, 141.0920, 124.0425	$C_{27}H_{44}O_7$	−4.6	25*R*-inokosterone	standards	0.90
10	9.51	525.3048	479.3033, 477.2902, 159.1055	$C_{27}H_{44}O_7$	−3.0	25*S*-inokosterone	standards	0.80

Table 4. *Cont.*

No.	t_R (min)	Detected Mass (*m*/*z*)	MS^2	Molecular Formula	Mass Error (ppm)	Compound (Name)	Identification	Peak Area Ratio
11	13.4	507.297	461.2921, 301.1889, 277.1472, 249.1255, 201.1278, 159.1051	$C_{27}H_{42}O_6$	2.4	Dehydroecdysone	standards	7.28
12	14.7	507.296	461.2861, 345.1842, 309.1479, 265.1693, 229.0488, 159.1066, 115.0768, 83.0535	$C_{27}H_{42}O_6$	1.0	Deoxykaladasterone	[26]	19.43
13	15.7	523.3109	477.3076, 459.2956, 361.2192, 317.1931, 159.1105	$C_{27}H_{42}O_7$	−1.5	Kaladasterone	[27]	15.25
14	16.4	507.2959	461.2989, 403.2529, 301.1815, 249.1308, 209.1000, 159.1058, 83.0489	$C_{27}H_{42}O_6$	0.2	Dacryhainansterone	[26]	0.58
15	17.5	187.0974	159.1022, 141.1016, 139.0442, 111.0111	$C_9H_{16}O_4$	2.1	Azelaic acid	standards	4.38
16	18.6	549.1589	341.1035, 311.0556, 295.0618, 268.0379, 255.0433, 252.0448	$C_{26}H_{30}O_{13}$	−4.4	Liquiritin apioside	standards	1.38
17	21.5	342.1357	327.1139, 190.0521, 178.0531, 148.0552, 135.0417, 134.0422	$C_{19}H_{21}NO_5$	4.7	*N-cis*-feruloyl-3-methoxytyramine	standards	0.26
18	24.4	312.1246	190.0584, 178.0544, 148.0556, 135.0491	$C_{18}H_{19}NO_4$	3.2	*N-trans*-feruloyltyramine	standards	0.81
19	25.4	342.1343	327.1171, 190.0533, 178.0529, 148.0544, 135.0460	$C_{19}H_{21}NO_5$	1.2	Feruloylmethoxytyramine	[28]	0.91

Table 4. *Cont.*

No.	t_R (min)	Detected Mass (*m*/*z*)	MS^2	Molecular Formula	Mass Error (ppm)	Compound (Name)	Identification	Peak Area Ratio
20	29.0	509.2856	494.1100, 489.6156, 466.2194, 421.8323, 410.1779, 387.1925, 384.8494, 358.5213, 350.2032, 342.0960, 333.4114	$C_{32}H_{46}O_5$	−1.8	Unknown	—	2.26
21	30.5	225.1124	214.0546, 209.0778, 196.8297, 187.0986, 171.8831, 158, 1576, 144.2472, 130.0387, 116.0739, 98.9125, 96.5702	$C_{11}H_{16}O_2$	−1.3	Dihydroactinidiolide	standards	0.65
22	32.1	507.2973	461.2921, 368.0941, 342.1358, 312.1195, 268.0684, 242.0818, 221.7878, 171.0949, 139.1091, 86.8855, 56.2580	$C_{27}H_{42}O_6$	4.9	Stachysterone D	standards	10.49
23	32.9	327.2175	229.1391, 221.1249, 211.1358, 171.1032, 85.0286	$C_{18}H_{32}O_5$	1.2	Corchorifatty acid F	[29]	0.87
24	33.4	327.2186	323.1074, 242.4109, 235.1030, 211.1431, 171.1025, 146.9710, 137.1052, 97.0710, 85.0290	$C_{18}H_{32}O_5$	4.6	9,12,13-Trihydroxy-10,15-octadecadienoic acid	[30]	0.93
25	35.8	329.2340	229.1482, 211.1358, 171.1053, 139.1160	$C_{18}H_{34}O_5$	3.6	9,12,13-Trihydroxy-10-octadecenoic acid	standards	0.51
26	36.3	329.2337	171.1043, 139.1148, 127.1147	$C_{18}H_{34}O_5$	2.7	9,10,13-TriHOME	[31]	0.61

Table 4. *Cont.*

No.	t_R (min)	Detected Mass (*m*/*z*)	MS^2	Molecular Formula	Mass Error (ppm)	Compound (Name)	Identification	Peak Area Ratio
27	39.3	329.2334	211.2371, 199.1361, 197.1192, 181.1294, 169.1259, 129.0907, 99.0819	$C_{18}H_{34}O_5$	1.8	Pinellic acid	[32]	0.71
28	43.2	329.2336	303.0145, 255.2305, 204.8009, 211.1199, 201.1148, 199.1152, 171.1017, 152.9974, 146.9705, 96.9641, 80.8853	$C_{18}H_{34}O_5$	2.4	Sanleng acid	standards	0.55
29	53.2	643.3685	610.5566, 588.5043, 569.3351, 531.3790, 512.9322	$C_{36}H_{54}O_{10}$	7.1	Unknown	—	0.80

Compounds **1**, **2**, and **3** showed the same [M−H]-ion at *m*/*z* 115, which corresponded to the formula of $C_4H_4O_4$. In comparison with standard compounds, compounds **1** and **3** were ambiguously identified as maleic acid and fumaric acid. Compound **4** at the retention time of 4.47 min displayed [M−H]-ion at *m*/*z* 204. Fragment ion at *m*/*z* 186 in the MS^2 spectrum was produced by losing one molecule of water. In comparison with the reference [22], compound **4** was unambiguously deduced as indolyllactic acid. Compound **5** showed the [M−H]-ion at *m*/*z* 235 and produced an ion at *m*/*z* 220 with the loss of the methyl group. By comparing with a related reference [23], compound **5** was deduced as methylflavone. Compound **6** at the retention time of 6.07 min showed [M + HCOO]-ion at *m*/*z* 541. Fragment ions at *m*/*z* 495 were produced by losing one molecule of HCOOH in the MS^2 spectrum. Compound **6** was deduced as polypodine B by comparing it with the fragmentation pathway in the literature [24]. The precursor ion of compound **7** at *m*/*z* 210 produced a fragment ion at *m*/*z* 177, which was attributed to the loss of both the H_2O and CH_3 groups. In comparison with the reference [25], compound **7** was assigned as methoxytyrosine. With the retention times of 8.43 min, 9.18 min, and 9.53 min, compounds **8**, **9**, and **10** displayed the same [M + HCOO]-ion at *m*/*z* 525, as well as the characteristic neutral loss of 46 Da yielding the product ion at *m*/*z* 479, which was ascribed to the loss of the HCOOH group. Compared with the standard substances, compounds **8**, **9**, and **10** were unambiguously deduced as ecdysterone, 25*R*-inokosterone, and 25*S*-inokosterone, respectively [33]. In the same way, compounds **11**, **12**, **14**, and **22** shared the same [M + HCOO]-ion at *m*/*z* 507. The fragmentation of compounds **11**, **12**, **14**, and **22** produced an ion at *m*/*z* 461 with a neutral loss of HCOOH in the MS^2 spectrum. Compared with the reference [26], compounds **11**, **12**, **14**, and **22** were tentatively deduced as dehydroecdysone, deoxykaladasterone, dacryhainansterone, and stachysterone D, respectively. The molecular formula of compound **13** is 16 Da more than that of compound **12**, whereas the retention time of compound **13** is a little later than that of compound **12**. Compared with the literature, compound **13** was plausibly assigned as kaladasterone [27]. Compound **15** showed the [M−H]-ion at *m*/*z* 187, which corresponded to the formula of $C_9H_{16}O_4$. Compound **15** was deduced to be azelaic acid by comparing it with the mass information in the literature [34]. Compound **16** showed the parent ion at *m*/*z* 547 and the daughter ion at m/z 255. Compared with the standard substance, compound **16** was unambiguously deduced as liquiritin apioside [35]. Compounds **17** and **19** shared the same [M−H]- ion at m/z 342 and yielded fragment ions at m/z 327, m/z 190, m/z 178, m/z 148, and m/z 135. The retention time of compound **19** was a little later than that of compound **17**. Compared with the standard substances, compounds **17** and **19** were unambiguously deduced as *N-cis*-feruloyl-3-methoxytyramine and feruloylmethoxytyramine [28]. Compound **18** showed the [M−H]- ion at m/z 312 and produced fragment ions at m/z 190, m/z 178, m/z 148, and m/z 135, which was consistent with the fragmentation pathway of *N-trans*-feruloyltyramine [36]. Compound **21** showed [M−H]- ion at m/z 225 with a formula of $C_{11}H_{16}O_2$. Compared with the literature [37], compound **21** was tentatively deduced as dihydroactinidiolide. Moreover, compounds **23** and **24** both shared the same [M−H]- ions at m/z 327, which corresponded to a formula of $C_{18}H_{32}O_5$. Compared with the references, compounds **23** and **24** were tentatively assigned as corchorifatty acid F [29] and 9,12,13-trihydroxy-10,15-octadecadienoic acid [30], respectively. Compounds **25**, **26**, **27**, and **28** were tentatively deduced as 9, 12, 13-trihydroxy-10-octadecenoic acid, 9,10,13-triHOME [31], pinellic acid [32], and sanleng acid [38], respectively.

3.3. Chemometric Analysis

Principal component analysis was first employed for the analysis of the data of the peak areas of 29 common peaks of nine batches of raw (S1, S2, S3, S4, S5, S6, S7, S8, S9) and salt-processed *A. bidentata* (Y1, Y2, Y3, Y4, Y5, Y6, Y7, Y8, Y9). The score and loading plots of the PCA model are displayed in Figure 3.

The samples of raw and salt-processed *A. bidentata* were well separated. Then, the supervised orthogonal partial least squares discriminant analysis method was used. The score and loading plots of the OPLS-DA model are shown in Figure 4. Q^2 of the OPLS-DA model is 97.6%, which indicates good prediction capability.

The score plot showed that the batches of raw *A. bidentata* were distinctly separated from the salt-processed batches. The VIP plot is sorted from high to low and shows confidence intervals for the VIP values, normally at the 95% level. VIP values larger than 1 indicate "important" X-variables, and values lower than 0.5 indicate "unimportant" X-variables. The importance projection of variables (VIP > 1) was used as the criteria to screen out the compounds that caused the difference. As depicted in Figure 5, methylflavone, indolyllactic acid, maleic acid, methoxytyrosine, ecdysterone, 5-deoxy kaladasterone, pinellic acid, kaladasterone, fumaric acid, dacryhainansterone, polypodine B, 9, 12, 13-trihydroxy-10, 15-octadecadienoic acid, 25*R*-inokosterone, liquiritin apioside, and *N-trans*-feruloyltyramine were identified as compounds with VIP values greater than 1. The VIP (Variable Importance for the Projection) plot summarizes the importance of the variables both to explain X and to correlate to Y. These compounds were considered important X-variables.

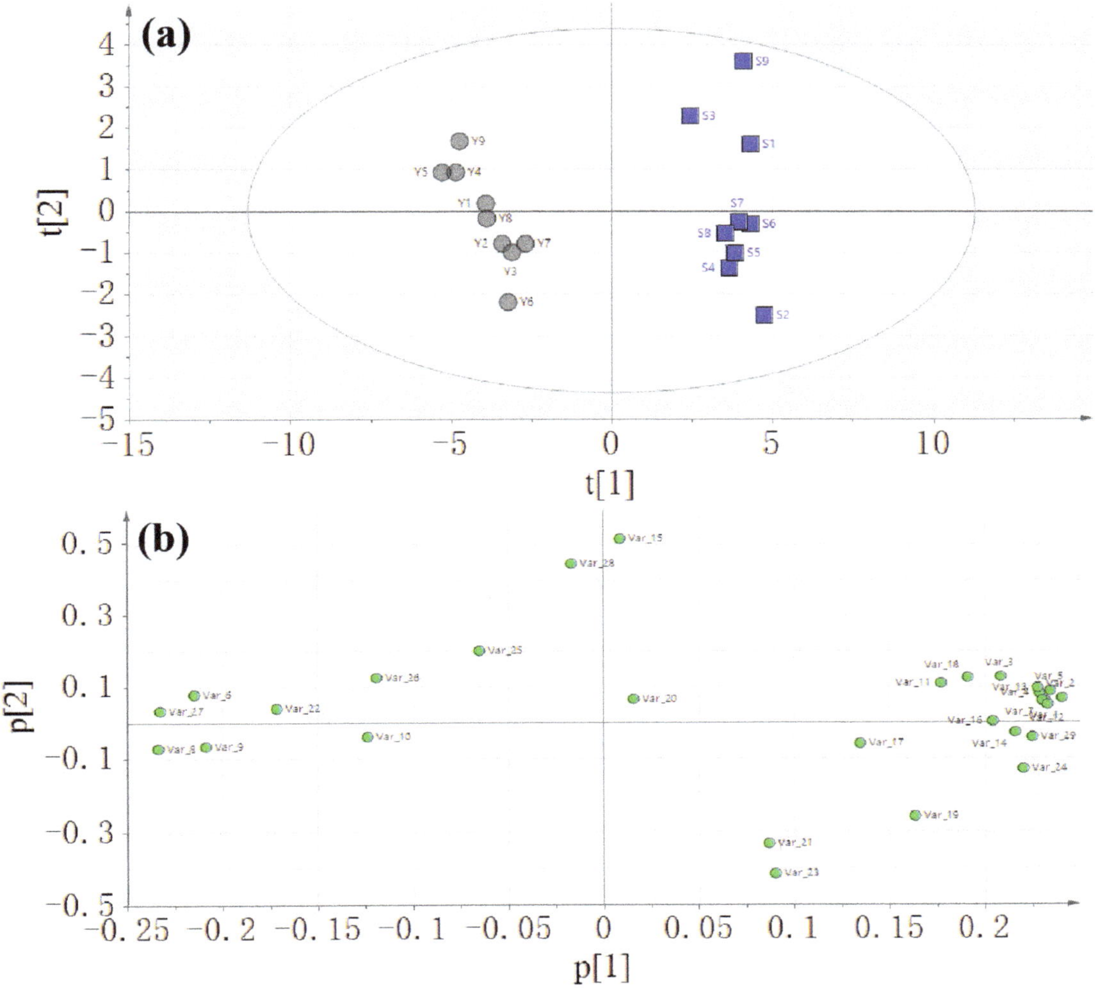

Figure 3. Score plot (**a**) and loading plot (**b**) of the principal component analysis.

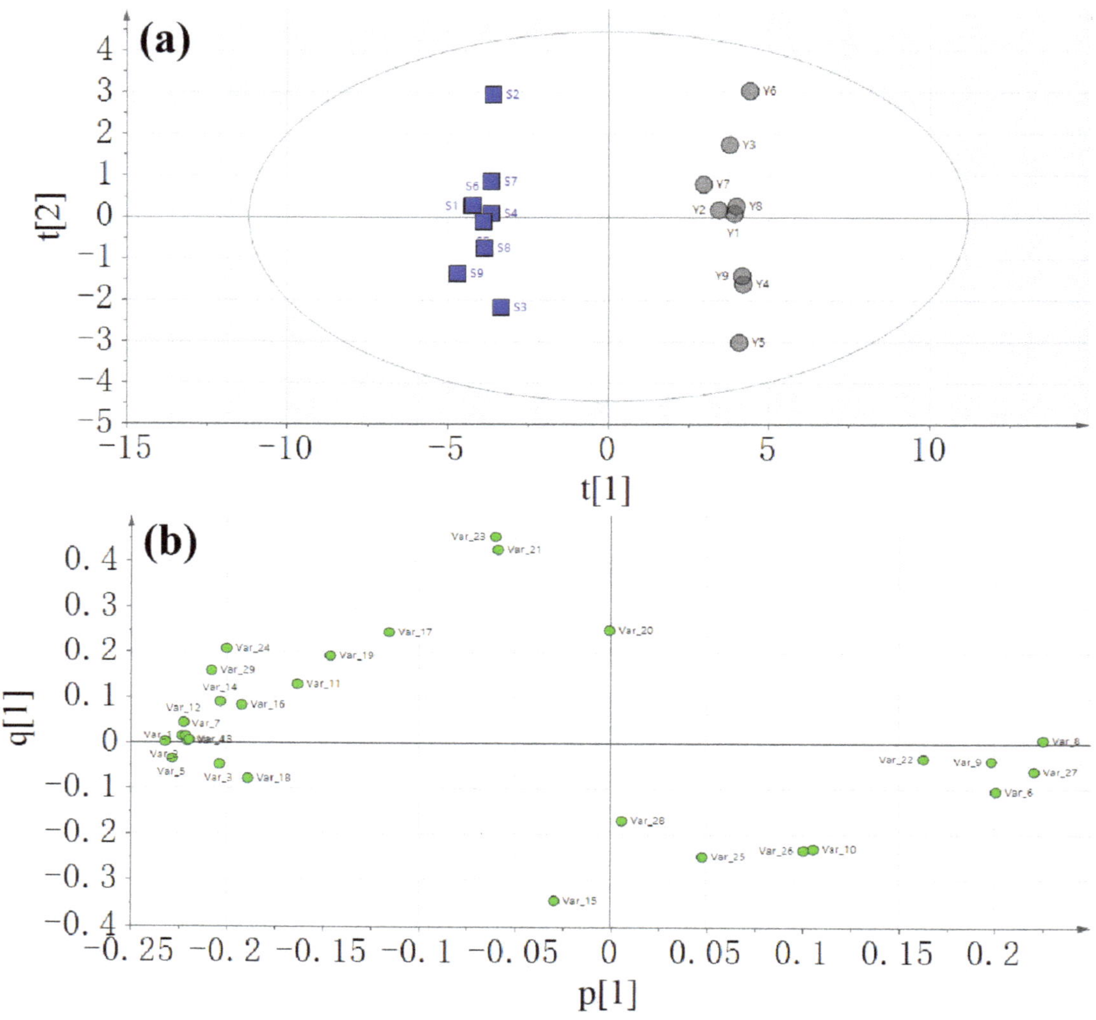

Figure 4. Score plot (**a**) and loading plot (**b**) of the OPLS-DA model.

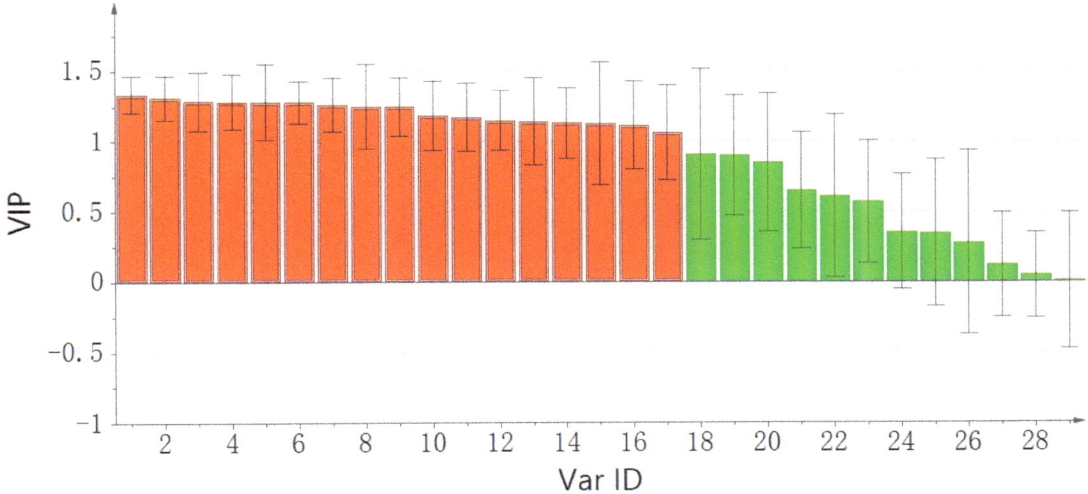

Figure 5. VIP plot of the OPLS-DA model. red indicated VIP value > 1,green indicated VIP value < 1.

3.4. Gray Correlation Analysis

The COX-2 inhibitory efficacy of nine batches of both raw and salt-processed *A. bidentata* samples was evaluated, with the findings presented in Table 5. Inhibition rates for raw *A. bidentata* spanned 58.6% to 73.5%, while those for salt-processed samples ranged from 79.5% to 88%. This implies that *A. bidentata* is strongly associated with COX-2 inhibition and has stronger osteoarthritis therapeutic activity. Correlation coefficients quantifying the relationship between individual compounds and COX-2 inhibition rates are detailed in Table 6. Eight compounds, including sanleng acid, stachysterone D, dihydroactinidiolide, *N-cis*-feruloyl-3-methoxytyramine, *N-trans*-feruloyltyramine, 9,12,13-trihydroxy-10-octadecenoic acid, azelaic acid, and dehydroecdysone, demonstrated a strong positive correlation with COX-2 inhibition. The data indicate a strong correlation between the components and the salt-processed *A. bidentata* in treating osteoarthritis. It is suggested that salt baking significantly contributes to the biological activity of *A. bidentata*. This finding also suggests that the therapeutic effects of salt-processed *A. bidentata* on osteoarthritis may be due to the combined action of multiple compounds.

Table 5. COX-2 inhibitory rate for nine batches of raw and salt-processed *Achyranthes bidentata*.

No.	Inhibitory Rate (%)	No.	Inhibitory Rate (%)
S1	68.5	Y1	81.1
S2	70.4	Y2	81.5
S3	73.5	Y3	79.5
S4	64.7	Y4	89.0
S5	65.6	Y5	81.7
S6	67.5	Y6	82.7
S7	70.2	Y7	82.1
S8	60.2	Y8	88.3
S9	58.6	Y9	83.4

Table 6. Correlation coefficient of compounds from raw and salt-processed *Achyranthes bidentata*.

No.	Compounds	Correlation Coefficient r
1	Maleic acid	0.7450
2	Unknown	0.7574
3	Fumaric acid	0.7326
4	Indolyllactic acid	0.8394
5	Methylflavone	0.8060
6	Polypodine B	0.8358
7	Methoxytyrosine	0.7346
8	Ecdysterone	0.8378
9	25*R*-inokosterone	0.7544
10	25*S*-inokosterone	0.8381
11	Dehydroecdysone	0.8506
12	Deoxykaladasterone	0.7291
13	Kaladasterone	0.7156
14	Dacryhainansterone	0.7267
15	Azelaic acid	0.8925
16	Liquiritin apioside	0.7818
17	*N-cis*-feruloyl-3-methoxytyramine	0.9175
18	*N-trans*-feruloyltyramine	0.9118
19	Feruloylmethoxytyramine	0.8186
20	Unknown	0.6732
21	Dihydroactinidiolide	0.9186
22	Stachysterone D	0.9291
23	Corchorifatty acid F	0.7514
24	9,12,13-Trihydroxy-10,15-octadecadienoic acid	0.8251
25	9,12,13-Trihydroxy-10-octadecenoic acid	0.9107
26	9,10,13-triHOME	0.7601
27	Pinellic acid	0.6485
28	Sanleng acid	0.9981
29	Unknown	0.7320

3.5. COX-2 Inhibitory Assay

The COX-2 inhibitory effects of sanleng acid, stachysterone D, dihydroactinidiolide, *N-cis*-feruloyl-3-methoxytyramine, *N-trans*-feruloyltyramine, 9,12,13-trihydroxy-10-octadecenoic acid, azelaic acid, and dehydroecdysone were evaluated. These compounds are the aforementioned substances that have a strong positive correlation with the rate of COX-2 inhibition. As shown in Figure 6, the positive control celecoxib showed an IC_{50} (half maximal inhibitory concentration) value of 25.64 nM, which is in agreement with the literature [39]. Sanleng acid, 9,12,13-trihydroxy-10-octadecenoic acid, and azelaic acid exhibited COX-2 inhibitory activity with IC_{50} values of 40.74 μM, 93.28 μM, and 86.36 μM, respectively. Stachysterone D and dehydroecdysone showed similar COX-2 inhibitory effects with IC_{50} values of 63.57 μM and 63.92 μM, respectively. Dihydroactinidiolide had an IC_{50} value of 70.66 μM. *N-cis*-feruloyl-3-methoxytyramine and *N-trans*-feruloyltyramine displayed IC_{50} values of 120.7 and 121.9 μM, respectively. However, the correlation strength between these compounds and the rate of COX-2 inhibition is not directly related to the corresponding compounds' IC_{50}. Of note, only the COX-2 inhibitory effect of *N-trans*-feruloyltyramine was reported in the literature [40], while the COX-2 inhibitory effect of seven other compounds, including sanleng acid, stachysterone D, dihydroactinidiolide, *N-cis*-feruloyl-3-methoxytyramine, 9,12,13-trihydroxy-10-octadecenoic acid, azelaic acid, and dehydroecdysone, was reported for the first time. The above eight compounds can be used as quality markers to focus on during the quality control of salt-processed *A. bidentata* for the treatment of osteoarthritis. Future studies will further consider the specific mechanisms of these compounds that are strongly associated with anti-COX-2 inhibition in osteoarthritis therapy.

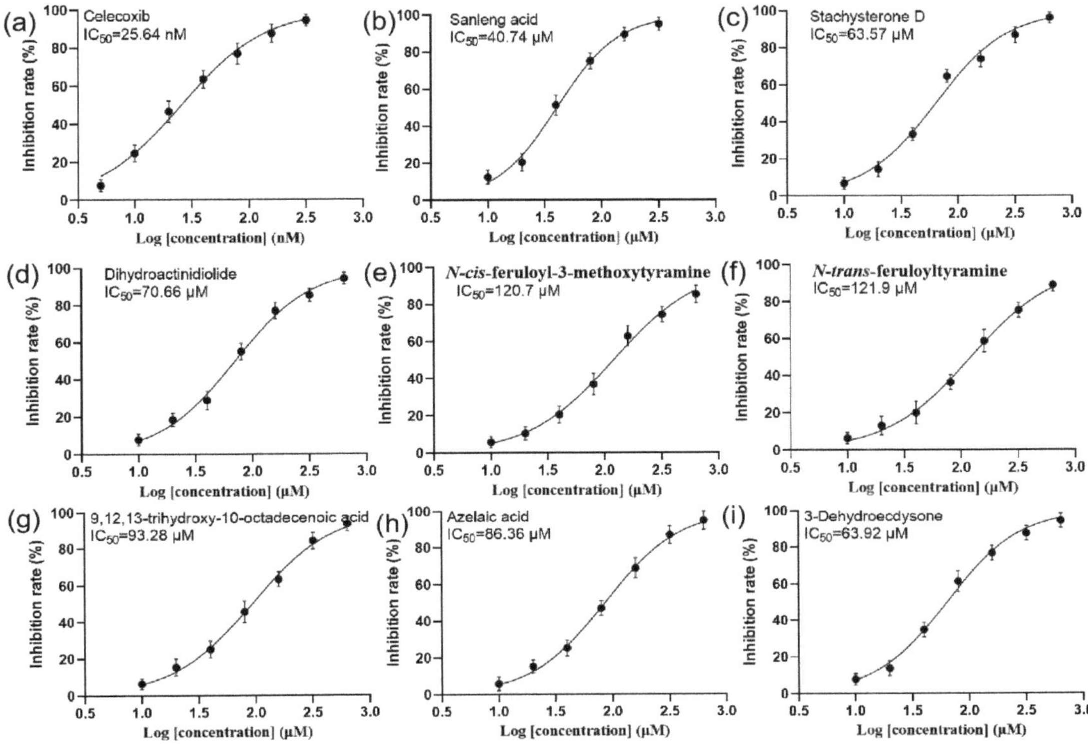

Figure 6. IC_{50} curves of celecoxib and other compounds. (**a**) celecoxib, (**b**) sanleng acid, (**c**) stachysterone, (**d**) dihydroactinidiolide, (**e**) *N-cis*-feruloyl-3-methoxytyramine, (**f**) *N-trans*-feruloyltyramine, (**g**) 9, 12, 13-trihydroxy-10-octadecenoic acid, (**h**) azelaic acid, and (**i**) dehydroecdysone.

4. Conclusions

A QbD-guided approach was successfully used to optimize the salt-processing parameters of *A. bidentata*. HPLC–Q-TOF-MS analysis combined with orthogonal partial least squares discriminant analysis facilitated the identification of the differential markers between raw and salt-processed *A. bidentate*. Notably, the COX-2 inhibitory effects of seven compounds were identified for the first time, which enhanced our understanding of the anti-inflammatory effects of salt-processed *A. bidentate*. In the future, these compounds can be considered critical product attributes of salt-processed *A. bidentate*.

Supplementary Materials: The following supporting information can be downloaded at: https://www.mdpi.com/article/10.3390/pr12030434/s1. Figure S1. Chemical structures of the identified compounds.

Author Contributions: Conceptualization, Y.T. and G.S.; methodology, J.Z. and L.S.; software, J.Z. and L.S.; writing—original draft preparation, J.Z. and L.S.; writing—review and editing, Y.T. and G.S.; supervision, Y.T. and G.S.; and funding acquisition, Y.T. and G.S. All authors have read and agreed to the published version of the manuscript.

Funding: This research was funded by the Hangzhou Municipal Bureau of Science and Technology, grant number 20220919Y164, and the Zhejiang Provincial Natural Science Foundation, grant number LY21H280008.

Data Availability Statement: Data are contained within the article.

Acknowledgments: The authors would like to thank the anonymous reviewers for their invaluable suggestions that helped improve the manuscript.

Conflicts of Interest: The authors declare no conflicts of interest.

References

1. Huang, P.; Tan, S.; Zhang, Y.X.; Li, J.S.; Chai, C.; Li, J.J.; Cai, B.C. The effects of wine-processing on ascending and descending: The distribution of flavonoids in rat tissues after oral administration of crude and wine-processed *Radix scutellariae*. *J. Ethnopharmacol.* **2014**, *155*, 649–664. [CrossRef]
2. Tao, Y.; Du, Y.; Li, W.; Cai, B.; Di, L.; Shi, L.; Hu, L. Integrating UHPLC-MS/MS quantification and DAS analysis to investigate the effects of wine-processing on the tissue distributions of bioactive constituents of herbs in rats: Exemplarily shown for *Dipsacus asper*. *J. Chromatogr. B* **2017**, *1055–1056*, 135–143. [CrossRef]
3. Zhu, T.; Liu, X.; Wang, X.; Cao, G.; Qin, K.; Pei, K.; Zhu, H.; Cai, H.; Niu, M.; Cai, B. Profiling and analysis of multiple compounds in rhubarb decoction after processing by wine steaming using UHPLC-Q-TOF-MS coupled with multiple statistical strategies. *J. Sep. Sci.* **2016**, *39*, 3081–3090. [CrossRef] [PubMed]
4. Li, W.; Ren, C.; Fei, C.; Wang, Y.; Xue, Q.; Li, L.; Yin, F.; Li, W. Analysis of the chemical composition changes of *Gardeniae Fructus* before and after processing based on ultra-high-performance liquid chromatography quadrupole time-of-flight mass spectrometry. *J. Sep. Sci.* **2021**, *44*, 981–991. [CrossRef] [PubMed]
5. Qin, K.; Wang, B.; Li, W.; Cai, H.; Chen, D.; Liu, X.; Yin, F.; Cai, B. Quality assessment of raw and processed *Arctium lappa* L. through multicomponent quantification, chromatographic fingerprint, and related chemometric analysis. *J. Sep. Sci.* **2015**, *38*, 1491–1498. [CrossRef] [PubMed]
6. He, X.; Wang, X.; Fang, J.; Chang, Y.; Ning, N.; Guo, H.; Huang, L.; Huang, X. The genus *Achyranthes*: A review on traditional uses, phytochemistry, and pharmacological activities. *J. Ethnopharmacol.* **2017**, *203*, 260–278. [CrossRef] [PubMed]
7. Li, Z.; Ma, D.; Peng, L.; Li, Y.; Liao, Z.; Yu, T. Compatibility of *Achyranthes bidentata* components in reducing inflammatory response through Arachidonic acid pathway for treatment of Osteoarthritis. *Bioengineered* **2022**, *13*, 1746–1757. [CrossRef] [PubMed]
8. Gawande, D.Y.; Goel, R.K. Pharmacological validation of in-silico guided novel nootropic potential of *Achyranthes aspera* L. *J. Ethnopharmacol.* **2015**, *175*, 324–334. [CrossRef] [PubMed]
9. Ishtiaq, A.; Muhammad, I.; Ullah, B.; Muhammad, N.; Muhammad, Z.; Ali, N. Pharmacognostic and hypoglycemic studies of *Achyranthus aspera* L. *J. Pharmacognosy Phytother.* **2013**, *5*, 127–131. [CrossRef]
10. Pan, R.; Hu, W.; Pan, J.; Huang, L.; Luan, C.; Shen, H. *Achyranthes bidentata* polypeptides prevent apoptosis by inhibiting the glutamate current in cultured hippocampal neurons. *Neural Regen. Res.* **2020**, *15*, 1086–1093.
11. Nazir, A.; Saleem, M.A.; Nazir, F.; Hussain, T.; Faizan, M.Q.; Usman, M. Comparison of UV Protection Properties of Cotton Fabrics Treated with Aqueous and Methanolic Extracts of *Achyranthes aspera* and *Alhagi maurorum* Plants. *Photochem. Photobiol.* **2016**, *92*, 343–347. [CrossRef]
12. Gawande, D.Y.; Druzhilovsky, D.; Gupta, R.C.; Poroikov, V.; Goel, R.K. Anticonvulsant activity and acute neurotoxic profile of *Achyranthes aspera* Linn. *J. Ethnopharmacol.* **2017**, *202*, 97–102. [CrossRef]
13. Zhang, S.; Zhang, Q.; Zhang, D.; Wang, C.; Yan, C. Anti-osteoporosis activity of a novel *Achyranthes bidentata* polysaccharide via stimulating bone formation. *Carbohyd. Polym.* **2018**, *184*, 288–298. [CrossRef]
14. Lee, T.G.; Hyun, S.W.; Jo, K.; Park, B.; Lee, I.S.; Song, S.J.; Kim, C.S. *Achyranthis radix* Extract Improves Urban Particulate Matter-Induced Dry Eye Disease. *Int. J. Environ. Res. Public Health* **2019**, *16*, 3229. [CrossRef]
15. Wang, S.; Zeng, M.; Li, B.; Kan, Y.; Zhang, B.; Zheng, X.; Feng, W. Raw and salt-processed *Achyranthes bidentata* attenuate LPS-induced acute kidney injury by inhibiting ROS and apoptosis via an estrogen-like pathway. *Biomed. Pharmacother.* **2020**, *129*, 110403. [CrossRef]
16. Nunavath, R.S.; Singh, M.T.; Jain, A.; Chakma, M.; Arivuselvam, R.; Azeeze, M. Quality by Design in Pharmaceuticals: A Review of its Impact on Regulatory Compliance and Product Quality. *Drug Res.* **2024**, *74*, 18–23.
17. Abramson, S.B. The role of COX-2 produced by cartilage in arthritis. *Osteoarthr. Cartil.* **1999**, *7*, 380–381. [CrossRef] [PubMed]
18. El-Miligy MM, M.; Al-Kubeisi, A.K.; El-Zemity, S.R.; Nassra, R.A.; Abu-Serie, M.M.; Hazzaa, A.A. Discovery of small molecule acting as multitarget inhibitor of colorectal cancer by simultaneous blocking of the key COX-2, 5-LOX and PIM-1 kinase enzymes. *Bioorg. Chem.* **2021**, *115*, 105171. [CrossRef] [PubMed]
19. Lv, Y.; Wu, H.; Hong, Z.; Wei, F.; Zhao, M.; Tang, R.; Li, Y.; Ge, W.; Li, C.; Du, W. Exploring active ingredients of anti-osteoarthritis in raw and wine-processed *Dipsaci Radix* based on spectrum-effect relationship combined with chemometrics. *J. Ethnopharmacol.* **2023**, *309*, 116281. [CrossRef] [PubMed]
20. Li, S.; Sun, Y.; Gao, Y.; Yu, X.; Zhao, C.; Song, X.; Han, F.; Yu, J. Spectrum-effect relationship analysis based on HPLC-FT-ICR-MS and multivariate statistical analysis to reveal the pharmacodynamic substances of Ling-Gui-Zhu-Gan decoction on Alzheimer's disease. *J. Pharm. Biomed. Anal.* **2024**, *237*, 115765. [CrossRef] [PubMed]
21. Alam, P.; Siddiqui, N.A.; Rehman, M.T.; Hussain, A.; Akhtar, A.; Mir, S.R.; Alajmi, M.F. Box–Behnken Design (BBD)-Based Optimization of Microwave-Assisted Extraction of Parthenolide from the Stems of *Tarconanthus camphoratus* and Cytotoxic Analysis. *Molecules* **2021**, *26*, 1876. [CrossRef] [PubMed]

22. Wen, S.; Tu, X.; Zang, Q.; Zhu, Y.; Li, L.; Zhang, R.; Abliz, Z. Liquid chromatography-mass spectrometry-based metabolomics and fluxomics reveals the metabolic alterations in glioma U87MG multicellular tumor spheroids versus two-dimensional cell cultures. *Rapid Commun. Mass Spectrom.* **2024**, *38*, e9670. [CrossRef]
23. Sun, J.; Song, Y.; Sun, H.; Liu, W.; Zhang, Y.; Zheng, J.; Zhang, Q.; Zhao, Y.; Xiao, W.; Tu, P.; et al. Characterization and quantitative analysis of phenolic derivatives in Longxuetongluo Capsule by HPLC-DAD-IT-TOF-MS. *J. Pharm. Biomed. Anal.* **2017**, *145*, 462–472. [CrossRef]
24. Yao, C.; Wang, Y.; Qu, H.; Li, J.; Hou, J.; Chen, X.; Zhang, J.; Wei, W.; Bi, Q.; Guo, D.A. Comparative identification of phytoecdysteroids in *Achyranthes bidentata* Blume and its three analogous species and application in differentiation between processing products from different species. *J. Pharm. Biomed. Anal.* **2023**, *227*, 115187. [CrossRef]
25. Zhang, L.; Yang, J.Q.; Luo, Y.; Shang, J.C.; Jiang, X.H. Simultaneous determination of eleven compounds related to metabolism of bioamines in rat cortex and hippocampus by HPLC-ECD with boron-doped diamond working electrode. *J. Pharm. Biomed. Anal.* **2016**, *118*, 41–51. [CrossRef]
26. Yadav, A.; Yadav, S.; Dabur, R. Higher plants exert interspecific effects on the phytoecdysteroids contents in *Tinospora cordifolia*. *Chem. Bio. Lett.* **2022**, *9*, 312.
27. Bourne, P.C.; Whiting, P.; Dhadialla, T.S.; Hormann, R.E.; Girault, J.P.; Harmatha, J.; Lafont, R.; Dinan, L. Ecdysteroid 7,9(11)-dien-6-ones as potential photoaffinity labels for ecdysteroid binding proteins. *J. Insect Sci.* **2002**, *2*, 11. [CrossRef] [PubMed]
28. Yang, L.; Jiang, H.; Wang, Q.H.; Yang, B.Y.; Kuang, H.X. A new feruloyl tyramine glycoside from the roots of *Achyranthes bidentata*. *Chin. J. Nat. Med.* **2012**, *10*, 16–19. [CrossRef]
29. Souza, L.; Oliveira, J.; Fernandes AD, S.; Macedo, A.F.; Araujo-Lima, C.F.; Felzenszwalb, I. UHPLC-MS metabolomic profile and in silico pharmacokinetic approach of *Kalanchoe daigremontiana* Raym.-Hamet & H. Perrier aqueous extracts. *J. Pharm. Biomed. Anal.* **2024**, *238*, 115827.
30. Hamberg, M.; Olsson, U. Efficient and specific conversion of 9-lipoxygenase hydroperoxides in the beetroot. Formation of pinellic acid. *Lipids* **2011**, *46*, 873–878. [CrossRef]
31. Fuchs, D.; Tang, X.; Johnsson, A.K.; Dahlén, S.E.; Hamberg, M.; Wheelock, C.E. Eosinophils synthesize trihydroxyoctadecenoic acids (TriHOMEs) via a 15-lipoxygenase dependent process. *Biochim. Et Biophys. Acta (BBA)-Mol. Cell Biol. Lipids* **2020**, *1865*, 158611. [CrossRef]
32. Khan, R.S.; Senthi, M.; Rao, P.C.; Basha, A.; Alvala, M.; Tummuri, D.; Masubuti, H.; Fujimoto, Y.; Begum, A.S. Cytotoxic constituents of *Abutilon indicum* leaves against U87MG human glioblastoma cells. *Nat. Prod. Res.* **2015**, *29*, 1069–1073. [CrossRef]
33. Mei, Y.; Zhang, X.; Hu, Y.; Tong, X.; Liu, W.; Chen, X.; Cao, L.; Wang, Z.; Xiao, W. Screening and characterization of xenobiotics in rat bio-samples after oral administration of Shen-Wu-Yi-Shen tablet using UPLC-Q-TOF-MS/MS combined with a targeted and non-targeted strategy. *J. Pharm. Biomed. Anal.* **2023**, *227*, 115286. [CrossRef]
34. Zhou, W.; Wang, P.G. Simultaneous determination of multi-class active pharmaceutical ingredients by UHPLC-HRMS. *J. Pharm. Biomed. Anal.* **2021**, *202*, 114160. [CrossRef]
35. Aly, S.H.; Elissawy, A.M.; Mahmoud, A.M.A.; El-Tokhy, F.S.; Mageed, S.S.A.; Almahli, H.; Al-Rashood, S.T.; Binjubair, F.A.; Hassab, M.A.E.; Eldehna, W.M.; et al. Synergistic Effect of *Sophora japonica* and *Glycyrrhiza glabra* Flavonoid-Rich Fractions on Wound Healing: In Vivo and Molecular Docking Studies. *Molecules* **2023**, *28*, 2294. [CrossRef] [PubMed]
36. Pan, L.; Li, L.; Xu, L.; Zhang, J.; Li, J.; Gao, M.; Yu, J.; Jin, L.; Lei, D. UHPLC-QTOF-MS/MS based characterization of anti-tumor constituents in *Ceratocarpus arenarius* L. and identification of EGFR-TK inhibitors by virtual screening. *Nat. Prod. Res.* **2022**, *36*, 6111–6115. [CrossRef] [PubMed]
37. Chang, Q.; Lan, L.; Gong, D.; Guo, Y.; Sun, G. Evaluation of quality consistency of herbal preparations using five-wavelength fusion HPLC fingerprint combined with ATR-FT-IR spectral quantized fingerprint: *Belamcandae rhizoma* antiviral injection as an example. *J. Pharm. Biomed. Anal.* **2022**, *214*, 114733. [CrossRef] [PubMed]
38. Yi-meng, Y.; Ke-xin, Y.; Yu-sheng, L.; Ge, C.; Han-wen, T.; Zhongying, L.; Zhiqiang, L.; Fengrui, S.; Zifeng, P. Characterization of Components in Huangying Kechuan Syrup by Ultra-high Liquid Chromatography Tandem Quadrupole-time-of-flight Mass Spectrometry. *Chin. J. App. Chem.* **2021**, *38*, 276.
39. Fadaly, W.A.A.; Nemr, M.T.M.; Zidan, T.H.; Mohamed, F.E.A.; Abdelhakeem, M.M.; Abu Jayab, N.N.; Omar, H.A.; Abdellatif, K.R.A. New 1,2,3-triazole/1,2,4-triazole hybrids linked to oxime moiety as nitric oxide donor selective COX-2, aromatase, B-RAF(V600E) and EGFR inhibitors celecoxib analogs: Design, synthesis, anti-inflammatory/anti-proliferative activities, apoptosis and molecular modeling study. *J. Enzym. Inhib. Med. Chem.* **2023**, *38*, 2290461.
40. Jiang, Y.; Yu, L.; Wang, M.H. N-trans-feruloyltyramine inhibits LPS-induced NO and PGE2 production in RAW 264.7 macrophages: Involvement of AP-1 and MAP kinase signalling pathways. *Chem. Biol. Interact.* **2015**, *235*, 56–62. [CrossRef] [PubMed]

processes

MDPI

Article

Anti-Glioblastoma Potential and Phenolic Profile of Berry Juices

Mirela Kopjar [1,*], Drazen Raucher [2], Mary Ann Lila [3] and Josip Šimunović [4,*]

1 Faculty of Food Technology, University of Josip Juraj Strossmayer in Osijek, Franje Kuhača 18, 31000 Osijek, Croatia
2 Department of Cell and Molecular Biology, University of Mississippi Cancer Institute, University of Mississippi Medical Center, 2500 North State Street, Jackson, MS 39216, USA; draucher@umc.edu
3 Plants for Human Health Institute, Department of Food Bioprocessing and Nutrition Sciences, North Carolina Research Campus, North Carolina State University, 600 Laureate Way, Kannapolis, NC 28081, USA; mlila@ncsu.edu
4 Department of Food, Bioprocessing and Nutrition Sciences, North Carolina State University, 116C Schaub Hall, Raleigh, NC 27695, USA
* Correspondence: mirela.kopjar@ptfos.hr (M.K.); simun@ncsu.edu (J.Š.)

Abstract: Glioblastoma is one of the most aggressive and lethal brain tumors. Due to the failure of conventional chemotherapies and targeted drugs pursuit of natural, less toxic agents is on the rise as well as their utilization in glioblastoma treatment. Consequently, this study explores the antiproliferative potential of selected berry juices (wild blackberry (*Rubus discolor*), dwarf elderberry (*Sambucus ebulus*), and raspberry (*Rubus idaeus*)) on glioblastoma cells (U87-MG and GBM43) in comparison to temozolomide. The juices were assessed for total phenolic content, proanthocyanins, polyphenol profiles, and antioxidant activity. Wild blackberry and dwarf elderberry juices exhibited higher total polyphenols, proanthocyanins, and monomeric anthocyanins compared to raspberry juice. HPLC analysis revealed distinctive anthocyanins, flavonoids, and phenolic acids in each juice. With the DPPH assay, the highest antioxidant potential had wild blackberry juice, while with other assays dwarf elderberry juice had the highest potential. Antiproliferative effects were dose-dependent, with wild blackberry juice demonstrating the highest potency, surpassing temozolomide in inhibiting GBM43 cell proliferation. In U87 cells, all juices exhibited antiproliferative effects, with wild blackberry showing the strongest impact. This study highlights the potential of wild blackberry juice as a potent natural agent against glioblastoma, suggesting its superiority over the conventional treatment.

Keywords: anthocyanins; antiproliferative potential; dwarf elderberry juice; raspberry juice; wild blackberry juice

1. Introduction

Nutritional guidelines have over the years highlighted the importance of daily consumption of fruits, especially berries. Berries are well known for their high content of flavonoids, phenolic acids, tannins, and especially anthocyanins, due to which they benefit from a "health halo". These phenolic compounds, through numerous in vitro, pre-clinical, clinical and epidemiological studies, have been linked to many beneficial health effects like improvement of the lipid balance, regulation of hyperglycemia, protection from cardiovascular disorders, possible regulation of diabetes, anticancer potential, anti-inflammation properties and antioxidant potential [1–6].

Another very important and popular aspect among the general population is the utilization of bioactive phytochemicals, among which anthocyanins have a crucial role, in order to substitute synthetic compounds used in chemotherapeutic or chemopreventive applications [7–9]. Anthocyanins exhibit numerous anticancer effects such as antioxidation, lipid peroxidation, antiproliferation, cytotoxicity anti-inflammation, cell cycle perturbations, epidermal growth factor receptor inhibition, and apoptosis [7].

Glioblastoma is one of the most aggressive and lethal brain tumors which is characterized by the existence of weakly differentiated anaplastic cells immersed in necrotic zones of brain tissue [10–12]. Despite all the efforts that were made through the years in the treatment of this tumor, it is still characterized by a low survival rate of only 13 months after diagnosis and even in the most positive situations, the majority of patients die within two years. In clinical trials, a 5-year survival rate of only 4–5% of patients has been achieved [11–16]. Conventional glioblastoma treatment comprises surgical resection followed by radiotherapy and chemotherapy including temozolomide [10,12,15]. However, the prognosis for these patients remains weak as already mentioned due to the high tumor recurrence [10,12,14–16]. These tumor cells are quite heterogeneous and grow rapidly, invading and infiltrating the nearby healthy brain tissues which makes the complete tumor resection challenging. Residual tumor cells cause the initiation of the occurrence of secondary glioblastoma lesions, which are even more resistant to therapy than rapidly proliferating primary tumor cells, leading to tumor recurrence [10]. In addition to the tumor recurrence and therapy resistance, additional challenges are the inability of drugs to cross the blood–brain barrier and pass through the blood–brain–tumor barrier which is formed during the later stages of tumor growth [10,17–19]. All those challenges and failure of conventional chemotherapies and targeted drugs govern scientist in the direction to find natural, less toxic agents and explore their utilization in glioblastoma treatment [20].

The most commonly used cell lines in cancer research, particularly in the study of glioblastoma multiforme (GBM), which is a type of malignant brain tumor, are U87-MG and GBM43. U87-MG cells are characterized by their ability to form tumors in experimental animals and are often used as a model system for studying glioblastoma. These glioblastoma cells are used to investigate various aspects of glioblastoma biology, including tumor growth, invasion, and response to therapies. They are also employed in drug testing and preclinical studies [21–23]. The specific characteristics of GBM43 cells may vary depending on factors such as the source and handling of the cells. They are likely to share features common to glioblastoma cells, such as rapid proliferation and invasiveness. These cells are used to gain insights into the biology of glioblastoma and to test potential therapeutic interventions [24–26].

The aim of this study was to evaluate the antiproliferative effect of selected berry juices (wild blackberry, dwarf elderberry and raspberry) on mentioned glioblastoma cells (U87-MG and GBM43). The effect of selected berry juices on glioblastoma cells was compared with temozolomide, the current drug of choice for treatment of glioblastoma. Additionally, berry juices were evaluated for their total phenolic content, proanthocyanidin content, individual polyphenols and antioxidant activity.

2. Materials and Methods

2.1. Chemicals

Products of Sigma-Aldrich (St. Louis, MO, USA) were 4-dimethylaminocinnamaldehyde, trolox, 2,2′-azino-bis(3-ethylbenzothiazoline-6-sulfonic acid) diammonium salt, 2,2-diphenyl-1-picrylhydrazyl and analytical standards of chlorogenic acid, gallic acid, ellagic acid, rutin and (-)-epicatechin. Standards of hyperoside, neochlorogenic acid and anthocyanins (cyanidin-3-glucoside, cyanidin-3-galactoside, cyanidin-3-rutinoside and cyanidin-3-sophoroside) were bought from Extrasynthese (Genay, France). From T.T.T. (Sveta Nedelja, Croatia) sodium carbonate was procured, while Folin–Ciocalteu reagent and potassium persulfate were from Kemika (Zagreb, Croatia). HPLC-grade orthophosphoric acid was product of Fisher Scientific (Loughborough, UK) and HPLC-grade methanol was from J.T. Baker (Deventer, The Netherlands). From Acros Organic (Geel, Belgium), cupric chloride, neocuproine and 2,4,6-tri(2-pyridyl)-s-triazine (TPTZ) were bought. From ThermoFisher Scientific (Waltham, MA, USA), Thiazolyl Blue tetrazolium bromide, 98% (2-(3,5-diphenyltetrazol-2-ium-2-yl)-4,5-dimethyl-1,3-thiazole;bromide), Temozolomide (3-methyl-4-oxoimidazol[5,1-d][1,2,3,5]tetrazine-8-carboxamide), DMEM media w/L-Glutamine, 4.5 g/L glucose and sodium pyruvate, Penicillin Streptomycin Solution (10,000 U/mL Penicillin, 10,000 μg/mL Streptomycin in 0.85% NaCl), and 0.25% Trypsin 0.1%

EDTA were obtained. U87-MB and GBM43 cells were acquired from Mayo Clinic (Rochester, MN, USA). Fetal bovine serum was bought from R&D Systems.

2.2. Preparation of Berry Juice

Wild blackberry (*Rubus discolor*), dwarf elderberry (*Sambucus ebulus*), and raspberry (*Rubus idaeus*) fruits were collected at location 46°18′39.7″ N 16°32′67.7″ E near Varaždin, Croatia. About 2 kg of each fruits were collected, washed and pressed. The obtained juice was filtered through cheesecloth and afterwards juice (300 mL) was thermally treated at 90 °C for two minutes in order to inactivate naturally present enzymes that could lead to degradation of polyphenols.

2.3. Spectrophotometric Analysis of Total Polyphenols, Monomeric Anthocyanins and Proanthocyanidins

The method described by Singleton and Rossi [27] was used for the determination of total polyphenols in samples. Folin–Ciocalteu reagent (7.5%) was prepared and 10 mL was mixed with a 0.2 mL of sample, 1.8 mL demineralized water and 8 mL of sodium carbonate solution (7.5%). This mixture was kept in the dark for 120 min prior to the measurement of the absorbance at 765 nm. The calibration curve was created for gallic acid. Results were expressed as g of gallic acid equivalents per L of the juice (g GAE/L). Each sample was analyzed in triplicate and all measurements were conducted using a UV/Vis spectrophotometer (Cary 60 UV-Vis, Agilent Technologies, Santa Clara, CA, USA).

Monomeric anthocyanins were determined by the pH differential method [28]. To conduct the analyses, it was necessary to prepare two different buffers (0.025 M KCl at pH 1 and 0.4 M sodium acetate at pH 4.5). Each buffer (2.8 mL) was mixed with the sample (0.2 mL). After these mixtures were kept in the dark for 15 min, absorbance was measured at 515 nm and 700 nm. The concentration of monomeric anthocyanins was expressed as mg of cyanidin-3-glucoside per L of the juice (mg cyanidin-3-glucoside/L).

To determine the concentration of proanthocyanidins, the DMAC method was applied [29]. Briefly, DMAC (4-dimethylaminocinnamaldehyde) reagent was prepared and mixed (1 mL) with sample and acidified ethanol. The absorbance of the samples was measured at 640 nm after the samples were kept in dark for 30 min. The calibration curve was set for procyanidin B2 and concentration of proanthocyanidins was expressed as mg of procyanidin B2 equivalent per L of the juice (mg B2E/L). All samples were analyzed in triplicate.

2.4. Evaluation of Antioxidant Activity—ABTS, DPPH, FRAP and CUPRAC Assays

Firstly, the ABTS assay by Arnao et al. [30] was conducted. After ABTS (2,2′-azino-bis(3-ethylbenzothiazoline-6-sulfonic acid) diammonium salt) reagent was prepared, 3.2 mL was mixed with 0.2 mL of the diluted samples and after 95 min, and absorbance was measured at 734 nm. Secondly, the DPPH (2,2-diphenyl-1-picrylhydrazyl) solution was prepared and 3 mL was mixed with 0.2 mL of the diluted samples. Absorbance was read at 517 nm after 15 min. This assay was described in detail elsewhere [31]. Thirdly, the ferric-reducing ability was determined using the method of Benzie and Strain [32]. FRAP reagent was prepared and 3 mL was mixed with 0.2 mL of the diluted samples and absorbance was read at 593 nm after 30 min. Finally, the CUPRAC assay was applied for the determination of cupric ion-reducing antioxidant capacity. This assay was described in detail by Apak et al. [33]. Copper chloride, neocuproine and ammonium acetate buffer (pH 7) solution were mixed in a ratio of 1:1:1 and 0.2 mL of the diluted samples were addes. After 30 min absorbance was read at 450 nm. All samples were analyzed in triplicate. Calibration curves were created using Trolox, and for all assays, results are expressed as μmol of Trolox equivalents per 100 mL of juice (μmol TE/100 mL).

2.5. Sample Preparation for High Performance Liquid Chromatography

Solid phase extraction was conducted using a commercial sorbent, Strata^{TM}-X 33 μm Polymeric Reversed Phase from Phenomenex (Torrance, CA, USA), in order to exclude impurities. After the cartridges were placed in a vacuum manifold operated at room temperture, they were preconditioned with HPLC-grade methanol. A solution of acetic acid (1% in water) was added next and then the sample was added by dripping it and allowing it to form a compact ring in the cartridges. Elution of polyphenols was performed with methanol after the cartridges were dry [34,35]. The eluents were collected and injected into the HPLC system.

2.6. Determination of Individual Polyphenols Using Reversed Phase HPLC

For the evaluation of individual polyphenols in samples, Agilent HPLC system 1260 Infinity II (Agilent Technology, Santa Clara, CA, USA) was used. This system consisted of a quaternary pump, a vial sampler, a column (Poroshell 120 EC C-18, 4.6 × 100 mm, 2.7 μm) and diode array detector that was recording in the range from 190 to 600 nm. The method used was previously published by Buljeta et al. [36]. Orthophosphoric acid (0.1%) was used as mobile phase A and HPLC-grade methanol was used as mobile phase B. Injected volume was 5 μL and the flow rate was 1 mL/min. The calibration curves for standards of cyanidin-3-glucoside, cyanidin-3-galactoside, cyanidin-3-rutinoside, cyanidin-3-sophoroside, rutin, hyperoside, ellagic acid, gallic acid, neochlorogenic acid, chlorogenic acid, and (-)-epicatechin were created and linearity was confirmed with $r^2 > 0.99$. All measurements were carried out in duplicate and concentrations of individual polyphenols were expressed as mg of polyphenol per L of the juice (mg/L).

2.7. Antiproliferative Effects of Berry Fruit Juices on Glioblastoma Cells

Glioblastoma cells (U87-MG and GBM43) were cultured in 96-well plates, a standard format for high-throughput screening and analysis. After 24 h, cells were exposed to fruit juices at concentrations of 1%, 2%, and 3% in the culture medium (DMEM-10% fetal bovine serum-1% penicillin/streptomycin). Untreated cells served as the negative control, while temozolomide, the current standard treatment for glioblastoma, was used as the positive control for inducing cell death. After 3 days of treatment, cell survival was assessed using the 3-(4,5-dimethylthiazol-2-yl)-2,5-diphenyl-2H-tetrazolium bromide (MTT) assay [37]. This colorimetric assay measures the metabolic activity of living cells. Viable cells convert the yellow MTT reagent into purple formazan crystals, quantified by absorbance at a specific wavelength. Higher absorbance values indicate greater cell viability, reflecting the effectiveness of the treatment. Percentage survival for each treatment group was calculated as follows:

$$\text{Percentage of survival} = (\text{Absorbance of treated cells}/\text{Absorbance of untreated control cells}) \times 100.$$

2.8. Statistical Analysis

Comparisons of the obtained results for polyphenols were carried out by analysis of variance (ANOVA) and Fisher's least significant difference (LSD), with the significance defined at $p < 0.05$. The software program STATISTICA 13.1 (StatSoft Inc., Tulsa, OK, USA) was applied for statistical analyses. Comparisons of the obtained results for antiproliferative effects of berry fruit juices on glioblastoma cells were graphed and statistically analyzed using one-way analysis of variance (ANOVA) to determine significant differences in survival percentages between groups. These analyses were performed using GraphPad Prism version 10.1.2.

3. Results

3.1. Phenolic Compounds and Antioxidant Activity of Berry Juices

Amounts of total polyphenols, proanthocyanidins and monomeric anthocyanins of selected berry juices are presented in Table 1. From the obtained results, it can be seen that wild blackberry and dwarf elderberry juices had approximately 3.2 g/L of total polyphenols while raspberry juice had a significantly lower amount, 1.67 g/L ($p < 0.05$). Investigated juices significantly differ in proanthocyanidins and monomeric anthocyanins amounts. The amount of pronanthocyanidins in juices declined in the following: order dwarf elderberry > wild blackberry

> raspberry juice (71.21 mg/L > 56.21 mg/L > 32.32 mg/L) ($p < 0.05$). A slightly different tendency was observed for the amounts of monomeric anthocyanins, and their amounts were ranked in the following order: wild blackberry > dwarf elderberry > raspberry juice (759.99 mg/L > 433.17 mg/L > 230.39 mg/L) ($p < 0.05$).

Table 1. Total polyphenols (TP), monomeric anthocyanins (MA) and proanthocyanidins (PAC) of investigated berry juices.

Juice	TP (g/L)	MA (mg/L)	PAC (mg/L)
DE	3.16 ± 0.02 ^{a}	433.17 ± 2.66 ^{b}	71.21 ± 0.81 ^{a}
WB	3.21 ± 0.01 ^{a}	759.99 ± 3.53 ^{a}	56.21 ± 1.96 ^{b}
RB	1.67 ± 0.02 ^{b}	230.39 ± 1.10 ^{c}	32.34 ± 0.27 ^{c}

DE—dwarf elderberry; WB—wild blackberry; RB—raspberry. Values in the same column, marked with different letters (a–c) are significantly different at $p \leq 0.05$.

HPLC profiles of individual polyphenols of investigated juices are presented in Table 2. It is evident from the results that juices differ in the type of anthocyanins inherent in their phytochemical profiles, as well as in phenolic acids and flavonoids. Dwarf elderberry juice contained cyanidin-3-galactoside (261.40 mg/L) while the other two investigated juices did not contain this anthocyanin. Wild blackberry juice contained high amounts of cyanidin-3-glucoside and cyanidin-3-rutinoside (389.60 and 269.72 mg/L, respectively). These two anthocyanins were also identified in raspberry juice but in significantly lower amounts (39.11 and 34.28 mg/L, respectively). The most abundant anthocyanin in raspberry juice was cyanidin-3-sophoroside (210.65 mg/L). Two flavonoids were detected in wild blackberry and dwarf elderberry juices, hyperoside and rutin. Dwarf elderberry and wild blackberry juice contained both of these flavonoids, but in different amounts. In dwarf elderberry juice, hyperoside was evaluated at 34.19 mg/L and rutin at 109.72 mg/L. The opposite trend was observed in wild blackberry juice, i.e., hyperoside was evaluated in the higher amount (73.27 mg/L) than rutin (3.37 mg/L). Raspberry contained only rutin in the amount of 14.70 mg/L.

Table 2. Polyphenols amount of investigated berry juices (mg/L of fresh weight).

	DE	WB	RB
Cyanidin-3-glucoside	nd	389.60 ± 10.98 ^{a}	39.11 ± 0.01 ^{b}
Cyanidin-3-rutinoside	nd	269.72 ± 8.52 ^{a}	34.28 ± 0.35 ^{b}
Cyanidin-3-galactoside	261.40 ± 1.41 ^{a}	nd	nd
Cyanidin-3-sophoroside	nd	nd	210.65 ± 0.23 ^{a}
Hyperoside	34.19 ± 0.09 ^{b}	73.27 ± 2.39 ^{a}	nd
Rutin	109.72 ± 0.05 ^{a}	34.37 ± 0.96 ^{b}	14.70 ± 0.41 ^{bc}
Gallic acid	4.58 ± 0.04 ^{a}	4.56 ± 0.03 ^{a}	nd
Ellagic acid	nd	5.14 ± 0.03 ^{a}	3.41 ± 0.00 ^{b}
Chlorogenic acid	160.02 ± 2.87 ^{a}	nd	nd
Neochlorogenic acid	27.18 ± 1.67 ^{a}	nd	nd

DE—dwarf elderberry; WB—wild blackberry; RB—raspberry; nd—not detected. Values in the same row, marked with different letters (a–c) are significantly different at $p \leq 0.05$.

Gallic, ellagic, chlorogenic and neochlorogenic acids were detected in the berry juices. Chlorogenic and neochlorogenic acids (160.02 and 27.18 mg/L, respectively) were identified only in dwarf elderberry juice in addition to gallic acid. Gallic acid was also identified in wild blackberry juice, and those two juices contained 4.57 mg/L of this phenolic acid. In raspberry and wild blackberry juices, ellagic acid was determined at 3.41 mg/L and 5.14 mg/L, respectively.

Results of the evaluation of antioxidant activities of berry juices are presented in Table 3. Antioxidant activities were evaluated using the following assays: DPPH (with DPPH radicals), ABTS (with ABTS cation radicals), FRAP (ferric reducing antioxidant power) and CUPRAC (cupric ion reducing capability) assays. Application of DPPH assay revealed that wild blackberry juice had the highest antioxidant potential, followed by the dwarf elderberry and raspberry juices (13.25, 9.86 and 6.95 μmol TE/100 mL, respectively) ($p < 0.05$). However, with other applied assays, a different trend was observed. Dwarf elderberry juice had the highest antioxidant potential, followed by raspberry and wild blackberry juices. For ABTS assay, these values were 26.18, 15.56 and 6.08 μmol TE/100 mL, respectively, ($p < 0.05$). The lowest values were obtained by FRAP (2.11, 1.38 and 0.59 μmol TE/100 mL, respectively) ($p < 0.05$) while the highest by CUPRAC assay (107.78, 64.00 and 27.37 μmol TE/100 mL, respectively) ($p < 0.05$). Comparing these results of antioxidant activity with individual polyphenols determined in berry juices, it is evident that antioxidant activity depends on the structure of polyphenols and the mechanism of action of each applied assay. Wild blackberry juice contained the highest amount of anthocyanins and had the highest antioxidant activity determined by a DPPH assay. For all other assays, dwarf elderberry juice possessed the highest antioxidant activity. It has to be noted that this juice contained, next to anthocyanins, a significant amount of flavonoids and phenolic acids in contrast to the other two berry juices. Even though raspberry juice contained a lower amount of anthocyanins and a much lower amount of flavonoids and phenolic acids than wild blackberry juice, through utilization of ABTS, FRAP and CUPRAC assay, a higher antioxidant activity was achieved than for wild blackberry juice. This can be explained by well-known phenomena of antagonistic or synergistic effects of phenolic compounds.

Table 3. Antioxidant activity (μmol TE/100 mL of fresh weight) of investigated berry juices.

Juice	DPPH	ABTS	FRAP	CUPRAC
DE	9.86 ± 0.01 [b]	26.18 ± 0.57 [a]	2.11 ± 0.01 [a]	107.78 ± 0.56 [a]
WB	13.25 ± 0.06 [a]	6.08 ± 0.01 [c]	0.59 ± 0.01 [c]	27.37 ± 0.37 [c]
RB	6.95 ± 0.03 [c]	12.59 ± 0.12 [b]	1.38 ± 0.00 [b]	64.00 ± 0.11 [b]

DE—dwarf elderberry; WB—wild blackberry; RB—raspberry; DPPH (2,2-diphenyl-1-picrylhydrazyl radicals assay); ABTS (2,2′-azino-bis(3-ethylbenzothiazoline-6-sulfonic acid) diammonium salt cation radicals assay); FRAP (ferric reducing antioxidant power assay); CUPRAC (cupric ion reducing capability assay). Values in the same column, marked with different letters (a–c) are significantly different at $p \leq 0.05$.

3.2. *Antiproliferative Effects of Berry Juices*

Results of the antiproliferative effect trials of berry juices on GBM43 and U87-MG cells are presented in Figure 1. Their potential for inhibition of proliferation of GBM43 and U87-MG cells was compared with temozolomide (TMZ), the drug of choice for the treatment of selected tumor cells. Juices were used in the amounts of 1%, 2% and 3%. The results demonstrated that berry juices exhibited antiproliferative effects on both GBM43 and U87-MG cells with different potency in a dose-dependent manner. The percentage of survival of U87-MG cells was 52.6% when the inhibition was conducted with TMZ. The potential for inhibition of proliferation of U87-MG cells decreased in the following order: wild blackberry > dwarf elderberry > raspberry. For the wild blackberry juice percentages of survival of U87 cells were 53.8%, 49% and 44%, respectively, when cells were treated with 1%, 2% and 3% of juice. The percentage of survival of U87 cells when treated with 1% of dwarf elderberry juice was 67.7%. 58.2% and approximately 60% when treated with higher concentrations. Raspberry juice was the least potent juice for this population of tumor cells, and percentages of survival were 75% when they were treated with 1% of juice and approximately 65% when treated with higher concentrations.

Remarkably, GBM43 cells, which are typically resistant to temozolomide, exhibited significantly higher inhibition when treated with berry juices compared to the temozolomide treatment. The percentage of survival of GBM43 cells was 73.7% when inhibition was conducted with TMZ. The potential for inhibition of proliferation of GBM43 cells decreased in the following order: wild blackberry > raspberry > dwarf elderberry. The percentages of survival of GBM43 cells were 43.3%, 40.2% and 30% when treated with 1%, 2% and 3% of wild blackberry juice. Treatment of GBM43 cells with 1%, 2% and 3% of raspberry juice resulted in survival rates of 64.5%, 54 and 47.2%, respectively. Percentages of survival rates of 66%, 62% and 51.2% were achieved when GBM43 cells were treated with 1%, 2% and 3% of dwarf elderberry juice.

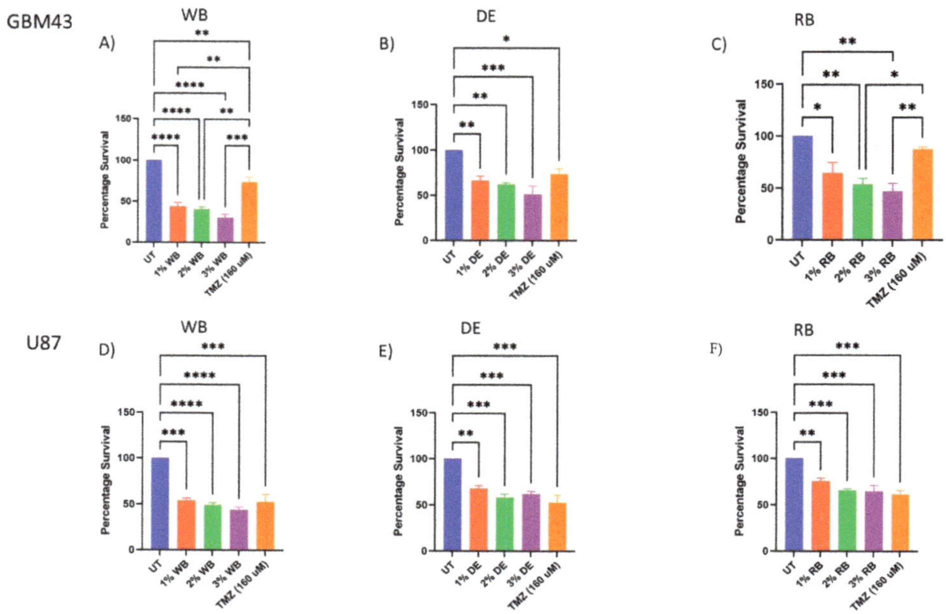

Figure 1. Antiproliferative effect of investigated berry juices on GBM43 and U87-MG cells. (p values * < 0.05, ** < 0.01, *** < 0.001, **** < 0.0001); DE—dwarf elderberry; WB—wild blackberry; RB—raspberry; UT—control; TMZ—Temozolomide.

These results of the in vitro study suggest that wild blackberry juice may be the most potent among the tested berry juices in combating these types of cancer. In both cell lines, wild blackberry juice demonstrated the most pronounced impact on the reduction in cell survival.

4. Discussion

In this study, we selected three berry juices, namely dwarf elderberry, raspberry and wild blackberry for the treatment of glioblastoma tumor cells (U87-MG and GBM43). Selected juices differ in their phenolic composition and antioxidant capacity. Wild blackberry and dwarf elderberry juices exhibited a higher content of total polyphenols, proanthocyanins, and monomeric anthocyanins compared to raspberry juice. HPLC analyses revealed distinctive anthocyanins, flavonoids, and phenolic acids in each juice. Comparisons of concentrations of individual phenolic compounds revealed that anthocyanins were the most prevalent flavonoid molecules in the profiles of all the juices. Wild blackberry and dwarf elderberry juice contained significant amounts of quercetin derivates, while the dwarf elderberry juice additionally contained significant amounts of phenolic acids. Wild blackberry juice had the highest potency of inhibition against both types of glioblastoma cells probably due to the highest concentration of anthocyanins, which is in agreement with other studies [7,11,16,38–46]. All berry juices selected for this study had cyanidin-based anthocyanins, but the base structure differed for each berry species which may account for their different potencies in inhibition of glioblastoma cells.

Other studies also highlighted anthocyanins among polyphenols as the compounds responsible for the inhibition of the proliferation of different tumor cells. Compared to other flavonoids, anthocyanins were more effective for the inhibition of direct cell growth [38]. Research on the potential for inhibition of migration of glioblastoma cells by anthocyanidins demonstrated that delphinidin, petunidin, and cyanidin could potentially inhibit these cancer cells. This property was ascribed to their structure, i.e., the number of hydroxyl groups on the B-ring. Delphinidin (with three hydroxyl groups) demonstrated the best inhibitory effect (83%) while the other two anthocyanidins (with two hydroxyl groups) demonstrated an inhibitory effect of 48% [16]. However, comparing the effectiveness of delphinidin and cyanidin in antiproliferative and apoptotic effects in MCF7 human breast cancer cells revealed that cyanidin was more effective [7]. Even though Lamy et al. [16] pointed out that aglycones are more potent in inhibition of tumor cell growth, Jing et al. [39] observed that the glycosides of anthocyanins might be more efficient since they can handicap glucose transport and cause inhibition of energy metabolism, which, in turn, may lead to mitochondrial damage and apoptosis of tumor cells.

In a study of mulberry anthocyanins (cyanidin 3-rutinoside and cyanidin 3-glucoside) on human lung cancer cells, an inhibitory effect on the migration and invasion of highly metastatic A549 cells was revealed. It was concluded that treatment of lung cancer cells with those anthocyanins could decrease the expression of matrix matallopro-tinase-2 (MMP-2) and urokinase-plasminogen activator (u-PA) on the one hand, and on the other increase the expression of tissue inhibitor of matrix matalloprotinase-2 (TIMP-2) and plasminogen activator inhibitor (PAI) [40]. Research with Mexican wild blackberries (*R. liebmannii* and *R. palmeri*) indicated that the anthocyanin fraction of extracts of these fruits mostly contained cyanidin 3-rutinoside and cyanidin 3-glucoside and it was determined that anthocyanins elicit apoptosis in C6 cell line and RG2 cell line. Both of these samples caused the arrest of the C6 cell lines in the G0/G1 phase (around 76.5% for *R. liebmannii* and 75.5% for *R. palmeri*), significantly higher compared to the control group [11]. Both of these anthocyanins were identified in our samples of wild blackberry and raspberry juices, with wild blackberry having these anthocyanins as dominant ones. Different berry ex-tracts (blackberry, black raspberry, red raspberry, blueberry, cranberry and strawberry) have been tested for the inhibition of growth of tumor cell lines like breast (MCF-7), colon (HT-29 and HCT116), prostate (LNCaP) and oral (KB and CAL27) cell lines. It was concluded that with the increase in berry extract concentration, the resulting inhibition of cell proliferation increased in all of the tested cells, with different degrees of potency between cell lines depending on the anthocyanin profiles of used berry extracts. A comparison of the effect of blackberry and raspberry extracts revealed that the blackberry extracts had a higher potency in the inhibition of tested cell lines than the raspberry extracts; however, their apoptosis effect was the same [41]. Comparison of the effects of different types of extracts (hexane, EtOAc and MeOH extracts) of blackberry and raspberry species (Jamaica-grown species: *Rubus jamaicensis*, *Rubus rosifolius* and *Rubus racemosus*, and of the Michigan-grown *Rubus acuminatus*, *Rubus idaeus* cv. Heritage and *Rubus idaeus* cv. Golden) resulted in the conclusion that their tumor cell proliferation inhibition can be attributed to anthocyanins, and the majority of species contained high amounts of cyanidin-3-glucoside. The most potent sample tested was the hexane extract of *Rubus jamaicensis*. It had the greatest overall capacity to inhibit the progression of tumor cell growth, inhibiting colon, breast, lung, and gastric human tumor cells by 50%, 24%, 54%, and 37%, respectively, [42].

We determined a dose-dependent behavior in the inhibition of tumor cells with the increase in applied berry juice amounts. That trend was also observed in previous studies where anthocyanins or phenolics showed a dose-dependent growth inhibition against breast, colon, stomach, central nervous system, and lung tumor cells [7,16,38,39,43–45].

A major mechanism of suppression of cancer is apoptosis or programmed cell death [41,46]. Generally, the growth rate of preneoplastic or neoplastic cells is higher than the growth of normal cells; thus, the initiation of apoptosis or cell cycle arrest can be a valuable mechanism for inhibition of the promotion and progression of carcinogenesis and consequently for the removal of genetically damaged, preinitiated, or neoplastic cells from the body [41,46]. Among berry phenolics, anthocyanins have been shown to possess apoptotic effects in human cancer cells [41,47,48]. In addition to anthocyanins, quercetin was also recognized for its apoptotic effects [46]. It can cause activation of caspases which are underexpressed in tumor cells leading to the apoptotic response [49].

Dwarf elderberry and wild blackberry juices contained derivatives of quercetin which could lead to induction of apoptosis in combination with anthocyanins. It has been demonstrated that quercetin and fruit extracts (strawberry and plum), in contrast to chlorogenic acid and (-)-epicatechin, caused the induction of apoptosis in HepG2 cells. Actually, quercetin and fruit extracts restrained the G1 phase in the progression of the cell cycle prior to apoptosis so they can contribute to the reduced cell viability in investigated tumor cells [47]. The anti-proliferative effect of quercetin on two breast cancer cell lines (MCF-7 and MDA-MB-231; cells that differ in hormone receptor) was also studied and it was concluded that quercetin had a significant cytotoxicity in MCF-7 cells, but not in MDA-MB-231 cells. In MCF-7 cells, quercetin also had an effect on the restriction of the G1 phase and caused effective suppression of the expression of CyclinD1, p21, Twist and phospho p38MAPK, which was not observed in the MDA-MB-231 cells [50]. Treatment of U87 and T98G cell lines with quercetin showed a significant decrease in the IL-6 mediated STAT3 activation [44]. Additionally, it caused an increase in the sensitivity of U87 and U251 cell lines to TMZ through suppression of Hsp27 known to confer drug resistance [51]. It was observed that quercetin can cause induction of mitochondria-mediated apoptosis in the resistant p53 mutant U373MG cell line [52]

Only rutin was detected in raspberry juice in significantly lower concentrations in comparison to the other two juices. The raspberry juice had a higher impact on inhibition of GBM43 glioblastoma cells than the dwarf elderberry, probably due to the already mentioned cyanidin 3-rutinoside and cyanidin 3-glucoside. Also, raspberry juice and wild blackberry juice contained ellagic acid, which is known for its anticarcinogenic properties. Both in vitro and in vivo studies have revealed that anticarcinogenic effects of ellagic acid were due to the inhibition of tumor cell proliferation, induction of apoptosis, breaking DNA binding to carcinogens, blocking virus infection, and disturbing inflammation, angiogenesis, and drug-resistance processes required for tumor growth and metastasis [53].

Asl et al. [45] also demonstrated the potential cytotoxic activities of ethyl-acetate and methanol extracts of leaf and the fruit of dwarf elderberry upon treatment of breast and stomach cancer cells (MCF-7 and AGS cells) and emphasized the importance of phenolic profile.

Ingestion of berry flavonoids, including anthocyanins, causes their degradation across the digestive system; thus, in circulation, they largely appear in the form of metabolites [1,54]. In plasma, anthocyanins can be found in their intact form, or as the corresponding phenolic acids and aldehydes, and conjugates (methyl, sulfate and glucuronyl conjugates) [54,55]. However, it was determined that flavonoids (including anthocyanins) and their metabolites can be found in the brain tissue passing through the blood–brain barrier [55–57], which is one of the challenges in effective glioblastoma treatment. Therefore, our results could be beneficial from this point of view. Additionally, since it has been determined that flavonoids can pass through the blood–brain barrier, our results can be beneficial from the aspect of formulation of delivery systems. Generally, drug delivery to glioblastoma cells is quite a challenging field. Significant efforts have been committed to the development of efficient delivery systems with the aim of overcoming the molecular and cellular heterogeneity of the tumor cells, its infiltrative nature, and the blood–brain barrier [17–19]. Since anthocyanins can pass the blood–brain barrier, they have the potential to be incorporated into delivery systems alone or, with TMZ, to boost their effect, which could be further explored.

In conclusion, our study highlights the diverse phenolic composition and antioxidant capacity of three berry juices—wild blackberry, dwarf elderberry, and raspberry—in their potential treatment of glioblastoma tumor cells (U87-MG and GBM43). The distinct anthocyanins, flavonoids, and phenolic acids present in each juice contribute to their varying inhibitory effects on glioblastoma cells. Wild blackberry juice, with its high concentration of anthocyanins, exhibited the most potent inhibition, aligning with findings from other studies.

Anthocyanins, particularly cyanidin types, emerged as the key compounds responsible for the inhibition of glioblastoma cells. The dose-dependent behavior observed in our study is consistent with previous research demonstrating the growth-inhibitory effects of anthocyanins on various tumor cell lines. Apoptosis, a major mechanism of cancer suppression, is induced by anthocyanins and other phenolic compounds, including quercetin. The combination of quercetin derivatives and anthocyanins in dwarf elderberry and wild blackberry juices suggests a potential synergistic effect in inducing apoptosis and cell cycle arrest.

Future studies should explore deeper into the specific mechanisms through which these berry juices exert their anticancer effects, investigating the interplay of individual compounds and potential synergies. Additionally, investigating the impact of different berry varieties and their specific phenolic profiles on various cancer types can provide valuable insights for the development of targeted therapies. In the realm of cancer research, the exploration of natural compounds, such as those found in berry juices, continues to offer promising venues for developing effective and potentially less toxic treatments for glioblastoma and other cancers.

5. Conclusions

Our results suggest that all three berry juices tested (wild blackberry, dwarf elderberry, and raspberry) have anticancer effects on both GBM43 and U87-MG glioblastoma cells. This is significant because GBM43 cells are resistant to temozolomide, the current drug of choice against glioblastomas. The antiproliferative effects of wild blackberry juice, specifically, against GBM43 cells are particularly interesting and warrant further investigation into its mechanism. Dwarf elderberry and raspberry also appear to significantly inhibit U87 and GBM43 cell growth, which warrants further study into the possible impacts of these juices as well. Overall, our study provides strong evidence that wild blackberry, dwarf elderberry, and raspberry juices have antiproliferative effects on glioblastoma cells. Further research is needed to investigate the mechanisms of action of these juices and to evaluate their potential as therapeutic agents for glioblastoma.

Author Contributions: Conceptualization, M.K., D.R., M.A.L. and J.Š.; methodology, M.K. and D.R.; formal analysis, M.K. and D.R.; investigation, M.K., D.R., M.A.L. and J.Š.; data curation, M.K. and D.R.; writing—original draft preparation, M.K. and D.R.; writing—review and editing, M.A.L. and J.Š.; funding acquisition, M.K., D.R., M.A.L. and J.Š. All authors have read and agreed to the published version of the manuscript.

Funding: This research was funded by Croatian Science Foundation; grant number IP-2019-04-5749.

Data Availability Statement: Data are contained within the article.

Conflicts of Interest: The authors declare no conflicts of interest.

References

1. Haizhou, W.; Oliveira, G.; Lila, M.A. Protein-binding approaches for improving bioaccessibility and bioavailability of anthocyanins. *Compr. Rev. Food Sci. Food Saf.* **2023**, *22*, 333–354.
2. Shen, Y.; Zhang, N.; Tian, J.; Xin, G.; Liu, L.; Sun, X.; Li, B. Advanced approaches for improving bioavailability and controlled release of anthocyanins. *J. Control. Release* **2022**, *341*, 285–299. [CrossRef] [PubMed]
3. Fallah, A.A.; Sarmast, E.; Fatehi, P.; Jafari, T. Impact of dietary anthocyanins on systemic and vascular inflammation: Systematic review and meta-analysis on randomised clinical trials. *Food Chem. Toxicol.* **2020**, *135*, 110922. [CrossRef] [PubMed]
4. Kimble, R.; Keane, K.M.; Lodge, J.K.; Howatson, G. Dietary intake of anthocyanins and risk of cardiovascular disease: A systematic review and meta-analysis of prospective cohort studies. *Crit. Rev. Food Sci. Nutr.* **2019**, *59*, 3032–3043. [CrossRef] [PubMed]
5. Krga, I.; Milenkovic, D. Anthocyanins: From sources and bioavailability to cardiovascular-health benefits and molecular mechanisms of action. *J. Agric. Food Chem.* **2019**, *67*, 1771–1783. [CrossRef]
6. Lila, M.A.; Burton-Freeman, B.; Grace, M.; Kalt, W. Unraveling anthocyanin bioavailability for human health. *Ann. Rev. Food Sci. Technol.* **2016**, *7*, 375–393. [CrossRef]
7. Tang, J.; Oroudjev, E.; Wilson, L.; Ayoub, G. Delphinidin and cyanidin exhibit antiproliferative and apoptotic effects in MCF7 human breast cancer cells. *Integr. Cancer Sci. Therap.* **2015**, 2, 82–86.

8. Hou, D.X.; Kai, K.; Li, J.J.; Lin, S.; Terahara, N.; Wakamatsu, M.; Fujii, M.; Young, M.R.; Colburn, N. Anthocyanidins inhibit activator protein 1 activity and cell transformation: Structure–activity relationship and molecular mechanisms. *Carcinogenesis* **2004**, *25*, 29–36. [CrossRef]
9. Zafra-Stone, S.; Yasmin, T.; Bagchi, M.; Chatterjee, A.; Vinson, J.A.; Bagchi, D. Berry anthocyanins as novel antioxidants in human health and disease prevention. *Mol. Nutr. Food Res.* **2007**, *51*, 675–683. [CrossRef]
10. Raucher, D. Tumor targeting peptides: Novel therapeutic strategies in glioblastoma. *Curr. Opin. Pharmcol.* **2019**, *47*, 14–19 [CrossRef] [PubMed]
11. Sánchez-Velázquez, O.A.; Cortés-Rodríguez, M.; Milán-Carrillo, J.; Montes-Ávila, J.; Robles-Bañuelos, B.; Santamaría del Ángel, A.; Cuevas-Rodríguez, E.O.; Rangel-López, E. Anti-oxidant and anti-proliferative effect of anthocyanin enriched fractions from two Mexican wild blackberries (*Rubus* spp.) on HepG2 and glioma cell lines. *J. Berry Res.* **2020**, *10*, 513–529. [CrossRef]
12. Thakkar, J.P.; Dolecek, T.A.; Horbinski, C.; Ostrom, Q.T.; Lightner, D.D.; Barnholtz-Sloan, J.S.; Villano, J.L. Epidemiologic and molecular prognostic review of glioblastoma. *Cancer Epidemiol. Biomark. Prev.* **2014**, *23*, 1985–1996. [CrossRef] [PubMed]
13. Rooprai, H.K.; Christidou, M.; Murray, S.A.; Davies, D.; Selway, R.; Gullan, R.W.; Pilkington, G.J. Inhibition of Invasion by Polyphenols from Citrus Fruit and Berries in Human Malignant Glioma Cells In Vitro. *Anticancer Res.* **2021**, *41*, 619–633 [CrossRef] [PubMed]
14. Ramírez-Expósito, M.J.; Martínez-Martos, J.M. The Delicate Equilibrium between Oxidants and Antioxidants in Brain Glioma *Curr. Neuropharmacol.* **2019**, *17*, 342–351. [CrossRef] [PubMed]
15. Stupp, R.; Mason, W.P.; van den Bent, M.J.; Weller, M.; Fisher, B.; Taphoorn, M.J.; Belanger, K.; Brandes, A.A.; Marosi, C.; Bogdahn, U.; et al. Radiotherapy plus concomitant and adjuvant temozolomide for glioblastoma. *N. Engl. J. Med.* **2005**, *352*, 987–996 [CrossRef] [PubMed]
16. Lamy, S.; Lafleur, S.; Bedard, V.; Moghrabi, A.; Barrete, S.; Gingras, D.; Béliveau, R. Anthocyanidins inhibit migration of glioblastoma cells: Structure-activity relationship and involvement of the plasmolytic system. *J. Cell. Biochem.* **2007**, *100*, 100–111 [CrossRef]
17. Dragojevic, S.; Mackey, R.; Raucher, D. Evaluation of Elastin-Like Polypeptides for Tumor Targeted Delivery of Doxorubicin to Glioblastoma. *Molecules* **2019**, *24*, 3242. [CrossRef]
18. Mathew, E.N.; Berry, B.C.; Yang, H.W.; Carroll, R.S.; Johnson, M.D. Delivering Therapeutics to Glioblastoma: Overcoming Biological Constraints. *Int. J. Mol. Sci.* **2022**, *23*, 1711. [CrossRef]
19. Raucher, D.; Dragojevic, S.; Ryu, J. Macromolecular Drug Carriers for Targeted Glioblastoma Therapy: Preclinical Studies, Challenges, and Future Perspectives. *Front. Oncol.* **2018**, *8*, 624. [CrossRef]
20. Vengoji, R.; Macha, M.A.; Batra, S.K.; Shonka, N.A. Natural products: A hope for glioblastoma patients. *Oncotarget* **2018**, *9*, 22194–22219. [CrossRef] [PubMed]
21. Allen, M.; Bjerke, M.; Edlund, H.; Nelander, S.; Westermark, B. Origin of the U87MG glioma cell line: Good news and bad news. *Sci. Transl. Med.* **2016**, *8*, 354re3. [CrossRef]
22. Pevna, V.; Wagnières, G.; Huntosova, V. Autophagy and Apoptosis Induced in U87 MG Glioblastoma Cells by Hypericin-Mediated Photodynamic Therapy Can Be Photobiomodulated with 808 nm Light. *Biomedicines* **2021**, *9*, 1703. [CrossRef] [PubMed]
23. Mousavi, M.; Koosha, F.; Neshastehriz, A. Chemo-radiation therapy of U87-MG glioblastoma cells using SPIO@AuNP-Cisplatin-Alginate nanocomplex. *Heliyon* **2023**, *9*, e13847. [CrossRef]
24. Di Cintio, F.; Dal Bo, M.; Baboci, L.; De Mattia, E.; Polano, M.; Toffoli, G. The Molecular and Microenvironmental Landscape of Glioblastomas: Implications for the Novel Treatment Choices. *Front. Neurosci.* **2020**, *14*, 603647. [CrossRef]
25. Nguyen, H.-M.; Guz-Montgomery, K.; Lowe, D.B.; Saha, D. Pathogenetic Features and Current Management of Glioblastoma. *Cancers* **2021**, *13*, 856. [CrossRef] [PubMed]
26. Tsai, C.-Y.; Ko, H.-J.; Huang, C.-Y.F.; Lin, C.-Y.; Chiou, S.-J.; Su, Y.-F.; Lieu, A.-S.; Loh, J.-K.; Kwan, A.-L.; Chuang, T.-H.; et al. Ionizing Radiation Induces Resistant Glioblastoma Stem-Like Cells by Promoting Autophagy via the Wnt/β-Catenin Pathway. *Life* **2021**, *11*, 451. [CrossRef] [PubMed]
27. Singleton, V.L.; Rossi, J.A. Colorimetry of total phenolics with phosphomolybdic-phosphotonutric acid reagents. *Am. J. Enol. Vitic.* **1965**, *16*, 144–158. [CrossRef]
28. Giusti, M.M.; Wrolstad, R.E. Characterization and Measurement of Anthocyanins by UV-Visible Spectroscopy. In *Current Protocols in Food Analytical Chemistry Current Protocols*; John Wiley & Sons Inc.: Hoboken, NJ, USA, 2001.
29. Prior, R.L.; Fan, E.; Ji, H.; Howell, A.; Nio, C.; Payne, M.J.; Reed, J. Multi-laboratory validation of a standard method for quantifying proanthocyanidins in cranberry powders. *J. Sci. Food Agric.* **2010**, *90*, 1473–1478. [CrossRef]
30. Arnao, M.B.; Cano, A.; Acosta, M. The hydrophilic and lipophilic contribution to total antioxidant activity. *Food Chem.* **2001**, *73*, 239–244. [CrossRef]
31. Brand-Williams, W.; Cuvelier, M.E.; Berset, C. Use of a free radical method to evaluate antioxidant activity. *LWT* **1995**, *28*, 25–30. [CrossRef]
32. Benzie, I.F.F.; Strain, J.J. The ferric reducing ability of plasma (FRAP) as a measure of "Antioxidant Power": The FRAP assay. *Anal. Biochem.* **1996**, *239*, 70–76. [CrossRef] [PubMed]
33. Apak, R.; Güçlü, K.; Ozyürek, M.; Karademir, S.E. Novel total antioxidant capacity index for dietary polyphenols and vitamins C and E, using their cupric ion reducing capability in the presence of neocuproine: CUPRAC method. *J. Sci. Food Agric.* **2004**, *52*, 7970–7981. [CrossRef]

34. Pambianchi, E.; Hagenberg, Z.; Pecorelli, A.; Grace, M.; Therrien, J.-P.; Lila, M.A.; Valacchi, G. Alaskan bog blueberry (*Vaccinium uliginosum*) extract as an innovative topical approach to prevent uv-induced skin damage. *Cosmetics* **2021**, *8*, 112. [CrossRef]
35. Može, Š.; Polak, T.; Gašperlin, L.; Koron, D.; Vanzo, A.; Poklar Ulrih, N.; Abram, V. Phenolics in Slovenian Bilberries (*Vaccinium myrtillus* L.) and Blueberries (*Vaccinium corymbosum* L.). *J. Agric. Food Chem.* **2011**, *59*, 6998–7004. [CrossRef] [PubMed]
36. Buljeta, I.; Pichler, A.; Šimunović, J.; Kopjar, M. Polyphenols and antioxidant activity of citrus fiber/blackberry juice complexes. *Molecules* **2021**, *26*, 4400. [CrossRef] [PubMed]
37. Mosmann, T. Rapid colorimetric assay for cellular proliferation and viability: Application to the proliferation and viability of murine macrophages. *J. Immunol. Methods* **1983**, *65*, 55–63. [CrossRef]
38. Kamei, H.; Kojima, T.; Hasegawa, M.; Koide, T.; Umeda, T.; Yukawa, T.; Terabe, K. Suppression of tumor cell growth by anthocyanins in vitro. *Cancer Investig.* **1995**, *13*, 590–594. [CrossRef]
39. Jing, N.; Song, J.; Liu, Z.; Wang, L.; Jiang, G. Glycosylation of anthocyanins enhances the apoptosis of colon cancer cells by handicapping energy metabolism. *BMC Complement. Med. Ther.* **2020**, *20*, 312. [CrossRef]
40. Chen, P.-N.; Chu, S.-C.; Chiou, H.-L.; Kuo, W.-H.; Chiang, C.-L.; Hsieh, Y.-S. Mulberry anthocyanins, cyanidin 3-rutinoside and cyanidin 3-glucoside, exhibited an inhibitory effect on the migration and invasion of a human lung cancer cell line. *Cancer Lett.* **2006**, *235*, 248–259. [CrossRef]
41. Seeram, N.P.; Adams, L.S.; Zhang, Y.; Lee, R.; Sand, D.; Scheuller, H.S.; Heber, D. Blackberry, black raspberry, blueberry, cranberry, red raspberry, and strawberry extracts inhibit growth and stimulate apoptosis of human cancer cells in vitro. *J. Agric. Food Chem.* **2006**, *54*, 9329–9339. [CrossRef]
42. Bowen-Forbes, C.S.; Zhang, Y.; Nair, M.G. Anthocyanin content, antioxidant, anti-inflammatory and anticancer properties of blackberry and raspberry fruits. *J. Food Compos. Anal.* **2010**, *23*, 554–560. [CrossRef]
43. Reddy, M.K.; Alexander-Lindo, R.L.; Nair, M.G. Relative inhibition of lipid peroxidation, cyclooxygenase enzymes, and human tumor cell proliferation by natural food colors. *J. Agric. Food Chem.* **2005**, *53*, 9268–9273. [CrossRef] [PubMed]
44. Michaud-Levesque, J.; Bousquet-Gagnon, N.; Beliveau, R. Quercetin abrogates IL-6/STAT3 signaling and inhibits glioblastoma cell line growth and migration. *Exp. Cell Res.* **2012**, *318*, 925–935. [CrossRef] [PubMed]
45. Asl, F.R.; Barzegar, A.; Ebrahimzadeh, M.A. Evaluating Cytotoxic Potential of the Fruit and the Leaf Extracts of *Sambucus ebulus* (L.) on MCF7 and AGS Cell Lines. *Res. Mol. Med.* **2021**, *9*, 11–20.
46. Pfeffer, C.M.; Singh, A.T.K. Apoptosis: A Target for Anticancer Therapy. *Int. J. Mol. Sci.* **2018**, *19*, 448. [CrossRef]
47. Ramos, S.; Alia, M.; Bravo, L.; Goya, L. Comparative effects of food-derived polyphenols on the viability and apoptosis of a human hepatoma cell line (HepG2). *J. Agric. Food Chem.* **2005**, *53*, 1271–1280. [CrossRef]
48. Heo, H.J.; Lee, C.Y. Strawberry and its anthocyanins reduce oxidative stress-induced apoptosis in PC12 cells. *J. Agric. Food Chem.* **2005**, *53*, 1984–1989. [CrossRef]
49. Schnekenburger, M.; Dicato, M.; Diederich, M. Plant-derived epigenetic modulators for cancer treatment and prevention. *Biotechnol. Adv.* **2014**, *32*, 1123–1132. [CrossRef]
50. Ranganathan, S.; Halagowder, D.; Sivasithambaram, N.D. Quercetin suppresses twist to induce apoptosis in MCF-7 breast cancer cells. *PLoS ONE* **2015**, *10*, e0141370. [CrossRef]
51. Sang, D.P.; Li, R.J.; Lan, Q. Quercetin sensitizes human glioblastoma cells to temozolomide *in vitro* via inhibition of Hsp27. *Acta Pharmacol. Sin.* **2014**, *35*, 832–938. [CrossRef]
52. Kim, H.; Moon, J.Y.; Ahn, K.S.; Cho, S.K. Quercetin induces mitochondrial mediated apoptosis and protective autophagy in human glioblastoma U373MG cells. *Oxid. Med. Cell. Longev.* **2013**, *2013*, 596496. [CrossRef] [PubMed]
53. Zhang, H.-M.; Zhao, L.; Li, H.; Xu, H.; Chen, W.-W.; Tao, L. Research progress on the anticarcinogenic actions and mechanisms of ellagic acid. *Cancer Biol. Med.* **2014**, *11*, 92–100. [PubMed]
54. Hribar, U.; Poklar Ulrih, N. The Metabolism of Anthocyanins. *Curr. Drug Metab.* **2014**, *15*, 3–13. [CrossRef] [PubMed]
55. Janle, E.M.; Lila, M.A.; Grannan, M.; Wood, L.; Higgins, A.; Yousef, G.G.; Rogers, R.B.; Kim, H.; Jackson, G.S.; Weaver, C. Method for evaluating the potential of 14C labeled plant polyphenols to cross the blood-brain barrier using accelerator mass spectrometry. *Nucl. Instrum. Methods Phys. Res.* **2010**, *268*, 1313–1316. [CrossRef] [PubMed]
56. Shimazu, R.; Anada, M.; Miyaguchi, A.; Nomi, Y.; Matsumoto, H. Evaluation of Blood-Brain Barrier Permeability of Polyphenols, Anthocyanins, and Their Metabolites. *J. Agric. Food Chem.* **2021**, *69*, 11676–11686. [CrossRef] [PubMed]
57. Youdim, K.A.; Dobbie, M.S.; Kuhnle, G.; Proteggente, A.R.; Abbott, N.J.; Rice-Evans, C. Interaction between flavonoids and the blood–brain barrier: *In vitro* studies. *J. Neurochem.* **2003**, *85*, 180–192. [CrossRef]

processes

MDPI

Article

Formulation and Characterization of Solid Lipid Nanoparticles Loaded with Troxerutin

Yahya F. Jamous [1,*], Najla A. Altwaijry [2,*], Mohamed T. S. Saleem [3], Aljoharah F. Alrayes [2], Sara M. Albishi [2] and Mashael A. Almeshari [2]

1 Vaccine and Bioprocessing Center, King Abdulaziz City for Science and Technology (KACST), Riyadh 12354, Saudi Arabia
2 Department of Pharmaceutical Sciences, College of Pharmacy, Princess Nourah Bint Abdulrahman University, Riyadh 11564, Saudi Arabia; 438001400@pnu.edu.sa (A.F.A.); 438001699@pnu.edu.sa (S.M.A.); 438001245@pnu.edu.sa (M.A.A.)
3 College of Pharmacy, Riyadh ELM University, Riyadh 13244, Saudi Arabia; tsmohamed.saleem@riyadh.edu.sa
* Correspondence: yjamous@kacst.edu.sa (Y.F.J.); naaltwaijry@pnu.edu.sa (N.A.A.)

Abstract: Troxerutin (TXR), a naturally derived compound with diverse therapeutic potential, faces limitations in clinical efficacy due to poor bioavailability and rapid plasma clearance. This study focuses on troxerutin-loaded solid lipid nanoparticles (TXR-SLNs) and their physicochemical properties, intending to enhance drug release. TXR-SLNs were prepared via high-shear homogenization followed by ultrasonication, yielding optimized nanoparticles with an average size of 140.5 ± 1.02 nm, a uniform distribution (polydispersity index: 0.218 ± 0.01), and a stable emulsion (zeta potential: 28 ± 8.71 mV). The formulation exhibited 83.62% entrapment efficiency, indicating improved drug-loading capacity and extended drug release. Spectroscopic and thermodynamic analyses confirmed component compatibility. Despite a decline in entrapment efficiency induced by temperature after one month of storage at 23 °C, the formulation may retain acceptable stability. This study provides insight into SLNs as effective carriers for enhancing troxerutin's release profile, motivating further in vivo investigations to optimize therapeutic interventions.

Keywords: troxerutin; flavonoids; solid lipid nanoparticles; high-shear homogenization; drug release

1. Introduction

Flavonoids represent a class of polyphenolic compounds found abundantly in an extensive array of plants, fruits, vegetables, and leaves [1]. In recent times, these naturally occurring phytochemicals have garnered significant attention within modern medicine, where they have been harnessed to engender innovative therapeutic agents. This choice stems from their diverse and expansive biological activities, coupled with their remarkable health benefits, all achieved with a markedly low proclivity for eliciting side effects and toxicity [2]. Their repertoire encompasses compelling attributes, including antioxidative, anti-inflammatory, antimutagenic, and anticarcinogenic properties [3]. Moreover, flavonoids are known to modulate key cellular enzyme functions, such as cyclo-oxygenase (COX), lipoxygenase, and xanthine oxidase (XO) [1].

Empirical evidence has progressively accumulated, underscoring the pharmacological prowess of flavonoids in the context of diabetes. Their impact on promoting glycemic control and ameliorating lipid levels has garnered attention. This dual effect translates into a protective shield against a spectrum of complications, spanning nephropathy, neuropathy, hepatic impairments, and cardiovascular maladies [4,5]. El-Shiekh et al. [6] notably documented an exploration of the antibacterial and antiviral potential embedded within a plant-derived flavonoid.

Furthermore, an assortment of investigations have converged to emphasize the profound antioxidant properties exhibited by flavonoids. This quality plays a pivotal role in mitigating oxidative stress, a factor intricately linked to the genesis of neurodegenerative diseases (NDDs) and neoplastic conditions [7–9].

Troxerutin (TXR), also known as vitamin P4, belongs to the trihydroxyethyl group and is derived from the natural bioflavonoid rutin [10–13]. TXR stands as a promising natural compound renowned for its multifaceted pharmacological effects across diverse conditions [14]. Notably, numerous investigations have substantiated TXR's capacity to confer protection to vital organs such as the kidneys, liver, and brain, primarily through its robust ability to scavenge reactive oxidative species (ROS). This property is underpinned by its dual role as an antioxidant and anti-inflammatory agent [14–18].

Furthermore, TXR emerges as a key player in the management of vascular disorders, displaying a beneficial impact on blood fibrinolysis and erythrocyte aggregation [19,20]. These attributes position it as a compelling therapeutic option for addressing chronic venous insufficiency (CVI), a condition characterized by abnormal increases in erythrocyte aggregation and diminished pro-fibrinolytic activity [20].

In general, the application of TXR is constrained by its limited reabsorption following biliary excretion and suboptimal cell membrane penetration, consequently leading to swift plasma clearance and diminished bioavailability [13,21]. Additionally, the chemical configuration of TXR (Figure 1) contributes to its reduced bioavailability due to the presence of polyphenolic groups, which are susceptible to substantial in vivo degradation [1,22]. Nevertheless, the employment of nanoparticles presents a promising avenue for augmenting TXR's bioavailability, thereby enhancing its potential for achieving an improved therapeutic effect.

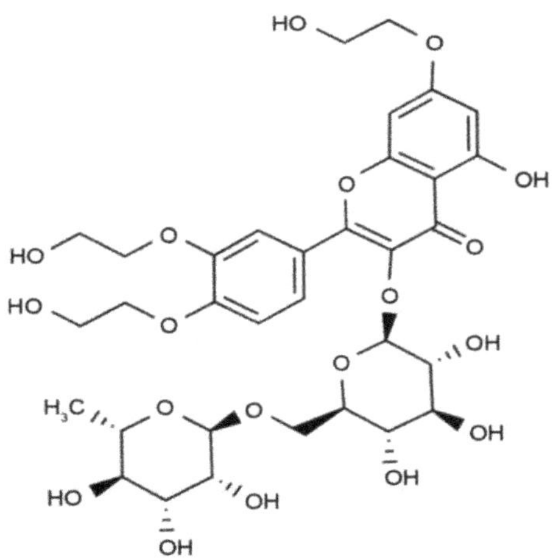

Figure 1. TXR's chemical structure.

In recent years, the realm of nanotechnology has experienced rapid expansion, emerging as a potent tool with extensive applications in medication delivery, diagnostics, cosmetics, and both biological and non-biological sciences [23]. However, the conventional delivery methods for therapeutic agents, including plant-derived flavonoids, are impeded by issues such as non-selectivity, low efficiency, and suboptimal bioavailability [24]. To surmount these limitations, the utilization of nanocarriers such as polymeric nanoparticles (NPs), liposomal NPs, and solid lipid NPs offers a promising avenue for optimizing drug delivery [24]. By enhancing the concentration of therapeutics at specific targets, these nanocarriers hold the potential to elevate efficacy, enhance biological system tolerability, and enable the utilization of lower dosages [24]. Furthermore, they offer the capacity to bolster the bioavailability of water-insoluble compounds [24]. The efficacy of nanocarriers hinges on a multitude of factors, encompassing their physicochemical attributes, drug loading efficiency, drug release kinetics, and minimal toxicity [23].

Solid lipid nanoparticles (SLNs) represent a promising drug delivery system composed of active drug molecules, solid lipids, surfactants, and/or co-surfactants [10]. These nanoparticles can be safely formulated for diverse routes of administration, including oral, injectable, and topical applications [25]. The scope of SLN applications is extensive, offering the potential to enhance the treatment of various diseases through judicious physicochemical modifications [26]. Their versatility extends to loading both lipophilic and hydrophilic drugs, thereby improving the drug characteristics, extending action duration, and prolonging drug release profiles [27]. Consequently, this enables a reduction in administration frequency while concurrently enhancing therapeutic efficacy [27]. Beyond their ability to enhance drug oral bioavailability and sustain drug release, the heightened precision of drug targeting is a paramount advantage of SLNs when compared to conventional delivery methods [28].

Targeted drug delivery serves as a ubiquitous platform for all solid lipid nanoparticles, facilitating precise drug delivery to specific tissues or cells across various pathological contexts [29]. Through the integration of drugs into nanocarriers, solid lipid nanoparticles offer a novel paradigm for targeted drug delivery strategies [29]. As such, solid lipid nanoparticles emerge as a propitious avenue for the development of controlled and targeted drug delivery systems [29].

Recent studies have reported the potential pharmacological benefits of troxerutin-loaded nanoparticles by using various polymers like chitosan and selenium [30,31]. In the present study, we aimed to establish the application of TXR loading within nanoparticles by adopting simple methods. This study aimed to meticulously characterize the physicochemical properties and assess the drug release kinetics of troxerutin-loaded solid lipid nanoparticles (TXR-SLNs) through investigations. The loading process was facilitated using a combination of high-shear homogenization followed by ultrasonication. The physicochemical attributes, encompassing shape, size, polydispersity index, and zeta potential, were scrutinized using a zeta sizer instrument (Malvern, UK). Moreover, a comprehensive assessment was conducted through techniques including differential scanning calorimetry (DSC), transmission electron microscopy (TEM), and Fourier transform infrared (FTIR) spectroscopy to ascertain any alterations in crystal structure and potential chemical interactions between constituents. Concurrently, the study evaluated parameters such as drug entrapment efficiency (EE) and drug release profiles to comprehensively gauge the performance of the TXR-SLN formulation.

2. Methodology

2.1. Chemicals and Materials

Troxerutin (purity: 98%), soy lecithin, Tween-80, and a dialysis bag were procured from Solarbio, Beijing, China. Glyceryl behenate was acquired from Kejian Chem, Beijing, China. Absolute ethanol was obtained from Fisher Scientific UK Limited, Leicestershire, UK. Sodium phosphate dibasic was sourced from Sigma Aldrich, Saint Louis, MO, USA, while sodium phosphate monobasic anhydrous was obtained from ICN Biomedicals, Inc., Santa Ana, CA, USA.

2.2. Preparation of the TXR-SLNs

The methodology employed for the preparation of the TXR-SLNs was adapted from Tan et al. [32] with slight modifications. The synthesis of the TXR-SLNs involved a two-step process comprising high-shear homogenization followed by ultrasonication. Initially, Tween-80 was dissolved in 25 mL of distilled water and heated to 73 degrees Celsius (°C) using a water bath, resulting in the formation of an aqueous phase. Meanwhile, the lipid phase was created by dissolving TXR, soy lecithin, and glyceryl behenate in 5 mL of ethanol, which was then heated to 73 °C.

Subsequently, the aqueous phase was carefully introduced into the lipid phase. The resulting mixture underwent homogenization at 7000 rpm for 10 min utilizing a high-speed Ultra Turrax D-500 homogenizer (IKA T25, Germany) to establish an emulsion. This emulsion was further subjected to ultrasonication for 15 min using an ultrasonic cell crusher (Ningbo Biological Technology Co., Ltd., Ningbo, China), leading to the formation of a TXR-SLN emulsion. The emulsion was then stored at 4 °C.

For the formulation of TXR-SLNs, three distinct compositions involving varying quantities of Tween-80, soy lecithin, and glyceryl behenate were prepared as delineated in Table 1.

Table 1. The ingredient quantities of each prepared TXR-SLN formulation.

Formulation	Active Ingredient	Surfactant	Lipid Polymer	
	TXR	**Tween-80**	**Glyceryl Behenate**	**Soy Lecithin**
TXR-SLN 1	10 mg	500 µL	100 mg	50 mg
TXR-SLN 2	10 mg	250 µL	50 mg	50 mg
TXR-SLN 3	10 mg	500 µL	50 mg	100 mg

2.3. Characterization

2.3.1. Estimation of Particle Size, Polydispersity, and Zeta Potential

A dynamic light scattering technique using a zeta sizer (Malvern, UK) was employed to assess the average particle size, polydispersity index (PDI), and zeta potential [32] in the context of this research study. For DLS particle sizing, the sample needs to be water clear to very slightly hazy. If the solution is white or too hazy, it should be diluted further before attempting a DLS size measurement. Thus, before the measurement of particle size, PDI analysis, and zeta potential, the samples were appropriately diluted with distilled water at a ratio of 1:10. If the sample is too concentrated, the measured size of the particles will be inaccurate due to multiple scattering or viscosity effects. The subsequent examination was conducted at a fixed angle of 90° for particle size and 25 °C for zeta potential.

The determination of the average particle size, PDI, and zeta potential was carried out in triplicate to ensure accuracy and consistency.

2.3.2. Imaging by Transmission Electron Microscopy (TEM)

The surface morphology analysis of all formulations was conducted using transmission electron microscopy (TEM). To prepare the samples, a 15 min ultrasonication process was performed using bath sonication at room temperature. Subsequently, a single drop of the prepared formulations was carefully pipetted onto a Formvar/Carbon 400 mesh copper grid (Ted Pella, CA). The samples were left to air dry for 24 h at room temperature.

Visual observation and imaging were accomplished using a TEM instrument (Hitachi HD-2300A, Chicago, IL, USA), operating at an acceleration voltage of 200 kV.

2.3.3. Determination of the Entrapment Efficiency (EE)

The determination of entrapment efficiency involved quantifying the concentration of an unentrapped substance within the aqueous solution after centrifugation. The prepared TXR-SLNs were subjected to centrifugation in a high-speed cooling centrifuge (Eppendorf, Hamburg, Germany) operating at 5000 rpm for 10 min, maintained at a temperature of 4 °C. Following centrifugation, the resultant sediment and supernatant liquid were carefully separated.

To assess the quantity of unentrapped drug present in the supernatant, a UV-spectrophotometer (PerkinElmer, Waltham, MA, USA) was employed, utilizing a wavelength of 360 nm. This measurement facilitated the estimation of the percentage of entrapment efficiency.

The calculation of the percentage entrapment efficiency (%EE) was executed using the subsequent formula:

$$\%EE = \text{Total drug content} - \text{free drug}/\text{Total drug content} \times 100$$

2.3.4. Differential Scanning Calorimetry (DSC) Analysis

Differential scanning calorimetry (DSC) analysis was conducted to assess the physical state of the TXR-SLNs (Seiko Instruments Inc, Chiba, Japan). The procedure involved placing samples of TXR, blank SLNs, and TXR-SLNs into individual aluminum pans. An empty pan was utilized as a reference or control. The analysis was performed with a continuous flow of nitrogen gas. The temperature was incrementally increased from 20 to 300 °C, following a controlled rate of 10 °C per minute [32].

2.3.5. Fourier Transform Infrared (FTIR) Analysis

The Fourier transform infrared (FTIR-8400S, Shimadzu, Nakagyo-ku, Kyoto, Japan) technique was employed to elucidate interactions among the drug, surfactant, and lipid components. A total of five samples were meticulously prepared, comprising three solids and two liquids. The solid samples encompassed TXR, soy lecithin, and glyceryl behenate, while the liquid samples included TXR-SLNs and SLNs without TXR. The FTIR spectra were acquired using the conventional KBr disc/pellet methodology.

For the solid samples, a procedure involving grinding with anhydrous KBr powder and subsequent compression into pellets was carried out. In contrast, the liquid samples were directly applied onto the KBr window. The FTIR spectrum became accessible after a comprehensive scan and underwent subsequent analysis through software (LabSolutions IR, Shimadzu, Nakagyo-ku, Kyoto, Japan).

The FTIR spectra were meticulously recorded with a resolution of 4 cm^{-1}, spanning a range of 4000–400 cm^{-1}, and involved 50 scans.

2.3.6. Quantification of the Drug Release from the TXR-SLNs

To investigate the drug release of the TXR-SLNs, the dialysis bag method was employed, as described in [32]. A dialysis bag (catalog no. YA1070, flat width: 10 mm, molecular weight cut-off: 8000–14,000 Da, length: 5 m, Solarbio) was employed to facilitate controlled drug release. The dialysis bag was loaded with the three different TXR-SLN formulations, enabling the gradual release of the free drug into the surrounding media. The bag was immersed in distilled water for approximately 12 h to facilitate initial hydration.

Subsequently, 1 mL of each TXR-SLN formulation was separately introduced into three dialysis bags, securely fastened with clamps. Each bag was then placed within a conical flask containing 100 mL of phosphate-buffered saline (PBS; pH 7.4). These three conical flasks were positioned within a thermostatic shaker, maintaining a consistent environment for 48 h. At predetermined intervals, 1 mL of the buffer solution was sampled, with an equivalent volume of fresh PBS replenished into the conical flasks. The drug release profiles were assessed using a UV-Vis spectrophotometer at a wavelength of 360 nm, with distilled water serving as the blank sample. A similar procedure was executed for the control group, which consisted of TXR alone, to facilitate comparative analysis of the drug release.

The cumulative release of the TXR-SLNs was calculated using the following equation:

$$Qn = Cn \times V0 + (C1 + C2 + C3 + \cdots + Cn - 1) \times V$$

where:

V0 represents the volume of the release medium;
Qn denotes the cumulative release at the n-th sampling point;
C signifies the solution concentration;
Cn indicates the concentration at the n-th sampling point;
V stands for the volume of each sample.

The drug concentration in the formulations was determined using a calibration curve of TXR. A stock solution of TXR was prepared by dissolving it in 100 mL of double-distilled water at a concentration of 1 mg/mL. Subsequent working standards were prepared through dilution with a phosphate buffer of pH 7.4, resulting in concentrations of 0.5, 1, 5, 10, and 15 mg/mL. UV-Vis spectrophotometry was employed at a wavelength of 360 nm to quantify the drug concentration using the calibration curve. The entire procedure was conducted in triplicate, and the average absorbance was calculated. The calibration equation and R^2 value were subsequently determined and reported.

Drug release kinetics

The dissolution data were fitted to various equations, including zero-order (cumulative quantity of drug released vs. time), first-order (log cumulative proportion of drug remaining vs. time), Higuchi (cumulative proportion of drug released vs. square root of time), Hixson–Crowell (the cubic root of percent drug release vs. time), and Korsmeyer–Peppas (log cumulative proportion of drug released vs. log time), to elucidate the rate and mechanism of TXR release from the formulated samples. The in-house developed PSP-DISSO V2 software (Developmed by BVDU's Poona College of Pharmacy, Pune, India) was employed to analyze the release kinetic parameters for the formulations in a phosphate buffer (pH 7.4). The correlation coefficients were calculated to assess data fit. The diffusion exponent (n) for different nanoparticles was determined, where $n \leq 0.45$ indicates Fickian (Case I) release, >0.45 but <0.89 suggests non-Fickian (anomalous) release, and >0.89 signifies super Case II release. In this context, Case II pertains to polymeric chain erosion, while anomalous transport (non-Fickian) refers to a combination of diffusion and erosion-controlled drug release [33].

2.3.7. Stability Studies

After measuring the initial particle size analysis of the three TXR-SLN formulations using the zeta sizer, the formulation samples were divided into two sets. One set was stored at 4 °C (in a refrigerator), and the second set was stored at room temperature (23 °C). All samples were kept in foil-sealed glass vials. Subsequently, the samples were retrieved at intervals of 5, 10, 20, and 30 days to assess particle size, polydispersity index, zeta potential, and entrapment efficiency.

3. Results

3.1. Drug Entrapment Efficiency (EE)

EE is a pivotal parameter for evaluating the drug-loading capacity of nanoparticles. The absorbance of each formulation was measured at 360 nm using a UV-Vis spectrophotometer and subsequently calculated through a calibration curve. As illustrated in Table 2, TXR-SLN 2 exhibited the highest EE, reaching 83.62%, in comparison to TXR-SLN 1 and TXR-SLN 3, with EEs of 76% and 77%, respectively.

Table 2. Entrapment efficiency (EE) of the TXR-SLN formulations.

S. No	Formulation Code	Entrapment Efficiency (%)
1	TXR-SLN 1	76.02
2	TXR-SLN 2	83.62
3	TXR-SLN 3	77.06

The entrapment was evaluated using a calibration curve depicted in Figure 2. The absorbance of each formulation was measured at 360 nm in phosphate-buffered saline (pH 6.8) using a UV-Vis spectrophotometer. A concentration–absorbance plot was generated, enabling the determination of the correlation coefficient and regression equation for TXR. Notably, the drug exhibited linearity across a concentration span of 0.5–15 mg/mL, with an R^2 value of 0.9992 for this concentration range.

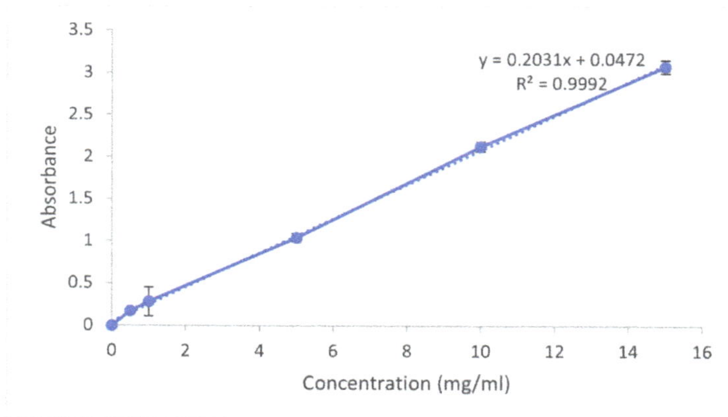

Figure 2. Calibration curve of TXR using a UV-Vis spectrophotometer (pH 6.8).

3.2. *Measurement of Particle Size, Polydispersity Index (PDI), and Zeta Potential*

The sizes of the three particles, PDI, and zeta potential were determined for the prepared samples. Among these, TXR-SLN 2 exhibited the smallest particle size, measuring at 140.5 ± 1.02 nm, and a PDI of 0.218 ± 0.01 (Table 3). The nanoparticle size and distribution of TXR-SLN 2 were notably favorable, as demonstrated in Figure 3. PDI values spanning from 0 to 0.5 indicate the presence of monodispersed and uniform materials. Furthermore, the zeta potential serves as an indicator of nanoparticle formulation stability by assessing surface charge and aggregation propensity. A zeta potential of ±20 mV signifies a stabilized formulation. Remarkably, TXR-SLN 2 exhibited a zeta potential of 28.6 ± 8.71 mV, effectively contributing to the stability of the system (Table 3).

Table 3. Particle size, polydispersity index, and zeta potential of the formulations.

S. No	Formulation	Particle Size (nm)	Polydispersity Index (PDI)	Zeta Potential (mV)
1	TXR-SLN 1	349.9 ± 7.67	0.387 ± 0.06	14.9 ± 4.87
2	TXR-SLN 2	140.5 ± 1.02	0.218 ± 0.01	28.6 ± 8.71
3	TXR-SLN 3	152.5 ± 2.35	0.350 ± 0.02	12.6 ± 4.52

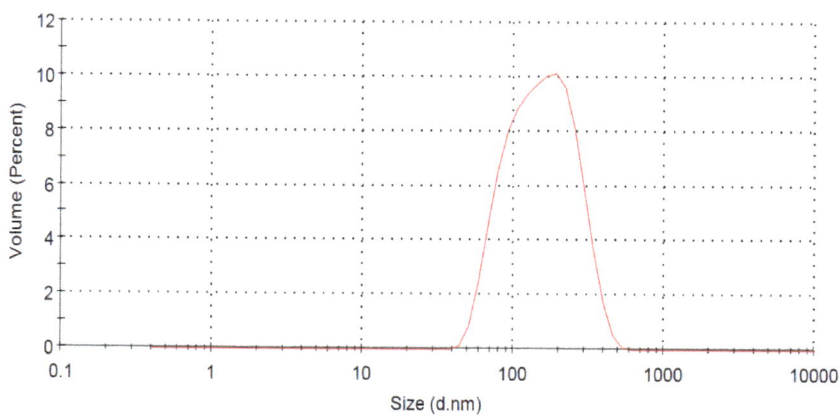

Figure 3. Particle size distribution of the TXR-SLN 2 formulation.

3.3. Transmission Electron Microscopy (TEM) Analysis

The shape and surface morphology of the optimal formulation were examined using TEM. The TEM micrograph depicts particles exhibiting a spherical shape, along with a smooth surface texture (Figure 4).

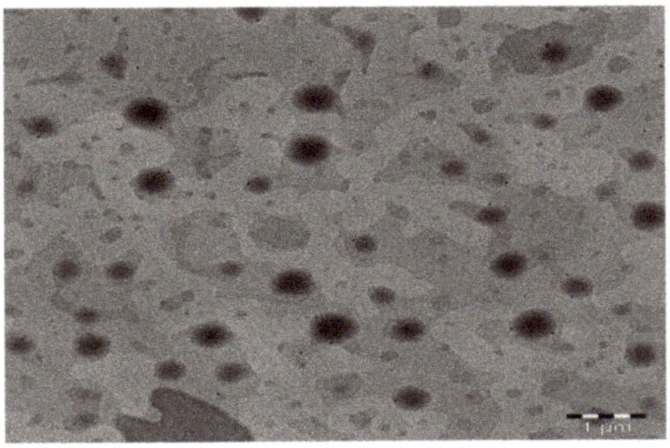

Figure 4. TEM micrograph of the TXR-SLN 2 formulation.

3.4. Fourier Transform Infrared Spectroscopy (FTIR)

A drug-excipient compatibility study was conducted using FTIR spectra, involving TXR and a TXR-excipient physical mixture. The outcomes are illustrated in Figure 5. The distinct peaks of troxerutin and those of its physical mixture reveal that the functional groups and characteristics of troxerutin remained unaltered. This observation indicates the absence of any interference from troxerutin solid lipid nanoparticles within the drug-excipient mixtures.

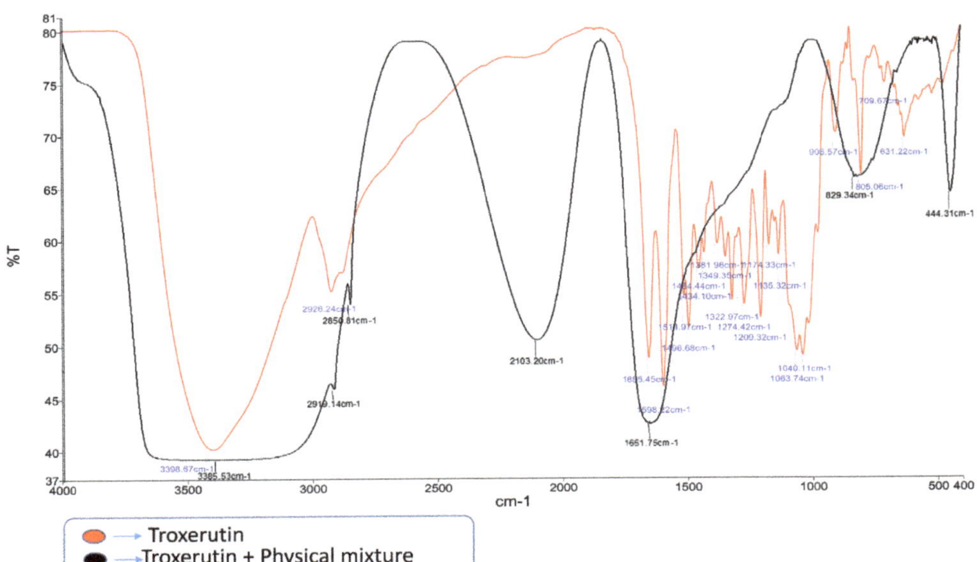

Figure 5. FTIR spectra of drug and excipient compatibility.

3.5. Differential Scanning Calorimetry (DSC)

The formulation of SLNs necessitated the validation of the expected physical state of the matrix lipid, which is of paramount importance. This validation can be achieved through DSC. When comparing the DSC thermograms of the bulk lipids with those of the corresponding SLNs, notable differences in the signals' positions and shapes were often observed. Specifically, the DSC curve for TXR alone exhibited a sharp exothermic peak at 177.07 °C (with a ΔH of −39.33 mJ), corresponding to its known melting point of 181 °C. For the lipid component, the peak was observed at 110 °C. In the case of the SLN formulation, two distinct peaks emerged at 135 and 177.07 °C (Figure 6). The shifting of the lipid peak in the solid lipid nanoparticle formulation may be due to the lipid polymorphism.

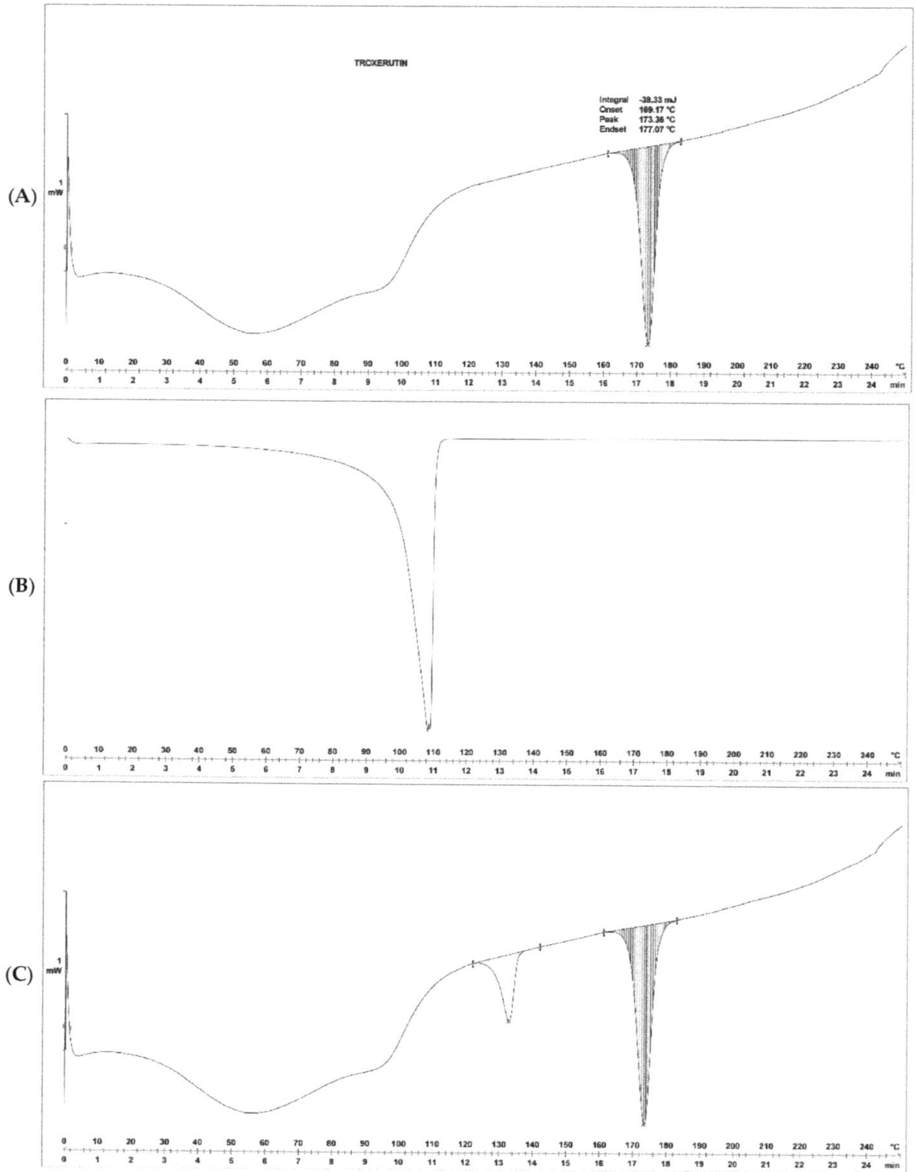

Figure 6. DSC spectra of the troxerutin (**A**), lipid (**B**), and SLN formulations (**C**).

3.6. Drug Release Study

The release profile of the TXR-SLNs, in comparison to TXR alone, is illustrated in Figure 7. All formulations demonstrated sustained drug release over 24 h in a phosphate-buffered saline medium. Within the same release medium, the cumulative drug release for all three formulations was more pronounced during the initial 4 h period. Notably, the TXR-SLN 2 formulation displayed the highest cumulative drug release at 24 h, reaching 82.47%. Conversely, TXR-SLN 1 exhibited a 56.33% drug release, and TXR-SLN 3 displayed a 66.51% release within the same timeframe. As for TXR alone, its maximum cumulative drug release occurred at 2 h. Comparatively, the TXR-SLNs exhibited elevated cumulative drug release over extended intervals in contrast to TXR alone. For a comprehensive overview, the dissolution data are presented in Table 4.

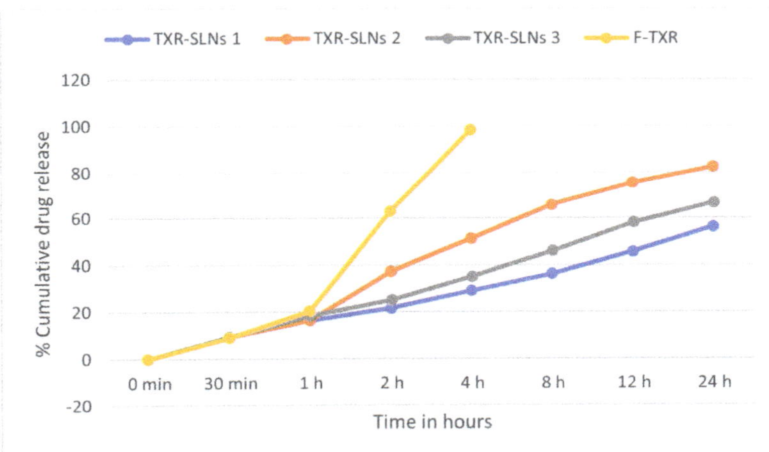

Figure 7. Percentage of cumulative drug release of the TXR-loaded SLNs.

Table 4. Percentage of cumulative drug release of the TXR-SLN dissolution data.

Time	% Cumulative Drug Released ± SD (*n* = 3)			
	TXR-SLN 1	**TXR-SLN 2**	**TXR-SLN 3**	**F-TXR**
0 min	0	0	0	0
30 min	9.36 ± 0.14	9.36 ± 0.83	9.36 ± 0.97	9.05 ± 0.64
1 h	16.46 ± 0.48	16.46 ± 0.93	18.45 ± 0.88	20.38 ± 0.37
2 h	21.65 ± 0.92	37.43 ± 0.47	25.05 ± 0.17	62.86 ± 0.29
4 h	29.17 ± 0.77	51.38 ± 0.84	35.19 ± 0.44	98.36 ± 0.47
8 h	36.47 ± 0.46	65.71 ± 0.64	46.02 ± 0.65	--
12 h	45.72 ± 0.13	75.45 ± 0.54	57.95 ± 0.66	--
24 h	56.33 ± 0.67	82.47 ± 0.16	66.51 ± 0.89	--

3.7. TXR Release Kinetic Study

The kinetics and mechanism of drug release were evaluated for the prepared SLNs. The release profiles of the formulations were scrutinized by fitting them to various equations, including zero-order, first-order, Higuchi, and Korsmeyer–Peppas. Among these, the optimal formulation underwent in-depth analysis to determine the underlying drug release mechanism. In these experiments, the release profile of TXR-SLN 2 exhibited substantial linearity, evident from a regression value of 0.916. This finding indicates that the drug release adhered to Higuchi's equation, thereby confirming diffusion as the primary driver of drug release. To validate the diffusion mechanism, the dissolution data were also subjected to fitting the Korsmeyer–Peppas equation, resulting in a slope (n) value of 0.95. A comprehensive presentation of these results is provided in Table 5 and Figure 8 below.

Table 5. Kinetic release study of TXR-SLN 2.

Zero-Order	First-Order	Higuchi	Korsmeyer–Peppas	
R^2	R^2	R^2	R^2	N
0.6993	0.8714	0.916	0.567	0.95

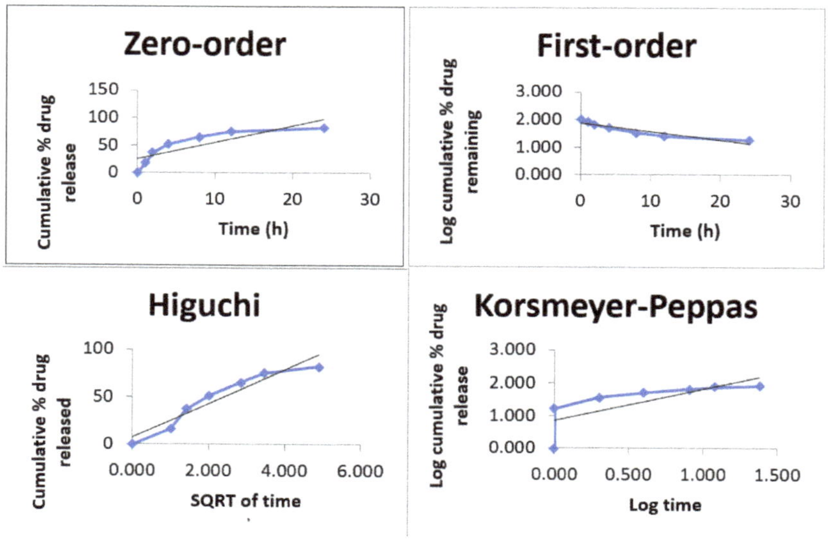

Figure 8. Release kinetic study of TXR-SLN 2.

3.8. Stability Studies

The results of the particle size, PDI, and zeta potential measurements for the three formulations after one month of refrigerated storage (4 °C) indicate minimal differences when compared to the measurements taken on the first day immediately after preparation. However, the particle size, PDI, and zeta potential for TXR-SLN 2 remained superior to those of TXR-SLN 1 and TXR-SLN 3, as illustrated in Table 6.

Table 6. Results of the particle size, polydispersity index (PDI), and zeta potential measurements for the three formulations after one month of refrigerated storage (4 °C).

S. No	Formulation	Particle Size (nm)	Polydispersity Index (PDI)	Zeta Potential (mV)
1	TXR-SLN 1	360.9 ± 6.10	0.391 ±0.05	16 ±3.48
2	TXR-SLN 2	151 ± 2.03	0.221 ± 0.03	28.4 ± 5.32
3	TXR-SLN 3	155.5 ± 2.05	0.356 ± 0.03	14.5 ± 4.31

The results of the particle size, PDI, and zeta potential measurements for the three formulations after one month of room-temperature storage (23 °C) indicate minimal differences when compared to the measurements taken on the first day immediately after preparation. However, the particle size, PDI, and zeta potential for TXR-SLN 2 remained superior to those of TXR-SLN 1 and TXR-SLN 3, as illustrated in Table 7.

Table 7. Results of the particle size, polydispersity index (PDI), and zeta potential measurements for the three formulations after one month of room-temperature storage (23 °C).

S. No	Formulation	Particle Size (nm)	Polydispersity Index (PDI)	Zeta Potential (mV)
1	TXR-SLN 1	382 ± 5.23	0.392 ±0.02	14 ±2.13
2	TXR-SLN 2	154 ± 3.01	0.221 ± 0.02	28.2 ± 3.21
3	TXR-SLN 3	162.5 ± 2.03	0.371 ± 0.04	13.1 ± 3.11

The results of the EE identification for the three TXR-SLN formulations after one month of refrigerated storage (4 °C) indicate a slight decrease in entrapment efficiency compared to the same formulations immediately after preparation on the first day, as illustrated in Table 8.

Table 8. Entrapment efficiency (EE) of the TXR-SLN formulations after one month of refrigerated storage (4 °C).

S. No	Formulation Code	Entrapment Efficiency (%)
1	TXR-SLN 1	71.23619
2	TXR-SLN 2	77.54231
3	TXR-SLN 3	73.23486

The results of the EE identification for the three TXR-SLN formulations after one month of room-temperature storage (23 °C) indicate a significant decrease in entrapment efficiency compared to the same formulations immediately after preparation on the first day. However, the EE of TXR-SLN 2 was still within the normal range for stability determination, as illustrated in Table 9.

Table 9. Entrapment efficiency (EE) of the TXR-SLN formulations after one month of room-temperature storage (23 °C).

S. No	Formulation Code	Entrapment Efficiency (%)
1	TXR-SLN 1	54.36215
2	TXR-SLN 2	71.32457
3	TXR-SLN 3	45.23615

4. Discussion

The impact of lipids on SLNs entails a highly intricate interplay. Notably, the presence of lipid molecules with longer chains and the heightened viscosity within the oil phase can culminate in the generation of larger particle sizes. It is essential to acknowledge that the imperfections stemming from the complex nature of lipid materials and their imperfect crystal structures can also contribute to EE [25,34].

In the context of our current study, we opted to employ glyceryl behenate as the solid lipid polymer. Characterized as a fusion of glycerol esters and behenic acid, glyceryl behenate predominantly comprises glyceryl behenate. This intricate lipid entity further encompasses an array of mono-, di-, and triglycerides, each marked by distinct chain lengths. This amalgamation of diverse components imparts an irregularity on the crystal lattice, thereby fostering an environment conducive to the assimilation of bioactive compounds.

Consequently, this intricate structural composition not only enhances compatibility, but also elevates the efficacy of the formed SLNs [25,34,35].

Surfactants play a pivotal role in shaping the ideal configuration of SLNs, wielding a range of beneficial effects by inducing surface property modifications. These modifications encompass a reduction in surface tension, facilitating particle dispersion, mitigating particle aggregation, and enhancing the stability of the SLN emulsion [36,37]. It is noteworthy that the influence of surfactants is not confined solely to surface properties; their impact extends to the degree of lattice order, contributing significantly to nanoparticle stability [25,36,37].

In the context of our investigation, we purposefully incorporated soy lecithin and Tween-80 as both surfactants and co-surfactants. This deliberate selection was underpinned by their demonstrated capability to yield a multitude of favorable results. Notably, this encompasses the achievement of reduced particle dimensions, elevated EE, and enhanced overall stability [25,36]. This judicious choice of surfactants not only seeks to fine-tune the system for optimal performance, but also harmonizes seamlessly with our overarching goal of engineering SLNs endowed with exceptional attributes.

The methodology employed in this study to design SLNs encompassed a two-step process: High-shear homogenization succeeded by ultrasonication. This combined approach stands out as a straightforward, dependable, and feasible technique renowned for SLN preparation [25]. The initial phase of homogenization yielded a stable and lucid dispersion of SLNs, characterized by an appropriate particle size. However, it is noteworthy that this process, due to its elevated speed, could inadvertently introduce a profusion of bubbles and atomization.

To circumvent such limitations, the subsequent step involved harnessing ultrasonication. This strategic inclusion was designed to mitigate the issues stemming from high-shear homogenization. The ultrasonication force proved instrumental in homogenizing the dispersion of the formulation, thereby rectifying the anomalies induced by the prior high-shear homogenization process [38]. This orchestrated combination of techniques not only yielded clear benefits in terms of stability and particle uniformity, but also highlighted the intricacies involved in optimizing SLN fabrication.

Physicochemical characteristics, such as particle size, PDI, and zeta potential, hold paramount importance as attributes of lipid nanocarriers. These characteristics exert substantial influence over multiple facets, including drug stability, entrapment efficiency, drug release kinetics, biodistribution, and cellular uptake. Notably, in the realm of nanoscale systems, size stability assumes heightened significance when juxtaposed with their microscale counterparts. This distinction arises from the vast specific surface area inherent to nanoscale systems, a factor that distinguishes them from their microscale counterparts [39].

As the radius of the sphere and particle size decrease, there is a consequential elevation in the surface area-to-volume ratio [40]. This fundamental geometric principle invariably underscores the fact that diminutive particle dimensions yield a proportionally larger surface area concerning the volume. It is within this interplay that the foundation for a host of nanoscale system properties is established [40].

In this study, the optimal formulation was TXR-SLN 2. The selection of TXR-SLN 2 was based on its particle size, PDI, and zeta potential values. The particle size measured 140.5 $\pm$ 1.02 nm, with a PDI of 0.218 $\pm$ 0.01 and a zeta potential of 28.6 $\pm$ 8.71. A PDI value of 0.3 or below in lipid-based nanocarriers is considered acceptable, indicating a homogeneous population [36]. The zeta potential range of 28.6 $\pm$ 8.71, which is close to 30, suggests good stability [38]. Maintaining a constant and narrow size distribution is crucial for achieving optimal clinical outcomes with nanocarrier formulations [39]. The size and polydispersity index in the present study was comparatively less than the recently published work with chitosan-loaded troxerutin nanoparticles, in which the size and polydispersity index was 692 d·nm and 0.355, respectively [30]. Both particle size and PDI were expected to increase with the addition of lipid components to the formulations [41,42]. TXR-SLN 2 contained fewer lipid polymers compared to the other formulations, resulting in a monodispersed distribution and smaller size.

The EE of the three formulations was evaluated by measuring absorbance using a UV-Vis spectrophotometer and then plotted on a calibration curve (Figure 2). The results further support the identification of TXR-SLN 2 as the best formulation, exhibiting the highest EE value of 83.62%. The entrapment efficiency percentage was increased as the sonication time increased from 5 to 15 min [43]. A higher entrapment efficiency (83.62%) was achieved when TXR-SLN 2 was processed by ultrasonication for 15 min, compared to the other two formulations. The optimal formulation displayed an increased EE attributed to the lower soy lecithin and surfactant concentrations, as well as fewer lipid polymers, all of which contributed to a smaller particle size.

FTIR is a technique used to elucidate the chemical composition of various organic chemicals, polymers, and other materials [44]. The measurements were conducted to identify the potential biomolecules responsible for capping and effectively stabilizing the TXR-SLNs [44]. The FTIR analysis in this study revealed characteristic peaks present in both TXR alone and the TXR-SLNs. The results demonstrated that there were no missing functional peaks in the formulated TXR-SLNs. Furthermore, the formulation exhibited no significant physicochemical interactions between the drug and lipid.

The DSC technique measures the disparity in heat flux delivered to a test sample and a reference sample at the same temperature [45]. This makes it possible to quantify the differences in heat flows resulting from sample melting, crystallization, chemical reactions, or evaporation [45]. The DSC curve of TXR alone displayed a sharp exothermic peak at 177.07 °C ($\Delta H = -39.33$ mJ), corresponding to its melting point of 181 °C. For the lipid, the peak was observed at 110 °C, while for the SLN formulation, two peaks were identified at 135 and 177.07 °C. Both TXR alone and the SLNs exhibited the same exothermic peak at 177.07 °C. DSC offers insights into the characteristics and temperature parameters of individual samples [45], thus enhancing the quality and physicochemical properties of the product during various technological processes involving TXR-SLNs [45].

The release profile of the TXR-SLNs was tested. Seven samples were taken from all three formulations, and their cumulative drug release was measured. All formulations successfully released the drug in a sustained manner over 24 h in a phosphate-buffered saline solution. Among the formulations, TXR-SLN 2 exhibited the highest cumulative drug release at 24 h, reaching 82.47%. In contrast, TXR alone displayed the maximum cumulative drug release within 2 h. The TXR-SLNs demonstrated higher cumulative drug release over extended periods compared to TXR alone.

The main lipid components used were glyceryl behenate, soy lecithin, and Tween-80 as surfactants. Tween-80 contributed to both the stability and improved bioavailability of the SLNs by acting as a permeability enhancer [46]. Furthermore, it can increase intestinal permeability while reducing the activity of intestinal efflux transporters, including P-glycoprotein efflux pump activity [46]. The primary factor influencing oral absorption and bioavailability is intestinal efflux by P-glycoprotein and other transporters [46].

The release kinetic study demonstrated the drug release behavior of the SLNs. The formulations were evaluated by fitting the drug release profiles to various equations, including the zero-order, first-order, Higuchi, and Korsmeyer–Peppas models [47]. The optimal formulation, TXR-SLN 2, exhibited a better fit to Higuchi's model, which characterizes drug release from polymeric systems and aids in understanding the release mechanism and comparing differences among formulations [48,49]. A release exponent (n) greater than 0.89 indicates super Case II transport, involving drug release through both diffusion and relaxation of the SLNs. For the optimized formulation, TXR showed a predominant mechanism of super Case II transport.

Additionally, an R^2 value of 0.916 came closest to 1. Both the SLN formulations and the drug release were time-dependent. Consequently, TXR-SLN 2 achieved the highest in vitro release concentration around 24 h. These findings are consistent with the drug release results.

TEM is employed in cell biology, involving the imaging of stained thin sections of plastic-embedded cells by passing an electron beam through the sample. This process causes the electrons to be absorbed and scattered, creating contrast and generating an image [50]. The results indicated that the TXR-SLNs were spherical with a smooth surface, indicative of the material's natural tendency to minimize its energy state [51].

The one-month stability study for the three formulations, measured by particle size, PDI, zeta potential, and EE, indicated minimal differences under both conditions (4 and 23 °C). However, a significant reduction in entrapment efficiency was observed after one month at 23 °C, suggesting a negative impact of temperature on the stability of the formulations over time. Notably, the stability study of TXR-SLN 2 remained within the acceptable stability range, distinguishing it from both TXR-SLN 1 and TXR-SLN 1.

This study has some potential limitations. First, there is a need for further optimization or factorial design. A factorial design could have assisted in examining how multiple factors affect a dependent variable, both individually and collectively. Evaluation of the toxicity of nanomaterials is essential to identify the effect of the nanomaterials used in the formulation during in vivo animal or human studies [52]. Additionally, this study employed a limited number of formulations, preventing us from anticipating potential variations in results across different formulations. Moreover, the use of a single surfactant and lipid polymer raises questions about the effectiveness of employing alternative surfactants or lipid polymers.

5. Conclusions

TXR-SLNs were prepared through high-shear homogenization followed by ultrasonication, and the formulation was optimized using CCD-RSM. The results of the drug release study indicated that the cumulative amount of drug released from the TXR-SLNs was significantly higher compared to that from TXR alone, reaching 82.47% within 24 h. This suggests that TXR-SLNs can enhance the drug release profile of TXR. Moreover, the kinetic investigation of drug release from TXR-SLN 2 revealed a combination of diffusion and relaxation mechanisms.

However, future plans involve in vivo drug release and efficacy studies for determining the performance of an optimal TXR-SLN formulation, along with toxicity evaluation of nanomaterials of the formulation.

Author Contributions: Conceptualization: Y.F.J., N.A.A. and M.T.S.S.; methodology: A.F.A., S.M.A. and M.A.A.; formal Analysis: Y.F.J., N.A.A. and M.T.S.S.; investigation: Y.F.J., N.A.A. and M.T.S.S.; writing, reviewing, and editing: Y.F.J., N.A.A., M.T.S.S., A.F.A., S.M.A. and M.A.A.; supervision: Y.F.J., N.A.A. and M.T.S.S. All authors have read and agreed to the published version of the manuscript.

Funding: This research was funded by grants from the Princess Nourah bint Abdulrahman University Researchers Supporting Project (grant number, PNURSP2023R98), Princess Nourah bint Abdulrahman University, Riyadh, Saudi Arabia.

Data Availability Statement: All of the data are included in the manuscript.

Acknowledgments: The authors would like to thank King Abdulaziz City for Science and Technology (KACST), Riyadh, Saudi Arabia.

Conflicts of Interest: The authors declare no conflict of interest.

Abbreviations

COX: Cyclo-oxygenase; CVI: Chronic venous insufficiency; DSC: Differential scanning calorimetry; EE: Entrapment efficiency; FTIR: Fourier transform infrared spectroscopy; NDDs: Neurodegenerative diseases; NPs: Nanoparticles; pH: Potential of hydrogen; Qn: Cumulative release at the n-th sampling point; ROS: Reactive oxidative species; R2: Coefficient of determination; rpm: Revolutions per minute; SLNs: Solid lipid nanoparticles; TEM: Transmission electron microscopy; TXR: Troxerutin; TXR-SLNs: Troxerutin-loaded solid lipid nanoparticles; UV: Ultraviolet; XO: Xanthine oxidase; °C: Degrees Celsius.

References

1. Ullah, A.; Munir, S.; Badshah, S.L.; Khan, N.; Ghani, L.; Poulson, B.G.; Emwas, A.H.; Jaremko, M. Important Flavonoids and Their Role as a Therapeutic Agent. *Molecules* **2020**, *25*, 5243. Available online: https://www.ncbi.nlm.nih.gov/pmc/articles/PMC7697716/ (accessed on 24 September 2022). [CrossRef]
2. Abbaszadeh, F.; Fakhri, S.; Khan, H. Targeting apoptosis and autophagy following spinal cord injury: Therapeutic approaches to polyphenols and candidate phytochemicals. *Pharmacol. Res.* **2020**, *160*, 105069. Available online: https://pubmed.ncbi.nlm.nih.gov/32652198/ (accessed on 25 September 2022). [CrossRef]
3. Kozłowska, A.; Szostak-Wegierek, D. Flavonoids—Food sources and health benefits. *Rocz. Panstw. Zakl. Hig.* **2014**, *65*, 79–85. Available online: https://pubmed.ncbi.nlm.nih.gov/25272572/ (accessed on 24 September 2022). [PubMed]
4. Ghorbani, A. Mechanisms of antidiabetic effects of flavonoid rutin. *Biomed. Pharmacother.* **2017**, *96*, 305–312. Available online: https://pubmed.ncbi.nlm.nih.gov/29017142/ (accessed on 25 September 2022). [CrossRef] [PubMed]
5. Mahmoud, A.M.; Hernández Bautista, R.J.; Sandhu, M.A.; Hussein, O.E. Beneficial Effects of Citrus Flavonoids on Cardiovascular and Metabolic Health. *Oxidative Med. Cell. Longev.* **2019**, *2019*, 5484138. Available online: https://www.ncbi.nlm.nih.gov/pmc/articles/PMC6431442/ (accessed on 6 August 2022). [CrossRef] [PubMed]
6. El-Shiekh, R.A.; Abdelmohsen, U.R.; Ashour, H.M.; Ashour, R.M. Novel Antiviral and Antibacterial Activities of Hibiscus schizopetalus. *Antibiotics* **2020**, *9*, 756. Available online: https://www.ncbi.nlm.nih.gov/pmc/articles/PMC7692239/ (accessed on 6 August 2022). [CrossRef]
7. Kopustinskiene, D.M.; Jakstas, V.; Savickas, A.; Bernatoniene, J. Flavonoids as Anticancer Agents. *Nutrients* **2020**, *12*, 457. Available online: https://www.ncbi.nlm.nih.gov/pmc/articles/PMC7071196/ (accessed on 6 August 2022). [CrossRef]
8. Hussain, G.; Zhang, L.; Rasul, A.; Anwar, H.; Sohail, M.U.; Razzaq, A.; Aziz, N.; Shabbir, A.; Ali, M.; Sun, T. Role of Plant-Derived Flavonoids and Their Mechanism in Attenuation of Alzheimer's and Parkinson's Diseases: An Update of Recent Data. *Mol. A J. Synth. Chem. Nat. Prod. Chem.* **2018**, *23*, 814. Available online: https://www.ncbi.nlm.nih.gov/pmc/articles/PMC6017497/ (accessed on 6 August 2022). [CrossRef]
9. de Andrade, R.B.T.; Diniz, T.C.; Pinto, T.C.C.; de Oliveira, R.G.; e Silva, M.G.; de Lavor, É.M.; Fernandes, A.W.C.; de Oliveira, A.P.; de Almeida, F.P.R.; da Silva, A.A.M.; et al. Flavonoids as Therapeutic Agents in Alzheimer's and Parkinson's Diseases: A Systematic Review of Preclinical Evidence. *Oxidative Med. Cell. Longev.* **2018**, *2018*, 7043213. Available online: https://www.ncbi.nlm.nih.gov/pmc/articles/PMC5971291/ (accessed on 6 August 2022). [CrossRef]
10. Rajpoot, K. Solid Lipid Nanoparticles: A Promising Nanomaterial in Drug Delivery. *Curr. Pharm. Des.* **2019**, *25*, 3943–3959. Available online: https://pubmed.ncbi.nlm.nih.gov/31481000/ (accessed on 13 July 2022). [CrossRef]
11. Shu, L.; Zhang, W.; Huang, C.; Huang, G.; Su, G. Troxerutin Protects Against Myocardial Ischemia/Reperfusion Injury Via Pi3k/Akt Pathway in Rats. *Cell Physiol. Biochem.* **2017**, *44*, 1939–1948. Available online: https://pubmed.ncbi.nlm.nih.gov/29241161/ (accessed on 12 July 2022). [CrossRef] [PubMed]
12. Yu, Z.P.; Yu, H.Q.; Li, J.; Li, C.; Hua, X.; Sheng, X.S. Troxerutin attenuates oxygen-glucose deprivation and reoxygenation-induced oxidative stress and inflammation by enhancing the PI3K/AKT/HIF-1α signaling pathway in H9C2 cardiomyocytes. *Mol. Med. Rep.* **2020**, *22*, 1351–1361. Available online: https://pubmed.ncbi.nlm.nih.gov/32626962/ (accessed on 24 September 2022). [CrossRef] [PubMed]
13. Xin, X.; Zhang, M.; Li, X.; Lai, F.; Zhao, G. Biocatalytic synthesis of acylated derivatives of troxerutin: Their bioavailability and antioxidant properties in vitro. *Microb. Cell Fact.* **2018**, *17*, 130. Available online: https://microbialcellfactories.biomedcentral.com/articles/10.1186/s12934-018-0976-x (accessed on 24 September 2022). [CrossRef] [PubMed]
14. Zamanian, M.; Bazmandegan, G.; Sureda, A.; Sobarzo-Sanchez, E.; Yousefi-Manesh, H.; Shirooie, S. The Protective Roles and Molecular Mechanisms of Troxerutin (Vitamin P4) for the Treatment of Chronic Diseases: A Mechanistic Review. *Curr. Neuropharmacol.* **2021**, *19*, 97. Available online: https://www.ncbi.nlm.nih.gov/pmc/articles/PMC7903491/ (accessed on 24 September 2022). [CrossRef]
15. Shan, Q.; Zheng, G.; Han, X.; Wen, X.; Wang, S.; Li, M.; Zhuang, J.; Zhang, Z.F.; Hu, B.; Zhang, Y.; et al. Troxerutin Protects Kidney Tissue against BDE-47-Induced Inflammatory Damage through CXCR4-TXNIP/NLRP3 Signaling. *Oxid. Med. Cell. Longev.* **2018**, *2018*, 9865495. Available online: https://www.ncbi.nlm.nih.gov/pmc/articles/PMC5932985/ (accessed on 24 September 2022). [CrossRef]
16. Shan, Q.; Zhuang, J.; Zheng, G.; Zhang, Z.; Zhang, Y.; Lu, J.; Zheng, Y. Troxerutin Reduces Kidney Damage against BDE-47-Induced Apoptosis via Inhibiting NOX2 Activity and Increasing Nrf2 Activity. *Oxid. Med. Cell. Longev.* **2017**, *2017*, 6034692. Available online: https://www.ncbi.nlm.nih.gov/pmc/articles/PMC5661100/ (accessed on 24 September 2022). [CrossRef]
17. Zhang, Z.F.; Shao-Hua, F.A.N.; Zheng, Y.L.; Jun, L.U.; Dong-Mei, W.U.; Shan, Q.U.N.; Hu, B. Troxerutin protects the mouse liver against oxidative stress-mediated injury induced by D-galactose. *J. Agric. Food Chem.* **2009**, *57*, 7731–7736. Available online: https://pubmed.ncbi.nlm.nih.gov/19722705/ (accessed on 24 September 2022). [CrossRef]
18. Zhang, Z.F.; Zhang, Y.Q.; Fan, S.H.; Zhuang, J.; Zheng, Y.L.; Lu, J.; Wu, D.M.; Shan, Q.; Hu, B. Troxerutin protects against 2,2′,4,4′-tetrabromodiphenyl ether (BDE-47)-induced liver inflammation by attenuating oxidative stress-mediated NAD+-depletion. *J. Hazard. Mater.* **2015**, *283*, 98–109. Available online: https://pubmed.ncbi.nlm.nih.gov/25262482/ (accessed on 24 September 2022). [CrossRef]

19. Casili, G.; Lanza, M.; Campolo, M.; Messina, S.; Scuderi, S.; Ardizzone, A.; Filippone, A.; Paterniti, I.; Cuzzocrea, S.; Esposito, E. Therapeutic potential of flavonoids in the treatment of chronic venous insufficiency. *Vasc. Pharmacol.* **2021**, *137*, 106825. Available online: https://www.sciencedirect.com/science/article/abs/pii/S153718912030330X?via%3Dihub (accessed on 24 September 2022). [CrossRef]
20. Boisseau, M.R.; Taccoen, A.; Garreau, C.; Vergnes, C.; Roudaut, M.F.; Garreau-Gomez, B. Fibrinolysis and Hemorheology in Chronic Venous Insufficiency: A Double Blind Study of Troxerutin Efficiency-PubMed. Available online: https://pubmed.ncbi.nlm.nih.gov/7593149/ (accessed on 24 September 2022).
21. Cui, X.; Zhang, M.; Guan, X.; Yin, L.; Sun, Y.; Fawcett, J.P.; Gu, J. LC-MS-MS determination of troxerutin in plasma and its application to a pharmacokinetic study. *Chromatographia* **2011**, *73*, 165–169. [CrossRef]
22. Troxerutin European Pharmacopoeia (EP) Reference Standard 7085-55-4. Available online: https://www.sigmaaldrich.com/SA/en/product/sial/y0000497?gclid=Cj0KCQjw1bqZBhDXARIsANTjCPLdKv45ZWE_h113biIEYAxhGAbl9AT0PMG9zg97_ThJBWSwAKd-ZrUaApiDEALw_wcB&gclsrc=aw.ds (accessed on 24 September 2022).
23. Najahi-Missaoui, W.; Arnold, R.D.; Cummings, B.S. Safe Nanoparticles: Are We There Yet? *Int. J. Mol. Sci.* **2020**, *22*, 385. Available online: https://pubmed.ncbi.nlm.nih.gov/33396561/ (accessed on 24 September 2022). [CrossRef] [PubMed]
24. Yetisgin, A.A.; Cetinel, S.; Zuvin, M.; Kosar, A.; Kutlu, O. Therapeutic Nanoparticles and Their Targeted Delivery Applications. *Molecules* **2020**, *25*, 2193. Available online: https://www.ncbi.nlm.nih.gov/pmc/articles/PMC7248934/ (accessed on 6 August 2022). [CrossRef] [PubMed]
25. Montoto, S.S.; Muraca, G.; Ruiz, M.E. Solid Lipid Nanoparticles for Drug Delivery: Pharmacological and Biopharmaceutical Aspects. *Front. Mol. Biosci.* **2020**, *7*, 319. Available online: https://pubmed.ncbi.nlm.nih.gov/33195435/ (accessed on 24 September 2022).
26. Souto, E.B.; Doktorovová, S. Chapter 6—Solid lipid nanoparticle formulations pharmacokinetic and biopharmaceutical aspects in drug delivery. *Methods Enzym.* **2009**, *464*, 105–129. Available online: https://pubmed.ncbi.nlm.nih.gov/19903552/ (accessed on 6 August 2022).
27. Tan, M.E.; He, C.H.; Jiang, W.; Zeng, C.; Yu, N.; Huang, W.; Gao, Z.G.; Xing, J.G. Development of solid lipid nanoparticles containing total flavonoid extract from Dracocephalum moldavica L. and their therapeutic effect against myocardial ischemia–reperfusion injury in rats. *Int. J. Nanomed.* **2017**, *12*, 3253–3265. Available online: https://www.dovepress.com/development-of-solid-lipid-nanoparticles-containing-total-flavonoid-ex-peer-reviewed-fulltext-article-IJN (accessed on 24 September 2022).
28. Zhao, T.; Wu, W.; Sui, L.; Huang, Q.; Nan, Y.; Liu, J.; Ai, K. Reactive oxygen species-based nanomaterials for the treatment of myocardial ischemia reperfusion injuries. *Bioact. Mater.* **2022**, *7*, 47–72. Available online: https://www.ncbi.nlm.nih.gov/pmc/articles/PMC8377441/ (accessed on 24 September 2022). [CrossRef]
29. Rostami, E.; Kashanian, S.; Azandaryani, A.H.; Faramarzi, H.; Dolatabadi, J.E.; Omidfar, K. Drug targeting using solid lipid nanoparticles. *Chem. Phys. Lipids.* **2014**, *181*, 56–61. Available online: https://pubmed.ncbi.nlm.nih.gov/24717692/ (accessed on 24 September 2022). [CrossRef]
30. Subbaraj, G.K.; Elangovan, H.; Chandramouli, P.; Yasam, S.K.; Chandrasekaran, K.; Kulanthaivel, L.; Pandi, S.; Subramanian, S. Antiangiogenic Potential of Troxerutin and Chitosan Loaded Troxerutin on Chorioallantoic Membrane Model. *Biomed. Res. Int.* **2023**, *2023*, 5956154. Available online: https://pubmed.ncbi.nlm.nih.gov/37260851/ (accessed on 24 September 2022). [CrossRef]
31. Saranya, T.; Kavithaa, K.; Paulpandi, M.; Ramya, S.; Winster, S.H.; Mani, G.; Dhayalan, S.; Balachandar, V.; Narayanasamy, A. The creation of selenium nanoparticles decorated with troxerutin and their ability to adapt to the tumour microenvironment have therapeutic implications for triple-negative breast cancer. *New J. Chem.* **2023**, *47*, 4565–4576. Available online: https://pubs.rsc.org/en/content/articlelanding/2004/yg/d2nj05671b/unauth (accessed on 1 October 2023). [CrossRef]
32. Zeng, C.; Jiang, W.; Tan, M.; Xing, J.; He, C. Improved Oral Bioavailability of Total Flavonoids of Dracocephalum moldavica via Composite Phospholipid Liposomes: Preparation, in-vitro Drug Release and Pharmacokinetics in Rats. *Pharmacogn. Mag.* **2016**, *12*, 313–318. Available online: https://pubmed.ncbi.nlm.nih.gov/27867275/ (accessed on 1 October 2023).
33. Janakiraman, A.K.; Sumathi, B.; Mohamed Saleem, T.; Ramkanth, S.; Odaya Kumar, P.; Venkatachalam, G. Design and evaluation of Carvedilol nanocrystals sustained release tablets. *J. App. Pharm. Sci.* **2017**, *7*, 61–68.
34. Kumar, V.V.; Chandrasekar, D.; Ramakrishna, S.; Kishan, V.; Rao, Y.M.; Diwan, P.V. Development and evaluation of nitrendipine loaded solid lipid nanoparticles: Influence of wax and glyceride lipids on plasma pharmacokinetics. *Int. J. Pharm.* **2006**, *335*, 167–175. Available online: https://europepmc.org/article/med/17161566 (accessed on 28 December 2022). [CrossRef] [PubMed]
35. Teeranachaideekul, V.; Souto, E.; Müller, R.; Junyaprasert, V.B. Physicochemical characterization and in vitro release studies of ascorbyl palmitate-loaded semi-solid nanostructured lipid carriers (NLC gels). *J. Microencapsul.* **2008**, *25*, 111–120. Available online: https://pubmed.ncbi.nlm.nih.gov/18246489/ (accessed on 19 April 2023). [CrossRef] [PubMed]
36. Hwang, K.M.; Byun, W.; Cho, C.H.; Park, E.S. Preparation and optimization of glyceryl behenate-based highly porous pellets containing cilostazol. *Pharm. Dev. Technol.* **2016**, *23*, 540–551. Available online: https://www.tandfonline.com/doi/abs/10.1080/10837450.2016.1245743 (accessed on 19 April 2023). [CrossRef] [PubMed]
37. Le, N.T.T.; Cao, V.D.; Nguyen, T.N.Q.; Le, T.T.H.; Tran, T.T.; Thi, T.T.H. Soy Lecithin-Derived Liposomal Delivery Systems: Surface Modification and Current Applications. *Int. J. Mol. Sci.* **2019**, *20*, 4706. Available online: https://pubmed.ncbi.nlm.nih.gov/31547569/ (accessed on 19 April 2023). [CrossRef] [PubMed]

38. Genç, L. Preparation and characterization of nocodazole-loaded solid lipid nanoparticles. *Pharm. Dev. Technol.* **2013**, *19*, 671–676. Available online: https://www.tandfonline.com/doi/abs/10.3109/10837450.2013.819017 (accessed on 19 April 2023). [CrossRef]
39. Kumar, A.J.; Ramkanth, S.; Lakshmana, S.P.; Gopal, V. Enhancement of saturation solubility and in vitro dissolution of carvedilol nanoparticles by high pressure homogenization technique. *Int. J. Curr. Pharm. Rev. Res.* **2015**, *6*, 269–273.
40. Danaei, M.; Dehghankhold, M.; Ataei, S.; Hasanzadeh, F.D.; Javanmard, R.; Dokhani, A.; Mozafari, M.R. Impact of Particle Size and Polydispersity Index on the Clinical Applications of Lipidic Nanocarrier Systems. *Pharmaceutics* **2018**, *10*, 57. Available online: https://pubmed.ncbi.nlm.nih.gov/29783687/ (accessed on 2 April 2023). [CrossRef]
41. Surface Area to Volume Ratio in Nanoparticles | Winner Science. Available online: https://winnerscience.com/surface-area-to-volume-ratio-in-nanoparticles/ (accessed on 10 May 2023).
42. The Effect of Rice Bran Wax on Physicochemical Properties of Curcuminoid-Loaded Solid Lipid Nanoparticles. Available online: https://www.researchgate.net/publication/305432628_The_effect_of_rice_bran_wax_on_physicochemical_properties_of_curcuminoidloaded_solid_lipid_nanoparticles#pf3 (accessed on 2 April 2023).
43. Satyanarayana, S.D.; Lila, A.S.A.; Moin, A.; Moglad, E.H.; Khafagy, E.S.; Alotaibi, H.F.; Obaidullah, A.J.; Charyulu, R.N. Ocular Delivery of Bimatoprost-Loaded Solid Lipid Nanoparticles for Effective Management of Glaucoma. *Pharmaceuticals* **2023**, *16*, 1001. [CrossRef]
44. Suhaimi, S.H.; Hasham, R.; Rosli, N.A. Akademia Baru Effects of Formulation Parameters on Particle Size and Polydispersity Index of Orthosiphon Stamineus Loaded Nanostructured Lipid Carrier. *J. Adv. Res. Appl. Sci. Eng. Technol.* **2015**, *1*, 36–39.
45. Devaraj, P.; Kumari, P.; Aarti, C.; Renganathan, A. Synthesis and characterization of silver nanoparticles using cannonball leaves and their cytotoxic activity against MCF-7 cell line. *J. Nanotechnol.* **2013**, *2013*, 598328. [CrossRef]
46. Musielak, E.; Feliczak-Guzik, A.; Nowak, I. Optimization of the Conditions of Solid Lipid Nanoparticles (SLN) Synthesis. *Molecules* **2022**, *27*, 2202. [CrossRef] [PubMed]
47. Hu, L.; Xing, Q.; Meng, J.; Shang, C. Preparation and Enhanced Oral Bioavailability of Cryptotanshinone-Loaded Solid Lipid Nanoparticles. *AAPS PharmSciTech* **2010**, *11*, 582. Available online: https://www.ncbi.nlm.nih.gov/pmc/articles/PMC2902353/ (accessed on 4 April 2023). [CrossRef] [PubMed]
48. Singh, S.; Kamal, S.S.; Sharma, A.; Kaur, D.; Katual, M.K.; Kumar, R. Formulation and In-Vitro Evaluation of Solid Lipid Nanoparticles Containing Levosulpiride. *Open Nanomed. J.* **2017**, *4*, 17–29. [CrossRef]
49. Shinde, G.; Shiyani, S.; Shelke, S.; Chouthe, R.; Kulkarni, D.; Marvaniya, K. Enhanced brain targeting efficiency using 5-FU (fluorouracil) lipid–drug conjugated nanoparticles in brain cancer therapy. *Prog. Biomater.* **2020**, *9*, 259–275. Available online: https://link.springer.com/article/10.1007/s40204-020-00147-y (accessed on 4 April 2023). [CrossRef] [PubMed]
50. Winey, M.; Meehl, J.B.; O'Toole, E.T.; Giddings, T.H. Conventional transmission electron microscopy. *Mol. Biol. Cell* **2014**, *25*, 319. Available online: https://www.ncbi.nlm.nih.gov/pmc/articles/PMC3907272/ (accessed on 16 April 2023). [CrossRef] [PubMed]
51. Ulusoy, U. A Review of Particle Shape Effects on Material Properties for Various Engineering Applications: From Macro to Nanoscale. *Minerals* **2023**, *13*, 91. Available online: https://www.mdpi.com/2075-163X/13/1/91/htm (accessed on 16 April 2023). [CrossRef]
52. Pagar, R.R.; Musale, S.R.; Pawar, G.; Kulkarni, D.; Giram, P.S. Comprehensive Review on the Degradation Chemistry and Toxicity Studies of Functional Materials. *ACS Biomater. Sci. Eng.* **2022**, *8*, 2161–2195. Available online: https://pubs.acs.org/doi/abs/10.1021/acsbiomaterials.1c01304 (accessed on 22 September 2022). [CrossRef]

processes

MDPI

Article

Metabolomic Profiling (LC–MS^2) of Flowers and Bee Honey of Dzidzilche (*Gymnopodium floribundum* Rolfe) and Jabin (*Piscidia piscipula* L. Sarg.) from Yucatán, México

Andrea Elizabeth Mendoza-Osorno [1], Kevin Alejandro Avilés-Betanzos [1], Alberto Uc-Varguez [1], Rommel Carballo-Castañeda [2], Aldo Moreno-Ulloa [2], Manuel Octavio Ramírez-Sucre [1] and Ingrid Mayanin Rodríguez-Buenfil [1,*]

[1] Centro de Investigación y Asistencia en Tecnología y Diseño del Estado de Jalisco A.C., Subsede Sureste, Tablaje Catastral 31264, Km. 5.5 Carretera Sierra Papacal-Chuburná Puerto, Parque Científico Tecnológico de Yucatán, Mérida 97302, Yucatán, Mexico; anmendoza_al@ciatej.edu.mx (A.E.M.-O.); keaviles_al@ciatej.edu.mx (K.A.A.-B.); auc@ciatej.mx (A.U.-V.); oramirez@ciatej.mx (M.O.R.-S.)

[2] Departamento de Innovación Biomédica, Centro de Investigación Científica y de Educación Superior de Ensenada (CICESE), Ensenada 22860, Baja California, Mexico; rommelomics@gmail.com (R.C.-C.); amoreno@cicese.mx (A.M.-U.)

* Correspondence: irodriguez@ciatej.mx

Abstract: Yucatan, Mexico, is renowned for its rich plant diversity, with ~40% melliferous plants. Yucatan bee honey (BH) constitutes ~15.83% of Mexico's annual BH production, giving high international value. Major melliferous families in Yucatan include Fabaceae, with *Piscidia piscipula* ("Jabin") as an example, and Polygonaceae, with *Gymnopodium floribundum* ("Dzidzilche"), crucial for BH production. This study aimed to profile the metabolome of Jabin and Dzidzilche flowers and their associated BH to identify metabolites for each flower coming from two regions (Tahdziu and Acanceh) of Yucatán. Liquid chromatography–tandem mass spectrometry (LC–MS^2), total polyphenol content (TPC), and antioxidant capacity (AC) were implemented. As many as 101 metabolites (69 in flowers, 55 in BH) were tentatively identified using spectral libraries and in silico predictions, predominantly flavonoids, which accounted for 50.7% of the total identified metabolites in flower and 16.4% in BH. Samples exhibited variations in TPC, AC, secondary metabolites, and chemical classes depending on geography and botanical origin. Dzidzilche flowers from Acanceh displayed the highest total polyphenol content (TPC, 1431.24 ± 15.38 mg GAE/100 g dry matter) and antioxidant capacity (AC, 93.63% inhibition). Among the metabolites detected in flowers (*Piscidia piscipula, Gymnopodium floribundum*), 50.7% were found to be part of the flavonoid chemical class, whereas in their respective honey samples, only 16.4% of the identified metabolites were categorized as flavonoids. Vanillin and vitexin were tentatively identified as potential markers for the botanical origin identification of honey from *Piscidia piscipula* and *Gymnopodium floribundum*, respectively. Recognizing botanical and geographic BH origin is important for product authentication, identification, and traceability. This study offers chemical insights that can be valuable and complementary to melissopalynology, aiding in determining the origin and quality of Yucatan BH.

Keywords: *Gymnopodium floribundum*; *Piscidia piscipula*; polyphenols; bee honey; untargeted metabolomics; Yucatán

1. Introduction

The Yucatán Peninsula in México has a diverse and abundant flora, with approximately 2329 reported taxa at the species level grouped in 956 genera and 161 native or wild families. In the case of Yucatán, there is a record of 1402 reported plant species, distributed in 120 families and 652 genera, of which 11 species are endemic. As part of the extensive range of documented plant species, the classification of melliferous flora has critical importance because it constitutes 40% of the reported species in Yucatan, representing about 849 species that serve as sources of nectar and pollen for bees. Bees, in turn, transform these resources into various products, like propolis, wax, and, primarily, honey [1]. The production of honey is one of the most important activities in Yucatan due to the abundance of a large quantity of plants (mentioned earlier). The honey production in Yucatan contributes 15.83% of Mexico's national production (9180 tons per year) [2]. The quality of Yucatecan BH is highly preferred in the international market; a total of 95% of the BH produced in Yucatán is destined for the markets of Germany, Switzerland, and England [3]. The quality of the BH is determined by the available floral composition for the foraging of *Apis mellifera*; in this sense, it is essential for apiaries to be located in areas with a high availability of nectar and pollen plant supply. It is reported that bees can fly between 2 and 5 km searching for floral resources [4].

According to melissopalynology studies characterizing honey in relation to pollen, it has been determined that certain taxonomic families exhibit a predominant presence in BH. Notably, among these families are Fabaceae and Polygonaceae [5]. Additionally, the presence of polyphenols in Yucatán's plants has been extensively studied. The scientific literature highlights the abundant presence of polyphenols in Yucatán's flora, showcasing significant antioxidant properties [6,7]. Alongside their antioxidant attributes, these plants yield a diverse range of secondary metabolites, carrying medicinal and therapeutic potential that has been harnessed by local cultures since ancient times. The metabolites present in the BH flora of Yucatan include flavonoids, terpenes, phenolic acids, and alkaloids, among others. This results in plants containing antioxidants, anti-inflammatory, antimicrobial, and analgesic characteristics. These plants have a history of use in traditional medicine for addressing a range of issues, including headaches, inflammation, infections, and gastrointestinal disorders. [8,9]. A study on the floristic knowledge of the Yucatan Peninsula reported that Fabaceae, commonly known as legumes, is one of the most diverse families in the region, with 230 species reported [10]. In these species, flavonoids, isoflavonoids, and quinones are reported as the prominent classes of secondary metabolites [11]. An important tree of Fabaceae species is the *Piscidia piscipula*, commonly known as "Jabin", a well-distributed tree within the Yucatán Peninsula, Mexico, that can reach up to 20 m in height, and its flowering occurs during the dry months of March and April. The inflorescences are generated on the branches, where up to 50 flowers develop. These flowers represent a food source for bees and many other insects [12] and play a significant ecological and socio-cultural role due to their use as a firewood (fuel) source or as poles, and it is considered by locals as a "godmother" plant or "mother of fire" [13]. The *Piscidia piscipula* honey has been classified as a potential source of monofloral honey [14] and contains several types of polyphenols, including flavonoids, such as genistein and quercetin, and phenolic acids, such as caffeic acid and chlorogenic acid; these metabolites have also been found in Jabin bark, with significant antioxidant anti-inflammatory properties [15,16]. Another plant family reported to be presented in Yucatecan BH is Polygonaceae, a diverse group of plants that include herbs, shrubs, trees, and vines; this family has alternate leaves, small flowers, and inflorescences with cymes form. A phytochemical study revealed the presence of tannins, saponins, flavonoids, alkaloids, and sesquiterpenes [17]. *Gymnopodium floribundum*, a prominent member of the Polygonaceae family in Yucatan, holds significant importance. Referred to as 'Dzidzilche' in the Mayan language, this species plays a vital role within the Yucatecan region, as it is utilized in commercial honey production [18] and traditional medicine for its anti-inflammatory and analgesic properties due to the high content of polyphenols in its leaves [19,20] as well as its tannins [21].

Speaking exclusively about flowers, these are a source of bioactive compounds of interest [22]; an example is the melliferous flora of Yucatán, which is important not only for honey production but also for its medicinal and therapeutic properties due to terpenes, polyphenols, and alkaloids, three metabolites present in these plants.

The polyphenol group is the most widely distributed and diverse category in plants from all around the world [23]. Polyphenols contribute to taste, color, odor, astringency, oxidative stability, bitterness, and chemical defense; they are synthesized by the phenylpropanoid and/or polyketide metabolic pathways derived from shikimic acid, and phenylalanine is the main compound needed for the synthesis of other polyphenols, such as flavonoids, phenolic acids, stilbenes, and lignans [15]. All the reported metabolites in *Gymnopodium floribundum* and *Piscidia piscipula* species from Yucatan are the result of biotic and abiotic stress in the region, mainly environmental conditions, such as climate and humidity. In this sense, the total content of polyphenols in flowers varies according to factors such as species, plant age, geographic origin, and environmental conditions [15,24].

A behavior similar to that in flower metabolites is observed in the chemical composition of honey, which is highly variable and mainly depends on the geographical and floral origin. Derived from this, honey is classified according to two characteristics: the region of origin and the flora of that region [25,26]. In a study conducted by Peilin et al. [27], the effects of different floral origins of honey (*Schisandra chinensis, Vicia dichroant, Astragalus sinicus, Robinia pseudoacacia,* and *Ziziphus jujuba*) on alcohol metabolism and their correlation with chemical composition were found to have significant differences in the polyphenol profile, especially in protocatechuic acid, p-Hydroxybenzoic acid, kaempferol, apigenin, and caffeic acid. Generally, it is recognized that evaluating the patterns of phenolic acids, phenolic esters, and aromatic carbonyl compounds can provide a good indication of the botanical origin of honey [28].

Some phenolic acids (e.g., ellagic acid) are used as floral markers of brezo honey (*Calluna vulgaris*), while the hydroxycinnamates (caffeic, p-coumaric, and ferulic acids) are used as floral markers of castaño honey (*Castanea sativa*) [29]. On the other hand, pinocembrin, pinobanksin, and chrysin are characteristic flavonoids found in most European honey [30], while abscisic acid was detected in rapeseed (*Brassica napus*), lime (*Citrus* sp.), and acacia (*Acacia* sp.) tropical honey [31].

Hence, the importance and interest in understanding the effect of geographical origin and different species of nectar-producing flowers are evident. Derived products, such as honey produced by *Apis mellifera* (hereinafter honeybee), have the potential to contain both the metabolites and bioactive properties that can be advantageous for human health. In this study, our objective was to investigate the impact of geographical location and flower species on metabolites identified using LC–MS2. This approach represents a novel and innovative process seeking to establish potential correlations between these factors and the honey produced by *Apis mellifera* from these local (Yucatán) floral species.

2. Materials and Methods

2.1. Collection of Plant Material

The collection of Dzidzilche (*Gymnopodium floribundum*) and Jabin (*Piscidia piscipula*) flowers was carried out in February, March, and April 2023 in two different apiaries in two locations in Yucatán, Mexico: Tahdziu and Acanceh. The selected sampling locations for plant material were near the apiaries (500 m approximately).

Flower collection consisted of taking a fragment of the complete terminal branch where flowering was present. Subsequently, the flowers were stored in resealable plastic bags in a Styrofoam container with ice for transportation (<4 h) to the Research and Technology Assistance Center of the State of Jalisco (CIATEJ), southeast unit.

The collected samples consisted of complete inflorescences of each species; however, only floral structures were exclusively utilized for the analyses. For the Jabin species (Figure 1a), the open flowers with the pedicel (3) and closed flowers with the pedicel (4) were employed, while the rachis (2) and developing fruits (1) were excluded. In the case of the Dzidzilche species (Figure 1b), we utilized the open flowers with the pedicel (1) and closed flowers with the pedicel (3); the rachis (2) was discarded.

Figure 1. (**a**) Floral structures of *Piscidia piscipula* (Jabin). (**b**) Floral structures of *Gymnopodium floribundum* (Dzidzilche). Figure (**a**) numeration: 1 = developing fruit; 2 = rachis; 3 = open flowers with pedicel; 4 = closed flowers with pedicel. Figure (**b**) numeration: 1 = open flower with pedicel; 2 = rachis; 3 = closed flower with pedicel.

2.2. Selection and Drying of Flowers

The selection of flower structures took place at the CIATEJ facilities for subsequent freeze-drying. Freeze-drying was performed according to Oney-Montalvo et al. [32] with some modifications. Before freeze-drying, the selected flowers were ultra-frozen (−32 °C, 24 h) in resealable plastic bags. Then, the frozen flowers were freeze-dried for 72 h at a temperature of −50 °C and a pressure of 240 mBar using the Freezone lyophilizer, LABCONCO® Brand, C to obtain a moisture content of ≤5%.

The Jabin flower sample obtained from the location of Tahdziu was used as a conventional dry control. It was dried according to Chel-Guerrero et al. [33], with some modifications, in a FELISA oven (model FE-292) at 50 °C for 48 h to obtain a moisture content of ≤5%.

Grinding and Sieving

Freeze-dried and oven-dried flowers were first ground using a coffee grinder (MasterChef®) until homogeneous powders (flower flours) were obtained. The flower flours were sieved using a No. 35 mesh (500 µm particle size). The final product was stored in resealable bags, labeled, and lined with aluminum foil at room temperature until use.

2.3. Determination of Moisture in Flowers and Flower Flours

The fresh flowers selected were subjected to moisture determination prior to the drying process (freeze-drying or oven-drying) according to the methodology reported by Avilés-Betanzos et al., 2023 with some modifications [34]. With an OHAUS® thermobalance (model MB90, NJ, USA), the moisture content of the samples was determined by placing 0.5 g of sample in an aluminum plate; the samples were kept in the equipment at a constant temperature of 105 °C until a constant weight was reached (weight loss < 1 mg in 60 s). The same procedure was followed with both the fresh flower and the flower flours. Moisture was the difference between the initial and final weights of the sample and was reported in percentage and in triplicate.

2.4. Extraction of Metabolites in Flower Flours

2.4.1. Obtainment of Total Polyphenol Content (TPC) and Antioxidant Capacity (AC)

The extraction process was carried out according to the methodology reported by Avilés-Betanzos et al. [34] with some modifications. A sample of flower flour was weighed (0.5 g); then, 2.5 mL of a methanol: water solution (80:20) was added. The solution was mixed with a vortex until a homogeneous solution was obtained. The solution was sonicated for 30 min at a frequency of 42 kHz using a BRANSON® sonicator (model 3510). After sonication, the samples were centrifuged at 4700× g rpm for 30 min at 4 °C. The supernatants were recovered and filtered using a 0.22 μm nylon filter. Finally, the sample was placed in chromatographic vials and refrigerated (5 °C) until use (<48 h).

2.4.2. Flower Extraction for LC–MS^2

We followed the methodology previously described by Contreras-Angulo et al. [35] with minor modifications. To 5 mg of freeze-dried flowers, 500 μL of a solvent mixture of methanol: acetonitrile (ACN): ethyl acetate (1:1:1) was added, and the mixture was vortexed for 1 min and sonicated for 30 min in ice-cold water. Then, the samples were centrifuged at 14,000× g rpm for 10 min at 4 °C, and 400 μL of the supernatant (top layer) was recovered, transferred to an Eppendorf tube, and dried in a SpeedVac system at room temperature. The dried extract was resuspended with 200 μL of a mixture of water: ACN (80:20), vortexed again for 1 min, and sonicated for 30 min in ice-cold water. Finally, extracts were centrifuged at 14,000 rpm for 10 min at 4 °C to obtain a particle-free supernatant suitable for LC–MS^2 data acquisition.

2.5. Extraction of Metabolites in Honey

2.5.1. Honey Dilution for Determination of CTP and AC

The dilution was carried out according to Al-Mamary et al. [36] with some modifications. A sample of honey (1 g) was weighed, and distilled water was added until a total weight of 10 g was reached. The sample was then vortexed to make a homogeneous solution to be used later to assess CTP and AC.

2.5.2. Honey Extraction for LC–MS^2 Data

A sample of 100 mg of freeze-dried honey was extracted with 500 μL of methanol: ACN: ethyl acetate (1:1:1) solution under sonication for 30 min in ice-cold water. Thereafter, the same steps described in Section 2.4.2 were followed. However, after the evaporation of solvents, the extracts were resuspended at 500 mg/μL using a mixture of water: ACN (80:20).

Then the same methodology as reported in Section 2.4.2 was subsequently followed.

2.6. Experimental Design

To evaluate the effects of the geographical zone on the total polyphenol content and antioxidant capacity in Dzidzilche and Jabin flowers and honey, a 2^3 factorial design was established. The levels of the main factors (botanical origin, geographical zone, and raw material) were presented at two levels (−1 for Tahdziu, Jabin, and flower and 1 for Acanceh, Dzidzilche, and honey), as shown in Table 1. The raw material factor (honey and flower) was also individually analyzed considering the factors of botanical origin and geographical origin.

Table 1. Experimental 2^3 design employed for the response variables total polyphenol content and antioxidant capacity.

Exp	Coded Values			Real Values			Variable Response	
	X_1	X_2	X_3	BO	GO	RM	TPC (mg GAE/100 g DM)	AC (% Inhibition)
1	−1	−1	−1	Jabin	Tahdziu	Flower	Y_1	Y_2
2	1	−1	−1	Dzidzilche	Tahdziu	Flower	Y_1	Y_2
3	−1	1	−1	Jabin	Acanceh	Flower	Y_1	Y_2
4	1	1	−1	Dzidzilche	Acanceh	Flower	Y_1	Y_2
5	−1	−1	1	Jabin	Tahdziu	Honey	Y_1	Y_2
6	1	−1	1	Dzidzilche	Tahdziu	Honey	Y_1	Y_2
7	−1	1	1	Jabin	Acanceh	Honey	Y_1	Y_2
8	1	1	1	Dzidzilche	Acanceh	Honey	Y_1	Y_2

Note: Exp = experiment; BO = botanical origin; GO = geographical origin; RM = raw material; TCP = total polyphenol content; AC = antioxidant capacity.

2.6.1. Determination of TPC in Dzidzilche and Jabin Flower Flours and Honey

The polyphenols contained in the extracts were evaluated using the Folin–Ciocalteu method with some modifications, following Singleton et al. [37]. A total of 25 µL of the sample (flower flour extract or honey dilution) was taken and diluted with distilled water (25 µL). Then, 3 mL of distilled water and 250 µL of Folin–Ciocalteu reagent (St. Louis, MO, USA) were added. After 5 min, 750 µL of 20% sodium carbonate (Na_2CO_3, 20%, St. Louis, MO, USA) and 950 µL of distilled water were added, and the mixture was incubated for 30 min. Finally, the samples were read at 765 nm using a quartz cell and UV–Vis spectrophotometer (THERMO SCIENTIFIC® Genesys 140 (Waltham, MA, USA)) to determine absorbance. A calibration curve (0.005−0.1 mg of gallic acid/mL) was prepared prior to sample analysis (flower and honey), obtaining an R^2 = 0.999 for both curves (the graphs can be observed in Figure S1).

2.6.2. Determination of AC in Dzidzilche and Jabin Flower Flours and Honey

The antioxidant capacity of the extracts was evaluated using the DPPH method, according to Brand et al. [38]. The 2,2-diphenyl-1- picrylhydrazyl (DPPH) reagent from Sigma Aldrich® (D9132-1G, St. Louis, MO, USA) was weighted (3.3 mg); then, reagent-grade methanol was added (100 mL). The obtained DPPH solution was adjusted (absorbance 0.700 ± 0.002) with a UV–Vis spectrophotometer (THERMO SCIENTIFIC® Genesys 140). To establish the antioxidant capacity, a sample volume of 100 µL (extract of Dzidzilche and Jabin flower flours and honey) was taken, and then a volume of 3.9 mL of DPPH-adjusted solution was added. The mixed solution was stirred and incubated for 30 min. Finally, the samples were read at 515 nm. The results of the extract were presented as the percentage of inhibition (%), according to Equation (1):

$$\%\ Inhibition = \left(\frac{DPPH\ adj - DPPH\ sam}{DPPH\ adj} \right) \quad (1)$$

DPPHadj = Absorbance of adjusted DPPH solution.

DPPHsam = Absorbance of sample (extracts of Dzidzilche and Jabin flower flours and honey).

2.7. Untargeted LC–MS² Data Acquisition

We followed the instrumentation and protocol previously described [35] with some modifications. We loaded 4 µL of the resuspended metabolite solutions into an Agilent 1260 Infinity LC (Agilent Technologies, Inc., Santa Clara, CA, USA). For the separation of molecules, we employed a ProtID-Chip-43 II column (C18, 43 mm × 75 µm, 300 Å, 5 µm particle size, equipped with a 40 nL enrichment column). The mobile phases were H_2O

with 0.1% formic acid (FA) as solution A and ACN with 0.1% FA as solution B. The gradient started at 5% B, was increased linearly to 40% B in 20 min and then to 100% B for 5 min, was held at 100% B for 5 min, returned to 5% B within 1 min, and was finally maintained at 5% B for 9 min before the next sample to ensure column re-equilibration. The total analysis time was 40 min at a flow rate of 300 nL/min. To minimize potential carryover, two blank samples (5 µL of mobile phases A and B at an 80:20 ratio) were run between experimental samples. The eluate was delivered to a 6530 Accurate-Mass Q-TOF mass spectrometer (Agilent Technologies, Inc., Santa Clara, CA, USA) using an HPLC-Chip Cube MS interface. Nanospray ionization under positive mode was utilized. Data-dependent acquisition was employed over the mass range 110–2000 *m/z* with a scan speed of 4 spectra/s. MS2 data were retrieved from the top 5 most intense precursor ions per cycle, reaching 150 cps over the mass range 50–2000 *m/z* at a rate of 3 spectra/s. The active exclusion option was enabled, set to 2 spectra, and disabled after 0.25 min. A ramped collision energy (CE) was used with slope and offset values of 6 and 4, respectively. Before data acquisition and every 24 h, the instrument was externally calibrated using an ESI-L low mix concentration tuning solution (Agilent Technologies, Inc., Santa Clara, CA, USA). The samples were randomly allocated in the autosampler for data acquisition.

LC–MS^2 Data Processing

The methodology followed was based on that reported previously by our group [35] with some modifications. The LC–MS^2 datasets were analyzed using a workflow that consisted of open-source software packages and free-access online platforms. First, datasets with Agilent, Santa Clara, CA, USA, proprietary format were converted to open-format files (.mzXML) using the MSConvert tool (version 3) from ProteoWizard [39]. The mzXML files were uploaded to the Global Natural Products Social Molecular Networking (GNPS) online platform (https://gnps.ucsd.edu (accessed on 25 august 2023)) [40] to perform classical molecular networking (CMN) and automated structural annotation (against GNPS spectral libraries) (Metabolomics Standards Initiative (MSI) classification level 2 [41]). The principal parameters used within GNPS were 0.02 Da for the mass tolerance of precursor and product ions, a minimum cosine score of 0.6, and at least 4 peaks matched at the MS^2 level. To enhance the putative identification of metabolites (MSI, level 3), we utilized MolDiscovery [42] and DEREPLICATOR+ (version 1.0.0) [43] in silico tools integrated into the GNPS platform. The clustered .mgf file generated by CMN was used for MolDiscovery analysis, and the DEREPLICATOR+ analysis was run through the "Annotate with DEREPLICATOR+" option on the results page of CMN. The precursor ion and product ion mass tolerances were set to 0.02 Da, while all other parameters were left as default. As an additional annotation tool, the CSI: FingerID algorithm from the SIRIUS software (version 4.9.12) [44] was used to assign structures found in the BioDatabases (MSI, level 3). First, a molecular formula was assigned with SIRIUS without databases [45] and reranked with the ZODIAC algorithm [46]. The chemical classes of the identified metabolites were assigned on the basis of ClassyFire chemical ontology using the web application http://classyfire.wishartlab.com (accessed on 25 August 2023) [47]. Spectra found in blank samples and annotations with >10 ppm mass accuracy and >5% FDR were filtered out from the results.

2.8. Statistical Analysis

The data presented are reported as means ± standard deviations. Data analysis for a factorial 2^3 design was employed to assess the total polyphenol content and antioxidant capacity of the factors (1) botanical origin (Dzidzilche or Jabin flowers), (2) geographical origin (Acanceh or Tahdziu locations), and (3) raw material (flower or honey) using an analysis of variance (ANOVA) to determine the effects and their interactions. These analyses were performed using the statistical software Statgraphics Centurion XVII.II-X64 (Statgraphics Technologies Inc., Virgin, UT, USA). XL-STAT software (version 2021.2.2, Addison, Paris, France) was utilized in the CATA analysis to obtain the principal components analysis of the metabolites obtained in the samples. The analysis of the linear correlation between the concentrations of total polyphenols and the antioxidant capacity obtained by the DPPH was carried out with a linear correlation utilizing Microsoft Excel (version 2019).

3. Results

3.1. Collection of Plant Material

The sampled apiaries (Acanceh and Tahdziu locations) were in two different areas of the state of Yucatan, Mexico, as indicated in Figure 2. The apiary located in Tahdziu (20°19′22.4″ N 089°24′38.6″ W) had scattered vegetation; it was surrounded by hectares with agricultural activity, such as livestock farming, agriculture, and other apiaries. On the contrary, the apiary located in Acanceh (20°45′28.2″ N 089°23′18.5″ W) had dense and closed vegetation, surrounded by unexploited lowland jungle unaffected by commercial activities.

Figure 2. Map of the state of Yucatan with the locations of the sampled apiaries.

The samples were labeled using codes for identification, which reflected their botanical origin (Jabin or Dzidzilche) and their geographical origin (Tahdziu or Acanceh). At the end of the plant material collection, 365 g of Dzidzilche material and 658 g of Jabin were obtained. The collected sample could contain additional structures other than flowers, principal stems, leaves, and secondary branching; however, these materials were discarded, as described in the following section.

3.2. Selection and Drying of Flowers

An initial weight of 365 g of Dzidzilche material was collected; after finalizing the process of classification and the selection of structures, 125 g remained that corresponded to flower structures. Similarly, Jabin decreased from the initial 658 g to 315 g of plant material corresponding to floral structures. After the drying and grinding process, 47 g and 39 g of floral flour were obtained from the Jabin and Dzidzilche species, respectively.

3.3. Moisture Determination in Flowers and Flower Flours

The Jabin species demonstrated a higher initial moisture percentage than the Dzidzilche species in both the Tahdziu (80.25 ± 0.19%) and Acanceh (72.26 ± 0.62%) apiaries. On the other hand, the Dzidzilche species showed the lowest initial moisture percentage in the Tahdziu locality (56.53 ± 1.09). Statistically significant differences ($p < 0.0001$) were found in the initial moisture between the species, resulting in three homogeneous groups: ACA-DZI and TAH-DZI, ACA-JA, and TAH-JA. Regarding the final moisture, a significant difference was also obtained ($p = 0.0082$), as shown in Table 2.

Table 2. Moisture percentages in Jabin and Dzidzilche flowers from Tahdziu and Acanceh apiaries.

	Moisture (%)	
Sample	**Initial**	**Final**
ACA-DZI	56.95 ± 1.30 [a]	2.91 ± 0.34 [bc]
TAH-DZI	56.53 ± 1.09 [a]	2.58 ± 0.18 [ab]
ACA-JA	72.26 ± 0.62 [b]	2.45 ± 0.27 [a]
TAH-JA	80.25 ± 0.19 [c]	3.33 ± 0.11 [c]

Note: ACA = Acanceh (geographical origin); TAH = Tahdziu (geographical origin); DZI = Dzidzilche (botanical origin); JA = Jabin (botanical origin). The values are the result of the average ± the standard deviation (n = 3). Different letters in each column show statistically significant differences (LSD, $p > 0.05$).

3.4. TPC and AC of Flower Flours and Honeys

The results of the total polyphenol content (TPC) and the antioxidant capacity (AC, with DPPH methodology) as response variables of the 2^3 factorial designs experiments are presented in Table 3. The data were obtained from two geographical locations (Tahdziu and Acanceh) and two different species (Jabin and Dzidzilche) using honey and flowers as raw materials.

Table 3. Results of total polyphenol content and antioxidant capacity for the experimental design (2^3).

Exp	Coded Values			Real Values			Variables Response	
	X_1	X_2	X_3	BO	GO	RM	TPC (mg GAE/100 g DM)	AC (% Inhibition)
1	−1	−1	−1	Jabin	Tahdziu	Flower	347.57 ± 1.33 [e]	83.64 ± 0.46 [a]
2	1	−1	−1	Dzidzilche	Tahdziu	Flower	294 ± 0.11.09 [d]	85.3 1± 0.87 [b]
3	−1	1	−1	Jabin	Acanceh	Flower	661.64 ± 1.98 [c]	92.39 ± 0.22 [c]
4	1	1	−1	Dzidzilche	Acanceh	Flower	1431.24 ± 15.38 [b]	93.63 ± 0.22 [d]
5	−1	−1	1	Jabin	Tahdziu	Honey	8.70 ± 0.51 [a]	13.68 ± 0.75 [e]
6	1	−1	1	Dzidzilche	Tahdziu	Honey	9.92 ± 0.56 [a]	18.52 ± 0.25 [f]
7	−1	1	1	Jabin	Acanceh	Honey	5.38 ± 0.14 [a]	21.60 ± 0.36 [g]
8	1	1	1	Dzidzilche	Acanceh	Honey	5.28 ± 0.47 [a]	20.32 ± 0.50 [h]

Note: The values are the result of the average ± the standard deviation (n = 3). Exp = experiment. BO = botanical origin. GO = geographical origin. RM = raw material. TCP = total polyphenol content AC = antioxidant capacity. Different letters in each column show statistically significant differences (LSD, $p > 0.05$).

The results showed that the highest concentration of total polyphenols (1431.24 ± 15.38 mg GAE/100 g dry matter) was obtained in the flower with the botanical origin of Dzidzilche and the geographical origin of Acanceh. In general, flowers exhibited higher values of total polyphenol content than the BH.

On the other hand, the BH contained a low concentration of polyphenols, in which the lowest TPC (5.28 ± 0.47 mg GAE/100 g dry matter) was obtained in the Dzidzilche sample from Acanceh.

Meanwhile, flowers developed a higher antioxidant capacity than the BH. Flowers with the botanical origin of Dzidzilche and the geographical origin of Acanceh exhibited the highest antioxidant capacity (93.63% inhibition) ($p < 0.05$). By contrast, the lowest antioxidant capacity was observed in the BH with Jabin botanical origin and Tahdziu geographical origin (13.68% inhibition).

Figure 3 demonstrates the significant effect of all factors and interactions on the total polyphenol content. However, the raw material factor, as expected, presented a greater statistical effect due to the inherent differences in the materials. The Pareto chart revealed that the main factors, raw material (flowers) and geographical origin (Acanceh), presented significant effects ($p < 0.05$) on the antioxidant capacity. The effect of all factors on the TPC and AC was confirmed by the interaction chart (Figure S2) and the main factors chart (Figure S3).

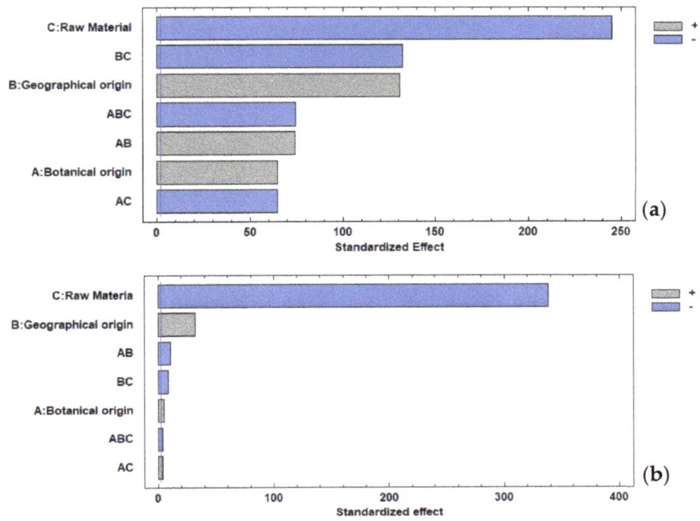

Figure 3. Pareto chart: (**a**) total polyphenol content, (**b**) antioxidant capacity.

Figure S4 shows the effects of the individual factors (botanical origin and geographical origin) and their interaction on AC and TPC in the flowers and their honeys. In this sense, the geographical origin (Tahdziu) had a greater statistical effect than the botanical origin (Jabin); a similar trend was observed for the AC. In addition, the geographical origin (Acanceh) had a greater statistical effect than the botanical origin (Dzidzilche).

In the individual analysis, the flower (Figure S5) showed a significant effect of all factors and interactions on the TPC. However, the geographical origin (Acanceh) presented a greater statistical effect than the botanical origin (Dzidzilche). For AC, the geographical origin (Acanceh) and the interaction had a significant effect, while the botanical origin did not.

3.5. Identification of Metabolites in Flower Flours and Honey of Dzidzilche and Jabin

From the entire dataset, we obtained 312 peaks containing MS^2. We focused solely on these peaks to obtain annotations at the molecular structure level through MS^2 spectral matching against spectral libraries or through in silico predictions. After filtering out contaminants or low-quality annotations, a total of 55 metabolites in the BH (Table S1) and 69 metabolites in the flower (Table S2) samples were putatively identified, with an identification rate of approximately 24%. Notably, the annotations reported herein are putative, and due to the limitations of the technique, we could not differentiate between enantiomers. Therefore, the reported molecular structures are those automatically retrieved from the automated spectral match and in silico predictions.

Raw materials (flowers and BH) were grouped to provide a comprehensive depiction of the chemical classes of the metabolites identified in both flowers and BH. A distinct chemical profile among flower and BH samples was observed. In flowers (Figure 4a), the 69 annotated compounds belonged to 15 chemical classes, and the flavonoid chemical class accounted for 50.7% of the total identified metabolites. By contrast, the 55 compounds annotated in BH (Figure 4b) spanned 16 chemical classes, with only one more chemical class than the flower repertoire. Despite this, the presence of flavonoids was lower in BH than in flowers, constituting 16.4% of BH metabolites. On the other hand, the percentage representation of carboxylic acids and their derivatives, organonitrogen compounds, and benzene and substituted derivative metabolites were more similar between flowers and BH.

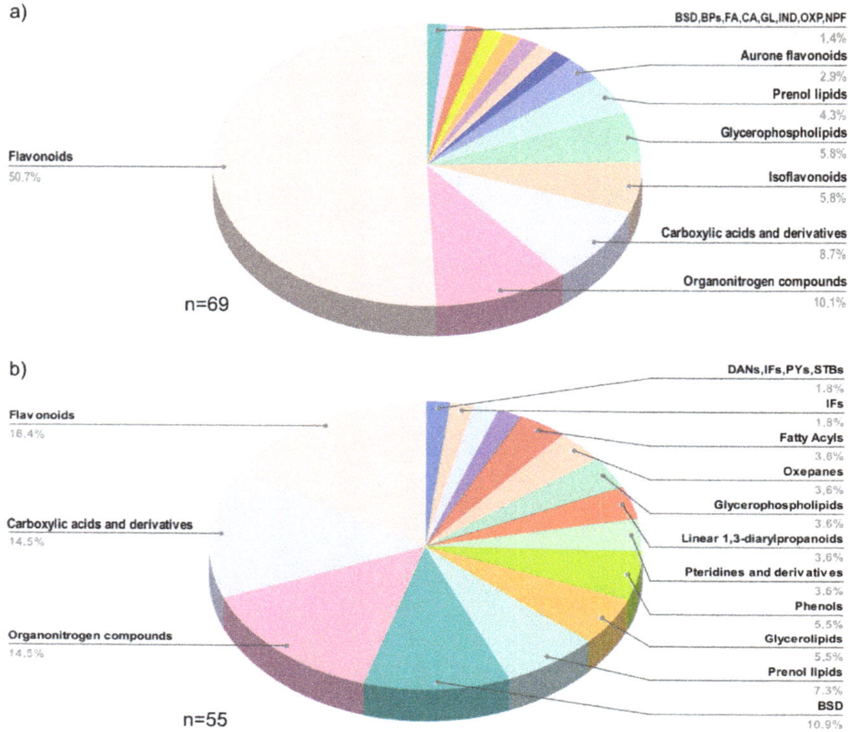

Figure 4. Distribution of chemical classes in (**a**) flower samples, BSD = benzene and substituted derivatives; BPs = benzopyrans; FA = fatty Acyls; CA = cinnamic acids and derivatives; GL = glycerolipids; IND = indoles and derivatives; OXP = oxepanes; NPF = naphthofurans; and (**b**) honey samples, BSD = benzene and substituted derivatives; DANs = diazanaphthalenes; IFs = isoflavonoids; PYs = pyrans; STBs = stilbenes.

Four chemical classes (aurone flavonoids, indoles and their derivatives, cinnamic acids and their derivatives, and benzopyrans) were only observed in flower samples. Moreover, the BH samples exhibited six chemical classes (pteridines and their derivatives, pyrans, stilbenes, diazanaphthalenes, linear 1,3-diarylpropanoids, and phenols) that were not present in the flower samples, as shown in Figure 5.

The comparative analysis of floral samples revealed a higher number of compounds (30 compounds) in Dzidzilche flowers from Acanceh compared with those from Tahdziu (27 compounds). Both Dzidzilche and Tahdziu samples displayed an equal number of chemical classes (n = 11). However, while cinnamic acids and their derivatives were present in Tahdziu, oxepanes were identified in the Acanceh samples (Figure 6a). Regarding botanical origin, Jabin flowers from Acanceh presented a higher number of compounds (50 compounds) than those from Tahdziu (32 compounds). These 50 metabolites were categorized into 12 different chemical classes, with the notable difference that fatty acyls were present in the Tahdziu samples and absent in the Acanceh samples (Figure 6b). For bee honey originating from Dzidzilche, a higher number of compounds were found in the Tahdziu location than in the Acanceh location (42 and 38 compounds, respectively). These compounds were classified into 15 chemical classes, wherein linear 1,3-diarylpropanoids were solely found in the Tahdziu location (Figure 6c). BH with Jabin botanical origin displayed a higher number of compounds in Tahdziu than Acanceh (48 and 43 compounds, respectively), categorized into 16 chemical classes. Among these isoflavonoids, linear

1,3-diarylpropanoids, oxepanes, and glycerophospholipids were present in the Tahdziu location (Figure 6d).

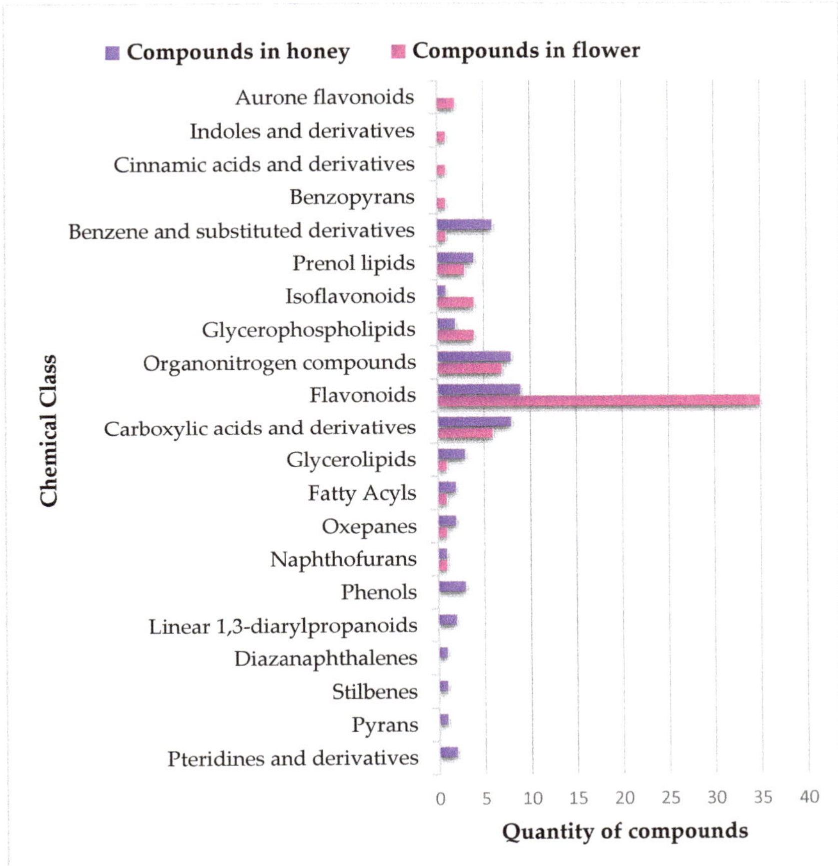

Figure 5. Comparative global analysis of the distribution of metabolites from various chemical classes among flower and honey samples.

Although the BH and flower samples shared botanical and geographical origin, differences at the chemical class level were observed. For instance, flowers of Dzidzilche botanical origin from the Tahdziu location exhibited the presence of cinnamic acids and their derivatives, aurone flavonoids, indoles and their derivatives, and naphthofurans (Figure 7a), while BH (associated with the same botany and geography) differentially presented linear 1,3-diarylpropanoids, pyrans, stilbenes, diazanaphthalenes, fatty acyls, phenols, and oxepanes. Bee honey associated with Acanceh flowers from Dzidzilche lacked aurone flavonoids, indoles and their derivatives, and naphthofurans but presented pyrans, stilbenes, diazanaphthalenes, fatty acyls, phenols, and pteridines and their derivatives (Figure 7b). Jabin flowers from Acanceh uniquely presented aurone flavonoids, indoles and their derivatives, naphthofurans, isoflavonoids, and glycerophospholipids, while BH (associated with the same botany and geography) uniquely presented pyrans, stilbenes, diazanaphthalenes, fatty acyls, and phenols (Figure 7c).

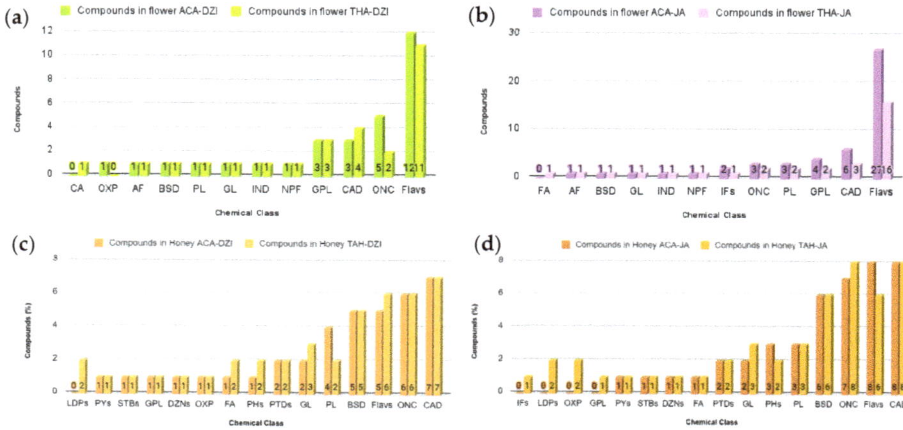

Figure 6. Comparative analysis of the distribution of metabolites from various chemical classes in flower and honey samples according to geography. (**a**) Dzidzilche flowers in Acanceh and Tahdziu. (**b**) Jabin flowers in Acanceh and Tahdziu. (**c**) Honey from Dzidzilche in Acanceh and Tahdziu. (**d**) Honey from Jabin in Acanceh and Tahdziu. BSD = benzene and substituted derivatives; BPs = benzopyrans; FA= fatty acyls; CA = cinnamic acids and derivatives; GL = glycerolipids; IND= indoles and derivatives; OXP = oxepanes; NPF = naphthofurans; DANs = diazanaphthalenes; IFs = isoflavonoids; PYs = pyrans; STBs = stilbenes; AF = aurone flavonoids; GPL = glycerophospholipids; CAD = carboxylic acids and derivatives; PL= prenol lipids; ONC = organonitrogen compounds; Flavs = flavonoids; DZNs = diazanaphthalenes; PHs = phenols; PTDs = pteridines and derivatives; LDPs = linear 1,3-diarylpropanoids. ACA = Acanceh (geographical origin); TAH = Tahdziu (geographical origin); DZI = Dzidzilche (botanical origin); JA = Jabin (botanical origin).

Jabin flowers from Tahdziu uniquely presented aurone flavonoids, indoles and their derivatives, and naphthofurans, while BH (associated with the same botany and geography) uniquely presented oxepanes, pyrans, stilbenes, diazanaphthalenes, pteridines and their derivatives, phenols, and linear 1,3-diarylpropanoids. Overall, aurone flavonoids, indoles and their derivatives, naphthofurans, isoflavonoids, and glycerophospholipids were found in flowers, while BH (associated with the same botany and geography) uniquely presented pyrans, stilbenes, diazanaphthalenes, fatty acyls, and phenols (Figure 7c). Naphthofurans were consistently solely found in flowers when compared with their BH counterpart (same botany and geography).

Principal Component Analysis (PCA) and Lineal Correlation

To examine similarities between flowers and BH considering their botany and geography, PCA was conducted. This analysis involved three distinct groups of chemical classes: (1) flavonoids, (2) benzenes and their derivatives, and (3) cinnamic acid and its derivatives. These groups were chosen due to their prominence as representative classes in honey and flowers.

In the PCA chart for flavonoids (Figure 8), the proximity between flower M (Tahdziu–Jabin) and N (Tahdziu–Jabin, oven-dried) suggests a shared similarity. Conversely, all honey samples exhibited close distances, implying common traits. The metabolite afzelin (O) demonstrated a close relationship between Dzidzilche flowers and honey from Acanceh and Tahdziu.

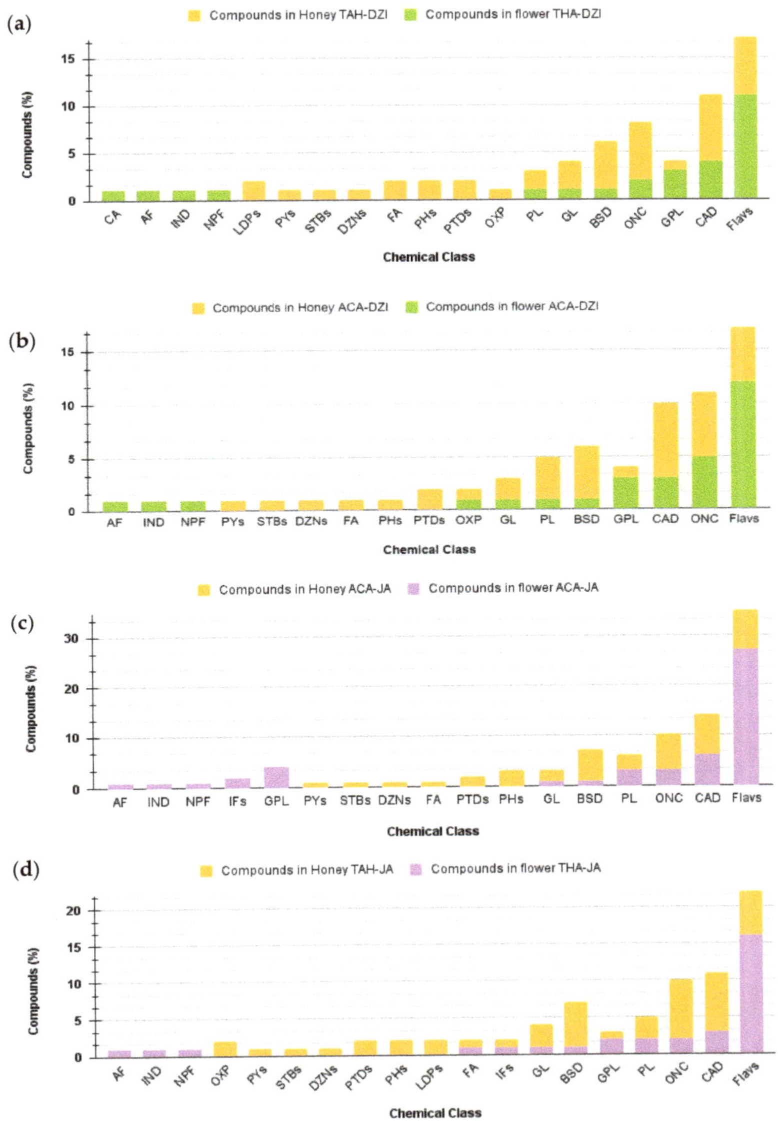

Figure 7. Comparative analysis of the distribution of metabolites from various chemical classes among flower and honey samples from the same botany and geography. (**a**) Dzidzilche flowers in Acanceh and Tahdziu. (**b**) Jabin flowers in Acanceh and Tahdziu. (**c**) Honey from Dzidzilche in Acanceh and Tahdziu. (**d**) Honey from Jabin in Acanceh and Tahdziu. Abbreviations: BPs = benzopyrans; FA = fatty Acyls; CA = cinnamic acids and derivatives; GL = glycerolipids; IND = indoles and derivatives; OXP = oxepanes; NPF = naphthofurans; DANs = diazanaphthalenes; IFs = isoflavonoids; PYs = pyrans; STBs = stilbenes; AF = aurone flavonoids; GPL = glycerophospholipids; CAD = carboxylic acids and derivatives; PL = prenol lipids; ONC = organonitrogen compounds; Flavs = flavonoids; DZNs = diazanaphthalenes; PHs = phenols; PTDs = pteridines and derivatives; LDPs = linear 1,3-diarylpropanoids. ACA = Acanceh (geographical origin); TAH = Tahdziu (geographical origin); DZI = Dzidzilche (botanical origin); JA = Jabin (botanical origin).

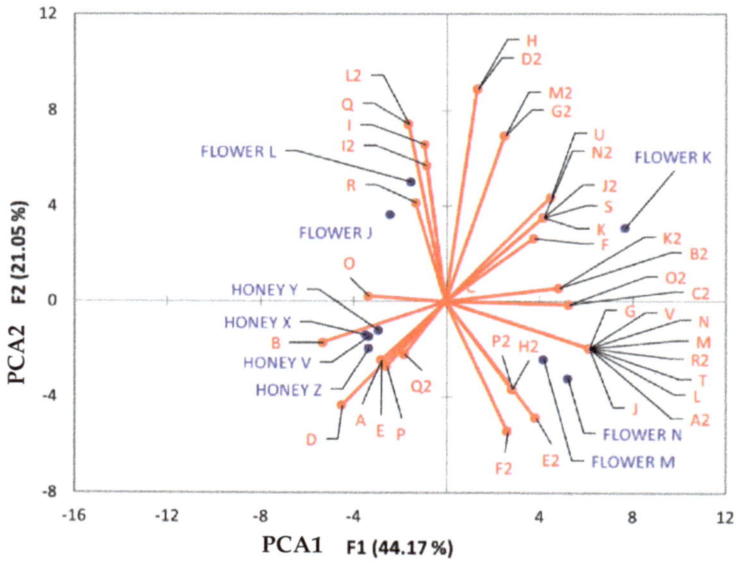

Figure 8. Principal component analysis (PCA) biplot of Dzidzilche and Jabin flower and honey samples based on the identified flavonoid metabolites. Abbreviations: A = naringin; B = vitexin; C = quercitrin; D = quercetin-3-O-rhamnoside; E = kaempferol-3-rhamninoside; F = kaempferol; G = aromadendrin; H = cianidanol; I = quercetin; J = taxifolin; K = isorhamnetin; L = 3,6,3′,4′-tetramethoxyflavone; M = retusin; N = tangeretin; O = afzelin; P = 3,5-dihydroxy-2-(4-hydroxyphenyl)-7-[(3,4,5-trihydroxy-6-methyltetrahydro-2H-pyran-2-yl) oxy]-4H-1-benzopyran-4-one; Q = reynoutrin; R = (-)-epicatechin gallate; S = luteolin-4′-O-glucoside; T = sinensetin; U = 3-[5-(1,2-dihydroxyethyl)-3,4-dihydroxyoxolan-2-yl] oxy-2-(3,4-dihydroxyphenyl)-5,7-dihydroxychromen-4-one; V = taxifolin-3-glucoside; A2 = isorhamnetin 3-galactoside; B2 = 2S,3S)-3,5,7-trihydroxy-2-[4-hydroxy-3-[(2S,3R,4S,5S,6R)-3,4,5-trihydroxy-6-(hydroxymethyl)oxan-2-yl]oxyphenyl]-2,3-dihydrochromen-4-one; C2 = luteolin 7-(6″ ″-malonylglucoside); D2 = procyanidin B2; E2 = 6″-O-(3-hydroxy-3-methylglutaroyl)astragalin; F2 = 3″-O-L-rhamnopyranosylastragalin; G2 = rutin; H2 = hesperidin; I2 = 2′-O-galloylhyperin; J2 = 3-[3,4-dihydroxy-6-(hydroxymethyl)-5-[3,4,5-trihydroxy-6-(hydroxymethyl)oxan-2-yl]oxyoxan-2-yl]oxy-2-(3,4-dihydroxyphenyl)-5,7-dihydroxychromen-4-one; K2 = rhamnetin 3-sophoroside; L2 = procyanidin B5, 3′-O-gallate; M2 = 2-(3,4-dihydroxyphenyl)-6,8-bis [2-(3,4-dihydroxyphenyl)-3,7,8-trihydroxy-3,4-dihydro-2H-chromen-5-yl]-3,4-dihydro-2H-chromene-3,5,7-triol; N2 = (2S,3S)-3,5,7-trihydroxy-2-[4-hydroxy-3-[(2S,3R,4S,5S,6R)-3,4,5-trihydroxy-6-(hydroxymethyl) oxan-2-yl]oxyphenyl]-2,3-dihydrochromen-4-one; O2 = 11-hydroxytephrosin; P2 = amorphigenin; Q2 = puerarin; R2 = 6-Hydroxysumatrol; FLOWER J = Acanceh–Dzidzilche; FLOWER K = Acanceh–Jabin; FLOWER L = Tahdziu–Dzidzilche; FLOWER M = Tahdziu–Jabin; FLOWER N = Tahdziu–Jabin oven-dried; HONEY V = Acanceh–Dzidzilche; HONEY X = Acanceh–Jabin; HONEY Y = Tahdziu–Dzidzilche; HONEY Z = Tahdziu–Jabin.

Compounds such as vitexin (B), pueratin (Q2), naringin (A), quercetin (D), kaempferol (E), and 3,5-dihydroxy-2-(4-hydroxyphenyl)-7-[(3,4,5-trihydroxy-6-methyltetrahydro-2H-pyran-2-yl) oxy]-4H-1-benzopyran-4-one (P) showed a clear association with all honey samples, irrespective of their botanical or geographical origins. Regarding the flowers, those stemming from Jabin demonstrated an association with phenolic compounds, including aromadendrin (G), taxifolin-3-glucoside (V), retusin (M), tangeretin (N), 3,6,3′,4′-tetramethoxyflavone (L), sinensetin (T), taxifolin (J), isorhamnetin 3-galactoside (A2), 6″-O-(3-hydroxy-3-methylglutaroyl) astragalin (E2), 6-hydroxysumatrol (R2), 3″-O-L-rhamnopyranosylastragalin (F2), amorphigenin (P2), and hesperidin (H2). On the other hand, flowers of Dzidzilche origin were associated with compounds like (-)-epicatechin gallate (R), reynoutrin (Q), 2′-O-galloylhyperin (I2), and procyanidin B5, and 3′-O-gallate (L2).

When analyzing benzenes and substituted derivatives (Figure 9), the analysis showed the following outcomes: compounds such as 3-methoxybenzoic acid (T2), benzylideneacetone (S2), tyramine (W2), and veratric acid (U2) were denoted as having a strong association with honey from Jabin, regardless of the geographical origin. Dzidzilche honey from Tahdziu showed an association with phenylalanine, whereas honey displayed an association with the compound 3-methylsalicylic acid (V2). Notably, no associations between flowers and compounds from this chemical class were found.

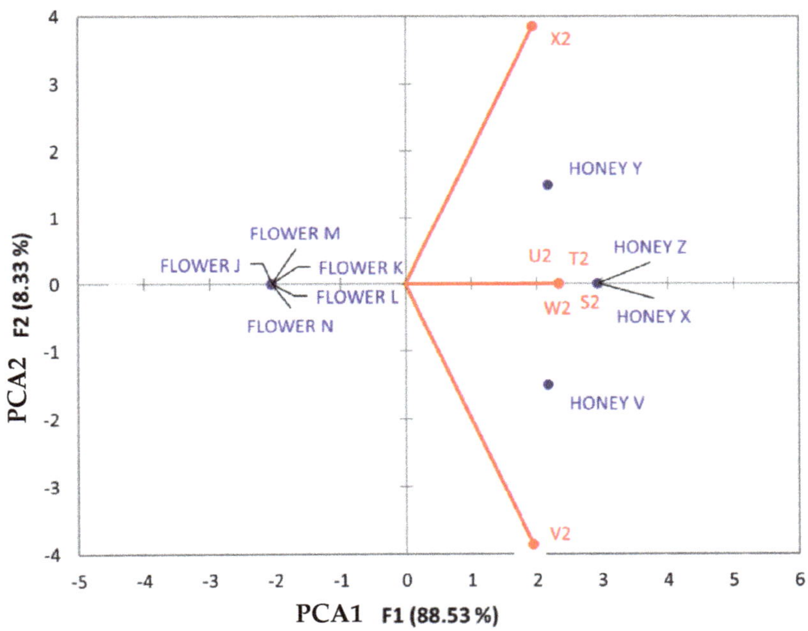

Figure 9. Principal Component Analysis (PCA) biplot of Dzidzilche and Jabin flowers and honey samples based on the identified benzenes and substituted derivate metabolites. Abbreviations: S2 = benzylideneacetone; T2 = 3-methoxybenzoic acid; U2 = veratric acid; V2 = 3-methylsalicylic acid; W2 = tyramine; X2 = n-(phenylacetyl)phenylalanine; FLOWER J = Acanceh–Dzidzilche; FLOWER K = Acanceh–Jabin; FLOWER L = Tahdziu–Dzidzilche; FLOWER M = Tahdziu–Jabin; FLOWER N = Tahdziu–Jabin, oven-dried; HONEY V = Acanceh–Dzidzilche; HONEY X = Acanceh–Jabin; HONEY Y = Tahdziu–Dzidzilche; HONEY Z = Tahdziu–Jabin.

As shown in Figure 10, various compounds (listed from M3-R3) linked to the carboxylic acid and derivative chemical class were associated with different honey samples, regardless of their botanical and geographical origins. Notably, asparagine (O3) was associated with the Jabin botanical origin, while proline (N3), 4-hydroxybenzoic acid (K3), and n-methyl-tryptophan (P3) were associated with Dzidzilche.

Furthermore, a lineal correlation was found between total polyphenols and antioxidant capacity in Dzidzilche and Jabin flowers ($p < 0.05$).

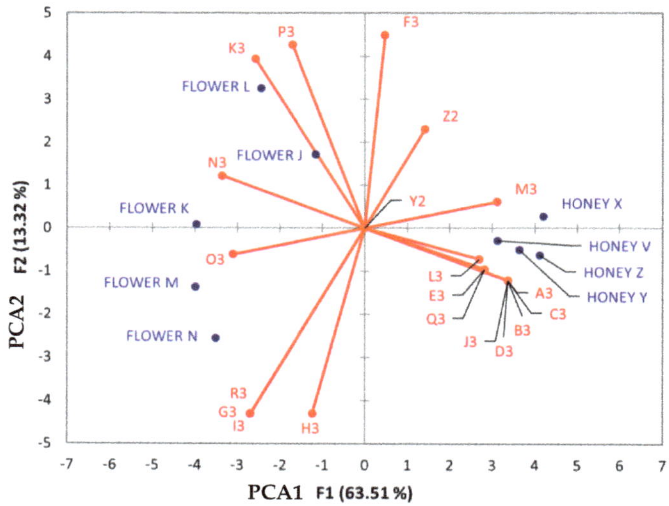

Figure 10. Principal component analysis (PCA) biplot of Dzidzilche and Jabin flower and honey samples based on the identified carboxylic acids and derivatives and phenols metabolites. Abbreviations: Y2 = phenylalanine; Z2 = tyrosine A3 = n-(phenylacetyl)phenylalanine; B3 = n-Fructosyl-isoleucine; C3 = n-fructosyl-phenylalanine; D3 = [6]-gingerol; E3 = vanillin; F3 = moupinamide; G3 = (2Z)-4,6-dihydroxy-2-[(4-hydroxy-3,5-dimethoxyphenyl) methylidene]-1-benzofuran-3-one; H3 = methyl (1S,3S,4S,4aS,9aR)-7-[(5R,6R,8R,8aS,10aS)-1,5,8,8a-tetrahydroxy-10a-methoxycarbonyl-6-methyl-9-oxo-5,6,7,8-tetrahydroxanthen-2-yl]-1,4,8,9a-tetrahydroxy-3-methyl-9oxo1,2,3,tetrahydroxanthene-4a-carboxylate; I3 = 3,4-dihydroxy cinnamic acid; J3 = resveratrol; K3 = 4-hydroxybenzoic acid; L3 = n-jasmonoyl-leucine; M3 = n-(1-deoxy-1-fructosyl) phenylalanine; N3 = proline; O3 = asparagine (0.93722378); P3 = n-methyl-tryptophan; Q3 = n-Acetyl-phenylalanine; R3 = chicoric acid. FLOWER J = Acanceh–Dzidzilche; FLOWER K = Acanceh–Jabin; FLOWER L = Tahdziu–Dzidzilche; FLOWER M = Tahdziu–Jabin; FLOWER N = Tahdziu–Jabin, oven-dried; HONEY V = Acanceh–Dzidzilche; HONEY X = Acanceh–Jabin; HONEY Y = Tahdziu–Dzidzilche; HONEY Z = Tahdziu–Jabin.

4. Discussion

The highest concentration of polyphenols (1431.24 ± 15.38 mg GAE/100 g DM) was observed in Dzidzilche flowers originating from Acanceh, which also exhibited the highest antioxidant capacity (93.63 ± 0.22% inhibition). By contrast, Jabin honey (botanical origin) had the lowest TPC concentration (5.38 ± 0.14 mg GAE/100 g DM) in Acanceh and the lowest AC (13.68 ± 0.75% inhibition) in Tahdziu. While there were differences in TPC between flowers and honey, their antioxidant capacity displayed a similar trend. This variation in TPC is mainly attributed to the higher percentage of flavonoids present in the flowers (50.7%) compared with that in honey (14.7%), consequently influencing the antioxidant capacity. According to Ghasemzadeh, et al. [48], these bioactive compounds demonstrate a significant positive correlation with the antioxidant capacity of plant matrix extracts, including flower extracts [49].

The antioxidant properties of flavonoids are derived from their ability to donate a hydrogen proton to an oxidizing molecule, thereby promoting molecular stability and inhibiting oxidative processes [50]. On the other hand, the higher percentage of phenolic compounds in flowers than in honey arises because these are secondary metabolites exclusively synthesized by plants [51], regardless of their botanical and geographical origins. By contrast, the concentration of phenolic compounds in honey is influenced by external factors, such as the availability of biological material for bees, the honey production process, and the amalgamation of flower or plant secretions with the bee's bodily fluids during the transport and storage of nectar [52].

Another crucial factor to consider when characterizing flowers and BH involves the edaphological conditions of the flower's geographical origin. In this study, Acanceh, a municipality situated in the central area of Yucatan, was selected, which is characterized by Mollic Leptosol soil (as per the World Reference Base for Soil Resources) or Box lu'um (according to the Mayan classification as described by Bautista et al.) [53]. These soils manifest distinctive features, including a black color, an intermediate percentage of gravel and stones (20–60%), a substantial organic matter content (greater than 10%), good drainage capacity, and a high concentration of exchangeable cations like calcium, phosphorus, sulfur, and sodium. On the other hand, the red soils (WRB) or Hay lu'um (Mayan classification) found in Tahdziu present a lower content of exchangeable cations than Mollic Leptosol soils, red color, clayey texture, reduced moisture retention, lower organic matter content (<15%), and occasional rocky outcrops [53,54].

The soil characteristics influence the polyphenol content of products such as plants, fruits, and their by-products. Numerous studies have evaluated the impact of black, brown, and red soils on the TPC and AC of habanero chili products from the Yucatan Peninsula. In one study, it was observed that plants cultivated in black soils presented a notably higher TPC (217.13 ± 28.04 mg GAE/100 g dry pepper) and AC (86.51 ± 0.82% inhibition) compared with those grown in other soil types, including red soil (135.17 ± 14.24 mg GAE/100 g dry pepper, with inhibition below 86%) and brown soil, which displayed intermediate behavior [54]. This study suggests that these differences were primarily attributed to variations in electrical conductivity, organic matter content, and increased nitrogen presence [55,56]. Overall, such factors are thought to impact nutrient availability and uptake in plants, ultimately modulating the polyphenol content and thereby the AC of the products.

Botanical origin emerged as a pivotal factor in assessing the TPC, AC, and metabolite chemical classes of honey. The Dzidzilche tree (*Gymnopodium floribundum*) has adapted to the climatic and soil conditions of the Yucatan Peninsula's tropical forest in Mexico and is notably abundant in the region [19]. This tree plays a crucial role in providing biological materials, such as flowers, nectar, and pollen to honeybees (*Apis mellifera*) [5]. It has also been identified as a producer of various secondary metabolites, mainly phenolic compounds, including flavonoids and tannins [19,57,58]. In this context, Ortíz-Ocampo et al. [19] conducted a study collecting biological materials (stems, leaves, and flowers) from Dzidzilche trees in the "Cuxtal" natural reserve from December 2017 to December 2018. They aimed to analyze changes in the concentration of phenolic compounds across the seasons (spring, summer, autumn, and winter). Their findings indicated significantly higher concentrations of phenolic compounds ($p < 0.05$) during the months of January, February, and March, coinciding with the same sampling period as the present study (March). During this period, Dzidzilche flowers presented a greater concentration of TPC than Jabin (another botanical origin). However, it is worth noting that this difference may also be attributed to the fact that Dzidzilche trees belong to the Polygonaceae family, which has demonstrated higher concentrations of phenolic compounds (1310 ± 100 mg GAE/100 g DM) than the Fabaceae family (713.3 ± 37.1 mg GAE/100 g DM), the family to which Jabin (*Piscidia piscipula*) belongs. This trend aligns with the current study's findings, in which Dzidzilche exhibited a significantly higher concentration of phenolic compounds than Jabin [57,59,60]. Furthermore, the concentrations of phenolic compounds exhibited a correlation with the AC in both Dzidzilche and Jabin flowers. For instance, Hallman [59] reported a significant linear correlation between the phenolic compound content in extracts of the Fabaceae family and their AC. Similarly, Feduraev et al. [57] reported a significant linear correlation between phenolic-rich extracts from plants of the Polygonaceae family and their AC. This same trend was observed in the present study when linearly correlating the phenolic compounds of Dzidzilche (Polygonaceae family) and Jabin (Fabaceae family) flower extracts with their respective antioxidant capacities, reaffirming the relevance of polyphenols in the antioxidant bioactivity of these plants.

Notably, the TCP and AC are not only affected by botanical origin but also by the presence of specific secondary metabolites. For example, within the Fabaceae family, plants are characterized by their repertoire of a set of compounds belonging to the chemical class of isoflavonoids. This type of compound is a subclass of the flavonoid family featuring a benzene ring B at the carbon 3 position of the pyran ring [61] and is synthesized de novo through the Shikimic acid pathway. At a certain stage, the activation of chalcone isomerase and isoflavone synthase is promoted, working as a defense mechanism in response to stressors, predominantly of microbiological origin [62]. Additionally, these compounds are regarded as pathogen resistance biomarkers [63]. On the other hand, plants belonging to the Polygonaceae family are renowned for their substantial content of flavonoids, particularly quercetin and epicatechin [51]. These metabolites are found in numerous genera within this family and are linked to pleiotropic effects, including a robust antioxidant capacity [64] along with antidiabetic, neuroprotective, cardioprotective, and anticancer effects [65,66].

In this study, the presence of 12 chemical classes of the samples analyzed was observed. However, it is worth noting that Jabin flowers contained two secondary metabolites classified as isoflavones, specifically 11-hydroxytephrosin and 6-hydroxysumatrol, which were absent in the Dzidzilche samples. Conversely, a notable association was observed between the flavanol catechin (or one of its stereoisomers) and the botanical origin of Dzidzilche. It is important to highlight that these phenolic compounds were also identified in honey. Although these metabolites are solely synthesized by plants, they can make their way into honey through the collection of nectar, propolis, and/or resins by honeybees, which are subsequently processed in the beehive [66].

Some authors, including Gašić et al. [67], Cheung et al. [68], and Kasiotis et al. [69], have proposed phenolic compounds as potential biomarkers for identifying the botanical and even geographical origin of honey, mainly monofloral honey, to ensure their authenticity in global markets. According to Lawag et al. [70], characteristic phenolic compounds found in honey sourced from plants of the Polygonaceae family across various geographical regions (China, Finland, Italy, Lithuania, Poland, Turkey, USA, etc.) encompass apigenin, luteolin, chrysin, vitexin, quercetin, kaempferol, rutin, phenylacetic acid, salicylic acid, p-coumaric acid, and chlorogenic acid, among others. On the other hand, representative phenolic compounds of the Fabaceae family identified from different geographical regions (China, Argentina, Italy, Turkey, Germany, Spain, Austria, USA, Poland, etc.) include vanillin, apigenin, luteolin, quercetin, quercitrin, rutin, myricetin, vanillic acid, p-coumaric acid, gallic acid, and protocatechuic acid. These reported results aligned with the findings of the present study. For example, vitexin (apigenin glycoside) was identified (Figure S6) in honey originating from Dzidzilche (Polygonaceae) but not in honey from Jabin (Fabaceae), while a characteristic metabolite of the Fabaceae family, vanillin, was only identified in honey with the botanical origin of Jabin (Figure S7).

Moreover, we detected numerous phenolic compounds of interest in both honeys (Dzidzilche and Jabin), including quercetin, kaempferol, naringenin, and quercitrin, with the latter being notably characteristic of the Fabaceae family. Although these honeys are considered monofloral, as they predominantly originate from a single plant species, it is important to acknowledge that metabolites from other plant species may be also present [71]. Additionally, during the honey production process, phenolic compounds can undergo hydrolysis or form glycosylated derivatives due to the activities of enzymes and microorganisms within the honeybee's digestive system [69,72]. Given the variability of phenolic compounds influenced by abiotic and biotic factors, there has been an exploration of alternative markers, such as nitrogenous compounds (e.g., essential amino acids), in honey [69,73–75]. However, no discernible differences in amino acids have been identified regarding botanical origin. For instance, Hermosín et al. [74] examined 31 Spanish honeys from various botanical (rosemary, eucalyptus, lavender, thyme, and orange blossom) and geographic (Alcarria, Cabañeros, Cáceres, Valencia, and Cantabria) origins and found that it was not feasible to definitively distinguish the botanical origin solely on the basis of the presence and concentration of amino acids. In this study, an association was found through PCA between tyrosine and honey sourced from Jabin collected from both geographic origins (Acanceh and Tahdziu). Moreover, samples from Dzidzilche and Tahdziu presented an association with phenylalanine. However, no such association with any amino acid was observed in honey originating from Dzidzilche or Acanceh. As highlighted by Kowalski et al. [76], the detection of amino acids from different botanical origins within the same geographic area may be possible due to environmental conditions and the proximity of flowering periods.

5. Conclusions

With the assistance of mass spectrometry and comprehensive chemoinformatics, a detailed analysis was conducted on the metabolome found in flowers and their associated honeys from Yucatan, Mexico. The aim was to uncover how the chemistry of these samples is influenced by their botanical and geographical origins. The flowers of the *Gymnopodium floribundum* and *Piscidia piscipula* (L.) genera presented a notable abundance of phenolic compounds, mainly flavonoids, and a remarkable AC, above 90% inhibition (DPPH). These characteristics were distinctly mirrored in the resulting honey; flavonoids and carboxylic acids and their derivatives were the main chemical classes detected, and they showed an AC range between 18 and 21% inhibition (DPPH). The intricate interplay of geographical and botanical origin factors emerged as a crucial determinant in shaping the presence of secondary metabolites (e.g., flavonoids and carboxylic acids and their derivatives), along with fostering a robust AC in both the flowers and the honey. Secondary metabolites previously suggested as biomarkers were identified by liquid chromatography–tandem mass spectrometry (LC–MS^2) in honey from *Gymnopodium floribundum* (vitexin) and *Piscidia piscipula* (L.) (vanillin). This substantiates and reinforces the ability to determine the botanical origin of Yucatan, México, honey through the presence of phenolic compounds, offering a valuable complement to melissopalynology. Such an approach not only serves as a potent tool for quality control but also substantiates the bioactive properties of honey sourced from this Mexican region, thereby enhancing its market value and streamlining export opportunities to global destinations.

Supplementary Materials: The following supporting information can be downloaded at: https://www.mdpi.com/article/10.3390/pr11103028/s1, Figure S1. (a) Total polyphenol content (TPC) calibration curve for flower sample analysis. (b) Total polyphenol content (TPC) calibration curve for honey sample analysis; Figure S2. Interaction chart; total polyphenol content (TPC) (a,b) antioxidant capacity (AC). Capital letters represent a main factor, where A: botanical origin; B: geographical origin; and C: raw material; Figure S3. Main effect plot; total polyphenol content (TPC) (a), antioxidant capacity (AC) (b). Capital letters represent a main factor, where BO: botanical origin; GO: geographical origin; and RM: raw material; Figure S4. Pareto chart in raw material honey: (a) total polyphenol content, (b) antioxidant capacity; Figure S5. Pareto chart in raw material flower: (a) total polyphenol content, (b) antioxidant capacity; Figure S6. Representative chromatogram used for the analysis of vitexin. (A) Extracted ion chromatogram for *Gymnopodium floribundum* flower from Acanceh; (B) Extracted ion chromatogram for *Gymnopodium floribundum* honey from Acanceh; (C) Extracted ion chromatogram for *Gymnopodium floribundum* flower from Tahdziu; (D) Extracted ion chromatogram for *Gymnopodium floribundum* honey from Tahdziu; Figure S7. Representative chromatogram used for the analysis of vanillin. (A) Extracted ion chromatogram for *Piscidia piscipula* flower from Acanceh; (B) Extracted ion chromatogram for *Piscidia piscipula* honey from Acanceh; (C) Extracted ion chromatogram for *Piscidia piscipula* flower from Tahdziu; (D) Extracted ion chromatogram for *Piscidia piscipula* honey from Tahdziu; Table S1. List of metabolites putatively annotated using spectral libraries and in silico predictions in honey; Table S2. List of metabolites putatively annotated using spectral libraries and in-silico predictions in flour flower.

Author Contributions: Conceptualization, I.M.R.-B. and A.E.M.-O.; methodology, I.M.R.-B., A.M.-U. and R.C.-C.; software, A.M.-U., R.C.-C., A.E.M.-O. and K.A.A.-B.; validation, I.M.R.-B., M.O.R.-S. and A.M.-U.; formal analysis, I.M.R.-B.; investigation, I.M.R.-B., A.E.M.-O., A.M.-U., R.C.-C., A.U.-V. and K.A.A.-B.; resources, I.M.R.-B.; data curation, I.M.R.-B., M.O.R.-S. and A.M.-U.; writing—original draft preparation A.E.M.-O. and K.A.A.-B.; writing—review and editing, I.M.R.-B., M.O.R.-S., A.M.-U.,

R.C.-C. and K.A.A.-B.; visualization, I.M.R.-B.; supervision, I.M.R.-B.; project administration, I.M.R.-B.; funding acquisition, I.M.R.-B. All authors have read and agreed to the published version of the manuscript.

Funding: The National Council of Humanities, Sciences, and Technologies of Mexico (CONAHCYT) which financed scholarship 1192863 for Andrea Elizabeth Mendoza-Osorno and scholarship 661099 for Kevin Alejandro Avilés-Betanzos. A.M.-U. thanks, CONAHCYT for grant F0001-2020-02-314964.

Data Availability Statement: The parameters and data for GNPS-derived analysis are available at the following links: (a) classical molecular networking/spectral matching, https://gnps.ucsd.edu/ProteoSAFe/status.jsp?task=2d10247511654d2eaaeef0ba768bde5f (accessed on 25 August 2023) (b) MolDiscovery, https://gnps.ucsd.edu/ProteoSAFe/status.jsp?task=ed9f60679c504868b7cfc7f119e867bd (accessed on 25 August 2023); (c) Dereplicator+, https://gnps.ucsd.edu/ProteoSAFe/status.jsp?task=ee920092b09a4b0cb5db9ee734727e89 (accessed on 25 August 2023).

Conflicts of Interest: The authors declare no conflict of interest.

References

1. Carnevali, F.C.G.; Tapia–Muñoz, J.L.; Duno de Stefano, R.; Ramírez Morillo, I. (Eds.) *Flora Ilustrada de la Penín-sula de Yucatán: Listado Florístico*; Centro de Investigación Científica de Yucatán, A.C: Mérida, Mexico, 2010; p. 328.
2. Gobierno de México. Yucatán Se Encuentra Entre Los Principales Productores de Miel Del País. Available online: https://www.gob.mx/agricultura/yucatan/articulos/yucatan-se-encuentra-entre-los-principales-productores-de-miel-del-pais?idiom=es (accessed on 14 August 2023).
3. Rogel, F.J.; Carlos, E.G.; Juan Manuel, R.; Álvarez, G. La Apicultura En La Península de Yucatán. Actividad de Subsistencia En un Entorno Globalizado. *Rev. Mex. Caribe* **2003**, *8*, 16.
4. Cairns, C.E.; Villanueva-Gutiérrez, R.; Koptur, S.; Bray, D.B. Bee Populations, Forest Disturbance, and Africanization in Mexico. *Biotropica J. Biol. Conserv.* **2005**, *37*, 686–692. [CrossRef]
5. Santiago, C.B.; Sosa, J.C.; Díaz, A.R.; Trejo, R.N.; May, D.C. Estudio de la flora presente en apiarios de tres municipios en el estado de Yucatán, México. *Polibotánica* **2022**, *3*, 1–15. [CrossRef]
6. Arteaga-Hernández, V.; González-Ávila, M.; Cauich-Rodríguez, J.V. Compuestos fenólicos y actividad antioxidante de plantas de la península de Yucatán. *Rev. U.D.C.A. Actual. Divulg. Científica* **2014**, *17*, 29–38.
7. Gómez-Caravaca, A.M.; Verardo, V.; Toselli, M.; Segura-Carretero, A.; Fernández-Gutiérrez, A.; Caboni, M.F. Actividad antioxidante y contenido fenólico total de extractos de hojas de doce especies de plantas seleccionadas de la Península de Yucatán, México. *Rev. Mex. Cienc. Forest.* **2011**, *2*, 7–18.
8. Stefano, R.D.; Morillo, I.M.; Tapia-Muñoz, J.L.; Hernández-Aguilar, S.; Can, L.L.; Cetzal-Ix, W.; Méndez-Jiménez, N.; Zamora-Crescencio, P.; Gutiérrez-Báez, C.; Carnevali-Fernaández-Concha , G. Aspectos generales de la flora vascular de la Península de Yucatán, México. *Bot. Sci.* **2018**, *96*, 512–532. [CrossRef]
9. Nowicka, P.; Wojdyło, A. Anti-Hyperglycemic and Anticholinergic Effects of Natural Antioxidant Contents in Edible Flowers. *Antioxidants* **2019**, *8*, 308. [CrossRef]
10. Pérez-Sarabia, J.E.; Duno de Stefano, R.; Carnevali Fernández-Concha, G.; Ramírez Morillo, I.; Méndez-Jiménez, N.; Zamo-ra-Crescencio, P.; Gutierrez-Báez, C.; Cetzal-Ix, W. El conocimiento florístico de la Península de Yucatán, México, actualización y colecciones botánicas. In Proceedings of the XX Congreso Mexicano de Botánica, Ciudad de, Mexico, Mexico, 4–9 September 2016.
11. Rosado-Vallado, M.; Brito-Loeza, W.; Mena-Rejón, G.; Quintero-Marmol, E.; Flores-Guido, J. Antimicrobial activity of Fabaceae species used in Yucatan traditional medicine. *Fitoterapia* **2000**, *71*, 570–573. [CrossRef]
12. Canché-Collí, C.; Jiménez, L.N.L.; Rodríguez, R.; Canto, A. El jabín y los secretos de su néctar. *Ecofronteras* **2022**, *26*, 2–5.
13. Zamora Crescencio, P.; Flores Guido, J.S.; Ruenes Morales, R. Flora útil y su manejo en el cono sur del estado de Yucatán, México. *Polibotánica* **2009**, *28*, 227–250.
14. Villanueva-Gutiérrez, R.; Moguel-Ordóñez, Y.B.; Echazarreta-González, C.M.; Arana-López, G. Monofloral honeys in the Yucatán Peninsula, Mexico. *Grana* **2009**, *48*, 214–223. [CrossRef]
15. Escobar-Rivera, P.; González-Mujica, F.; Rodríguez-Amado, J.; lejandro Cruz Sánchez, T. Composición química, actividades antioxidantes y antiinflamatorias de la corteza de *Piscidia piscipula* L. *Aliment. Chem. Toxicol.* **2019**, *123*, 476–483. [CrossRef]
16. Hernández-Moreno, L.V.; Salazar, J.R.; Pabón, L.C.; Hernández-Rodríguez, P. Antioxidant activity and quantification of phenols and flavonoids of Colombian plants used in urinary tract infections. *Rev. U.D.C.A Actual Divulg. Cient.* **2022**, *25*, 1–7. Available online: http://www.scielo.org.co/pdf/rudca/v25n1/0123-4226-rudca-25-01-e1690.pdf (accessed on 23 August 2023).
17. Mohtashami, L.; Amiri, M.S.; Ayati, Z.; Ramezani, M.; Jamialahmadi, T.; Emami, S.A.; Sahebkar, A. Ethnobotanical uses, phytochemistry and pharmacology of different Rheum Species (Polygonaceae): A. Review. In *Pharmacological Properties of Plant-Derived Natural Products and Implications for Human Health*; Barreto, G.E., Sahebkar, A., Eds.; Advances in Experimental Medicine and Biology Series; Springer: Cham, Switzerland, 2021; Volume 1308, pp. 1–22. [CrossRef]
18. Moguel-Ordóñez, Y.; Echazarreta-Gonzalez, C.; Mora-Escobedo, R. Fisicoquímica de la miel de abeja Apis mellifera producida en el estado de Yucatán durante diferentes etapas del proceso de producción y tipos de floración. *Técnica Pecu. México* **2005**, *43*, 323–334.

19. Ortíz-Ocampo, G.I.; Sandoval-Castro, C.A.; González-Pech, P.G.; Mancilla-Montelongo, G.; Ventura-Cordero, J.; Castañeda-Ramírez, G.S.; Tun-Garrido, J.; Torres-Acosta, J.F.d.J. Month of Harvest and Leaf Age Impact the Bromatological Composition and Polyphenol Content of *Gymnopodium floribundum* Rolfe Leaves. *Agriculture* **2022**, *12*, 1110. [CrossRef]
20. Cuevas-Glory, L.; Sosa-Moguel, O.; Ortiz-Vázquez, E.; Sauri-Duch, E.; Pino, A. Volatile constituents of tzizilché flower (*Gymnopodium* floribundum Rolfe) from Yucatán Peninsula, Mexico. *J. Essent. Oil Res.* **2012**, *24*, 359–361. [CrossRef]
21. Ortíz-Ocampo, G.I.; Sandoval-Castro, C.A.; González-Pech, P.G.; Mancilla-Montelongo, G.; Ventura-Cordero, J.; Castañeda-Ramírez, G.S.; Pérez, J.I.C.; Leal, C.C.; Torres-Acosta, J.F.d.J. El Mes de Cosecha y la Edad de la Hoja Impact. *Rev. Mex. Cienc. Pecu.* **2020**, *12*, 1289–1303. [CrossRef]
22. Yang, L.; Wen, K.-S.; Ruan, X.; Zhao, Y.-X.; Wei, F.; Wang, Q. Response of Plant Secondary Metabolites to Environmental Factors. *Molecules* **2018**, *23*, 762. [CrossRef]
23. Cheynier, V.; Comte, G.; Davies, K.M.; Lattanzio, V.; Martens, S. Plant phenolics: Recent advances on their biosynthesis, genetics, and ecophysiology. *Plant Physiol. Biochem.* **2013**, *72*, 1–20. [CrossRef]
24. García-Lara, S.; Dzib, G.; Cetzal-Ix, W. Flora Melífera de la Península de Yucatán, México: Estrategia Para Incrementar la Pro-ducción de Miel en los Periodos de. CICY. Available online: https://www.cicy.mx/Documentos/CICY/Desde_Herbario/2019/2019-09-05-Cetzal-Noguera-Martinez-Flora-melifera-de-PY.pdf (accessed on 25 August 2023).
25. Wang, J.; Qing, X.L. *Chapter 3—Chemical Composition, Characterization, and Differentiation of Honey Botanical and Geo-Graphical Origins*; Steve Taylor, L., Ed.; Advances in Food and Nutrition Research; Academic Press: New York, NY, USA, 2011; Volume 62, pp. 89–137. [CrossRef]
26. Anklam, E. A review of the analytical methods to determine the geographical and botanical origin of honey. *Food Chem.* **1998**, *63*, 549–562. [CrossRef]
27. Guo, P.; Deng, Q.; Lu, Q. Anti-alcoholic effects of honeys from different floral origins and their correlation with honey chemical compositions. *Food Chem.* **2019**, *286*, 608–615. [CrossRef] [PubMed]
28. Kortesniemi, M.; Rosenvald, S.; Laaksonen, O.; Vanag, A.; Ollikka, T.; Vene, K.; Yang, B. Sensory and chemical profiles of Finnish honeys of different botanical origins and consumer preferences. *Food Chem.* **2018**, *246*, 351–359. [CrossRef] [PubMed]
29. Cherchi, A.; Spanedda, L.; Tuberoso, C.; Cabras, P. Solid-phase extraction and high-performance liquid chromatographic determination of organic acids in honey. *J. Chromatogr. A* **1994**, *669*, 59–64. [CrossRef]
30. Toma´s-Barbera´n, F.A.; Martos, I.; Ferreres, F.; Radovic, B.S.; Anklam, E. HPLC flavonoid profiles as markers for the botanical origin of European unifloral honeys. *Sci. Food Agric.* **2001**, *81*, 485–496. [CrossRef]
31. Toma´s-Barbera´n, F.A.; Garcı´a-Viguera, C.; Vit-Olivier, P.; Ferreres, F.; Toma´s-Lorente, F. Flavonoids in honey of different geographical origin. *Z. Lebensm. Unters. Forsch.* **1993**, *196*, 38–44. [CrossRef]
32. Oney-Montalvo, J.E.; Avilés-Betanzos, K.A.; Ramírez-Rivera, E.J.; Ramírez-Sucre, M.O.; Rodríguez-Buenfil, I.M. Polyphenols Content in *Capsicum chinense* Fruits at Different Harvest Times and Their Correlation with the Antioxidant Activity. *Plants* **2020**, *9*, 1394. [CrossRef] [PubMed]
33. Chel-Guerrero, L.D.; Oney-Montalvo, J.E.; Rodríguez-Buenfil, I.M. Phytochemical Characterization of By-Products of Habanero Pepper Grown in Two Different Types of Soils from Yucatán, Mexico. *Plants* **2021**, *10*, 779. [CrossRef] [PubMed]
34. Avilés-Betanzos, K.A.; Cauich-Rodríguez, J.V.; Ramírez-Sucre, M.O.; Rodríguez-Buenfil, I.M. Optimization of Spray-Drying Conditions of Microencapsulated Habanero Pepper (*Capsicum chinense* Jacq.) Extracts and Physicochemical Characterization of the Microcapsules. *Polibotánica* **2023**, *11*, 1238. [CrossRef]
35. Contreras-Angulo, L.A.; Moreno-Ulloa, A.; Carballo-Castañeda, R.A.; León-Felix, J.; Romero-Quintana, J.G.; Aguilar-Medina, M.; Ramos-Payán, R.; Heredia, J.B. Metabolomic Analysis of Phytochemical Compounds from Agricultural Residues of Eggplant (*Solanum melongena* L.). *Molecules* **2022**, *27*, 7013. [CrossRef]
36. Al-Mamary, M.; Al-Meeri, A.; Al-Habori, M. Antioxidant Activities and total phenolics of different types of honey. *Nutr. Res.* **2002**, *22*, 1041–1047. [CrossRef]
37. Singleton, V.L.; Orthofer, R.; Lamuela-Raventos, R.M. Analysis of total phenols and other oxidation substrates and antioxidants by means of Folin-Ciocalteu reagent. *Method Enzym.* **1999**, *299*, 152–178. [CrossRef]
38. Brand-Williams, W.; Cuvelier, M.E.; Berset, C. Use of a free radical method to evaluate antioxidant activity. *LWT Food Sci. Technol.* **1995**, *28*, 25–30. [CrossRef]
39. Holman, J.D.; Tabb, D.L.; Mallick, P. Employing ProteoWizard to convert raw mass spectrometry data. *Curr. Protoc. Bioinform.* **2014**, *46*, 13.24.1–13.24.9. [CrossRef] [PubMed]
40. Aron, A.T.; Gentry, E.C.; McPhail, K.L.; Nothias, L.-F.; Nothias-Esposito, M.; Bouslimani, A.; Petras, D.; Gauglitz, J.M.; Sikora, N.; Vargas, F.; et al. Reproducible molecular networking of untargeted mass spectrometry data using GNPS. *Nat. Protoc.* **2020**, *15*, 1954–1991. [CrossRef] [PubMed]
41. Schymanski, E.L.; Jeon, J.; Gulde, R.; Fenner, K.; Ruff, M.; Singer, H.P.; Hollender, J. Identifying Small Molecules via High Res-olution Mass Spectrometry: Communicating Confidence. *Env. Sci Technol* **2014**, *48*, 2097–2098. [CrossRef] [PubMed]
42. Cao, L.; Guler, M.; Tagirdzhanov, A.; Lee, Y.-Y.; Gurevich, A.; Mohimani, H. MolDiscovery: Learning mass spectrometry fragmentation of small molecules. *Nat. Commun.* **2021**, *12*, 3718. [CrossRef] [PubMed]
43. Mohimani, H.; Gurevich, A.; Shlemov, A.; Mikheenko, A.; Korobeynikov, A.; Cao, L.; Shcherbin, E.; Nothias, L.-F.; Dorrestein, P.C.; Pevzner, P.A. Dereplication of microbial metabolites through database search of mass spectra. *Nat. Commun.* **2018**, *9*, 4035. [CrossRef] [PubMed]

44. Dührkop, K.; Shen, H.; Meusel, M.; Rousu, J.; Böcker, S. Searching molecular structure databases with tandem mass spectra using CSI:FingerID. *Proc. Natl. Acad. Sci. USA* **2015**, *112*, 12580–12585. [CrossRef]
45. Feunang, Y.D.; Eisner, R.; Knox, C.; Chepelev, L.; Hastings, J.; Owen, G.; Fahy, E.; Steinbeck, C.; Subramanian, S.; Bolton, E.; et al. ClassyFire: Automated Chemical Classification with a Comprehensive, Computable Taxonomy. *J. Chemin.* **2016**, *8*, 61. [CrossRef]
46. Dührkop, K.; Fleischauer, M.; Ludwig, M.; Aksenov, A.A.; Melnik, A.V.; Meusel, M.; Dorrestein, P.C.; Rousu, J.; Böcker, S. SIRIUS 4: A rapid tool for turning tandem mass spectra into metabolite structure information. *Nat. Methods* **2019**, *16*, 299–302. [CrossRef]
47. Ludwig, M.; Nothias, L.-F.; Dührkop, K.; Koester, I.; Fleischauer, M.; Hoffmann, M.A.; Petras, D.; Vargas, F.; Morsy, M.; Aluwihare, L.; et al. Database-independent molecular formula annotation using Gibbs sampling through ZODIAC. *Nat. Mach. Intell.* **2020**, *2*, 629–641. [CrossRef]
48. Ghasemzadeh, A.B.; Ghasemzadeh, N.C.D. Flavonoids and phenolic acids: Role and biochemical activity in plants and human. *J. Med. Plants Res.* **2011**, *5*, 6697–6703. [CrossRef]
49. Zheng, J.; Yu, X.; Maninder, M.; Xu, B. Total phenolics and antioxidants profiles of commonly consumed edible flowers in China. *Int. J. Food Prop.* **2018**, *21*, 1524–1540. [CrossRef]
50. Panche, A.N.; Diwan, A.D.; Chandra, S.R. Flavonoids: An overview. *J. Nutr. Sci.* **2016**, *5*, e47. [CrossRef] [PubMed]
51. Shen, N.; Wang, T.; Gan, Q.; Liu, S.; Wang, L.; Jin, B. Plant flavonoids: Classification, distribution, biosynthesis, and antioxidant activity. *Food Chem.* **2022**, *383*, 132531. [CrossRef] [PubMed]
52. Becerril-Sánchez, A.L.; Quintero-Salazar, B.; Dublán-García, O.; Escalona-Buendía, H.B. Phenolic compounds in honey and their relationship with antioxidant activity, botanical origin, and color. *Antioxidants* **2021**, *10*, 1700. [CrossRef] [PubMed]
53. Bautista, F.; Zinck, J.A. Construction of an Yucatec Maya soil classification and comparison with the WRB framework. *J. Ethnobiol. Ethnomed.* **2010**, *6*, 7. [CrossRef] [PubMed]
54. Bautista-Zúñiga, F.; Jiménez-Osornio, J.; Navarro-Alberto, J.; Manu, A.; Lozano, R. Microrelieve y color del suelo como propiedades de diagnóstico en leptosoles cársticos. *Tierra Latinoame-Ricana* **2003**, *21*, 1–11. Available online: http://www.redalyc.org/pdf/573/57321101.pdf (accessed on 25 August 2023).
55. Oney-Montalvo, J.; Uc-Varguez, A.; Ramírez-Rivera, E.; Ramírez-Sucre, M.; Rodríguez-buenfili, I.M. influence of soil composition on the profile and content of polyphenols in habanero peppers (capsicum Chinese Jacq). *Agronomy* **2020**, *10*, 1234. [CrossRef]
56. Montalvo, J.E.O.; Madrigal, A.C.d.S.; Sucre, M.O.R.; Rodríguez-Buenfil, I.M. Effect of the Soil and Ripening Stage in *Capsicum chinense* var. Jaguar on the Content of Carotenoids and Vitamins. *Horticulturae* **2021**, *7*, 442. [CrossRef]
57. Feduraev, P.; Skrypnik, L.; Nebreeva, S.; Dzhobadze, G.; Vatagina, A.; Kalinina, E.; Pungin, A.; Maslennikov, P.; Riabova, A.; Krol, O.; et al. Variability of Phenolic Compound Accumulation and Antioxidant Activity in Wild Plants of Some *Rumex* Species (*Polygonaceae*). *Antioxidants* **2022**, *11*, 311. [CrossRef]
58. Hussein, S.; El-Magly, U.; Tantawy, M.; Kawashty, S.; Saleh, N. Phenolics of selected species of Persicaria and Polygonum (Polygonaceae) in Egypt. *Arab. J. Chem.* **2017**, *10*, 76–81. [CrossRef]
59. Hallmann, E. Quantitative and qualitative identification of bioactive compounds in edible flowers of black and bristly locust and their antioxidant activity. *Biomolecules* **2020**, *10*, 1603. [CrossRef] [PubMed]
60. Tungmunnithum, D.; Drouet, S.; Lorenzo, J.M.; Hano, C. Characterization of Bioactive Phenolics and Antioxidant Capacity of Edible Bean Extracts of 50 Fabaceae Populations Grown in Thailand. *Foods* **2021**, *10*, 3118. [CrossRef] [PubMed]
61. Desta, K.T.; El-Aty, A.M.A. Millettia isoflavonoids: A comprehensive review of structural diversity, extraction, isolation, and pharmacological properties. *Phytochem. Rev.* **2023**, *22*, 275–308. [CrossRef] [PubMed]
62. Foudah, A.I.; Abdel-Kader, M.S. Isoflavonoids. In *Flavonoids—From Biosynthesis to Human Health*; InTech: Mexico City, Mexico, 2017. [CrossRef]
63. Araya-Cloutier, C.; den Besten, H.M.W.; Aisyah, S.; Gruppen, H.; Vincken, J.-P. The position of prenylation of isoflavonoids and stilbenoids from legumes (Fabaceae) modulates the antimicrobial activity against Gram positive pathogens. *Food Chem.* **2017**, *226*, 193–201. [CrossRef] [PubMed]
64. Palma, M.; Robert, P.; Holgado, F.; Velasco, J.; Márquez-Ruiz, G. Antioxidant Activity and Kinetics Studies of Quercetin, Epicatechin and Naringenin in Bulk Methyl Linoleate. *J. Am. Oil Chem. Soc.* **2017**, *94*, 1189–1196. [CrossRef]
65. Shay, J.; Elbaz, H.A.; Lee, I.; Zielske, S.P.; Malek, M.H.; Hüttemann, M. Molecular mechanisms and therapeutic effects of (-)-epicatechin and other polyphenols in cancer, inflammation, diabetes, and neurodegeneration. In *Oxidative Medicine and Cellular Longevity*; Hindawi Publishing Corporation: London, UK, 2015. [CrossRef]
66. Shabir, I.; Pandey, V.K.; Shams, R.; Dar, A.H.; Dash, K.K.; Khan, S.A.; Bashir, I.; Jeevarathinam, G.; Rusu, A.V.; Esatbeyoglu, T.; et al. Promising bioactive properties of quercetin for potential food applications and health benefits: A review. *Front. Nutr.* **2022**, *9*, 999752. [CrossRef]
67. Gašić, U.M.; Milojković-Opsenica, D.M.; Tešić, L. Polyphenols as possible markers of botanical origin of honey. *J. AOAC Int.* **2017**, *100*, 852–861. [CrossRef]
68. Cheung, Y.; Meenu, M.; Yu, X.; Xu, B. Phenolic acids and flavonoids profiles of commercial honey from different floral sources and geographic sources. *Int. J. Food Prop.* **2019**, *22*, 290–308. [CrossRef]
69. Kasiotis, K.M.; Baira, E.; Iosifidou, S.; Bergele, K.; Manea-Karga, E.; Theologidis, I.; Barmpouni, T.; Tsipi, D.; Machera, K. Characterization of Ikaria Heather Honey by Untargeted Ultrahigh-Performance Liquid Chromatography-High Resolution Mass Spectrometry Metabolomics and Melissopalynological Analysis. *Front. Chem.* **2022**, *10*, 924881. [CrossRef] [PubMed]

70. Lawag, I.L.; Lim, L.-Y.; Joshi, R.; Hammer, K.A.; Locher, C. A Comprehensive Survey of Phenolic Constituents Reported in Monofloral Honeys around the Globe. *Foods* **2022**, *11*, 1152. [CrossRef] [PubMed]
71. Schievano, E.; Finotello, C.; Uddin, J.; Mammi, S.; Piana, L. Objective Definition of Monofloral and Polyfloral Honeys Based on NMR Metabolomic Profiling. *J. Agric. Food Chem.* **2016**, *64*, 3645–3652. [CrossRef]
72. Deadman, B.J. The Flavonoid Profile of New Zealand Manuka Honey. Master Thesis, The University of Waikato, Hamilton, New Zealand, 2009; pp. 1–270. Available online: https://hdl.handle.net/10289/5443 (accessed on 1 March 2023).
73. Kuś, P.; Jerković, I.; Jakovljević, M.; Jokić, S. Extraction of bioactive phenolics from black poplar (*Populus nigra* L.) buds by supercritical CO_2 and its optimization by response surface methodology. *J. Pharm. Biomed. Anal.* **2018**, *152*, 128–136. [CrossRef] [PubMed]
74. Hermosín, I.; Chicón, R.M.; Cabezudo, M.D. Free amino acid composition and botanical origin of honey. *Food Chem.* **2003**, *83*, 263–268. [CrossRef]
75. Dimins, F.; Cinkmanis, I.; Radenkovs, V.; Augspole, I.; Valdovska, A. Analysis of 18 Free Amino Acids in Honeybee and Bumblebee Honey from Eastern and Northern Europe and Central Asia Using HPLC-ESI-TQ-MS/MS Approach Bypassing Derivatization Step. *Foods* **2022**, *11*, 2744. [CrossRef] [PubMed]
76. Kowalski, S.; Kopuncová, M.; Ciesarová, Z.; Kukurová, K. Free amino acids profile of Polish and Slovak honeys based on LC–MS/MS method without the prior derivatisation. *J. Food Sci. Technol.* **2017**, *54*, 3716–3723. [CrossRef]

processes

MDPI

Article

Plant-Derived Essential Oils and Aqueous Extract as Potential Ingredients for a Biopesticide: Phytotoxicity in Soybean and Activity against Soybean Mosaic Virus

María Evangelina Carezzano [1,2,3], Pablo Gastón Reyna [4,5], Efrén Accotto [1], Walter Giordano [2,3], María de las Mercedes Oliva [1,2], Patricia Rodriguez Pardina [4,5] and María Carola Sabini [1,6,*]

1 Departamento de Microbiología e Inmunología, Universidad Nacional de Río Cuarto (UNRC), Ruta 36 Km 601, Río Cuarto CP 5800, Argentina; ecarezzano@exa.unrc.edu.ar (M.E.C.); efrenaccotto93@gmail.com (E.A.); moliva@exa.unrc.edu.ar (M.d.l.M.O.)
2 Instituto de Biotecnología Ambiental y Salud, INBIAS-CONICET, Río Cuarto CP 5800, Argentina; wgiordano@exa.unrc.edu.ar
3 Departamento de Biología Molecular, Universidad Nacional de Río Cuarto, Ruta 36 Km 601, Río Cuarto CP 5800, Argentina
4 Instituto Nacional de Tecnología Agropecuaria, Instituto de Patología Vegetal, Córdoba CP 5020, Argentina; pablogreyna@hotmail.com (P.G.R.); patorpar@gmail.com (P.R.P.)
5 Consejo Nacional de Investigaciones Científicas y Técnicas, Unidad de Fitopatología y Modelización Agrícola, Cordoba CP 5020, Argentina
6 Instituto de Investigaciones en Ciencias de la Salud (INICSA)—Consejo Nacional de Investigaciones Científicas y Técnicas, Facultad de Ciencias Médicas, Universidad Nacional de Córdoba, Cordoba CP 5016, Argentina
* Correspondence: csabini@exa.unrc.edu.ar

Abstract: Soybean mosaic disease, caused by the soybean mosaic virus (SMV), is responsible for major losses in yield and seed quality worldwide. Although resistant cultivars are used for its prevention and control, an alternative strategy could consist of applying environmentally friendly antimicrobial agents, such as extracts and essential oils (EOs) of aromatic plants. This study assessed an extract of *Achyrocline satureioides* and EOs of *Minthostachys verticillata*, *Origanum vulgare*, and *Thymus vulgaris* in terms of their phytotoxicity in soybean. Since all the concentrations tested were found to be safe, the activity of each product against SMV was then assayed in vivo, i.e., in experimentally infected soybean plants. The parameters measured were plant height, wet weight, and virus titer. All the treated plants had a greater height and weight than those in the viral control group. The EOs of *M. verticillata* (0.80 mg/mL) and *T. vulgaris* (0.71 mg/mL) inhibited the production of viral antigens, as determined by an ELISA test. These findings could encourage further studies aimed at developing an effective biopesticide against SMV.

Keywords: Potyvirus; *Glycine max*; Biocontrol

1. Introduction

Soybean (*Glycine max* (L.) Merry), a legume thought to have originated in northern/central China, is now grown around the world. Argentina is a leading producer: 16.1 hectares were cultivated in the 2021/2022 campaign, during which the average national yield amounted to around 27.7 qq/ha [1]. The seeds are medium-sized and rich in proteins and oils. They have a good balance of essential amino acids, especially lysine and leucine [2], which makes them a valuable addition to human and animal diets.

However, soybean cultivation is often affected by uneven weather, low germination percentages, inadequate growing time and planting space, poor seed irrigation, weeds, and diseases caused by microorganisms (fungi, bacteria, viruses). Among the latter, some of the most important are bacterial blight (the result of infection by *Pseudomonas savastanoi* pv. *glycinea*) [3], bacterial pustules (caused by *Xanthomonas axonopodias* pv. *glycinea*) [4], brown spot (caused by *Septoria glycines*) [5], root and stem rot (for which *Rhizoctonia solani* is responsible) [6], and soybean leaf blight (whose causal agent is *Cercospora kikuchii*) [7]. Such diseases damage around 8–10% of the annual global production. They not only limit yields but can also deform the seeds and thus reduce their quality [3].

Viral diseases, more specifically, can be responsible for yield losses between 10 and 30%, though these numbers can range from 50 to 100% in the case of serious epidemics. One of the most frequent and destructive pathologies is soybean mosaic disease (SMD), which was first reported in 1915 in the US but is now detected in all soybean-producing countries. Caused by the SMV, it usually reduces yield by between 8 and 35%. Nevertheless, losses amounting to 90% have also been reported, since the damage depends on the soybean genotype and the SMV strain [8].

SMV, a positive-sense, single-stranded RNA virus, belongs to the family *Potyviridae* and to the genus *Potyvirus*. Its hosts are commonly from the Fabaceae family, but it can also infect other plants like *Passiflora* spp., *Pinellia ternata*, *Senna occidentalis*, and *Vigna angularis* [8,9]. It is transmitted by more than 30 types of aphids, including the Asian soybean aphid (*Aphis glycines*). The main source of inoculum is infected seeds. The transmission rates ascribed to seeds (secondary spread) can reach 75% but generally do not exceed 5% [10–14].

The virus enters the plant through natural openings or wounds caused by environmental factors or insect vectors. If the host cannot recognize and counteract its effectors, the virus establishes itself in the cells and tends to cause systemic infection. Its ability to decimate proteins and reduce the host's immune response explains why the disease can be so severe and difficult to control [12,15].

The symptoms vary according to the soybean cultivar, the plant's age at the time of infection, the virus strain, and the weather conditions (infections are more serious in cold temperatures). The leaves become mottled, wrinkled, distorted, and necrotic. They may display vein clearing, chlorotic areas, and shortened internodes. Additionally, the plant may be stunted, and the pods and seeds reduced in size and number [15]. The infected seeds also show mottling.

In general, diseases caused by phytopathogenic viruses are difficult to manage, since viruses evolve quickly, and pests are often involved as vectors. In the case of SMD, prevention and control strategies currently include the following: using cultivars that have been genetically engineered to be resistant or immune, planting virus-free seeds, rotating between alternative crop hosts, interrupting transmission routes (prophylaxis), controlling vectors through pesticides, eliminating the primary source of inoculum and infected plants in the field, and managing weeds. As with any other disease, the success of these practices may vary depending on the aggressiveness of the virus and/or the vector, the vulnerability of the plants, and different environmental factors. Better results are obtained when they are combined as part of integral management programs. Moreover, their appropriate implementation relies heavily on farmers having sufficient knowledge about epidemiology and the disease cycle.

Although resistant cultivars have proven particularly efficient in fighting SMV, they are associated with concerns about the insertion of foreign genes. In addition, several SMV strains can bypass their resistance and infect them [8,12,15]. For this reason, alternative methods for sustainable control are necessary. Natural antimicrobial agents derived from medicinal plants, such as essential oils and extracts, could prove useful as control tools for plant pathogenic viruses [3,16–20].

The strategies used at present for SMD prevention and control include good agricultural practices, breeding and genetic engineering of resistant cultivars, vector control, elimination of the primary source of inoculum and infected plants in the field, rotation of alternative hosts, and weed management. Even though resistant cultivars are a particularly effective and environmentally friendly approach, several SMV strains are still able to infect them [8,12,15]. Alternative methods for sustainable control are thus necessary. This is where natural antimicrobial agents derived from medicinal plants, such as essential oils and extracts, could come in handy [3,16].

Among such plants, *Achyrocline satureioides* (marcela), *Minthostachys verticillata* (peperina), *Origanum vulgare* (oregano), and *Thymus vulgaris* (thyme) stand out for their antimicrobial, antiviral, antioxidant, and immunomodulatory properties [16,21–32]. The EOs that they synthesize as secondary metabolites are volatile compounds with potential uses in the pharmaceutical, food, agricultural, and cosmetic industries. The same is true for extracts from these plants. The EOs of *T. vulgaris* and *O. vulgare*, for instance, have shown bactericidal and phytotoxin-inhibitory activity against phytopathogens that infect soybean [16]. A study on soybean seeds infected with *Pseudomonas syringae* found that the application of *T. vulgaris* EO reduced the disease caused by the bacterium [3]. On the other hand, the EO of *M. verticillata* and pulegone, one of its main components, inhibited Suid herpesvirus 1 (SuHV-1) [33] and Herpes simplex virus type 1 [23]; and extracts of *A. satureioides* inhibited the Western equine encephalitis virus (WEEV). However, little is known about the effect of these natural products on phytopathogenic viruses.

In earlier studies, we characterized the cold aqueous extract of *A. satureioides* by HPLC-ESI-MS/MS, and detected the presence of quercetin, chlorogenic acid, luteolin, 5,7,8-trimethoxyflavone, 3-O-methylquercetin, and caffeic acid [26,34]. A GC–FID analysis by Escobar et al. (2012) [35] found that the most important compounds in the EO of peperina are pulegone (60.5%), menthone (18.2%), and limonene (3.76%); those in the EO of oregano are γ-terpinene (22.7%), carvacrol (19.7%), and cis-sabinene hydrate (19.7%); and in thyme EO they are carvacrol (29.5%) and p-cymene (31.5%). The last two EOs, moreover, contain low percentages of thymol [36].

The present study assessed the same extract and EOs in terms of their phytotoxicity in soybean and their activity against SMV in vivo, with the aim of finding evidence that may contribute to the ulterior development of a biopesticide against SMD.

2. Materials and Methods

2.1. Collection of Plant Material

Between 80 and 100 specimens of *Achyrocline satureioides* (Lam) DC were collected in the area of Villa Jorcoricó, in the southern mountains of the province of Córdoba (32°41′26″ S, 64°43′16″ W 800 m ASL). Collection was performed by hand during the flowering stage (in autumn, April–May). Only the shoots were cut; the roots were left in place to allow the plants to regrow. The collected specimens were taxonomically identified by Prof. Luis A. del Vitto at the National University of San Luis. One of them was deposited in that university's herbarium under exsiccate number 6362. Before the preparation of the extract, the plants were washed to remove dirt, insects, and foreign elements, and then left to dry in the open air at room temperature.

The same number of specimens of *Minthosthachys verticillata* (Griseb.) Epling (peperina) (leaves, shoots, and stems) were collected in Santa Rosa, Córdoba (32°04′08″ S, 64°32′10″ W 598 m ASL). They were taxonomically identified at the National University of Río Cuarto. One full specimen was stored in that university's herbarium of vascular plants under exsiccate number 1955. Collection was carried out in the same manner as for A. satureioides, and the plants were also subjected to washing and drying before they were used to obtain the extract.

Finally, 20 kg of leaves and stems of different cultivars of *Origanum vulgare* L. (oregano) and *Thymus vulgaris* L. (thyme) were purchased at a farm called "Los Molles", in the province of San Luis (32°31′34″ S, 64°58′43″ W 1820 m ASL).

2.2. Preparation of the Cold Aqueous Extract (CAE) of *A. satureioides*

The plant material was dried and ground, and 15 g were weighed and extracted with 700 mL of distilled water at room temperature for 2 days. The mixture was filtered through a cloth, and the resulting liquid was labeled "CAE". Then, the extract was lyophilized [37]. Its yield was determined by considering the weight of the initial material vs. the weight of the final lyophilized powder. After that, the extract was resuspended in phosphate-buffered saline (PBS, pH 7), filtered through Whatman No. 2 paper, and sterilized through 0.22-μm pore cellulose acetate filters.

2.3. *Obtention of the Essential Oils (EOs) of M. verticillata, O. vulgare, and T. vulgaris*

The EOs were extracted from the aromatic plants by hydrodistillation in a Clevenger-type apparatus [38]. Briefly, an extraction column was filled with the plant material and placed on a grid. Water was boiled for 2 h in a 1-L balloon; upon passing through the plant-filled column, the steam carried away the volatile components to a condenser. Two phases were obtained by decantation: EO and water [39,40]. The EO was dried with anhydrous sodium sulfate and stored at 4 °C until it was used.

The concentration of each EO (mg/mL) was calculated based on the density of the oily compound, with the formula $\delta = M\ (g)/V\ (mL)$, where M = mass; V = volume. The mass of each EO was obtained from the average weight of five 1 mL fractions.

2.4. *Phytotoxicity Assay*

2.4.1. Soybean Seeds

This assay (as well as the antiviral activity assay, see Section 2.5) was carried out with soybean seeds (cultivar Don Mario 4800 or DM 4800), donated by Dr Rodríguez Pardina from IPAVE-INTA, Córdoba. They were soaked in sodium hypochlorite (3%) for 3 min, rinsed five times with sterile distilled water, and left to dry on absorbent paper for 1 h in a sterile room [3].

2.4.2. Determination of the Phytotoxicity of the Natural Products (NPs) in Soybean

Soybean seedlings (V2) were exposed to different concentrations of each natural product (NP). Different groups of plants (consisting of 10 specimens each) received the following treatments: CAE of *A. satureioides* (0.5, 1, and 1.1 mg/mL); EO of *M. verticillata* (0.4 and 0.8 mg/mL); EO of *O. vulgare* (0.46 and 0.92 mg/mL); or EO of *T. vulgaris* (0.35 and 0.71 mg/mL). Two control groups were included: a negative control (treated with water), and a DMSO control (treated with dimethyl sulfoxide 1/8 in water). The plants were sprayed at the beginning (day 0) and on days 7 and 14. Any observable changes during this time were recorded. When the treatments concluded (on day 21), the plants were evaluated again in terms of morphology, leaf color, height, and wet weight [3].

2.5. *Antiviral Activity Assay*

2.5.1. Source of Soybean Mosaic Virus

The virus used in this assay was an SMV-MJ isolate [41], whose inoculum was recovered from a freeze-dried sample.

2.5.2. Inhibitory Activity of the NPs against SMV Evaluated In Vivo by Applying Koch's Postulates

Soybean plants were grown under greenhouse conditions from DM 4800 seeds (see Section 2.4.1). They were then mechanically inoculated with the virus following Camelo García (2010) [42], with modifications. More precisely, the SMV-MJ isolate was inoculated with potassium phosphate buffer (0.05 M, pH 7.6) in the first emerged trifoliate leaf. Groups of 10 plants were sprayed 7 and 14 days post-inoculation (dpi) with the different NPs: CAE of *A. satureioides* (1.1 mg/mL); EO of *M. verticillata* (0.8 mg/mL); EO of *O. vulgare* (0.92 mg/mL); or EO of *T. vulgaris* (0.71 mg/mL), in independent trials. Symptoms were monitored daily until day 21 after inoculation. The NPs were considered to be effective when symptoms were absent or reduced.

A positive control (10 plants infected with SMV, left untreated), a negative control (10 uninfected plants sprayed with sterile distilled water), and a DMSO control (10 plants infected with SMV and treated with DMSO 1/8) were included.

2.6. Determination of Viral Replication Inhibition through Indirect ELISA

Twenty-one dpi, the inhibition of viral replication was indirectly measured by analyzing the virus titer with PTA-ELISA (Plate Trapped Antibody-Enzyme Linked Immunosorbent Assay) [43]. The primary antibody was a rabbit polyclonal anti-SMV serum, produced at IPAVE-INTA (unpublished data). The secondary antibody was a goat anti-rabbit IgG conjugated with alkaline phosphatase (BIO-RAD, Hercules, CA, USA). The reaction was detected by adding 0.75 mg/mL of p-Nitrophenyl Phosphate, Disodium Salt (PNPP) (Agdia Inc, Elkhart, IN, USA). Six healthy samples and one SMV-positive sample per plate were used as controls. The reactions were quantified in a Thermo Labsystem MultisKan MS spectrophotometer. A sample was considered positive when the absorbance at 405 nm (A_{405}) was higher than the mean of the healthy controls plus three times the standard deviation (cut-off), or 0.100.

2.7. Statistical Analysis

The data were statistically analyzed (ANOVA) on GraphPad Prism v6.0. Values were expressed as the mean ± standard deviation (SD). The data from the phytotoxicity assays and antiviral studies were compared with the parametric *t*-test. The level of significance was established at $p < 0.05$.

3. Results

3.1. Yield of NPs

The yields of the different products were as follows: CAE of *A. satureioides* 4.76% (*w*/*w*); EO of *M. verticillata* 5.1% (*w*/*v*); EO of *O. vulgare* 0.5% (*w*/*v*); and EO of *T. vulgaris* 1.9% (*w*/*v*).

3.2. Phytotoxicity Assay

The following concentrations of each product were tested for their phytotoxicity in soybean: 0.5, 1, and 1.1 mg/mL for the CAE of *A. satureioides;* 0.4 and 0.8 mg/mL for the EO of *M. verticillata;* 0.46 and 0.92 mg/mL for the EO of *O. vulgare;* and 0.35 and 0.71 mg/mL for the EO of *T. vulgaris*. A negative control group was included, which consisted of plants that did not receive any of the NPs. In those experiments where DMSO was used as a diluent, a solvent control was added. The measurements obtained for plant height (including shoots and roots) can be seen in Figure 1a,c,e,g. Those for wet weight are shown in Figure 1b,d,f,h.

Figure 1 shows that there were no statistically significant differences in plant height or wet weight between the negative control group and the treated plants, regardless of the NP concentration tested. This means that these concentrations are not phytotoxic, i.e., they do not affect the normal development of soybean plants. Phytotoxicity of NPs in soybean. (a) Plant height and (b) wet weight after exposure to CAE of *A. satureioides;* (c) plant height and (d) wet weight after exposure to EO of *M. verticillata;* (e) plant height and (f) wet weight after exposure to EO of *O. vulgare;* (g) plant height and (h) wet weight after exposure to EO of *T. vulgaris.* In all cases, significant differences with respect to the control were determined through the *t*-test, ANOVA ($p \leq 0.05$).

3.3. In Vivo Evaluation of the Inhibitory Activity of the NPs against SMV

The concentrations ascertained to be safe in the previous step were then assessed in vivo on a greenhouse scale, to determine their inhibitory activity against SMV. Soybean seedlings were experimentally infected with the virus and spray-treated independently with the different NPs. Once again, their total height and wet weight were individually measured for each treatment (Figures 2 and 3).

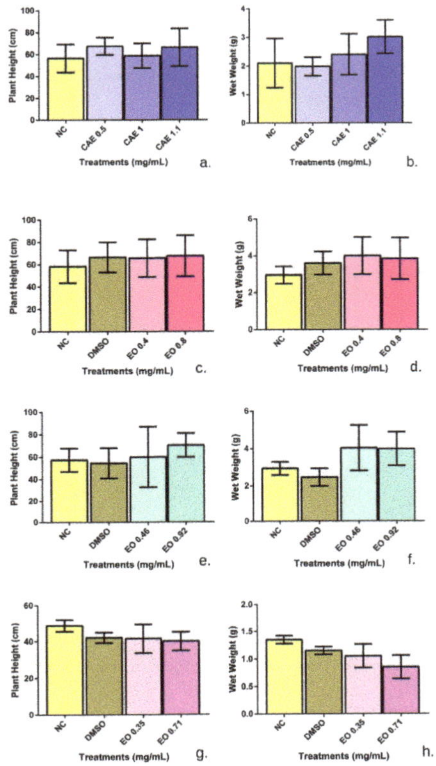

Figure 1. Phytotoxicity of NPs in soybean. (**a**) Plant height and (**b**) wet weight after exposure to CAE of *A. satureioides*; (**c**) plant height and (**d**) wet weight after exposure to EO of *M. verticillata*; (**e**) plant height and (**f**) wet weight after exposure to EO of *O. vulgare*; (**g**) plant height and (**h**) wet weight after exposure to EO of *T. vulgaris*. In all cases, significant differences with respect to the control were determined through the *t*-test, ANOVA ($p \leq 0.05$).

As seen in Figure 2, the untreated plants infected with SMV (positive control) were shorter than those in the negative control group ($p < 0.05$). This indicates that the virus significantly affects the normal growth of soybean seedlings. In turn, all the treated plants were taller than those in the positive control group. Statistically significant differences, however, were only registered for those plants treated with the CAE of *A. satureioides* ($p < 0.05$) and the EO of *M. verticillata* ($p < 0.01$), so these NPs seem to protect soybean plants from damage caused by the virus.

Moreover, the plants in the untreated positive control group weighed less than those in the negative control group ($p < 0.01$), which is further evidence of how the virus affects normal growth (Figure 3). The treated plants weighed more than the viral controls, in particular those sprayed with the EO of peperina ($p < 0.001$), which had the highest weight.

The photographs in Figure 4 show the leaves of treated plants and their negative and positive controls. A significant reduction in the spots and other symptoms produced by SMV was observable in the plants treated with the EO of *T. vulgaris*. For their part, the plants treated with the EO of *M. verticillata* did not exhibit any symptoms at all. These results are consistent with the statistical data, according to which the EO of *M. verticillata* is the most efficient among the NPs tested here in counteracting the effects of SMV.

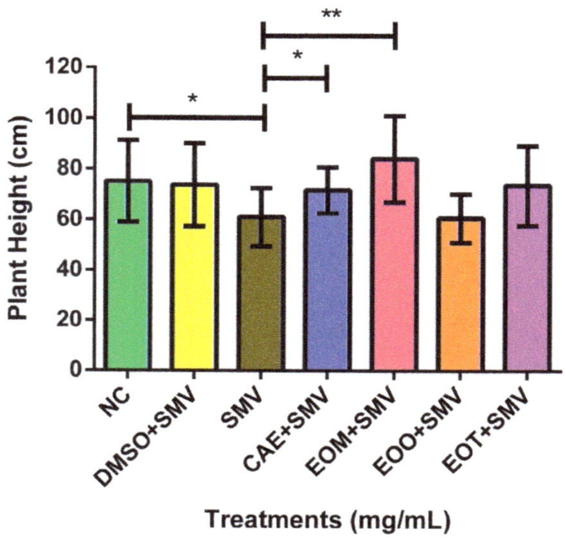

Figure 2. Inhibitory activity of NPs against SMV in vivo, as indicated by plant height (cm). NC: negative control; DMSO: dimethyl sulfoxide (solvent control); SMV: soybean mosaic virus; CAE: cold aqueous extract of *A. satureioides;* EOM: essential oil of *M. verticillate;* EOO: essential oil of *O. vulgare;* EOT: essential oil of *T. vulgaris*. Means with different * indicate statistically significant differences ($p < 0.05$). In all cases, significant differences with respect to the control were determined through *t*-test, ANOVA ($p \leq 0.05$). Ref.: * $p \leq 0.05$, ** $p \leq 0.01$.

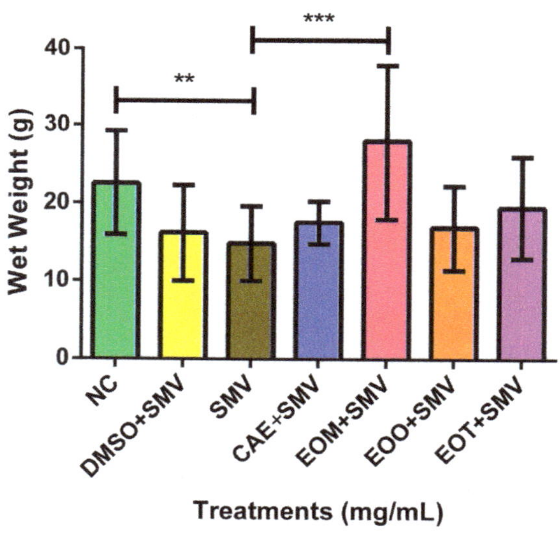

Figure 3. Inhibitory activity of NPs against SMV in vivo, as indicated by wet weight (g). NC: negative control; DMSO: dimethyl sulfoxide (solvent control); SMV: soybean mosaic virus; CAE: cold aqueous extract of *A. satureioides;* EOM: essential oil of *M. verticillate;* EOO: essential oil of *O. vulgare;* EOT: essential oil of *T. vulgaris*. Means with different * indicate statistically significant differences ($p < 0.05$). In all cases, significant differences with respect to the control were determined through *t*-test, ANOVA ($p \leq 0.05$). Ref.: ** $p \leq 0.01$, *** $p \leq 0.001$.

Figure 4. (**a**) Negative control; (**b**) positive control (SMV+DMSO); (**c**) SMV+EO of *T. vulgaris*; (**d**) SMV+EO *M. verticillate*.

3.4. Inhibition of Viral Replication Determined through Indirect ELISA

Indirect ELISA, a technique that determines the amount of viral antigen (Ag), was used to assess viral inhibition after the application of the NPs. In other words, it was performed to detect the presence of the virus in the treated plants and the controls. The results of this test are shown in Figure 5.

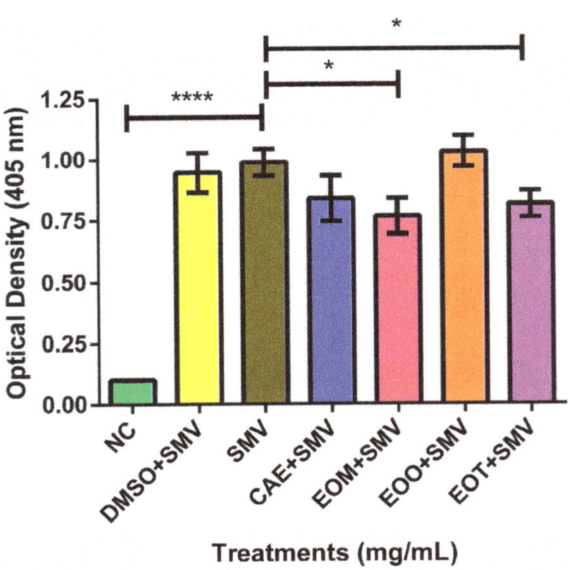

Figure 5. Inhibitory activity of NPs against SMV in vivo, as determined by indirect ELISA (405 nm). NC: Negative control; DMSO: dimethyl sulfoxide; SMV: soybean mosaic virus; CAE: cold aqueous extract of *A. satureioides*; EOM: essential oil of *M. verticillate*; EOO: essential oil of *O. vulgare*; EOT: essential oil of *T. vulgaris*. Means with different * indicate statistically significant differences ($p < 0.05$). In all cases, significant differences with respect to the control were determined through *t*-test, ANOVA ($p \leq 0.05$). Ref.: * $p \leq 0.05$, **** $p < 0.0001$.

As expected, a statistically significant difference ($p < 0.0001$) was found between the negative control and the untreated control infected with SMV. With respect to the latter, the EO of *M. verticillata* (0.80 mg/mL) and the EO of *T. vulgaris* (0.71 mg/mL) significantly inhibited viral antigen production or relative virus concentration. This concentration was not significantly altered by the other two products (CAE of *A. satureioides* at 1.1 mg/mL or EO of *O. vulgare* at 0.92 mg/mL).

4. Discussion

Among the microbial diseases that affect soybean, one of the most devastating is SMD, which causes significant economic losses around the world [12,13].

SMV, the virus responsible for this disease, spreads rapidly through insect vectors and seeds. Planting virus-free seeds and resistant cultivars, maintaining good agricultural practices, and striving to detect cases early on are some of the control strategies implemented at present [12,44]. Insecticides (which are sometimes applied to control the vectors) are not only toxic to the environment but also quite unsuccessful in reducing the incidence of SMV [13].

Given the demand for sustainable strategies to mitigate the harmful impact of microorganisms on agriculture, a lot of research currently focuses on exploring the antimicrobial properties of natural products. Many medicinal plants have antibacterial, antifungal, and antiviral activity, and can inhibit toxin production, biofilm formation, swarming, swimming, and other virulence factors [16,45–48]. However, any new substance with potential applications in medicine, agriculture, or the food sector must first be assessed in terms of its toxicity.

The present study found that the CAE of *A. satureioides* is not toxic for soybean at the concentrations tested. Although no other research had previously looked into the phytotoxicity of this product, an assay performed on cultures of mouse marrow cells demonstrated it was not genotoxic. In addition, high concentrations of the extract (50 mg/kg p.c.) were safe for mice [34].

When we used this CAE (1.1 mg/mL) to treat soybean plants infected with SMV, both plant height and wet weight improved with respect to the untreated infected specimens, but only the first parameter was significantly different. An aqueous extract of *A. satureioides* was previously reported to act against the Western Equine Encephalitis virus (WEEV) at the intracellular replication stage [26]. This alphavirus, which infects humans and horses and for which there are no other effective antivirals, has a similar genome to that of SMV: the two of them are positive-sense, single-stranded RNA viruses, and may thus be inhibited through similar mechanisms. The antiviral properties of the extract are likely related to the action of quercetin, one of its components, which was effective on its own against other RNA and naked viruses such as poliovirus type 1 [49] and respiratory syncytial virus [50].

Marcela has also been successful against different bacterial species. A hexanic extract showed inhibitory and acidic properties in vitro against mechanisms related to the pathogenicity of *Paenibacillus larvae* (biofilm formation, swarming, synthesis of proteases). *P. larvae* attacks honeybee hives, and the extract was assessed to be safe for larvae and adult bees [32]. Other bacteria which the extract has been observed to antagonize include *Enterococcus faecalis*, *Escherichia coli*, *Proteus mirabilis*, *Pseudomonas aeruginosa*, *Salmonella enterica* serovar *enteritidis*, and *Staphylococcus aureus* [51]. A decoction of *A. satureiodes* not only inhibited *Staphylococcus* spp. almost completely, but also immunostimulated T CD8+ lymphocytes in humans [23].

The application of EO in aqueous media is limited due to their hydrophobicity and volatile nature, so DMSO is used in low proportion to aid dissolution in water in both in vivo and in vitro experiments. In all our EO assays, emulsions were prepared with this solvent plus sterile water. In our phytotoxicity studies, a solvent control was performed evaluating the possible toxic effect of DMSO: water (at a concentration of 12.5%) on soybean plants, and no toxic damage (in any of the parameters evaluated) was observed in the tested plants.

Regarding previous studies of DMSO toxicity, Kloverpris et al. (2010) [52] exposed in vitro human PBMC to a concentration of 10% for 1 h not affecting the viability. Similar results were obtained in vivo when a dose of 0.22 g DMSO/kg/day was administered in NOD mice not producing diarrhea or weight loss [53]. Other trials demonstrated that minimal side effects were induced during clinical applications [54] only when concentrations of intravenous infusion greater than 40% were tested. They produced intravascular hemolysis [55]. These investigations demonstrated that low concentrations of DMSO in animal cell cultures and in animal species were safe. Therefore, this solvent can be used for a variety of clinical in vitro or in vivo treatments.

Our study is the first report on the toxicity of DMSO in plants; a concentration of 12.5% was shown to be safe for soybean plants. Similarly, antiviral studies included a DMSO control group, infected with SMV and treated with DMSO:water (12.5%) in which no statistically significant difference from the untreated viral control was observed.

On the other hand, EOs and their components have attracted significant scientific interest thanks to their beneficial properties for general health and their lack of toxicity. Their antimicrobial and antioxidant abilities have been explored for their incorporation into food [48], but the present study investigated three of them in relation to their potential benefits for the protection of soybean.

According to our findings, the EO of *M. verticillata* is not toxic for soybean. Earlier research had proven its safety in murine and bovine models [56,57], in human peripheral blood mononuclear cells (PBMC), mouse bone marrow cells [35], porcine and equine blood cells [58], and even brine shrimp (*Artemia salina*) [59,60]. In short, the EO of *M. verticillata* appears to be a safe natural product that could be used on agriculturally important crops destined for human and animal consumption. Several studies compared the toxicity of this EO as a whole against that of some of its components on their own. Rossi et al. (2012) [61] reported that the EO was less toxic than pulegone and menthone for houseflies (*Musca domestica*). Similarly, Vogt et al. (2010) [33] observed that the 50% cytotoxic concentration (CC_{50}) values for the oil were higher than for its major components in HEp-2 cells.

In terms of its protective activity, the EO of peperina (0.8 mg/mL) was effective in protecting plants against SMV infection in the present study, as shown by the plants' significantly greater height and weight compared to infected specimens that were not treated. Furthermore, the virus titer or concentration was significantly reduced after this product was applied, and none of the characteristic symptoms of SMD appeared in the plants sprayed with it. Previous studies had confirmed its activity against viruses that infect animal hosts, such as Suid herpesvirus 1 (SuHV1) [33], and Herpes simplex virus type I (HSV-1) [23]. As with *A. satureiodes*, this activity might be due to the action of specific components in *M. verticillata*. Pulegone was observed to inhibit SARS-CoV-2 [62], and limonene, a monoterpene, reduced the infectivity of HSV-1 by 100%, by exerting antiherpetic activity during the early phase of viral multiplication [63].

In addition to inhibiting viruses, this EO showed protective power against the damage caused by aflatoxin B1 (AFB1), a mycotoxin. More precisely, it improved the histomorphometry of intestinal villi and reduced DNA damage in bone marrow cells of male Wistar rats in which AFB1 toxicity had been induced [64]. Evidence of its activity against enteropathogenic bacteria (*S. aureus*, *S. faecalis*, *Bacillus cereus*, *E. coli*, and *Salmonella typhi*) supports its popular use for digestive troubles [35]. Its ability to interfere with growth and biofilm formation in bacteria that cause bovine mastitis has been described as well [47].

The safety of *O. vulgare* EO for soybean was also confirmed in our study. Twenty-one days after being sprayed with 0.46 and 0.92 mg/mL of the oil, none of the plants displayed alterations in their growth parameters. This agrees with the results by Gonçalves et al. (2021) [65], who treated tomato seeds (*Solanum lycopersicum*) with 1.2 mg/mL of oregano EO and recorded no phytotoxicity.

Less is known about its antiviral potential. Polar extracts of *O. vulgare* (infusion, decoction, and hydroalcoholic extract) completely inhibited bovine alphaherpesvirus 1 (BoHV-1), responsible for bovine infectious rhinotracheitis [66]. An aqueous extract was active against Equine Arteritis Virus (EAV), and aqueous and ethanolic extracts were effective against canine distemper virus (CDV) [67]. Nevertheless, we were unable to reliably demonstrate that the EO of this plant can inhibit SMV at the concentration tested (0.92 mg/mL), since there were no significant differences after treatment in plant height, wet weight, or virus titer. Further studies should perhaps try out other concentrations, or test this EO in combination with other NPs.

As with the other two EOs, that of *T. vulgaris* was not toxic for soybean at 0.35 and 0.71 mg/mL, since there were no significant differences in growth between treated plants and the untreated control. Other researchers similarly recorded no statistically significant modifications in growth after treatment with *T. vulgaris* EO in soybean (at 1.76 mg/mL) [3], and potato (0.170 mg/mL) [68]. In contrast, the oil was deemed highly toxic for brine shrimp and for cancer cell lines in vitro [69]. Its vapors slightly affected the germination of wheat seeds [70], and produced some scalding in apple tissues [71]. This likely means that the toxicity of this EO depends on the concentration and the sensitivity of the species on which it is applied.

We observed that when sprayed at 0.71 mg/mL, it improved height and weight in infected plants with respect to the control. Moreover, it reduced symptoms and viral antigen production in a statistically significant manner. Although several studies have focused on its antimicrobial properties [16,24,72–74], this is the first time it has proven to be effective against a phytopathogenic virus.

Still, there is evidence of its ability to antagonize animal viruses. For instance, it affected the replication of retroviruses like human immunodeficiency virus 1 (HIV-1) [75], and other RNA viruses such as the feline infectious peritonitis virus (*Coronaviridae*). In the last case, a higher concentration than the one tested here (27 μg/mL) reduced virus concentration by 2 log10 TCID50/50 μL [29]. When tested against HSV-1 at 25 μg/mL, it achieved 45% inhibition, while 25 μg/mL of p-cymene (one of its main terpene components) achieved around 70% inhibition [76]. The antiviral activity observed in our assay might therefore be related to the action of this compound. Likewise, an aqueous extract of thyme exerted antiviral activity against HSV-1 and HSV-2 [77].

As far as bacteria are concerned, the EO of *T. vulgaris* has been reported to successfully antagonize *P. syringae* isolated from soybean and reference phytopathogenic strains in vitro [45]. Its effectiveness against *P. syringae* (as well as against *P. savastanoi pv. Glycinea* B076) was then confirmed in vivo in soybean [3] and explained in terms of an ability to interfere with phytotoxin production, biofilm formation [16], swarming, and swimming in these bacteria [31]. Microencapsulated in maltodextrin and hydroxypropyl methylcellulose, thyme EO inhibited *Streptomyces scabiei*, the causative agent of common potato scab [68]. Its antifungal activity, on the other hand, has been documented on *Fusarium avenaceum*, *Botrytis cinerea*, *Penicillium expansum*, *Neofabraea vagabunda* [71], *P. infestans*, *R. solani* [78], *Drechslera* [70], *Candida albicans*, and *Candida tropicalis* [79]. Finally, it had insecticidal activity on the larvae of the *Aedes aegypti* mosquito, the main vector of urban arboviruses [80], and reduced the survival and longevity of bean weevil adults (*Acanthoscelides obtectus*). It also retracted ovipositioned females of this species [81].

In short, the EO of *T. vulgaris* has a broad spectrum of action, which comprises viruses, bacteria, fungi, and even insects. Nevertheless, more information is necessary about the antiviral capacities of its specific components.

In terms of the specific mechanisms through which these and other NPs fight viruses, several possibilities are feasible. Monoterpenes might bind to viral proteins involved in host cell penetration, and thus prevent the virus from entering the cell [82]. As was mentioned earlier for an aqueous extract of *A. satureioides*, viral replication inside the cell may also be interrupted through molecular interactions between the virus and the natural agents. Lastly, although the present study did not explore this, EOs and extracts may not act directly against the virus, but rather on the host plant by enhancing its immunity. When tomato plants infected with the tobacco mosaic virus were treated with zinc oxide (ZnO) nanoparticles (synthesized with an aqueous extract of *Mentha spicata*), systemic acquired resistance (SAR) was induced, and disease symptoms were diminished [83–85].

5. Conclusions

All the natural products assessed in this study appear to be safe for soybean and had positive effects on plants infected with SMV. The most relevant improvement after the application of the CAE of *A. satureioides* was observed in plant height. The EO of *T. vulgaris* also improved plant height and weight and managed to inhibit the virus, as demonstrated by the statistically significant reduction in the virus titer. The best results, both in terms of plant growth and viral inhibition, were obtained with the EO of *M. verticillata*. Plants sprayed with this product showed no symptoms of disease, i.e., the EO was able to control infection. The only product for which no statistically significant effects could be demonstrated was the EO of *O. vulgare*. Although these findings are conclusive, further research should corroborate them by working with larger sampling sizes and under less or no controlled conditions, e.g., in the greenhouse and in the field. Furthermore, the NPs could also be tested for their ability to inhibit other plant pathogenic viruses. Given the promising performance of the EO of *M. verticillata* in the assays described here, its potential applications may also deserve more in-depth exploration. Finally, a combination of NPs could be assayed, with the ultimate aim of formulating a "biopesticide" to manage SMD sustainably.

Author Contributions: Conceptualization, M.E.C. and M.C.S.; methodology, M.E.C., M.C.S., P.G.R., E.A., W.G. and M.d.l.M.O.; formal analysis, M.E.C., M.C.S. and P.R.P.; investigation, M.E.C. and M.C.S.; writing, reviewing, and editing, M.E.C., P.R.P., W.G., M.d.l.M.O. and M.C.S.; supervision, M.E.C. and M.C.S.; funding acquisition, M.E.C., P.G.R., E.A., M.d.l.M.O., P.R.P. and M.C.S. All authors have read and agreed to the published version of the manuscript.

Funding: This research was funded by grants from Consejo Nacional de Investigaciones Científicas y Técnicas, Argentina (CONICET), SeCyT Resolution 083/20 from Universidad Nacional de Río Cuarto. W.G., M.d.l.M.O., R.P.P. and M.C.S. are Career Members of CONICET.

Institutional Review Board Statement: Not applicable.

Informed Consent Statement: Not applicable.

Data Availability Statement: All of the data are included in the manuscript.

Acknowledgments: The authors wish to thank Candela Winter for his expert help with the illustrations; F. Sgarlatta for proofreading the manuscript's English.

Conflicts of Interest: The authors declare no conflict of interest.

References

1. Available online: https://www.bcr.com.ar/es/mercados/gea/estimaciones-nacionales-de-produccion/estimaciones (accessed on 20 March 2023).
2. Scott, W.O.; Aldrich, S.R. *Producción Moderna De La Soja*; Hemisferio Sur: Buenos Aires, Argentina, 1975; pp. 67–99.
3. Sotelo, J.P.; Oddino, C.; Giordano, D.F.; Carezzano, M.E.; Oliva, M.d.l.M. Effect of *Thymus vulgaris* essential oil on soybeans seeds infected with *Pseudomonas syringae*. *Physiol. Mol. Plant Pathol.* **2021**, *116*, 101735. [CrossRef]
4. Zhang, J.; Pang, B.; Wang, Y.; Jiang, H.; Herrera-Balandrano, D.D.; Jin, Y.; Wang, Y.; Laborda, P. Exogenous genistein enhances soybean resistance to *Xanthomonas axonopodis* pv. *glycines*. *Pest Manag. Sci.* **2022**, *78*, 3664–3675. [CrossRef]
5. Lin, H.A.; Mideros, S.X. The Effect of *Septoria glycines* and fungicide application on the soybean phyllosphere mycobiome. *Phytobiomes J.* **2023**. [CrossRef]
6. Akber, M.A.; Mubeen, M.; Sohail, M.A.; Khan, S.W.; Solanki, M.K.; Khalid, R.; Abbas, A.; Divvela, P.K.; Zhou, L. Global distribution, traditional and modern detection, diagnostic, and management approaches of *Rhizoctonia solani* associated with legume crops. *Front. Microbiol.* **2023**, *13*, 1091288. [CrossRef] [PubMed]
7. Kashiwa, T.; Suzuki, T. High-quality genome assembly of the soybean fungal pathogen *Cercospora kikuchii*. *G3-Genes Genomes Genet.* **2021**, *11*, jkab277. [CrossRef]

8. Usovsky, M.; Chen, P.; Li, D.; Wang, A.; Shi, A.; Zheng, C.; Shakiba, E.; Lee, D.; Canella Vieira, C.; Lee, Y.C.; et al. Decades of genetic research on soybean mosaic virus resistance in soybean. *Viruses* **2022**, *14*, 1122. [CrossRef]
9. Hajimorad, M.R.; Domier, L.L.; Tolin, S.A.; Whitham, S.A.; Saghai Maroof, M.A. Soybean mosaic virus: A successful potyvirus with a wide distribution but restricted natural host range. *Physiol. Mol. Plant Pathol.* **2018**, *19*, 1563–1579. [CrossRef]
10. Rupe, J.C.; Luttrell, R.G. Effect of pests and diseases on soybean quality. In *Soybeans: Chemistry, Production, Processing, and Utilization*; Johnson, L.A., White, P.J., Galloway, R., Eds.; AOCS Press: Urbana, IL, USA, 2008; pp. 93–116. [CrossRef]
11. Sweets, L. Virus diseases may begin to show up in missouri wheat fields. *Integr. Pest Crop Manag.* **2011**, *4*, 1–12.
12. Yang, Q.; Jin, H.; Yu, X.; Fu, X.; Zhi, H.; Yuan, F. Rapid Identification of soybean resistance genes to soybean mosaic virus by SLAF-seq bulked segregant analysis. *Plant Mol. Biol. Rep.* **2020**, *38*, 666–675. [CrossRef]
13. Rehman, F.U.; Kalsoom, M.; Adnan, M.; Naz, N.; Nasir, T.A.; Ali, H.; Shafique, T.; Murtaza, G.; Anwar, S.; Arshad, M.A. Soybean mosaic disease (SMD): A review. *Egypt. J. Basic Appl. Sci.* **2021**, *8*, 12–16. [CrossRef]
14. Available online: https://www.sinavimo.gob.ar/plaga/soybean-mosaic-virus (accessed on 25 April 2023).
15. Widyasari, K.; Alazem, M.; Kim, K.H. Soybean resistance to soybean mosaic virus. *Plants* **2020**, *9*, 219. [CrossRef]
16. Carezzano, M.E.; Sotelo, J.P.; Primo, E.; Reinoso, E.B.; Paletti Rovey, M.F.; Demo, M.S.; Giordano, W.; Oliva, M.d.l.M. Inhibitory effect of *Thymus vulgaris* and *Origanum vulgare* essential oils on virulence factors of phytopathogenic *Pseudomonas syringae* strains. *Plant Biol.* **2017**, *19*, 599–607. [CrossRef]
17. Rubio, L.; Galipienso, L.; Ferriol, I. Detection of plant viruses and disease management: Relevance of genetic diversity and evolution. *Front. Plant Sci.* **2020**, *11*, 539737. [CrossRef]
18. Jones, A.C.; Naidu, R.A. Global dimensions of plant virus diseases: Current status and future perspectives. *Annu. Rev. Virol.* **2019**, *6*, 387–409. [CrossRef]
19. Jones, R.A. Global plant virus disease pandemics and epidemics. *Plants* **2021**, *10*, 233. [CrossRef]
20. Shahzad, G.; Passera, A.; Maldera, G.; Casati, P.; Marcello, I.; Bianco, P.A. Biocontrol potential of endophytic plant-growth-promoting bacteria against phytopathogenic viruses: Molecular interaction with the host plant and comparison with chitosan. *Int. J. Mol. Sci.* **2022**, *23*, 6990. [CrossRef]
21. Bettega, J.M.R.; Teixeira, H.; Bassani, V.L.; Barardi, C.R.M.; Simões, C.M.O. Evaluation of the antiherpetic activity of standardized extracts of *Achyrocline satureioides*. *Phytother. Res.* **2004**, *18*, 819–823. [CrossRef]
22. Calvo, D.; Cariddi, L.N.; Grosso, M.; Demo, M.S.; Maldonado, A.M. *Achyrocline satureioides* (LAM.) DC (Marcela): Antimicrobial activity on *Staphylococcus* spp. and immunomodulating effects on human lymphocytes. *Rev. Latinoam. Microbiol.* **2006**, *48*, 247–255.
23. Sabini, M.C.; Cariddi, L.N.; Escobar, F.M.; Aguilar, J.J.; Tonn, C.E.; Contigiani, M.S.; Sabini, L.I. Action of extracts obtained with organic solvents from *Minthostachys verticillata* (Griseb.) Epling on viability of *Herpes simplex Type 1* virus (HSV-1). *Mol. Med. Chem.* **2010**, *21*, 84–87.
24. Borugă, O.; Jianu, C.; Mişcă, C.; Goleţ, I.; Gruia, A.T.; Horhat, F.G. *Thymus vulgaris* essential oil: Chemical composition and antimicrobial activity. *J. Med. Life* **2014**, *7*, 3.
25. Zhang, X.L.; Guo, Y.S.; Wang, C.H.; Li, G.Q.; Xu, J.J.; Chung, H.Y.; Ye, W.C.; Li, Y.L.; Wang, G.C. Phenolic compounds from *Origanum vulgare* and their antioxidant and antiviral activities. *Food Chem.* **2014**, *152*, 300–306. [CrossRef] [PubMed]
26. Sabini, M.C.; Cariddi, L.N.; Escobar, F.M.; Mañas, F.; Comini, L.; Iglesias, D.; Larrauri, M.; Núñez Montoya, S.; Sereno, J.; Contigiani, M.S.; et al. Potent inhibition of *Western Equine Encephalitis virus* by a fraction rich in flavonoids and phenolic acids obtained from *Achyrocline satureioides*. *Rev. Bras. Farmacogn.* **2016**, *26*, 571–578. [CrossRef]
27. Han, X.; Parker, T.L. Anti-inflammatory, tissue remodeling, immunomodulatory, and anticancer activities of oregano (*Origanum vulgare*) essential oil in a human skin disease model. *Biochim. Open* **2017**, *4*, 73–77. [CrossRef] [PubMed]
28. Micucci, M.; Protti, M.; Aldini, R.; Frosini, R.; Corazza, I.; Marzetti, C.L.B.; Mattioli, L.B.; Tocci, G.; Chiarini, A.; Mercolini, L.; et al. *Thymus vulgaris* L. essential oil solid formulation: Chemical profile and spasmolytic and antimicrobial effects. *Biomolecules* **2020**, *10*, 860. [CrossRef]
29. Catella, C.; Camero, M.; Lucente, M.S.; Fracchiolla, G.; Sblano, S.; Tempesta, M.; Martella, V.; Buonavoglia, C.; Lanave, G. Virucidal and antiviral effects of *Thymus vulgaris* essential oil on feline coronavirus. *Res. Vet. Sci.* **2021**, *137*, 44–47. [CrossRef]
30. Montironi, I.D.; Campra, N.A.; Arsaute, S.; Cecchini, M.E.; Raviolo, J.M.; Vanden Braber, N.; Barrios, B.; Montenegro, M.; Correa, S.; Grosso, M.C.; et al. *Minthostachys verticillata* Griseb (Epling.) (Lamiaceae) essential oil orally administered modulates gastrointestinal immunological and oxidative parameters in mice. *J. Ethnopharmacol.* **2022**, *290*, 115078. [CrossRef]
31. Carezzano, M.E.; Paletti Rovey, M.F.; Sotelo, J.P.; Giordano, M.; Bogino, P.; Oliva, M. dlM.; Giordano, W. Inhibitory potential of *Thymus vulgaris* essential oil against growth, biofilm formation, swarming, and swimming in *Pseudomonas syringae* isolates. *Processes* **2023**, *11*, 933. [CrossRef]
32. Paletti Rovey, M.F.; Sotelo, J.P.; Carezzano, M.E.; Huallpa, C.; Oliva, M.d.l.M. Hexanic extract of *Achyrocline satureioides*: Antimicrobial activity and in vitro inhibitory effect on mechanisms related to the pathogenicity of *Paenibacillus larvae*. *Vet. Res. Commun.* **2023**, 1–13. [CrossRef]
33. Vogt, M.V.; Sutil, S.B.; Escobar, F.M.; Sabini, M.C.; Cariddi, L.N.; Torres, C.V.; Zanon, S.M.; Sabini, L.I. *Minthostachys verticillata* essentials oil and its major components: Antiherpetic selective action in HEp-2 cells. Instituto de Investigación de las Ciencias Exactas Físicas y Naturales. *Mol. Med. Chem.* **2010**, *21*, 117–120.

34. Sabini, M.C.; Cariddi, L.N.; Escobar, F.M.; Mañas, F.; Comini, L.; Reinoso, E.; Sutil, S.B.; Acosta, A.C.; Núñez Montoya, S.; Contigiani, M.S.; et al. Evaluation of the cytotoxicity, genotoxicity and apoptotic induction of an aqueous extract of *Achyrocline satureioides* (Lam.) DC. *FCT* **2013**, *60*, 463–470. [CrossRef]
35. Escobar, F.M.; Cariddi, L.N.; Sabini, M.C.; Reinoso, E.; Sutil, S.B.; Torres, C.V.; Zanon, S.M.; Sabini, L.I. Lack of cytotoxic and genotoxic effects of *Minthostachys verticillata* essential oil: Studies in vitro and in vivo. *FCT* **2012**, *50*, 3062–3067. [CrossRef]
36. Oliva, M.M.; Carezzano, E.; Gallucci, N.; Freytes, S.; Zygadlo, J.A.; Demo, M. Growth inhibition and morphological alterations of *Staphylococcus aureus* caused by the essential oil of *Aloysia triphylla*. *Bol. Latinoam. Caribe Plant. Med. Aromat.* **2015**, *14*, 83–91.
37. Sabini, M.C.; Escobar, F.M.; Tonn, C.E.; Zanon, S.M.; Contigiani, M.S.; Sabini, L.I. Evaluation of antiviral activity of aqueous extracts from *Achyrocline satureioides* against *Western equine encephalitis* virus. *Nat. Prod. Res.* **2012**, *26*, 405–415. [CrossRef]
38. Beoletto, V.G.; De Las Mercedes Oliva, M.; Marioli, J.M.; Carezzano, M.E.; Demo, M.S. Antimicrobial natural products against bacterial biofilms. In *Antibiotic Resistance: Mechanisms and New Antimicrobial Approaches*; Kon, K., Rai, M., Eds.; Elsevier: London, UK, 2016; pp. 290–307.
39. Bamba, D.; Bessiere, J.M.; Pélissier, Y.; Fourasté, I. Essential oil of *Eupatorium odoratum*. *Planta Médica* **1993**, *59*, 184.
40. De Feo, V.; Ricciardi, A.; Biscardi, D.; Senatore, F. Chemical composition and antimicrobial screening of the essential oil of *Minthostachys verticillata* (Griseb). Epl. (*Lamiaceae*). *J. Essent. Oil Res.* **1998**, *10*, 61–65.
41. Maugeri Suarez, M.; Rodriguez, M.S.; Bejerman, N.; Laguna, I.G.; Rodriguez Pardina, P. Biological, molecular and phiysiological characterization of four soybean mosaic virus isolates present in Argentine soybean crops. *BioRxiv* **2021**, *10*, 447356. [CrossRef]
42. Camelo García, V. Detección e Identificación de Los Virus Patógenos de Cultivos de Gulupa (*passiflora edulis sims*) en la Región del Sumapaz (Cundinamarca). Master's Thesis, Universidad Nacional de Colombia, Bogotá, Colombia, 2010.
43. Converse, R.H.; Martin, R.R. ELISA methods for plant viruses. In *Serological Methods for Detection and Identification of Viral and Bacterial Plant Pathogens. A Laboratory Manual*; Hampton, R., Ball, E., de Boer, S., Eds.; APS Press: St. Paul, MN, USA, 1990; pp. 179–196.
44. Ren, R.; Wang, T.; Gao, L.; Song, P.; Yang, Y.; Zhi, H.; Li, K. Development of comprehensive serological techniques for sensitive, quantitative and rapid detection of Soybean mosaic virus. *Int. J. Mol. Sci.* **2022**, *23*, 9457. [CrossRef]
45. Oliva, M.d.l.M.; Carezzano, M.E.; Giuliano, M.; Daghero, J.; Zygadlo, J.; Bogino, P.; Giordano, W.; Demo, M. Antimicrobial activity of essential oils of *Thymus vulgaris* and *Origanum vulgare* on phytopathogenic strains isolated from soybean. *Plant Biol. J.* **2015**, *17*, 758–765. [CrossRef]
46. Gallucci, M.N.; Carezzano, M.E.; Oliva, M.M.; Demo, M.S.; Pizzolitto, R.P.; Zunino, M.P.; Zygadlo, J.A.; Dambolena, J.S. In vitro activity of natural phenolic compounds against fluconazole-resistant Candida species: A quantitative structure-activity relationship analysis. *J. Appl. Microbiol.* **2014**, *116*, 795–804. [CrossRef]
47. Reinoso, E.B.; Oliva, M.M.; Cariddi, L.N.; Carezzano, M.E.; Marioli, J.M.; Beoletto, V. Antimicrobial activity of essential oils on pathogenic strains. In *Understanding Microbial Pathogens: Current Knowledge and Educational Ideas on Antimicrobial Research*; Torres-Hergueta, E., Méndez-Vilas, A., Eds.; Formatex Research Center: Bajadoz, Spain, 2018; pp. 36–42.
48. Lambir Jacobo, A.J.; Carezzano, M.E.; Quiroga, P.R.; Grosso, N.R. Fractions of laurel essential oil obtained by molecular distillation with greater antioxi-dant and antimicrobial activities. *Agriscientia* **2022**, *39*, 105–116. [CrossRef]
49. Sharma, N.; Raut, P.W.; Baruah, M.M.; Sharma, A. Combination of quercetin and 2-methoxyestradiol inhibits epithelial-mesenchymal transition in PC-3 cell line via Wnt signaling pathway. *Future Sci. OA* **2021**, *7*, FSO747. [CrossRef] [PubMed]
50. Cushnie, T.P.; Lamb, A.J. Antimicrobial activity of flavonoids. *Int. J. Antimicrob. Agents* **2005**, *26*, 343–356. [CrossRef] [PubMed]
51. Joray, M.B.; del Rollán, M.R.; Ruiz, G.M.; Palacios, S.M.; Carpinella, M.C. Antibacterial activity of extracts from plants of central argentina-isolation of an active principle from *Achyrocline satureioides*. *Plant Med.* **2011**, *77*, 95–100. [CrossRef] [PubMed]
52. Kloverpris, H.; Fomsgaard, A.; Handley, A.; Ackland, J.; Sullivan, M.; Goulder, P. Dimethyl sulfoxide (DMSO) exposure to human peripheral blood mononuclear cells (PBMCs) abolish T cell responses only in high concentrations and following coincubation for more than two hours. *J. Immunol. Methods* **2010**, *356*, 70–78.
53. Lin, G.J.; Sytwu, H.K.; Yu, J.C.; Chen, Y.W.; Kuo, Y.L.; Yu, C.C.; Chang, H.M.; Chan, D.C.; Huang, S.H. Dimethyl sulfoxide inhibits spontaneous diabetes and autoimmune recurrence in non-obese diabetic mice by inducing differentiation of regulatory T cells. *Toxicol. Appl. Pharmacol.* **2015**, *282*, 207–214.
54. Jacob, S.W.; de la Torre, J.C. Pharmacology of dimethyl sulfoxide in cardiac and CNS damage. *Pharmacol. Rep.* **2009**, *61*, 225–235.
55. Waller, F.T.; Tanabe, C.T.; Paxton, H.D. Treatment of elevated intracranial pressure with dimethyl sulfoxide. *Ann. N. Y. Acad. Sci.* **1983**, *411*, 286–292. [CrossRef]
56. Montironi, I.D.; Cariddi, L.N.; Reinoso, E.B. Evaluation of the antimicrobial efficacy of *Minthostachys verticillata* essential oil and limonene against *Streptococcus uberis* strains isolated from bovine mastitis. *Rev. Argen. Microbiol.* **2016**, *48*, 210–216. [CrossRef]
57. Montironi, I.D.; Reinoso, E.B.; Croce Paullier, V.; Siri, M.I.; Pianzzola, M.J.; Moliva, M.; Campra, N.; Bagnis, G.; Ferreira LaRocque-de-Freitas, I.; Decote-Ricardo, D.; et al. *Minthostachys verticillata* essential oil activates macrophage phagocytosis and modulates the innate immune response in a murine model of *Enterococcus faecium* mastitis. *Res. Vet. Sci.* **2019**, *125*, 333–344. [CrossRef]
58. Cecchini, E.; Paoloni, C.; Campra, N.; Picco, N.; Grosso, M.C.; Soriano Perez, M.L.; Alustiza, F.; Cariddi, N.; Bellingeri, R. Nanoemulsion of *Minthostachys verticillata* essential oil. In-vitro evaluation of its antibacterial activity. *Heliyon* **2021**, *7*, e05896. [CrossRef]
59. Sutil, S.B.; Astesano, A.; Vogt, V.; Torres, C.V.; Zanon, S.M.; Sabini, L.I. *Minthostachys verticillata*: Toxicity of its essential oil and major constituents to *Artemia salina* and cell lines. *Mol. Med. Chem.* **2006**, *10*, 41–42.

60. Oliva, M. dlM.; Gallucci, N.; Zygadlo, J.A.; Demo, M.S. Cytotoxic activity of argentinean essential oils on *Artemia salina*. *Pharm. Biol.* **2007**, *45*, 259–262. [CrossRef]
61. Rossi, Y.E.; Canavoso, L.; Palacios, S.M. Molecular response of *Musca domestica* L. to *Mintostachys verticillata* essential oil, (4R)+-pulegone and menthone. *Fitoterapia* **2012**, *83*, 336–342. [CrossRef]
62. Ćavar Zeljković, S.; Schadich, E.; Džubák, P.; Hajdúch, M.; Tarkowski, P. Antiviral activity of selected *Lamiaceae* essential oils and their monoterpenes against SARS-CoV-2. *Front. Pharmacol.* **2022**, *13*, 893634. [CrossRef]
63. Astani, A.; Schnitzler, P. Antiviral activity of monoterpenes beta-pinene and limonene against *herpes simplex virus* in vitro. *Iran. J. Microbiol.* **2014**, *6*, 149–155. [PubMed]
64. Escobar, F.M.; Magnoli, A.; Sabini, M.C.; Cariddi, L.N.; Bagnis, G.; Soltermann, A.; Cavaglieri, L. *Minthostachys verticillata* essential oils as potential phytogenic additives and chemoprotective strategy on aflatoxin B1 toxicity. *J. Appl. Anim. Res.* **2019**, *47*, 217–222. [CrossRef]
65. Gonçalves, D.C.; Tebaldi de Queiroz, V.; Vidal Costa, A.; Pinheiro Lima, W.; Belan, L.L.; Moraes, W.B.; Lopes Pontes Póvoa Iorio, N.; Cardoso Corrêa Póvoa, H. Reduction of Fusarium wilt symptoms in tomato seedlings following seed treatment with *Origanum vulgare* L. essential oil and carvacrol. *Crop Prot.* **2021**, *141*, 105487. [CrossRef]
66. Picoli, T.; Waller, S.B.; Hoffmann, J.F.; Peter, C.M.; da Silva Barcelos, L.; Lopes, M.G.; de Faria, R.O.; Cleff, M.B.; Hübner, S.d.O.; Lima, M.; et al. Antiviral and virucidal potential of *Origanum vulgare* Linn. (oregano) extracts against Bovine alpha herpesvirus 1 (BoHV-1). *Res. Soc. Dev.* **2021**, *10*, e28410514979. [CrossRef]
67. Blank, D.; Hübner, S.; Alves, G.; Cardoso, C.; Freitag, R.; Cleff, M. Chemical composition and antiviral effect of extracts of *Origanum vulgare*. *Adv. Biosci. Biotechnol.* **2019**, *10*, 188–196. [CrossRef]
68. Prieto, M.C.; Camacho, N.M.; Dell Inocenti, F.; Mignolli, F.; Lucini, E.; Palma, S.; Bima, P.; Grosso, N.R.; Asensio, C.M. Microencapsulation of *Thymus vulgaris* and *Tagetes minuta* essential oils: Volatile release behavior, antibacterial activity and effect on potato yield. *J. Saudi Soc. Agric. Sci.* **2022**, *22*, 195–204. [CrossRef]
69. Niksic, H.; Becic, F.; Koric, E.; Gusic, I.; Omeragic, E.; Muratovic, S.; Miladinovic, B.; Duric, K. Cytotoxicity screening of *Thymus vulgaris* L. essential oil in brine shrimp nauplii and cancer cell lines. *Sci. Rep.* **2021**, *11*, 13178. [CrossRef]
70. Bota, V.; Sumalan, R.M.; Obistioiu, D.; Negrea, M.; Cocan, I.; Popescu, I.; Alexa, E. Study on the sustainability potential of thyme, oregano, and coriander essential oils used as vapours for antifungal protection of wheat and wheat products. *Sustainability* **2022**, *14*, 4298. [CrossRef]
71. Di Francesco, A.; Aprea, E.; Gasperi, F.; Parenti, A.; Placì, N.; Rigosi, F.; Baraldi, E. Apple pathogens: Organic essential oils as an alternative solution. *Sci. Hortic.* **2022**, *300*, 111075. [CrossRef]
72. Rota, M.C.; Herrera, A.; Martínez, R.M.; Sotomayor, J.A.; Jordán, M.J. Antimicrobial activity and chemical composition of *Thymus vulgaris*, *Thymus zygis* and *Thymus hyemalis* essential oils. *Food Control* **2008**, *19*, 681–687. [CrossRef]
73. Imelouane, B.; Amhamdi, H.; Wathelet, J.P.; Ankit, M.; Khedid, K.; El Bachiri, A. Chemical composition of the essential oil of thyme (*Thymus vulgaris*) from Eastern Morocco. *Int. J. Agric. Biol.* **2009**, *11*, 205–208.
74. Gedikoğlu, A.; Sökmen, M.; Çivit, A. Evaluation of *Thymus vulgaris* and *Thymbra spicata* essential oils and plant extracts for chemical composition, antioxidant, and antimicrobial properties. *Food Sci. Nutr.* **2019**, *7*, 1704–1714. [CrossRef]
75. Wani, A.R.; Yadav, K.; Khursheed, A.; Rather, M.A. An updated and comprehensive review of the antiviral potential of essential oils and their chemical constituents with special focus on their mechanism of action against various influenza and coronaviruses. *Microb. Pathog.* **2021**, *152*, 104620. [CrossRef]
76. Astani, A.; Reichling, J.; Schnitzler, P. Comparative study on the antiviral activity of selected monoterpenes derived from essential oils. *Phytother. Res.* **2010**, *24*, 673–679. [CrossRef]
77. Nolkemper, S.; Reichling, J.; Stintzing, F.C.; Carle, R.; Schnitzler, P. Antiviral effect of aqueous extracts from species of the *Lamiaceae* family against Herpes simplex virus type 1 and type 2 in vitro. *Planta Med.* **2006**, *72*, 1378–1382. [CrossRef]
78. Aksit, H.; Bayar, Y.; Simsek, S.; Ulutas, Y. Chemical composition and antifungal activities of the essential oils of thymus species (*Thymus pectinatus*, *Thymus convolutus*, *Thymus vulgaris*) against plant pathogens. *J. Essent. Oil-Bear. Plants* **2022**, *25*, 200–207. [CrossRef]
79. Jafri, H.; Ahmad, I. *Thymus vulgaris* essential oil and thymol inhibit biofilms and interact synergistically with antifungal drugs against drug resistant strains of *Candida albicans* and *Candida tropicalis*. *J. Mycol. Med.* **2020**, *30*, 100911. [CrossRef] [PubMed]
80. Dias Maia, J.; La Corte, R.; Martinez, J.; Ubbink, J.; Prata, A.S. Improved activity of thyme essential oil (*Thymus vulgaris*) against *Aedes aegypti* larvae using a biodegradable controlled release system. *Ind. Crops Prod.* **2019**, *136*, 110–120. [CrossRef]
81. Lazarević, J.; Jevremović, S.; Kostić, I.; Kostić, M.; Vuleta, A.; Manitašević Jovanović, S.; Šešlija Jovanović, D. Toxic, oviposition deterrent and oxidative stress effects of *Thymus vulgaris* essential oil against *Acanthoscelides obtectus*. *Insects* **2020**, *11*, 563. [CrossRef] [PubMed]
82. Mieres-Castro, D.; Ahmar, S.; Shabbir, R.; Mora-Poblete, F. Antiviral activities of eucalyptus essential oils: Their effectiveness as therapeutic targets against human viruses. *Pharmaceuticals* **2021**, *14*, 1210. [CrossRef]
83. Abdelkhalek, A.; Al-Askar, A.A. Green synthesized ZnO nanoparticles mediated by *Mentha spicata* extract induce plant systemic resistance against tobacco mosaic virus. *Appl. Sci.* **2020**, *10*, 5054. [CrossRef]

34. Mollaei, S.; Ghanavi, Z. Green Tea (*Camellia sinensis*) Extract induces systemic acquired resistance against witches' broom diseases of *Citrus aurantifolia*. *JMPB* **2020**, *9*, 7–16. [CrossRef]
35. Ben-Jabeur, M.; Ghabri, E.; Myriam, M.; Hamada, W. Thyme essential oil as a defense inducer of tomato against gray mold and Fusarium wilt. *Plant Physiol. Biochem.* **2015**, *94*, 35–40. [CrossRef]

processes

Communication

Indoor Space Disinfection Effect and Bioactive Components of *Chamaecyparis obtusa* Essential Oil

Seung-Yub Song [1,2,†], Dae-Hun Park [3,†], Sung-Ho Lee [1,2], Chul-Yung Choi [4], Jung-Hyun Shim [1,2], Goo Yoon [1], Jin-Woo Park [1,2], Min-Suk Bae [5] and Seung-Sik Cho [1,2,*]

1 Department of Pharmacy, College of Pharmacy, Mokpo National University, Muan-gun 58554, Republic of Korea; tgb1007@naver.com (S.-Y.S.); tjdgh0730@naver.com (S.-H.L.); s1004jh@gmail.com (J.-H.S.); gyoon@mnu.ac.kr (G.Y.); jwpark@mnu.ac.kr (J.-W.P.)
2 Biomedicine, Health & Life Convergence Sciences, BK21 Four, College of Pharmacy, Mokpo National University, Muan-gun 58554, Republic of Korea
3 College of Oriental Medicine, Dongshin University, Naju-si 58245, Republic of Korea; dhj1221@hanmail.net
4 Department of Biomedical Science, College of Natural Science, Chosun University, Gwangju-city 61452, Republic of Korea; blockstar@chosun.ac.kr
5 Department of Environmental Engineering, College of Engineering, Mokpo National University, Muan-gun 58554, Republic of Korea; minsbae@hotmail.com
* Correspondence: sscho@mokpo.ac.kr; Tel.: +82-61-450-2687; Fax: +82-61-450-2689
† These authors contributed equally to this work.

Abstract: *Chamaecyparis obtusa* (Siebold & Zucc.) Endl. (Cupressaceae) is known to produce a variety of antimicrobial substances. In the present study, components of three lots of essential oil from *C. obtusa* were analyzed by GCMS. It was confirmed that thujopsene and pinene were common markers. In addition, we report indoor space disinfectant effects of products containing *C. obtusa* essential oil (PO100, PO500, PO1000). It was confirmed that PO100 and PO500 could effectively remove airborne microorganisms in indoor spaces. Results of our study suggest that *C. obtusa* essential oil is effective in reducing contamination by infectious microorganisms in confined spaces.

Keywords: *Chamaecyparis obtuse*; essential oil; indoor space disinfection

1. Introduction

In the global pandemic situation, the prevention of pathogenic microbes in indoor air spaces and on various surfaces attract significant attention. Recently, as interest in disinfection methods or antibacterial materials increases, interest in antibacterial materials as natural disinfectants is also increasing [1]. The disinfectants should not only be effective in reducing the microbes present in the air but should also not be toxic to humans, allowing their continuous application [2].

Bacteria can be released into the air in a variety of natural and anthropogenic environments. When infectious bacteria are released into the air, they can be easily transferred to susceptible individuals via bioaerosols. Representative microorganisms among airborne bacteria include Legionella, Mycobacterium, Escherichia, Enterobacter, Pseudomonas, Aspergilius, and Acinetobacter species. These infectious microorganisms cause diseases such as legionellosis, pneumonia, tuberculosis, and leprosy [3].

Essential oils are mixtures of volatile and non-volatile compounds and can be obtained from natural plants by organic solvent extraction, hydrodistillation, and compression [4]. Essential oils have been shown to have antimicrobial, antifungal, and antiviral activities [5]. The antimicrobial effects of essential oils are explained by their composition and cytotoxic effects, which cause cell membrane damage. For example, essential oils containing carvacrol, cinnamaldehyde, and thymol showed antibacterial activity against *E. coli* and *S. aureus* in vitro and exhibited antibacterial activity through inhibition of cell membrane formation. The cytotoxicity of the essential oils tested was lower than that of chlorhexidine.

Therefore, carvacrol, cinnamaldehyde, and thymol appear to be natural agents that can be used effectively against *E. coli* and *S. aureus*-borne infections [6].

Several scientific studies on disinfection methods using the antibacterial effects of essential oils have been reported. The potential use of *Melaleuca alternifolia* (Tea tree) oil as a disinfectant has been shown in a previous report for control bacteria [7] due to terpinen-4-ol (35–45%) and 1,8-cineole (1–6%) [8,9].

Eucalyptus oil is usually rich in sesquiterpenes that show antimicrobial activities against gram-positive and negatives [10]. Tea tree oil and eucalyptus oil have been reported to have significant activity against airborne pathogenic bacteria in a short period of time [11].

Several papers have reported that the antimicrobial efficacy of essential oil vapor is more effective than essential oils in liquid form [12]. Therefore, empirical experiments on the indoor space control effect of various essential oil components are needed, and the need to present the basis for the development of various natural disinfectants based on empirical data is increasing.

The antimicrobial effect of an essential oil mixture in a small chamber was reported by Chaoet et al. [13]. Chaoet et al. evaluated the antimicrobial activity of blended essential oils of cinnamon, clove, rosemary, eucalyptus, and lemon in a closed hood with an area of 0.4 m^3. For further empirical studies, Lanzerstorfer et al. reported that a mixture of *Citrus limon* essential oil and *Abies alba* essential oil removed 40% of bacteria using an ultrasonic vaporizer in a hospital room [14]. There have been reports on demonstration tests of various essential oil mixtures using ultrasonic vaporizers; however, there have been no demonstration tests on *Chamaecyparis obtusa* essential oil.

In 1926, *Chamaecyparis obtusa* was adopted as an afforestation species in southern regions of Korea. In 1960, it began to be cultivated in Jeju Island, Gyeongnam, Jeollanam-do, Korea. Among them, the *C. obtusa* forest in Jangheung is the oldest. In Jeollanam-do, *C. obtusa* occupies about 78,000 ha, with 65% of the nation's reforestation concentrated in this area. *C. obtusa* is a species of growing interest in Korea as it has excellent effects on environmental and infectious diseases [15,16]. Previously, we have determined the efficacy and components of *C. obtusa* essential oil to confirm the value of Jangheung *C. obtusa*. Some studies have been reported on the effect of *C. obtusa* essential oil on the human body and animal inhalation experiments. One clinical trial of *C. obtusa* essential oil was reported. Chen et al. extracted *C. obtusa* essential oil and *C. formosensis* essential oil. Physiological and psychological effects were investigated by inhaling essential oils for 5 min in 16 healthy adults. After inhalation of *C. obtusa* essential oil, systolic blood pressure, heart rate, and parasympathetic activity decreased while sympathetic nervous system activity increased. Additionally, in the profile of mood states test, *C. obtusa* essential oil stimulated a pleasant mood state. Chen's results indicated that *C. obtusa* essential oil was helpful in controlling sympathetic nervous system dysfunction.

As a result, it showed excellent effects on general microorganisms and antibiotic-resistant bacteria [15]. Although the antimicrobial effect of *C. obtusa* has been reported previously, there has been no empirical study on the antimicrobial effect of the essential oil of *C. obtusa*. Thus, the objective of the present study was to analyze the bioactive components of three lots of essential oil. The antibacterial excellence of a product containing essential oil was identified through an indoor space disinfection test.

2. Materials and Methods

2.1. Essential Oil Preparation and Chemical Analysis

C. obtusa essential oil was produced three times (June 2014, December 2014, and April 2015) from leaves of *C. obtusa* and supplied by Earthmate Jungamjin Co (Jangheung, Republic of Korea). Its active ingredient was measured by GCMS. GCMS analysis analyzed the components of essential oil using the method used by Bae et al. [15]. Agilent 7890 gas chromatography (GC) and Agilent 5975 quadrupole mass spectrometry (MS) systems (Agilent Technologies, Palo Alto, CA, USA) were utilized to analyze molecular mass fragments (50–550 amu) of the Lot 1~3, with Agilent HP-5MS fused silica capillary column (30 mm l. × 0.25 mm i.d., 0.25 μm film thickness, Agilent Technologies, Palo Alto, CA, USA). The mass fragments were ionized under electron ionization (EI) conditions. A GC oven was isothermally programmed at 65 °C for 10 min at 10 to 300 min^{-1} with helium (He) as a carrier gas. All the data were compared with the system library (NIST 2017). Three lots were analyzed for essential oil components, and common components between lots were selected.

2.2. Space Disinfectant Preparation and Space Disinfectant Test

A space disinfectant test sample (HINOPHY PO) was provided by Earthmate Jungamjin (Jangheung, Republic of Korea) by diluting *C. obtusa* essential oil (LOT3) 100 times, 500 times, and 1000 times (Product code, PO100-001, PO500-001, PO1000-001) to make test product. The product did not contain any synthetic ingredients other than essential oil. For the disinfectant test, an actual space was selected to confirm the degree of control efficiency of essential oil-containing products. For actual space selection, the space (average 8 h, 20–40 actual users) was selected so that it could be applied to offices and public institutions. The area of the test space was 60 m^2 for Room 1 (R1) and 90 m^2 for Room 2 (R2) and Room 3 (R3). PO100, PO500, and PO1000 were supplied to the selected space, and 0.25 L per space was evenly sprayed over 15 min. For measuring falling bacteria, 4 points were selected for each space. At each point, a solid medium (nutrient agar, BD DIFCO. Co., Franklin Lakes, NJ, USA) in a 10 cm petri dish (SPL life science, Seoul. Republic of Korea) was exposed to the space for 8 h to recover the falling bacteria. The recovered strain was cultured at 37 °C for 24–48 h, and the number of falling cells was counted. The falling bacteria were visually counted by the number of colonies before and after the treatment of the PO series. We only considered the collection of falling bacteria when the petri dish was left in the test space after sample processing. Indoor temperature and humidity were not recorded (Figure 1).

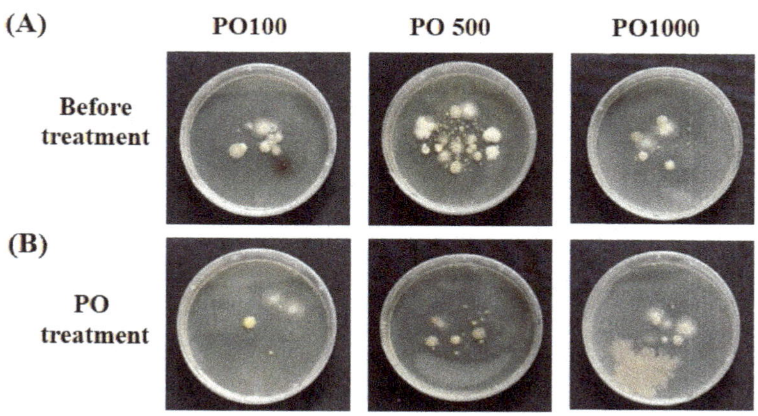

Figure 1. Visualization of falling bacterial strain. Example, (**A**) Falling bacteria pattern before PO treatment; (**B**) Falling bacteria pattern after space disinfection.

3. Results and Discussion

3.1. *C. obtusa* Essential Oil Component Analysis

In lot 1, several components, such as limonene (6%), thujopsene (3%), alpha-pinene (2.3%), and phellandrane (1.7%), were detected (Table 1). In lot 2, thujopsene (5.2%), limonene (4%), (+)-Epi-bicyclosesquiphellandrene (2.4%), cedrol (2%), and alpha-pinene (1.5%) were detected. In lot 3, alpha-pinene (1.5%), phellandrane (3.3%), (+)-4-Carene (2.1%), cedrol (1.55%), thujopsene (5.1%), and (+)-Epi-bicyclosesquiphellandrene (2.6%) were detected. Lots 2 and 3 also showed similar results. Since the manufacturing date of each source is different, there are slight differences in some components; however, the main components are similar.

Table 1. Example of GC-MS analysis of essential oil from *C. obtusa* (Lot 1).

No	RT (min)	% in Sample	Hit Name
1	8.753	1.299	Bicyclo[3.1.0]hex-2-ene, 2-methyl-5-(1-methylethyl)-
2	9.019	2.27	1R-alpha-Pinene
3	9.551	0.542	Camphene
4	10.638	1.756	.beta.-Phellandrene
5	10.711	0.256	.beta.-Pinene
6	12.439	3.212	Bicyclo[4.1.0]hept-2-ene, 3,7,7-trimethyl-
7	12.747	1.482	Benzene, 1-methyl-3-(1-methylethyl)-
8	13.007	6.004	Limonene
9	14.233	5.735	1,4-Cyclohexadiene, 1-methyl-4-(1-methylethyl)-
10	14.445	0.09	Terpineol, cis-.beta.-
11	15.321	2.191	Cyclohexene, 1-methyl-4-(1-methylethylidene)-
12	15.647	0.064	Terpineol, cis-.beta.-
13	15.768	0.049	1,6-Octadien-3-ol, 3,7-dimethyl-
14	16.288	0.051	1-Octen-3-yl-acetate
15	16.547	0.058	2-Cyclohexen-1-ol, 1-methyl-4-(1-methylethyl)-, trans-
16	17.254	0.05	2-Cyclohexen-1-ol, 1-methyl-4-(1-methylethyl)-, cis-
17	17.387	0.071	Bicyclo[2.2.1]heptan-2-one, 1,7,7-trimethyl-, (1R)-
18	20.269	0.076	Bicyclo[2.2.1]heptan-2-ol, 1,3,3-trimethyl-, acetate, (1S-exo)-
19	21.574	0.206	1,6-Octadien-3-ol, 3,7-dimethyl-, 2-aminobenzoate
20	22.746	8.374	Bicyclo[2.2.1]heptan-2-ol, 1,7,7-trimethyl-, acetate, (1S-endo)-
21	25.006	13.384	3-Cyclohexene-1-methanol, alpha.,.alpha.,4-trimethyl-, acetate
22	25.961	0.131	Di-epi-alpha.-cedrene-(I)
23	26.045	0.046	2H-2,4a-Methanonaphthalene, 1,3,4,5,6,7-hexahydro-1,1,5,5-tetramethyl-, (2S)-
24	26.178	0.195	Cyclohexane, 1-ethenyl-1-methyl-2,4-bis(1-methylethenyl)-, [1S-(1.alpha.,2.beta.,4.beta.)]-
25	26.831	0.722	1H-3a,7-Methanoazulene, 2,3,4,7,8,8a-hexahydro-3,6,8,8-tetramethyl-, [3R-(3.alpha.,3a.beta.,7.beta.,8a.alpha.)]-
26	27.06	0.614	1H-3a,7-Methanoazulene, octahydro-3,8,8-trimethyl-6-methylene-, [3R-(3.alpha.,3a.beta.,7.beta.,8a.alpha.)]-
27	27.471	3.012	Thujopsene
28	27.58	0.137	1,3-Cyclohexadiene, 1-methyl-4-(1-methylethyl)-
29	28.021	0.084	beta-Guaiene
30	28.112	0.098	1,4,7,-Cycloundecatriene, 1,5,9,9-tetramethyl-, Z,Z,Z-
31	28.202	0.078	1,6,10-Dodecatriene, 7,11-dimethyl-3-methylene-, (E)-
32	28.438	1.476	(+)-Epi-bicyclosesquiphellandrene

Table 1. *Cont.*

No	RT (min)	% in Sample	Hit Name
33	28.523	0.034	Di-epi-.alpha.-cedrene
34	29.115	0.216	Naphthalene, 1,2,3,4,4a,5,6,8a-octahydro-4a,8-dimethyl-2-(1-methylethenyl)-, [2R-(2.alpha.,4a.alpha.,8a.beta.)]-
35	29.562	2.748	Tricyclo[5.4.0.0(2,8)]undec-9-ene, 2,6,6,9-tetramethyl-
36	29.665	0.12	Spiro[5.5]undeca-1,8-diene, 1,5,5,9-tetramethyl-, (R)-
37	29.755	0.731	Benzene, 1-methyl-4-(1,2,2-trimethylcyclopentyl)-, (R)-
38	29.979	0.28	Naphthalene, 1,2,4a,5,6,8a-hexahydro-4,7-dimethyl-1-(1-methylethyl)-
39	30.275	1.979	Naphthalene, 1,2,3,5,6,8a-hexahydro-4,7-dimethyl-1-(1-methylethyl)-, (1S-cis)-
40	30.68	0.148	1H-Cycloprop[e]azulene, 1a,2,3,4,4a,5,6,7b-octahydro-1,1,4,7-tetramethyl-, [1aR-(1a.alpha.,4.alpha.,4a.beta.,7b.alpha.)]-
41	31.272	0.07	gamma-Elemene
42	31.441	0.087	1,6,10-Dodecatrien-3-ol, 3,7,11-trimethyl-, [S-(Z)]-
43	32.607	1.522	Cedrol
44	32.673	0.042	2-Naphthalenemethanol, 1,2,3,4,4a,5,6,8a-octahydro-alpha,alpha,4a,8-tetramethyl-, (2.alpha.,4a.alpha.,8a.alpha.)-
45	33.09	0.138	2-Naphthalenemethanol, 1,2,3,4,4a,5,6,7-octahydro-alpha,alpha,4a,8-tetramethyl-, (2R-cis)-
46	49.627	0.041	2-Phenanthrenol, 4b,5,6,7,8,8a,9,10-octahydro-4b,8,8-trimethyl-1-(1-methylethyl)-, (4bS-trans)-

To summarize representative active ingredients of lots 1 to 3, thujopsene and pinene are common ingredients in the analysis. Thujopsene has shown potent antibacterial activity, for example, against *Phytophthora ramorum* at 2.0~3.0 ppm [17]. Limonene also can be utilized as one of the solutions to the problem of antimicrobial resistance. Advantageous contributions have been made by various research groups in the study of the antimicrobial properties of limonene. Previous studies have shown that limonene inhibits disease-causing pathogenic microbes [18]. In addition, it was observed positive results for the isolated antibacterial action of α-pinene. Of these, the strains that worked were *Escherichia coli* ATCC, *Staphylococcus aureus* ATCC and *Salmonella enterica*, revealing their susceptibility to the α-pinene [19]. The recently extracted lot 2 and lot 3 showed almost similar component detection, such as thujopsene, limonene, and α-pinene and content patterns. It seems desirable to set the standard for quality control based on active ingredients of lots 1~3 in the future (Figure 2).

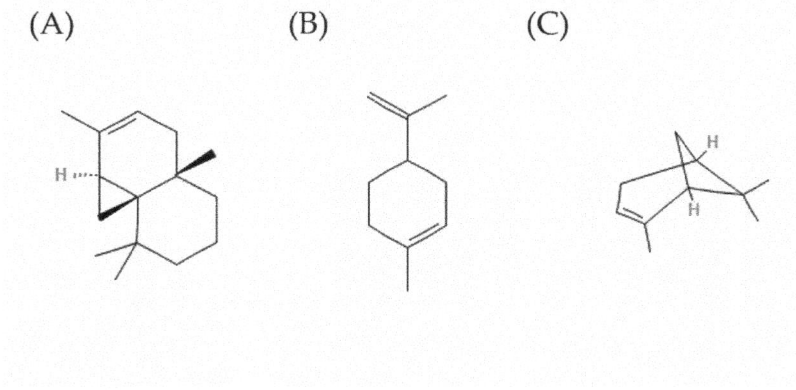

Figure 2. Biomarkers of *C. obtusa* essential oil. (**A**) thujopsene, (**B**) limonene, and (**C**) α-pinene.

3.2. Indoor Space Disinfectant Test

As shown in Figure 3, we measured the number of falling bacteria before and after spraying PO100, PO500, and PO1000 into three spaces. Since the distribution of dropping bacteria varied by location, 4~6 points were designated per room to measure falling bacteria. The average number of falling bacteria per room was calculated. On average, falling bacteria in R1 and R2 before PO treatment were 50~60 CFU/plate. When PO100 and 500 were used for treatment, a statistically significant reduction in falling bacteria was shown in R1 and R2. When treated with PO100, falling bacteria were reduced by 12 to 65% in R1 and 10 to 55% in R2. In R3, 42 to 67% of falling bacteria were reduced. The area of R2 is the same as that of R3. The reason for the slightly different decrease in falling bacteria is thought to be that the distribution of airborne bacteria is different depending on the number of people and their activities. When PO500 was used for treatment, the number of falling bacteria decreased by 21 to 63.6% in R1, by 11.3 to 43.8% in R2, and by 3.5 to 31.8% in R3. When PO1000 was used for treatment, the number of falling bacteria decreased by 44 to 61% in R1, by 6 to 25% in R2, and by 8.4 to 68.2% in R3. The statistical significance of the PO1000 treatment group was not observed. Experimental results revealed that PO100 and PO500 showed statistically significant indoor space disinfectant effects, probably due to the volatilization of essential oil components in PO products and the antibacterial power of non-volatile substances. In particular, non-volatile substances that remain in the indoor space are presumed to inhibit the growth of microorganisms distributed in the space. Previously, we reported the study on the essential oil and non-volatile residues derived from *C. obtusa* leaves showed broad antimicrobial activities. Thus, both essential oil and non-volatile residues may be eco-friendly disinfectants for the prevention of skin disorders caused by infectious bacteria [15]. There have been no studies on the control of disinfectants containing *C. obtusa* essential oils in real indoor spaces. For effective product development in the future, research on deriving the optimal blending ratio between the existing indoor space disinfectant and *C. obtusa* essential oil and a community study of controlled microbes should be conducted.

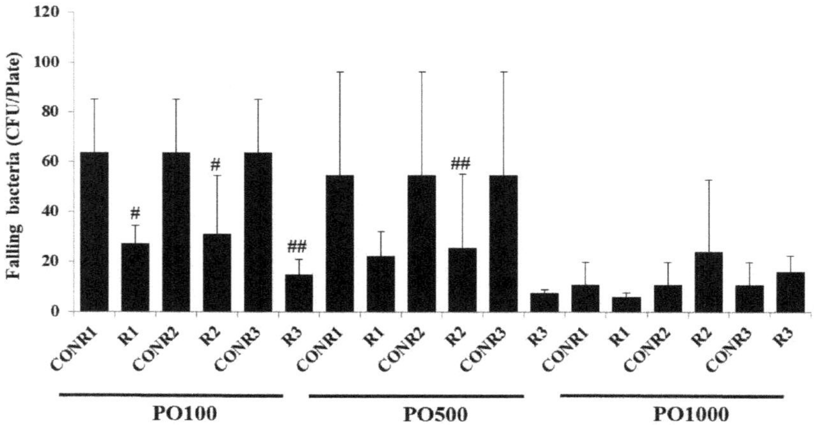

Figure 3. Inhibition of falling bacteria of PO100, PO500, PO1000 containing *C. obtusa* essential oil. CONR1~3 means the colonies of falling bacteria before not treated with PO series in the experimental space (R1~3), *p* value, # < 0.05 and ## < 0.02.

4. Conclusions

In the present study, components and indoor space disinfectant efficacy of *C. obtusa* essential oil were evaluated. As a result of the analysis, it was found that common ingredients such as thujopsene, limonene, and α-pinene were identified for lots 1~3. Anti-bacterial efficacy was also confirmed. In the case of the GCMS analysis method, if 2~3 markers were selected, it was considered appropriate as an analysis method for quality control. In addition, through an indoor space disinfectant test of products containing *C. obtusa* essential oil, it was thought that essential oil could efficiently remove airborne microorganisms. In conclusion, our study showed the possibility of practical use of *C. obtusa* essential oil by demonstrating that *C. obtusa* essential oil could inhibit the growth of various microorganisms and reduce human exposure to infectious microorganisms.

Author Contributions: Conceptualization, S.-S.C. and D.-H.P.; methodology and investigation, S.-Y.S., S.-H.L., J.-H.S., G.Y., J.-W.P., C.-Y.C. and M.-S.B.; writing and editing, S.-S.C. All authors have read and agreed to the published version of the manuscript.

Funding: This work was supported by the National Research Foundation of Korea (NRF) grant funded by the Korean government (MSIT) (No. 2022R1A5A8033794), and this work was supported by a grant from the National Institute of Environment Research (NIER), funded by the Ministry of Environment (MOE) of the Republic of Korea (NIER-2021-03-03-007).

Data Availability Statement: Data available on request.

Conflicts of Interest: The authors declared no potential conflict of interest with respect to the research, authorship, and/or publication of this article.

References

1. Mirskaya, E.; Agranovski, I.E. Control of Airborne Microorganisms by Essential Oils Released by VaxiPod. *Atmosphere* **2021**, *12*, 1418. [CrossRef]
2. Musee, N.; Ngwenya, P.; Motaung, L.K.; Moshuhla, K.; Nomngongo, P. Occurrence, effects, and ecological risks of chemicals in sanitizers and disinfectants: A review. *Environ. Chem. Ecotoxicol.* **2023**, *5*, 62–78. [CrossRef]
3. Stetzenbach, L.D. Airborne Infectious Microorganisms. *Encycl. Microbiol.* **2009**, *52*, 175–182. [CrossRef]
4. Rai, M.; Zacchino, S.; Derita, M. *Essential Oils and Nanotechnology for Treatment of Microbial Diseases*; CRC Press: Boca Raton, FL, USA, 2017.
5. Tariq, S.; Wani, S.; Rasool, W.; Shafi, K.; Bhat, M.A.; Prabhakar, A.; Shalla, A.H.; Rather, M.A. A comprehensive review of the antibacterial, antifungal and antiviral potential of essential oils and their chemical constituents against drug-resistant microbial pathogens. *Microb. Pathog.* **2019**, *134*, 103580. [CrossRef] [PubMed]

6. García-Salinas, S.; Elizondo-Castillo, H.; Arruebo, M.; Mendoza, G.; Irusta, S. Evaluation of the Antimicrobial Activity and Cytotoxicity of Different Components of Natural Origin Present in Essential Oils. *Molecules* **2018**, *23*, 1399. [CrossRef] [PubMed]
7. Wilkinson, J.M.; Cavanagh, H.M. Antibacterial activity of essential oils from Australian native plants. *Phytother. Res. PTR* **2005**, *19*, 643–646. [CrossRef]
8. May, J.; Chan, C.H.; King, A.; Williams, L.; French, G.L. Time-kill studies of tea tree oils on clinical isolates. *J. Antimicrob. Chemother.* **2000**, *45*, 639–643. [CrossRef] [PubMed]
9. Carson, C.F.; Hammer, K.A.; Riley, T.V. Melaleuca alternifolia (Tea Tree) oil: A review of antimicrobial and other medicinal properties. *Clin. Microbiol. Rev.* **2006**, *19*, 50–62. [CrossRef] [PubMed]
10. Barbosa, L.C.A.; Filomeno, C.A.; Teixeira, R.R. Chemical variability and biological activities of Eucalyptus spp. essential oils. *Molecules* **2016**, *21*, 1671. [CrossRef]
11. Usachev, E.V.; Pyankov, O.V.; Usacheva, O.V.; Agranovski, I.E. Antiviral activity of tea tree and eucalyptus oil aerosol and vapour. *J. Aerosol Sci.* **2013**, *59*, 22–30. [CrossRef]
12. Tian, F.; Lee, S.Y.; Chun, H.S. Comparison of the Antifungal and Antiaflatoxigenic Potential of Liquid and Vapor Phase of Thymus vulgaris Essential Oil against Aspergillus flavus. *J. Food Prot.* **2019**, *82*, 2044–2048. [CrossRef] [PubMed]
13. Chao, S.C.; Young, D.G.; Oberg, C.J. Effect of a diffused essential oil blend on bacterial bioaerosols. *J. Essent. Oil Res.* **1998**, *10*, 517–523. [CrossRef]
14. Lanzerstorfer, A.; Hackl, M.; Schlömer, M.; Rest, B.; Deutsch-Grasl, E.; Lanzerstorfer, C. The influence of air-dispersed essential oils from lemon (Citrus limon) and silver fir (Abies alba) on airborne bacteria and fungi in hospital rooms. *J. Environ. Sci. Health Part A* **2019**, *54*, 256–260. [CrossRef] [PubMed]
15. Bae, M.S.; Park, D.H.; Choi, C.Y.; Kim, G.Y.; Yoo, J.C.; Cho, S.S. Essential Oils and Non-volatile Compounds Derived from Chamaecyparis obtusa: Broad Spectrum Antimicrobial Activity against Infectious Bacteria and MDR(multidrug resistant) Strains. *Nat. Prod. Commun.* **2016**, *11*, 693–694. [CrossRef] [PubMed]
16. Kim, H.-S.; Han, S.-R.; Yang, M. Evaluations on the Deodorization Effect and Antibacterial Activity of Chamaecyparis obtusa Essential Oil. *J. Odor Indoor Environ.* **2009**, *8*, 101–107.
17. Manter, D.K.; Kelsey, R.G.; Karchesy, J.J. Antimicrobial activity of extractable conifer heartwood compounds toward Phytophthora ramorum. *J. Chem. Ecol.* **2007**, *33*, 2133–2147. [CrossRef] [PubMed]
18. Gupta, A.; Jeyakumar, E.; Lawrence, R. Journey of Limonene as an Antimicrobial Agent. *J. Pure Appl. Microbiol.* **2021**, *15*, 1094–1110. [CrossRef]
19. Borges, M.F.d.A.; Lacerda, R.d.S.; Correia, J.P.d.A.; de Melo, T.R.; Ferreira, S.B. Potential Antibacterial Action of α-Pinene. *Med. Sci. Forum* **2022**, *12*, 11.

processes

MDPI

Review

Recent Developments in *Citrus aurantium* L.: An Overview of Bioactive Compounds, Extraction Techniques, and Technological Applications

Joaquín Fernández-Cabal [1], Kevin Alejandro Avilés-Betanzos [1], Juan Valerio Cauich-Rodríguez [2], Manuel Octavio Ramírez-Sucre [1] and Ingrid Mayanin Rodríguez-Buenfil [1,*]

1 Centro de Investigación y Asistencia en Tecnología y Diseño del estado de Jalisco A.C., Subsede Sureste, Tablaje Catastral 31264, Km. 5.5 Carretera Sierra Papacal-Chuburná Puerto, Parque Científico Tecnológico de Yucatán, Mérida C.P. 97302, Yucatán, Mexico; jofernandez_al@ciatej.edu.mx (J.F.-C.); keaviles_al@ciatej.edu.mx (K.A.A.-B.); oramirez@ciatej.mx (M.O.R.-S.)

2 Centro de Investigación Científica de Yucatán, Unidad de Materiales, Calle 43 No. 130 x 32 y 34, Colonia Chuburná de Hidalgo, Mérida C.P. 97205, Yucatán, Mexico; jvcr@cicy.mx

* Correspondence: irodriguez@ciatej.mx

Abstract: This review provides an overview of recent developments in *Citrus aurantium* L. (sour or bitter orange), focusing on its bioactive compounds, innovative extraction techniques, and technological applications. *C. aurantium* is rich in bioactive compounds such as flavonoids (naringin, hesperidin, kaempferol, quercetin), essential oils (β-pinene, limonene), and vitamin C, which represents significant biological activities including antioxidant, antimicrobial, anti-inflammatory, and anticancer effects. The review discusses traditional extraction methods, such as solvent extraction and hydrodistillation, alongside newer, eco-friendly approaches like ultrasound-assisted extraction, microwave-assisted extraction, supercritical fluid extraction, and natural deep eutectic solvents. It also highlights cutting-edge techniques, including molecular imprinting polymer-based extraction, which enable the more efficient enrichment and purification of specific compounds like synephrine. Finally, the review examines the diverse industrial applications of these bioactive compounds in sectors such as foods, pharmaceuticals, and cosmetics, while emphasizing the growing need for sustainable and efficient extraction technologies.

Keywords: *Citrus aurantium* (bitter orange); bioactive compounds; polyphenols; extraction techniques; biological activities

1. Introduction

Citrus fruits constitute one of the most highly regarded categories of fruits due to their significant nutritional and pharmacological value. They also rank among the most extensively cultivated crops worldwide [1]. The botanical name for citrus is Citrus L., a genus within the family Rutaceae, which comprises 28 recognized species. This genus encompasses a diverse range of cultivated natural hybrids, including oranges, lemons, grapefruits, limes, mandarins, and citrons. They are widely cultivated in areas classified as tropical or subtropical, as well as in many other non-tropical areas, which together produce more than 100 million tons annually worldwide [2–5].

Citrus fruits are globally cultivated in 168 countries. More than 50% of global production is concentrated in six countries. The main citrus-producing countries are China, leading with 22% of the total production; Brazil with 10%; India with 7%; Mexico and the United States, both with 4.4%; and Spain with 3% [6]. Citrus fruits are essential to Mexico's economy, comprising 34.89% of the national agricultural Gross Domestic Product (GDP). In 2016, oranges represented the perennial crop with the largest cultivated area, covering 335,336 hectares. Citrus fruits are not only widely consumed nationally, but they also contribute to exports, positioning Mexico as the world leader in lemon (*Citrus aurantifolia swingle*) exports [7].

C. aurantium L., commonly known as sour or bitter orange, is a hybrid of *Citrus maxima* (pomelo) and *Citrus reticulata* (mandarin) [8]. It is an evergreen tree characterized by fragrant, small, white flowers, is capable of growing up to 5 m height, and is believed to have originated in East Africa and Syria [9]. It can be found in several varieties, including *amara* [10], *dulcis* [11], and *sinensis* [12], among others. Its fruits are utilized across various industries, including food, pharmaceuticals, and perfumery, due to their acidifying and flavor properties. However, *C. aurantium* is not typically consumed fresh, unlike other citrus fruits, due to a pronounced bitterness [9,13].

The fruit, flowers, leaves, and other components of *C. aurantium* are utilized as rich sources of bioactive compounds [14], including flavonoids [15–17], alkaloids [18–20], essential oils [21–23], coumarins [24], terpenoids [25–27], tannins [28–30], as well as essential vitamins and minerals [9,26,31]. These metabolites exhibit a wide range of health-promoting properties, such as antioxidant [32–34], anti-inflammatory [27,34,35], antidiabetic [36–38], anxiolytic [39–41], antimicrobial [42–44], and anticancer effects [45–47], among others (Figure 1).

These compounds can be extracted using various methodologies, from traditional techniques such as maceration, Soxhlet extraction, and organic solvent-based methods, including infusions [48–50], to achieve more innovative and environmentally friendly approaches. These advanced methods include ultrasound-assisted extraction (UAE) [20,51, 52], microwave-assisted extraction (MAE) [21,53], supercritical fluid extraction (SFE) [54,55] techniques, nanotechnology-based processes [56,57], and the application of deep eutectic solvents (DESs) [58,59] and their natural derivatives, Natural Deep Eutectic Solvents (NADESs) [60,61]. Furthermore, studies have demonstrated that the combination of these techniques can enhance the efficiency and yield of metabolite extraction [62,63]. For example, the combination of MAE and NADESs works synergistically to maximize the extraction of phenolic compounds from agro-industrial residues. The energy generated from microwaves penetrates the matrix of the agro-residues and directly heats the polar solvents, such as the water present in NADES. This rapid heating increases the internal pressure within the plant matrix cells, causing the rupture of the cell walls and release of the phenolic compounds into the solvent. MAE accelerates the extraction process by providing quick and efficient heating, reducing processing time, and minimizing the degradation of heat-sensitive compounds. Meanwhile, NADESs enhance the solubility and stability of phenolic compounds, improving both the extraction efficiency and the quality of the final extract [64]. Once extracted, these compounds are used across various industries, including food [65], pharmaceuticals [66], and cosmetics [67] because a green non-toxic solvent is used.

This review focuses on the bioactive compounds of *C. aurantium*, their potential as therapeutic agents for the prevention of various diseases, the advanced extraction techniques for obtaining these compounds, and their potential applications in the food, pharmacological, and cosmetic industries.

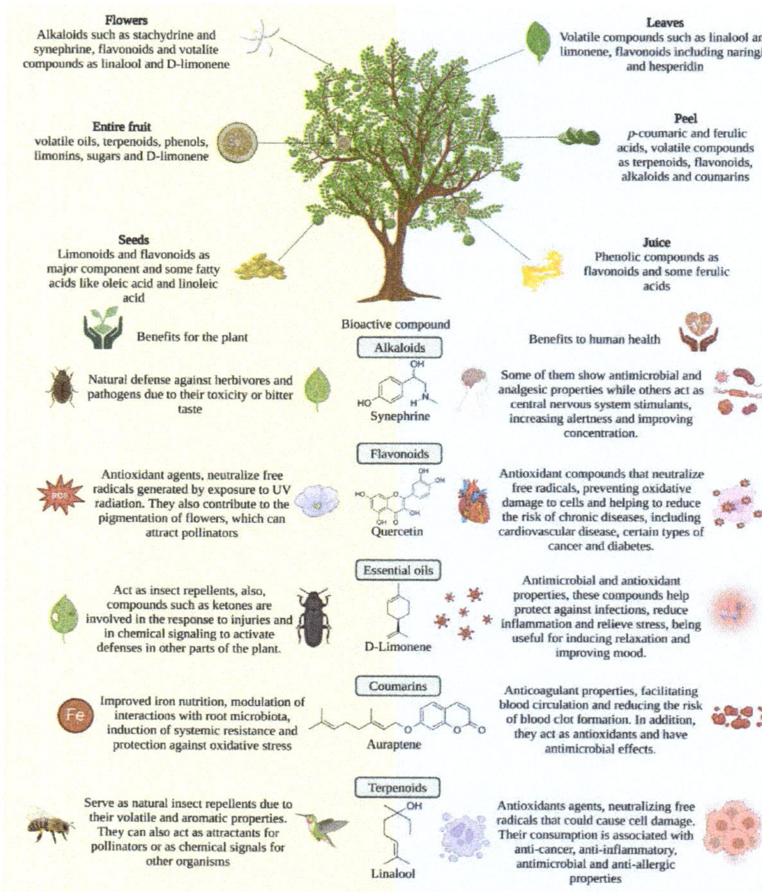

Figure 1. *C. aurantium* tree. Main bioactive compounds present in the fruit, peel, seeds, leaves, and flowers of *C. aurantium*, the benefits provided to the plant as defense mechanisms against biotic and abiotic stresses, and the benefits of their consumption. Created with Biorender.com.

2. Bioactive Compounds in *C. aurantium*

Secondary metabolites are organic molecules produced through the secondary metabolism of plants [68]. While they are not essential for plant growth and reproduction, they play a crucial role in adapting plants to their environment. These compounds function as defense mechanisms against herbivores and microbial pathogens or as pollinator attractors. This defense mechanism works through direct toxicity to herbivores or the inhibition of microbial growth. Some secondary metabolites, known as phytoalexins, are produced in response to infection, while others, called phytoanticipins, are stored in plant tissues and activated after tissue damage. For instance, glucosinolates, found in many cruciferous plants, hydrolyze into toxic compounds while plant tissues are damaged by insects [69]. Secondary metabolites benefit human health by participating in human metabolic processes, regulating physiological activities, scavenge free radicals in the body, and exhibit antibacterial and anticancer properties. They are particularly suitable for these purposes due to their ability to cross cell membranes and exert biological activity [70–72].

Primary metabolites are essential precursors for secondary metabolite synthesis. Pathways such as the shikimic acid pathway and the Krebs cycle generate key intermediates necessary for the synthesis of phenolic compounds and other secondary metabolites (Figure 2) [73].

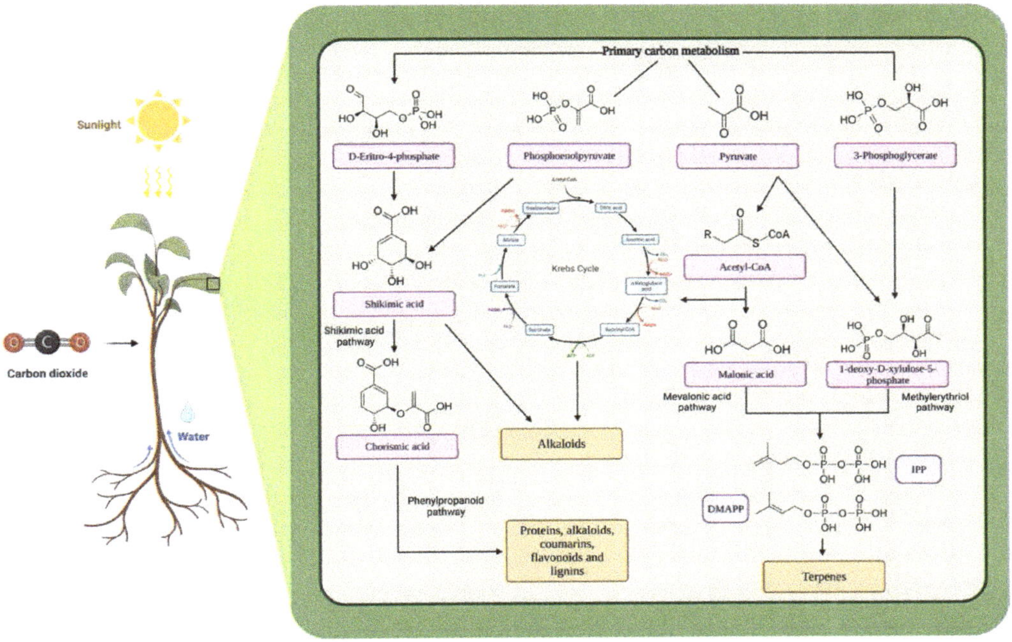

Figure 2. Simplified description of the biosynthesis of secondary metabolites starting from intermediates generated in glycolysis (phosphoenolpyruvate, pyruvate, and 3-phosphoglycerate) and the pentose phosphate pathway (D-erythrose-4-phosphate). These intermediates form the basis for secondary metabolic pathways (violet boxes), such as the shikimic acid or the malonic acid pathways in plants, leading to the production of a wide variety of secondary compounds essential for plant defense and adaptation (yellow boxes). IPP, isopentenyl diphosphate; DMAPP, dimethylallyl diphosphate [68,74,75]. Created with Biorender.com.

Citrus fruits contain a variety of secondary metabolites, such as phenolic acids, alkaloids, coumarins, flavonoids, limonoids, carotenoids, and essential oils, among others. These compounds demonstrate numerous biological activities beneficial to human health, including antioxidants, anti-inflammatory, anticancer, cardiovascular, and neuroprotective effects (Table 1). Below, a description of the main bioactive compounds present throughout the morphology of *C. aurantium* is presented.

2.1. *Alkaloids*

Alkaloids are a diverse class of natural compounds characterized by cyclic structures containing one or more basic nitrogen atoms, which confer their alkaline properties and contribute to their pharmacological effects. Alkaloids are typically biosynthesized from amino acids such as tyrosine, lysine, ornithine, phenylalanine, and tryptophan, are extracted mainly from plant sources, although they are also present in fungi and certain animal species. In contemporary pharmaceutical research, alkaloids are chemically modified to enhance their bioactivity, and in some cases, are fully synthesized to optimize therapeutic effectiveness [76–78].

Citrus species contain abundant bioactive alkaloids with notable antioxidant properties. These alkaloids play an indirect role in the growth, reproduction, and metabolism of citrus plants, with *C. aurantium* showing particularly high levels of antioxidant alkaloids compared to other citrus species [79].

In *C. aurantium*, various types of alkaloids have been identified across different parts of the plant. For instance, eight types of alkaloids (stachydrine, caffeine, cathine, synephrine, hordenine, cytisine, N-methylcytisine, and choline) have been reported in ethanolic extracts from *C. aurantium* flowers, inhibiting the α-glucosidase and pancreatic lipase activities by more than 80% [20]. Additionally, synephrine and stachydrine have been detected in aqueous extracts derived from *C. aurantium* residues (Changshanhuyou) [19]. Stachydrine has shown unique benefits in preventing and treating cardiovascular disease by modulating various disease-related signaling pathways and molecular targets [80], while synephrine is recognized for its stimulant effects and has been studied for its potential in weight loss and athletic performance [81].

Moreover, synephrine has been identified in whole *C. aurantium* fruit using liquid chromatography–mass spectrometry (LC-MS), a technique that separates components based on polarity, with polar compounds like synephrine eluting first [18]. Other studies also mention the presence of synephrine in *C. aurantium*; certain properties are reported, specifically, for p-synephrine, that may increase fat oxidation rates during exercise, particularly at low and moderate intensities, and reduce carbohydrate oxidation rates during exercise [82].

Alkaloids are present at every part of *C. aurantium* plant, like leaves, where the anti-inflammatory and antioxidant activities have been reported through in vitro studies. This study focused on the metabolic pathways of cyclooxygenase (COX) and 5-lipoxygenase (5-LOX), both of which play roles in inflammatory processes. The extract inhibited the COX-1, COX-2, and 5-LOX enzymes in vitro, suggesting its potential to reduce the production of prostaglandins and leukotrienes, which are key mediators of inflammation [83].

2.2. Flavonoids

Flavonoids are a group of secondary metabolites present in plants [84] and are involved in the regulation of auxin transport, male fertility, pollination, seed development, flower coloration, and allelopathy [85]. Their basic structure consists of two phenol rings (A and B), linked by a linear three-carbon chain that forms a heterocyclic pyran ring (C) with an oxygen atom [86]. Their importance lies on their various pharmacological effects, such as immunomodulatory [87], hypoglycemic [88], and antibacterial [89] as well as in the prevention of many diseases such as cancer [90], Alzheimer's [91], diabetes [92], among others.

Plant flavonoids are generated through various biosynthetic pathways beginning with the shikimate pathway for the biosynthesis of shikimic acid, which begins with an aldol condensation of phosphoenolpyruvic acid and D-erythrose 4-phosphate, to obtain chorismic acid, which is converted into the amino acid phenylalanine through the action of the enzymes prephenate-aminotransferase (PhAT) and arogenate-dehydratase. Finally, it enters the phenylpropanoid pathway and flavonoid biosynthesis, where the compound 4-coumaroyl-CoA, a product of the action of the enzyme cinnamate-4-hydroxylase (C4L), plays a crucial role in the biosynthesis of flavonoids through the phenylpropanoid pathway, the production of the coumarin skeleton, and the initiation of the flavonoid pathway [93].

Flavonoids are classified into several types according to their chemical structure, degree of unsaturation, and oxidation of the carbon ring, including flavones, flavanones, isoflavones, flavonols, chalcones, flavanols, and anthocyanins [84].

Flavonoids and phenolic acids are the primary classes of phenolic compounds present in citrus fruits. Typically, the fruit peel has a higher concentration of antioxidant compounds compared to other parts of the fruit. Flavanones are the most important flavonoids on citrus species, with hesperidin, narirutin, naringin, and eriocitrin being the main ones [94].

In *C. aurantium*, numerous flavonoids are distributed throughout the entire tree. For instance, compounds such as naringin, narirutin, hesperidin, neohesperidin, hesperetin, and cimbitin have been extracted from daidaihua (dried fruit of *C. aurantium*) through a multistep process. After extracting essential oils via steam distillation, sodium hydroxide is added to the solution, and its pH is adjusted with hydrochloric acid to precipitate and isolate these flavonoids [16].

Further studies examined the influence of harvest season, fruit ripeness, solvent, and the interactions on flavonoid content and antioxidant activity in extracts from *C. aurantium* post-juice extraction residue. In these analyses, naringin, neohesperidin, quercetin, and (+)-catechin were extracted and identified using high-performance liquid chromatography (HPLC). Notably, methanolic extracts from spring-harvested *C. aurantium* exhibited the highest antioxidant activity, assessed through 2,2-diphenyl-1-picrylhydrazyl (DPPH) and 2,2′-azino-bis (3-ethylbenzothiazoline-6-sulfonic acid) (ABTS) assays [17].

Researchers obtained a flavonoid-enriched extract from the peel of *C. aurantium* L. var. *amara* Engl. using the UAE technique, followed by purification steps that included evaporation, precipitation, and liquid extraction. Although the specific flavonoids identified were not detailed, flavanones are noted as the predominant class in *C. aurantium* peel, including compounds such as hesperetin, naringenin, hesperidin, neohesperidin, naringin, and narirutin. Other flavonoid types, such as nobiletin and tangeretin, were also present. The extract showed significantly enhanced biological activity, with a greater DPPH radical scavenging capacity compared to a traditional extract obtained via maceration. This outcome suggests that the optimized extract achieved a higher concentration of antioxidant compounds, attributable to the efficiency of the UAE method and the subsequent purification steps [30].

2.3. Essential Oils

Essential oils (EOs) are volatile, aromatic liquids with low molecular weight, produced through the secondary metabolism of aromatic plants. They are extracted from various plant parts, including bark, seeds, flowers, peels, buds, fruits, roots, leaves, wood, berries, and are named according to their plant of origin. The term "essential" refers to their role as key components of the plant. These oils are complex mixtures of organic compounds derived from various plant sources that contain multiple active constituents, such as alkaloids, tannins, steroids, glycosides, resins, phenols, volatile oils, and flavonoids, conferring their distinctive fragrances [95–98].

Citrus flavedo is extensively used to produce essential oils, widely applied in the food and cosmetic industries. These citrus essential oils contain over 2000 organic compounds, primarily comprising monoterpenes and sesquiterpenes, as well as their oxygenated derivatives, aliphatic aldehydes, alcohols, and esters. The concentration of these constituents in hydrodistillates varies depending on environmental factors, including climate, irrigation practices, ripening stage, and soil composition [98,99].

The presence of essential oils of *C. aurantium* has been widely reported, for example, Badalamenti et al. [32] studied seven essential oils (EOs) obtained from the flavedo of different *C. aurantium* cultivars from Sicily: Canaliculata (C1), Consolei (C2), Crispifolia (C3), Fasciata (C4), Foetifera (C5), Listata (C6), and Bizzaria (C7), which were extracted by hydrodistillation and analyzed with GC-MS.

D-limonene was the main component in all EOs, although its concentration varied between 33.35% and 89.17% [68]. The cultivar Crispifolia (C3) stood out for a lower amount of D-limonene and a higher percentage of oxygenated monoterpenes, such as β-linalool, α-terpineol, bergamol, and geranyl acetate. The antioxidant activity of individual EOs and their combinations (1:1 v/v) was evaluated by DPPH, ABTS, FRAP, and β-carotene decolorization assays. Fasciata extract (C4) presented the highest antioxidant potential among individual EOs, while the combination of Canaliculata + Bizzaria (C1C7) was the most active among the combinations. The results suggest that the antioxidant activity is not attributed to D-limonene alone but is also affected by other minor compounds.

Other studies, such as Oulebsir et al. [100] focused on analyzing the essential oil extracted from the leaves of *C. aurantium* L. The research group employed hydrodistillation to obtain the oil, achieving a yield of 0.57%. Chemical analysis showed a composition predominantly of linalool (30.62%), linalyl acetate (33.01%), and α-terpineol (9.57%), classifying the oil within the linalool/linalyl acetate chemotype. In terms of biological activity, the essential oil exhibited a low total phenol content and limited antioxidant activity. However, it demonstrated a significant inhibition of elastase and collagenase enzymes, indicating a potential anti-aging effect—an innovative finding for the leaf essential oil of this species.

Elhawary et al. [21] evaluated the essential oil extracted from the leaves of *C. aurantium* using three distinct distillation techniques: hydrodistillation (HD), steam distillation (SD), and microwave-assisted distillation (MV). Thirty-five volatile components were identified, with monoterpenes as the predominant class, followed by sesquiterpenes. The oil's composition varied according to the extraction method, with steam distillation yielding a distinct chemical profile.

The essential oils demonstrated antioxidant activity, with the SD-derived oil showing the highest potency. Additionally, the oils exhibited neuroprotective properties by inhibiting the enzymes butyrylcholinesterase (BChE) and acetylcholinesterase (AChE), both relevant in Alzheimer's treatment, as their inhibition can support cognitive function in patients with the disease. The oils also displayed antidiabetic effects by inhibiting α-amylase and α-glucosidase enzymes; SD oil showed the highest potency in α-amylase inhibition, while MV oil was most effective in inhibiting α-glucosidase.

2.4. Coumarins

Coumarins are a major class of phytochemicals belonging to the benzopyrone family. Their name originates from *Coumarouna odorata* Aubl. (*Dipteryx odorata*), the species from which coumarins were first isolated. The core structure of coumarins is benzo-α-pyrone (2H-1-benzopyran-2-one), comprising an aromatic benzene ring in a conjugated system fused with an α-pyrone (lactone ring). This electron-rich structure can interact with various molecular targets, such as enzymes and receptors, contributing to their potent medicinal properties. Coumarins are fundamental compounds of the broader group of phenolic compounds bearing the same name [101–103].

Natural coumarins are categorized into distinct classes according to their chemical structure and complexity: simple coumarins (e.g., esculetin, scopoletin, umbelliferone), furanocoumarins (e.g., psoralen, bergapten, angelicin), pyranocoumarins (e.g., xanthyletin, seselin), and coumarins substituted on the pyrone ring, such as biscoumarins (e.g., dicumarol) and benzocoumarins (e.g., urolithins) [103].

Coumarins exhibit a broad spectrum of biological activities, including antioxidant [104], anti-inflammatory [105], antibacterial [106], antifungal [107], and therapeutic effects against neurodegenerative [108] and chronic metabolic diseases [109].

Citrus essential oils comprise over 200 compounds and are typically extracted by cold pressing or the steam distillation of citrus plant materials. The major fraction (85% to 99%) of these essential oils consists of volatile compounds, primarily aromatic hydrocarbons and their oxygenated derivatives, such as linear hydrocarbons, esters, ketones, aldehydes, and alcohols. However, elevated levels of coumarin derivatives are also common in citrus oils, especially those obtained by the cold pressing of citrus peel. This method facilitates the co-extraction of relatively polar, non-volatile coumarins [110,111].

Wen et al. [23] identified eight coumarins in extracts from *C. aurantium* fruits: scoparone, 4-(2,3-dihydroxy-3-methylbutoxy)-7H-furo [3,2-g] chromen-7-one, 7-methoxy-4-methylcoumarin, meranzine, isomeranzine, bergapten, ostrutin, and auraptene. The levels of these coumarins varied significantly between the two cultivars studied, Morocco sour orange (MSO) and United sour orange (USO). In MSO, meranzine and isomeranzine were the predominant coumarins, whereas in USO, auraptene was the major component, with a higher (about 209 times) concentration than MSO. The study also analyzed auraptene's fragmentation, detailing the loss of specific chemical groups from its quasimolecular ion.

2.5. Terpenoids

Terpenoids represent the largest and most structurally diverse class of natural phytochemicals, comprising multiple isoprene units (a gaseous hydrocarbon with the molecular formula (C5H8). These compounds are abundantly present in nature and contribute to the fragrance, flavor, and pigmentation of plants. They are classified into monoterpenoids (C10), sesquiterpenoids (C15), diterpenoids (C20), and triterpenoids (C30). Terpenoid biosynthesis occurs via two primary pathways: the mevalonate (MVA) pathway, primarily in the cytoplasm of eukaryotic cells, producing sesquiterpene, triterpene, and steroid precursors; and the 2-C-methyl-D-erythritol-4-phosphate (MEP) pathway, predominantly in the plastids of prokaryotes and plants, forming monoterpenes, diterpenes, and tetraterpenes [112,113].

Over 90,000 terpenoid compounds have been characterized, initially identified as constituents of plant essential oils. They are now acknowledged for their role in plant growth. Carotenoids, for example, play a crucial part in photosynthesis. Additionally, terpenoids function as defense mechanisms, providing resistance to pathogens through antimicrobial, antifungal, and insect-repellent properties [114,115].

Terpenoids are of significant interest due to their widespread occurrence and extensive medicinal applications [116]. Numerous terpenoids exhibit diverse biological activities, including antioxidant [117], anti-inflammatory [118], anticancer [113], and antiallergic effects [119]. These activities have practical applications, in creams, lotions, and face oils development [120].

Citrus essential oils are composed primarily of terpenes, with up to 95% of the oil's content. Despite their high concentration, terpenes are easily degraded by oxidation while slightly contributing to the aroma of the oil. Conversely, terpenoids, which represent less than 5% of the crude oil, are key to its aromatic qualities. For this reason, deterpenation is essential to enhance terpenoid concentration and achieve high-quality essential oils [121].

Essential oils from *C. aurantium* contain a diverse array of terpenoids. Okla et al. [122] used hydrodistillation with a Clevenger-type apparatus to extract essential oils from the leaves/twigs, small branches, branch wood, and bark of *C. aurantium*, analyzing their chemical composition and antibacterial properties. They identified various terpenoids, including 4-terpineol, D-limonene, 4-carvomenthenol, linalool, γ-terpinene, (−)-β-fenchol, and dimethyl anthranilate, with concentrations varying by plant part. Essential oils from leaves/twigs and small branches demonstrated notable antibacterial activity against the phytopathogens *Agrobacterium tumefaciens*, *Dickeya solani*, and *Erwinia amylovora*, while oils from bark and branch wood showed no activity against *A. tumefaciens*. Leaf/twig oils exhibited the highest overall activity, correlating the increased antibacterial efficacy to the concentration of essential oil applied. The study concluded that differences in essential oil composition and antibacterial activity might be influenced by the extraction method, plant part, soil type, and local climate conditions.

Nidhi et al. [123] analyzed essential oil from *C. aurantium* leaves (CAEO) extracted via hydrodistillation with a Clevenger apparatus. GC-MS analysis identified ten main compounds, including 2-β-pinene (100%), δ-3-carene (84%), and D-limonene (28%). CAEO exhibited potent antifungal activity against *Candida albicans*, proving comparable or superior to the antibiotics fluconazole and amphotericin B. Beyond inhibiting fungal growth, CAEO effectively reduced fungus growth. Additionally, CAEO showed a synergistic effect with both antibiotics, enhancing their efficacy, a significant finding for addressing drug resistance in *C. albicans*. In silico analyses confirmed that the three primary terpenoids in CAEO interact with key antifungal targets in *C. albicans*: NMT (an enzyme vital for fungal survival) and CYP51 (involved in ergosterol biosynthesis, essential for fungal cell membranes). These terpenoids also met the "Lipinski rule of five", indicating favorable oral bioavailability, while ADMET (absorption, distribution, metabolism, excretion, and toxicity) analysis suggested a strong safety profile. The study underscores CAEO's potential as both an antifungal agent and antibiotic enhancer for treating candidiasis.

Table 1. Bioactive compounds in *C. aurantium* and their biological activities.

Class	Isolated Bioactive Compound	Biological Activity	Reference
Alkaloids	Citrusinina-I Citracridona-I 5-Hidroxinoracronicin Natsucitrina-I Glicofolinina Citracridona-III	In vitro Cytotoxic activity against four human cancer cell lines: MCF7 (breast cancer) HepG2 (liver cancer) A549 (lung cancer) PC3 (prostate cancer)	[124]
	p-synephrine	In vivo Decrease in weight gain. Low subchronic toxicity in mice and alteration of oxidative metabolism.	[125]
Flavonoids	Extract of 14 flavonoids, with nobiletin, naringin, and hesperidin as the main flavonoids.	In vitro Inhibition of cell motility, decreased MMP-2 activity, and induction of morphological changes related to apoptosis in A549 cells. In vivo Inhibition of A549 cell infiltration into the lungs of mice; increased expression of activated caspase-3 and p-p53; decreased levels of DDX3X and ANP32B proteins.	[126]

Table 1. *Cont.*

Class	Isolated Bioactive Compound	Biological Activity	Reference
Essential Oils	Essential oils from the peel, leaves, and flowers. The main components were limonene (87.02%), linalyl acetate (53.76%), and linalool (39.74%), respectively.	In vitro Antifungal activity against *P. italicum* and *P. digitatum* in vitro, reduction of mycelial growth, spore germination, and fungal mycelium weight. In vivo Reduction of infection rate by *Penicillium italicum* and *P. digitatum* in inoculated oranges.	[127]
Class	Isolated bioactive compound	Results/Biological activity	Reference
Coumarins	8,3′-β-D-glucopiranosiloxi-2′-hidroxi-3′-metilbutil-7-methoxycoumarin, merazin hydrate, and isomerazin	In vitro The extract did not significantly inhibit platelet aggregation induced by ADP, epinephrine, collagen, and arachidonic acid in human plasma. In vivo Anti-inflammatory activity by significantly inhibiting the increase in vascular permeability induced by histamine and dextran in rats.	[128]
Terpenoids	Terpenes such as limonene, valencene, β-caryophyllene, α-bisabolene, γ-terpinene, α-terpineol, β-elemene, β-farnesene, among others.	In vitro Antioxidant activity (ABTS assay; IC50 of 3.39 mg/mL), antimicrobial activity (inhibited the growth of 14 out of 15 tested bacterial strains), and cytotoxic activity (dose-dependent reduction in the viability of Hep-2 cells).	[12]
	D-limonene and α-terpineol	In vivo Mice that received a single oral dose of 2000 mg/kg of bitter orange peel extract showed no clinical symptoms of acute toxicity or mortality during the 14-day test period.	[37]

3. Extraction Techniques

With knowledge of the diverse compounds present in plant materials, scientists have developed various methodologies over time to isolate and purify these compounds from plants and foods. The extraction of biological active compounds depends on numerous factors, including the extraction method, raw materials, and choice of solvent. Consequently, technologies for extracting bioactive compounds from foods must prioritize product quality while balancing efficiency, cost-effectiveness, and sustainability [129,130]. Below are the most common extraction methodologies, along with recent innovations specifically applied to *C. aurantium* for isolating its therapeutic components.

3.1. Conventional Extraction Methods

Conventional extraction techniques, though effective at isolating target compounds, present significant limitations. These methods (such as maceration and Soxhlet) rely on the solvent's extraction capacity and often require agitation and/or heat, which can lead to high energy consumption and large solvent volumes, resulting in potentially hazardous waste [129,131,132]. The application of these traditional techniques for extracting bioactive compounds from *C. aurantium* is briefly outlined below, followed by an introduction to innovative, eco-friendly methodologies.

In maceration, raw material is cut, soaked in a solvent, and left to extract, with occasional shaking to improve solvent penetration. After a set time, filtration completes the process [133].

Abdallah et al. [30] compared UAE with traditional maceration for isolating polyphenols and flavonoids from *C. aurantium* peel. The optimized UAE method produced an extract with polyphenol yields of 33.76 ± 5.63 mg of gallic acid equivalent per gram of dry matter (DM), flavonoids of 75.50 ± 5.59 mg of naringin equivalent per gram of DM, and tannins of 19.85 ± 1.67 mg of tannic acid equivalent per gram of DM, showing significantly higher yields of polyphenols and flavonoids. This also leads to higher DPPH inhibition activity (IC50 = 30.7 ± 3.21 μg/mL), compared to the conventional extract prepared by maceration.

Other studies focused on lipophilic compounds, used the Soxhlet method with hexane, a non-polar solvent, demonstrating its effectiveness for the continuous and efficient extraction of essential oils and other non-polar constituents by solvent recirculation through the plant material. In this sense, extraction with hexane produced a 1.5% yield of non-polar extract, while extraction with methanol produced a 35% yield of polar extract [10]. However, while effective, Soxhlet method drawbacks include prolonged extraction times, low yield, and high solvent usage, which may lead to target compound evaporation and degradation [129].

Other studies carried out on *C. aurantium* using the maceration technique, such as Sharma et al. [96] who obtained a yield of 15.33% when extracting bioactive compounds from the *C. aurantium* flower to evaluate the antioxidant and inhibitory effects of the extract on acetylcholinesterase activity and the production of amyloid nanobiofibrils from bovine serum albumin (BSA). In the studies carried out by Pasandideh et al. [33] they used maceration with water as a solvent to extract bioactive compounds in the form of essential oils from the inner core of sour orange seeds, obtaining a diverse profile of fatty acids and showing a high DPPH inhibition capacity (almost 90%).

3.2. Green Extraction Techniques

Green chemistry is based on 12 fundamental principles aimed at harnessing chemistry to prevent pollution by optimizing product and process designs to reduce or, ideally, eliminate the use and generation of hazardous substances [134,135]. Therefore, green extraction aims to design and implement extraction processes that lower energy consumption, employ alternative solvents, yield renewable natural products, and ensure that the resulting extracts or products are both safe and of high quality [132].

Green extraction offers several advantages, including enhanced efficiency, lower solvent use, and the preservation of valuable bioactive compounds, addressing the limitations of conventional methods like low yields, high solvent consumption, and degradation risks for heat-sensitive compounds [131]. Examples of green extraction techniques include SFE, UAE, MAE, NADES (Table 2), among others. Below are modern green extraction methods for isolating bioactive compounds from *C. aurantium*.

Table 2. Advantages and limitations of novel extraction techniques.

Extraction Technique	Principle (Graphical Illustration)	Factors Influencing Extraction Efficiency	Advantages	Limitations
UAE [136]	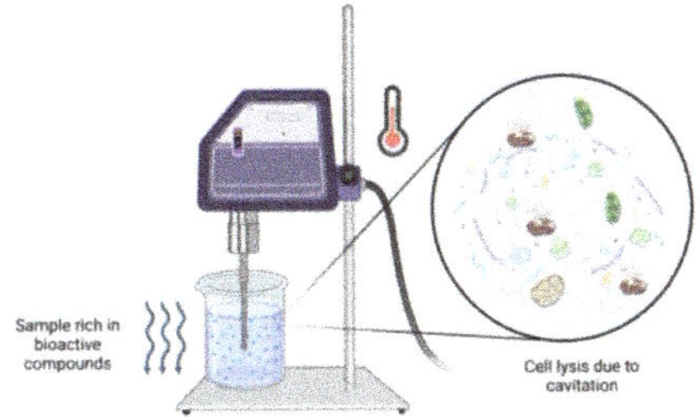	• Extraction time • Temperature • Solvent selection • Ultrasonic power • Ultrasonic frequency • Solvent/solid ratio	Higher extraction yield, shorter extraction time, lower solvent consumption, preservation of thermosensitive compounds, increased bioaccessibility of compounds, and improved extract quality.	Degradation of compounds, complex optimization, equipment cost, scalability, limitations in the extraction of certain compounds, and effects on the functional properties of the compounds.
MAE [137]	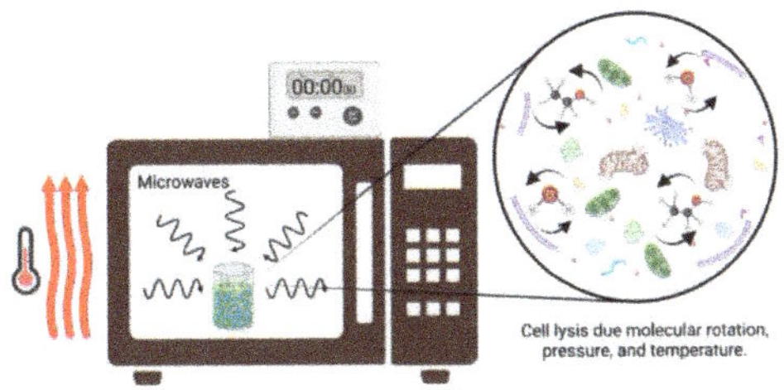	• Solvent/solid ratio • Irradiation time • Effect of agitation • Microwave power and temperature • Plant sample characteristics and water content • Microwave energy density	Reduction of extraction time, lower solvent consumption, higher extraction yield, better selectivity, and reproducibility.	Degradation of heat-sensitive compounds, limitations in sample size, equipment cost, parameter optimization, and appropriate safety precautions when operating the equipment.
SFE [138]	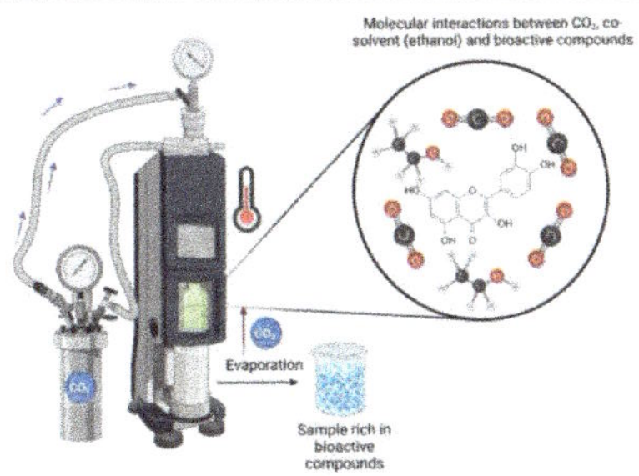	• Temperature • Pressure • Co-solvent • Raw matrix • Extraction time	High selectivity and efficiency, short extraction time, reduction in solvent usage, environmental protection, suitability for thermolabile compounds, and solvent-free extracts.	Weaknesses in the extraction of polar compounds, effect of particle size, initial investment cost, scalability, degradation of sensitive compounds, and potential safety issues.

Table 2. *Cont.*

Extraction Technique	Principle (Graphical Illustration)	Factors Influencing Extraction Efficiency	Advantages	Limitations
DES/NADES [139]	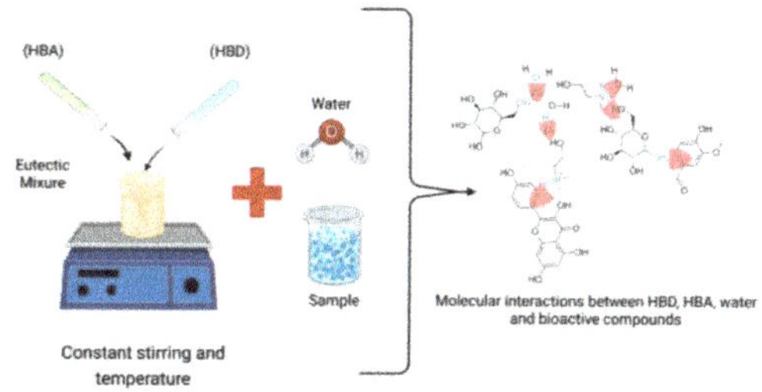	• Type of HBA and HBD • Molar ratio • Water content • Temperature • Viscosity • pH	Low cost and easy synthesis, biodegradability and low toxicity, low volatility and flammability, high adjustability, and high solubility for a variety of compounds.	High viscosity, low conductivity, water sensitivity, and limited information on toxicity and biodegradability.

3.2.1. Ultrasound Assisted Extraction

UAE is a method that utilizes solvent volumes ranging from 20 to 200 mL and typically lasts between 2 to 20 min [133]. This technique employs ultrasound energy with solvents to extract target compounds from various plant matrices [129].

The primary mechanism underlying UAE is acoustic cavitation, which involves the formation and subsequent collapse of cavitation bubbles. The collapse of these bubbles, along with the propagation of sound waves, can induce several phenomena, including fragmentation, localized erosion, pore formation, shear forces, and enhanced absorption and swelling rates within the plant cell matrix. As the cavitation bubbles collapse, they generate shock waves that promote accelerated particle collisions, leading to the fragmentation of the cellular structure. This rapid fragmentation results in the solubilization of bioactive components within the solvent, driven by reduced particle size, increased surface area, and elevated mass transfer rates at the solid matrix boundary. Consequently, this enhances the contact area between the bioactive compounds and the solvent, thereby accelerating the mass transfer process [129,136,140].

UAE has been employed with both polar and non-polar solvents to enhance the extraction of bioactive compounds in *C. aurantium*. Cai et al. [20] applied this method to isolate and optimize alkaloids from CAVA flower (*C. aurantium* L. var. *amara* Engl.) with the goal of maximizing alkaloid yield and examining their impact on lipid metabolism. Optimal conditions were achieved with a 72 min ultrasonication, an ethanol concentration of 78%, and a liquid-to-solid ratio of 30 mL/g, yielding a maximum alkaloid content of 5.66%. These extracts exhibited significant effects on both in vivo and in vitro lipid regulation, including reductions in lipid accumulation, triglycerides (TGs), and total cholesterol (TCHO). Additionally, they modulated key lipid metabolism genes, such as fatty acid synthase (FAS), peroxisome proliferator-activated receptor γ (PPARγ), uncoupling protein 2 (UCP2), and retinol-binding protein (RBP4) in Hep G2 cells and genes including fat-5, fat-6, fat-7, sbp-1, cebp-2, nhr-49, nhr-80, and mdt-15 in *C. elegans*.

Yan et al. [51] applied an ultrasound-assisted aqueous biphasic extraction (UA-ATPE) system to isolate synephrine, naringin, and neohesperidin from *C. aurantium* fruits. After testing five ethanol/salt systems, they identified ethanol/K_2CO_3 as the optimal system for achieving the highest extraction yields for synephrine and neohesperidin. They then optimized key parameters for UA-ATPE: K_2CO_3 concentration at 20.60% (*w*/*w*), ethanol concentration at 27% (*w*/*w*), a solvent-to-material ratio of 45.17:1 (g:g), fruit powder at 120 mesh, extraction temperature at 50 °C, an extraction time of 30 min, and an ultrasonic power of 80 W. Under these conditions, yields were 11.17 $\pm$ 0.18 mg/g for synephrine, 7.39 $\pm$ 0.13 mg/g for naringin, and 89.27 $\pm$ 1.70 mg/g for neohesperidin. The yield of neohesperidin was eight times greater than that achieved with conventional UAE, while doubled total target compound yield. Additionally, extracts obtained through UA-ATPE had significantly lower impurity levels compared to UAE. The study concludes that UA-ATPE is an efficient and promising technique for extracting these bioactive compounds from *C. aurantium*.

Other researchers like Wang et al. [141] employed an optimized UAE process using water as a solvent to obtain bioactive compounds from *C. aurantium* leaves. Researchers assessed the effects of temperature, extraction time, liquid-to-solid ratio, and ultrasound power on the total phenolic content (TPC) of the extract. Under optimized conditions, *C. aurantium* leaf extract (CALE) exhibited multiple beneficial biological activities, including significant antioxidant capacity (notably ABTS radical scavenging), antityrosinase activity with potential skin-whitening effects, anti-aging properties by inhibiting collagenase, elastase, and MMP-1 enzymes involved in skin aging, and antimicrobial effects against *Staphylococcus aureus, Escherichia coli, and Pseudomonas aeruginosa*, potentially indicating it as an antiseptic. Additionally, CALE was safe for skin cells and showed high transdermal absorption, supporting its potential for topical applications. Key active compounds, including linalool, linalyl acetate, limonene, and α-terpineol, contribute to these bioactivities. This study is noted as the first to explore the multifunctional bioactivities of CALE on a pilot scale, suggesting its applications in the food, pharmaceutical, and cosmetics industries.

Table 3 shows a summary of the aforementioned information and additionally shows other relevant studies on the use of UAE to obtain bioactive compounds in *C. aurantium*. Cui et al. [142] analyzed the peel extract of *C. aurantium* Changshan-huyou (CACH) obtained by ultrasound-assisted n-hexane extraction; a total of 544 volatile constituents (VOCs) were identified using gas chromatography-mass spectrometry (GC-MS). These VOCs were classified into 12 different chemical classes, including terpenoids, alkanes, alkenes, alicyclic hydrocarbons, aromatic compounds, alcohols, ethers, aldehydes, ketones, fatty acids, esters, and others.

3.2.2. Microwave-Assisted Extraction

MAE integrates traditional solvent extraction with microwave energy to enhance extraction efficiency. MAE operates through two key mechanisms: dipolar rotation and ionic conduction. In dipolar rotation, microwaves induce analyte molecules to align with the electric field, generating heat and causing water molecules to evaporate. Ionic conduction, on the other hand, generates heat through the resistance of the solution to ion flow, producing friction. Together, these effects disrupt plasma membranes and cell walls, enhancing the release of target compounds and making extraction more effective [143–145].

In a comparative study, García-Martín et al. [53] evaluated MAE with Soxhlet extraction (SE) for isolating phenolic compounds from *C. aurantium* residues left after industrial neohesperidin extraction. Using aqueous solutions of ethanol and ketone at concentrations of 50%, 75%, and 100% (*v*/*v*), they applied a Box–Behnken design to optimize MAE parameters, such as temperature, extraction time, and solvent concentration. Results demonstrated that MAE yielded higher extraction efficiency in shorter durations compared to SE. Specifically, MAE using 50% acetone at 75 °C for 15 min produced a yield of 16.7 g/100 g, while 50% ethanol under similar conditions reached 20.2 g/100 g. UHPLC analysis identified primary flavonoids, including naringin, neohesperidin, hesperidin, and naringenin. The findings indicate that MAE is an efficient technique for flavonoid extraction from orange residues, positioning these residues as promising by-products for obtaining high-value compounds.

On the other hand, Agustin et al. [146] employed microwave hydrodistillation (MHD), a technique that combines water distillation with microwave heating, to extract essential oils from *C. aurantium* leaves. They varied microwave power (150, 300, and 450 W), extraction time (30, 60, and 90 min), and fresh material/solvent (F/S) ratios (0.375, 0.5, and 0.625 g/mL) to identify optimal extraction conditions. The highest essential oil yield (0.373%) was obtained with a microwave power of 300 W, an extraction time of 90 min, and an F/S ratio of 0.625. GC-MS analysis indicated that the primary constituents were Germacrene D (29.51%), Alpha-Copaene (18.47%), CIS-Caryophyllene (15.42%), 9-Eicosene (E) (8.58%), and Limonene (5.51%). The study demonstrated that microwave power, extraction time, and F/S ratio significantly impact essential oil yield, concluding that the MHD method is an efficient approach for essential oil extraction from *C. aurantium* leaves.

Comparative studies such as the one carried out by Lakache et al. [36] evaluated the essential oil obtained through microwave-assisted hydrodistillation (MAHD) from *C. aurantium* peels, which was characterized by a high percentage of limonene (99.23%) and a low diversity of compounds compared to steam distillation (SD) and hydrodistillation

(HD). This method produced an oil yield of 0.7 mL from 250 g of dried peels, which only presented one additional oxygenated compound, linalool (0.06%). Although MAHD oil presented the lowest antioxidant activity in most of the assays, it showed notable values in the Cupric Reducing Antioxidant Capacity (CUPRAC) and Ferric Reducing Antioxidant Power (FRAP) assays. Regarding enzymatic inhibition, MAHD oil demonstrated a high inhibition of acetylcholinesterase (AChE) comparable to HD, and the highest inhibition of butyrylcholinesterase (BChE) and tyrosinase.

All this information can be found in Table 3, where the principal compounds and their concentration are summarized.

3.2.3. Supercritical Fluid Extraction

A supercritical fluid (SCF) is a substance at a temperature and pressure above its critical point, where differences between liquid and gas phases disappear, giving it intermediate properties, such as density, viscosity, and diffusion rate, between those of a gas and a liquid [147]. SFE leverages this state by introducing raw material into an extraction chamber where the solvent, upon reaching its supercritical point under controlled pressure and temperature, extracts bioactive compounds from the material [129].

The SFE process involves three main steps: (a) the dissolution of analytes in the SCF (often CO_2), (b) the movement of analytes through the fluid, and (c) the capture of the extracted analytes. Carbon dioxide (CO_2) is the most used SCF solvent due to its non-toxic, environmentally friendly properties, including non-flammability, high availability, low viscosity, high diffusivity, supercritical conditions (31.1 °C and 1070.38 Psi) and low cost. A key advantage of CO_2 in SCFE is its easy removal from the extract after decompression, as it reverts to a gaseous state, leaving no solvent residue. In essence, SCFE is a technique that yields high-purity extracts considered safe (GRAS), meeting both quality and environmental safety standards [148–150]. Below is detailed information presented in Table 3 on the use of extracts obtained through SFE in *C. aurantium*.

Trabelsi et al. [54] applied SFE to isolate bioactive compounds from bitter orange peels (*C. aurantium* L. var *amara*) using supercritical CO_2 and ethanol as a co-solvent. The extraction parameters, specifically pressure, static time, and CO_2 flow rate, were optimized through a central composite design and response surface methodology, showing a significant quadratic effect of pressure and CO_2 flow on extraction yield. Optimal conditions for maximum yield were identified as 170 bar, a static time of 53 min, and a CO_2 flow rate of 2.9 kg/h. Extract analysis revealed various compounds, such as fatty acid esters, terpenes, and coumarins, with isogeijerin as the major component, followed by osthole, squalene, and hexadecane. The highest osthole content (47%) was achieved at 170 bar, 2.7 kg/h CO_2 flow, and a static time of 50 min. Sovová's model was also used to predict and understand extraction behavior on a larger scale, successfully aligning with experimental data.

Jerković et al. [55] conducted a study utilizing SFE with CO_2 at 10 MPa and 40 °C to obtain extracts from *C. aurantium* peels and three varieties of *Citrus sinensis* for comparative analysis. The findings indicated that supercritical CO_2 extraction yielded enriched profiles of volatiles and semi-volatiles, including oxygenated monoterpenes, sesquiterpenes, and coumarin derivatives, as opposed to the simple mechanical pressing of peels. Limonene was identified as the predominant compound across all extracts, consistent with profiles seen in essential oils from citrus peels obtained by conventional methods. However, supercritical CO_2 extracts highlighted varietal differences; for example, *C. aurantium* peel oil contained higher percentages of linalool, linalyl acetate, and geranyl acetate, along with notable amounts of sabinene and isogeijerin. The study concludes that supercritical CO_2 extraction is an effective technique for obtaining enriched citrus peel extracts and that the resulting profiles can offer valuable insights into the biodiversity of different citrus varieties.

Extracts obtained using this extraction technology have already been tested in in vivo models, as performed by Jiang et al. [151] who investigated the properties and sleep-promoting effect of a nanoemulsion of *C. aurantium* L. var. *amara* Engl. essential oil (CAEO) extracted with supercritical fluid (SFE-CAEO-NE), analyzing the serum concentrations of neurotransmitters, such as 5-hydroxytryptamine (5-HT), γ-aminobutyric acid (GABA), and norepinephrine (NE) in a mouse model of insomnia. The results showed that SFE-CAEO-NE significantly improved insomnia-like behavior in mice, increasing 5-HT and GABA levels and decreasing NE levels in serum. A restorative effect on the function of damaged neurons in the hippocampus was also observed. The authors concluded that SFE-CAEO-NE has a good sleep-promoting effect and potential for application in aromatherapy.

3.2.4. Deep Eutectic Solvents

DESs are formed by mixing two or more pure compounds in a specific ratio to create an eutectic system with non-ideal thermodynamic behavior. This behavior is due to strong interactions between components, acting as hydrogen bond donors (HBDs) and acceptors (HBAs), which form a dense molecular network. This structure confers a DES of low volatility, high thermal stability, a low melting point, and a strong ability to dissolve a wide variety of compounds. Additionally, these physical and chemical properties can be modified by selecting different DES components [152].

DESs have numerous advantageous characteristics that make them preferable to conventional solvents, including low cost, safety, low toxicity, non-volatility, thermal stability, non-flammability, sustainability, tunability, and biodegradability [153].

Edrisi et al. [13] utilized a series of DESs based on choline chloride (ChCl) with different hydrogen bond donors (HBDs) to extract polyphenolic compounds (PCs) from *C. aurantium* L. peels. Among these, ChCl-1,4-butanediol (DES 7) showed the highest extraction efficiency, overcoming other DESs and conventional solvents (70% ethanol and water). UAE was used to enhance the process. Response surface methodology (RSM) and central composite design (CCD) optimized the extraction conditions, including sonication time, temperature, liquid-to-solid ratio, and water content. Optimal conditions were found to be 30 min at 40 °C, 49.95% water, and a liquid-to-solid ratio of 18.97:1 (mL/g), yielding 7.85 mg of gallic acid equivalent per gram of peel. The study concluded that this method, conducted with an aqueous two-phase system (ATPS) for purification, is efficient and sustainable for extracting PC from *C. aurantium* peels. The high extraction efficiency, short extraction times, and low temperatures help to preserve polyphenol quality.

He et al. [58] conducted screening studies, synthesizing fourteen different DESs using various hydrogen bond donors (HBDs) and acceptors (HBAs). They selected a DES composed of choline chloride and ethylene glycol in a 1:3 molar ratio with 30% water for extracting the flavonoids naringin (NAR) and neohesperidin (NEO) from *Jiang Fructus aurantii* (dried fruit of *C. aurantium*, JFA) using UAE. The DES-UAE method proved to be more effective than traditional solvents like 95% ethanol, methanol, and water, as well as conventional extraction methods like cold extraction, agitation-assisted extraction, and high temperature extractions. Using a Box–Behnken design (BBD), the extraction conditions were optimized at a 27% water content in DESs, a liquid-to-solid ratio of 16 mL/g, an extraction time of 72 min, and an extraction temperature of 62 °C, yielding extraction rates of 48.18 mg/g for NAR and 34.50 mg/g for NEO. Fourier-transform infrared spectroscopy (FT-IR) confirmed the presence of flavonoids in the DES-UAE extracts, while scanning electron microscopy (SEM) revealed that DES-1 exhibited higher penetration and erosion capability on the plant cell wall compared to traditional solvents, explaining the high extraction efficiency. In conclusion, the DES-UAE protocol developed in this study represents a promising, sustainable, and efficient approach for flavonoid extraction from JFA and potentially other traditional Chinese medicinal herbs.

Zhou et al. [59] used an innovative DES called DES-14, composed of choline chloride as the hydrogen bond acceptor and diethanolamine as the hydrogen bond donor in a 1:5 molar ratio, to extract hesperidin from *C. aurantium* L. They added 40% water to DES-14 and optimized other parameters such as the solid–liquid ratio, extraction temperature, and extraction time. Under optimal conditions, they achieved a hesperidin extraction yield of 6.26 ± 0.05%. This yield overcomes traditional extraction methods with organic solvents or alkali and acid precipitation. They concluded that DES-14 is an effective solvent for hesperidin extraction due to its high solubility, the formation of hydrogen bonds, and van der Waals forces between DES-14 and hesperidin, and its ability to disrupt the cell structure of *C. aurantium*, enhancing extraction efficiency.

Liu et al. [154] focused on finding a green and efficient method to extract flavonoids from *C. aurantium* fruit using eutectic solvents (DESs) as an alternative to traditional organic solvents. The researchers evaluated 24 different DESs, with choline chloride and betaine as hydrogen bond acceptors (HBAs), combined with various hydrogen bond donors (HBDs). The DES composed of betaine and ethanediol (DES-19) in a molar ratio of 1:4 demonstrated the highest extraction efficiency for the four flavonoids studied: narirutin, naringin, hesperidin, and neohesperidin. Extraction conditions, including the HBA/HBD ratio, water content, extraction temperature, solid/liquid ratio, and extraction time, were optimized to maximize yield. Under the optimal conditions, DES-19 improves the extraction (narirutin: 8.39 ± 0.6, naringin: 83.98 ± 1.92, hesperidin: 3.03 ± 0.35, and neohesperidin: 35.94 ± 0.63 mg of flavonoid per gram of dry mass) conducted with methanol in the extraction of the four flavonoids confirming its potential as a green and efficient solvent for the extraction of bioactive compounds from medicinal plants. Finally, a simple and effective method was implemented to recover flavonoids from DES extract using column chromatography with C18 adsorbent, achieving a recovery higher than 93% for the four compounds. This study highlights the importance of research into sustainable extraction methods and opens new possibilities for obtaining bioactive compounds with lower environmental impact.

All information regarding the extraction of bioactive compounds from *C. aurantium* using DESs can be found in the summary in Table 3.

3.2.5. Natural Deep Eutectic Solvents

NADESs are a subclass of DESs, distinguished by their composition of naturally occurring primary metabolites. These metabolites, which plants utilize for physiological processes and survival, include sugars, organic acids, organic bases, and amino acids. They are considered less harmful with significantly lowered melting points due to specific interactions between their components. Their adaptable physicochemical properties make NADESs widely used in the pharmaceutical and food industries, as they align with many principles of green chemistry [155,156].

Ramírez-Sucre et al. [60] studied NADESs based on choline chloride and glucose, combined with UAE, used to extract polyphenols from citrus peels like bitter orange (*C. aurantium*), sweet orange (*Citrus sinensis*), and lemon (*Citrus limon*). They used various molar ratios of choline chloride/glucose and different water content percentages in NADESs to evaluate their impact on the polyphenol profile, total polyphenol content (TPC), and antioxidant capacity (Ax) of the extracts.

The results showed that *C. aurantium* peel, extracted with a 1:1 molar ratio (ChCl:Glu) and 60% added water, exhibited the highest concentration of hesperidin (2003.37 ± 10.91 mg/100 g dry mass). The highest TPC (96.23 ± 0.83 mg GAE/100 g dry mass) was obtained using a 1:0.5 molar ratio and 60% added water. Furthermore, *C. aurantium* extracts prepared with a 1:2 molar ratio and either 60% or 70% water demonstrated the highest antioxidant capacity, achieving a total inhibition in the DPPH assay. The linear correlation analysis revealed a strong relationship between antioxidant capacity and certain individual polyphenols in the extracts from each citrus variety. The study concluded that NADESs present a promising, environmentally friendly alternative for extracting phenolic compounds from citrus peels. Authors propose that, by optimizing extraction conditions, specific compound-rich extracts can be obtained, showing the potential for *C. aurantium and C. sinensis* in functional foods, nutraceuticals, and pharmaceuticals.

Finally, Du et al. [61] employed a NADES, choline chloride-levulinic acid (Ch-Lea), to extract flavonoids, specifically naringin and neohesperidin, from *Fructus aurantii* (dried *C. aurantium* fruit) using UAE. A theoretical analysis with the COSMO-RS model predicted the high solubility of naringin in Ch-Lea, and the subsequent extraction experiments confirmed the solvent's efficiency. The optimal levels for each extraction factor (molar ratio, water content, solid–liquid ratio, and extraction time) were identified by performing individual extraction experiments while keeping other variables constant. Ch-Lea achieved a flavonoid extraction yield of 122.68 of dry mass mg/g, overcoming traditional solvents like methanol and ethanol (115.01 and 112.81 mg/g of dry mass respectively).

Sustainability evaluation using the Eco-Scale methodology demonstrated that Ch-Lea was more environmentally friendly compared to conventional solvents. The extraction mechanism analysis revealed that hydrogen bonding interactions, especially O-H bonds between Ch-Lea and naringin, were key to the extraction process. In conclusion, Ch-Lea proved to be a green, efficient, and sustainable solvent for extracting flavonoids from *Fructus aurantii*.

Table 3 provides a summarized overview of the main bioactive compounds found in various parts of *C. aurantium*, including the fruit, leaves, and flowers.

Table 3. Main bioactive compounds in *C. aurantium* and their biological activities.

Extraction Methods		Plant Part	Extraction Conditions	Main Classes of Bioactive Compounds Quantification or Identification Studied	Reference
Technology	Solvent				
Maceration	Ketone and water (2:1)	Peel	Solid–liquid ratio of 2.4 g/100 mL, acetone concentration of 80%, and sonication time of 34.7 min	TPC (20.36 ± 3.04 GAE/g dry mass), F=flavonoids (37 ± 2.48 mg of NE/g dry mass), and tannins (20.14 ± 1.70 TAE/g dry mass)	[30]
	Water	Seeds	n-hexane as solvent in a 1:8 (powder/solvent) ratio for 4 h and ultrasound bath at 65 °C for 2 h	Fatty acids: palmitic acid, 35.70%; oleic acid, 31.27%; linoleic acid, 27.65%; stearic acid, 3.08%; and linolenic acid, 0.92%	[157]
	Ethanol 80%	Flower	80% ethanol at 100 °C for 3 h.	Choerospondin	[158]
Soxhlet	Hexane	Peel	Extraction with hexane for 5 h	45 components, classified into 5 chemical groups; 23 terpenes, 7 alkanes, 6 sterols, 4 flavones, and saturated and polyunsaturated fatty acids	[12]
UAE	Ethanol	Flowers	Ultrasonic time 72 min, ethanol concentration 78%, and liquid/solid ratio 30 mL/g	Nine alkaloid compounds: stachydrine, synephrine, cathine, hordenine, cytisine, N-methylcytisine, coniine, (R,S)-anatabine, and caffeine	[20]
	Ethanol and K_2CO_3	Fruit	K_2CO_3 concentration 20.60% (*w*/*w*), ethanol concentration 27% (*w*/*w*), solvent/material ratio 45.17:1 (g:g), extraction temperature 50 °C, extraction time 30 min, and ultrasonic power 80 W	11.17 ± 0.18 mg of synephrine /g dry mass, 7.39 ± 0.13 mg of naringin/g dry mass, and 89.27 ± 1.70 mg of neohesperidin/g dry mass	[51]
	Water	Leaves	Temperature of 68.5 °C, extraction time of 31.8 min, liquid/solid ratio of 22.1 mL/g, and extraction power of 302.6 W	17 compounds; principal compounds identified were linalool (30.46%); linalyl acetate (13.18%); limonene (9.28%); and α-terpineol (9.25%)	[141]
	Chromatographic n-hexane	Pericarp	60 min at an ultrasonic power of 500 W and 50 °C.	Volatile components: 127 alkanes, 91 terpenoids, 68 alcohols, 52 esters, 47 aromatic compounds, 40 alicyclic hydrocarbons, 44 alkenes, 24 ketones, 22 aldehydes, 15 ethers, 11 fatty acids, and 3 other components	[142]
MAE	Water	Peel	35 min at 80% radiation	Limonene as main component (98.68%)	[22]
	Ethanol and ketone	Fruit waste	Sample/solvent ratio of 1/20 *m*/*v*, temperature of 75 °C, time of 14 min, ketone/water ratio 50% (*v*/*v*), power of 500 W	Primary flavonoids as naringin, neohesperidin, hesperidin, and naringenin	[53]
	Water	Leaves	Power of 300 W, 90 min, and a mass/solvent ratio of 0.625 g/mL	Germacrene D (29.51%), lpha-Copaene (18.47%), CIS-Caryophyllene (15.42%), 9-Eicosene (8.58%), and limonene (5.51%)	[146]

Table 3. *Cont.*

Extraction Methods		Plant Part	Extraction Conditions	Main Classes of Bioactive Compounds Quantification or Identification Studied	Reference
Technology	Solvent				
SFE	CO_2 (97%) and ethanol as co-solvent (3%)	Peel	Pressure (130–210 bar), static time (30–70 min), and CO_2 flow rate (2.1–3.3 kg·h^{-1})	Fatty acid esters, terpenes, and coumarins, with isogeijerin as the major component, followed by osthole, squalene, and hexadecane	[54]
	CO_2		Pressure of 10 MPa, temperature of 40 °C, and CO_2 mass flow rate of 1.76 kg/h	Limonene was identified as the predominant compound and percentages of linalool, linalyl acetate, and geranyl aetate, sabinene and isogeijerin	[55]
DES	DES based on choline chloride (ChCl) with different HBDs	Peel	Liquid/solid ratio of 18.97 mL/g, extraction temperature of 40 °C, extraction time of 30 min, and 49.95% water content	TPC of 7.85 mg GAE/g peel	[13]
	Choline chloride as HBA ethylene glycol as HBD	Fruit	27% water content, liquid/solid ratio of 16 mL/g, extraction time of 72 min, and extraction temperature of 62 °C	48.18 mg of naringin per gram of dry mass and 34.50 mg of nehoesperidin per gram of dry mass	[58]
	DES based on Choline chloride (HBA) and diethanolamine (HBD)		40% water content, solid/liquid ratio of 1:12.5 (g/mL), extraction temperature of 75 °C, and extraction time of 40 min	Hesperidin extraction yield of 6.26 ± 0.05%	[59]
	DES with choline chloride and betaine (HBA) combined with various HBDs		Betaine/ethanediol ratio of 1:4, 40% water content, extraction temperature of 60 °C, extraction time of 30 min, and solid-to-liquid ratio of 1:100 g/mL	Narirutin, 8.39 ± 0.6; naringin, 83.98 ± 1.92; hesperidin, 3.03 ± 0.35; and neohesperidin, 35.94 ± 0.63 mg of flavonoid per gram of dry mass	[154]
NADES	NADES based on choline chloride (HBA) and glucose (HBD)	Peel	Ultrasonic bath for 30 min at 42 kHz with three molar ratio levels of 1:0.5, 1:1, and 1:2 mol/mol of Choline Chloride/Glucose, and three water levels of 50%, 60%, and 70%	Total phenolic content of 96.23 ± 0.83 mg GAE/100 g of dry mass, being hesperidin (96.23 ± 0.83 mg GAE/100 g dry mass), which showed the highest concentration	[60]
	NADES based on choline chloride (HBA) and levulinic acid (HBD)	Fruit	Molar ratio (Choline chloride:Levulinic acid) of 1:2, water content of 30%, solid-to-liquid ratio of 1/60 g/g, and extraction time of 15 min	Flavonoid extraction yield of 122.68 mg/g dry mass	[61]

Note: TPC = total polyphenol content; GAE = gallic acid equivalent; NE = naringenin equivalent; TAE = tanin equivalent; HBA = hydrogen bond acceptor; HBD = hydrogen bond donor.

4. Technological Applications of *C. aurantium*

In the pursuit of sustainable solutions to critical global challenges such as food security [159], reliance on synthetic drugs [160], and environmental pollution [161], plant-derived materials have emerged as essential resources across a variety of technological domains. From the food industry to biomedical applications, plant by-products like peels, leaves, and roots not only support circular economy initiatives but also open new avenues for innovation in sectors such as renewable energy and advanced biomaterials [162,163].

The following section highlights the key technological uses of *C. aurantium* extracts, alongside Table 4, which provides examples of commercially available products based on these extracts.

Table 4. Commercial and cosmetics with ingredients derived from *C. aurantium*.

Product	Plant Part	Brand	Type of Product	Beneficial Effect	Reference
Essential oil	Peel	Pranarom	Dietary supplement	Relaxing	[164]
Perfume	-	Lacura	Perfume	-	[165]
Perfume	-	L'Occitane en Provence	Perfume	-	[166]
Tea	Peel	Chás do Mundo	Naturist	Anti-inflammatory properties beneficial for relieving indigestion, constipation, flatulence, and headaches	[167]
Essential oil	Flowers	Cremas caseras	Cosmetic	Soothing effect for irritated, reddened, or inflamed skin	[168]
Capsules	-	Biform	Pharmacological	Fat elimination	[169]
Powder	Peel	Droguería Cosmopolita	Cosmetic	Exfoliant and antiseptic properties	[170]
Powder	Fruit	Naturmed Scientific	Pharmacological	Treats fungal infections, helps with wound healing, weight loss, and diabetes control	[171]
Cream	-	Benditaluz	Cosmetic	Skin hydration	[172]
Jizhi Syrup	Fruit	999	Pharmacological	Anti-inflammatory effect, effective in treating acute bronchitis	[173]
Hesperidin powder	Peel	PURE	Dietary supplement	Improvement of cardiovascular health and anti-inflammatory effects	[174]
Tonic	-	Aesop	Cosmetic	Hydration	[175]

Note: "-" indicates that the information is not available.

4.1. Nanotechnology

Nanotechnology is of crucial importance in the use of *C. aurantium* extracts in various industries. Several studies have shown that the application of nanotechnology increases the efficacy of plant bioactive compounds, reduces side effects, and promotes sustainability [42,176]. The synthesis of nanoparticles from *C. aurantium* peel extract (NPs-CA) and the nanoemulsion of the flower essential oil (CABE-NE), has resulted in particle sizes of 94–373 nm and 76.9 nm, respectively. The zeta potentials indicate good stability, which is a crucial parameter to assess the stability of colloidal dispersions; a high zeta potential, either positive or negative, indicates a strong repulsion between particles, preventing their aggregation and sedimentation while a zeta potential close to zero suggests low repulsion, favoring agglomeration and dispersion instability. NPs-CA showed an enhanced cytotoxic effect against breast cancer cells, while CABE-NE exhibited cytotoxicity against lung cancer cells and induced apoptosis. CABE-NE also presented potential as a natural and safe cell-dependent apoptosis inducer as no notable cell death was detected in the liver, kidney, and jejunum biopsies of CABE-NE-treated mice [45,177]. Furthermore, the environmentally friendly synthesis of copper nanoparticles using *C. aurantium* extract was effective against mosquito eggs and larvae, suggesting its potential as a biopesticide [178]. The biosynthesis of zinc ferrite nanoparticles also offers promising applications in medicine, including the diagnosis and treatment of cancers. In the future, nanotechnology is expected to open new possibilities for the use of *C. aurantium* in the medical, agricultural, and food industries [179]. However, challenges remain to be solved, such as the evaluation of nanoparticle toxicity, the development of cost-effective methods for large-scale production, and the improvement of the long-term stability of nanoemulsions and nanoparticles [45,177]. Other studies have used nanotechnology to produce biodiesel from the essential oils of *C. aurantium*. This oil was subjected to a transesterification process, which converts the triglycerides present in the oil into methyl esters, which are the main components of biodiesel. For this, green zirconium oxide (ZrO_2) nanoparticles were used, synthesized with the aqueous extract of *Alternanthera pungens* leaves, as a catalyst. The results of the research indicate a maximum biodiesel yield of 94% under optimal reaction conditions, including a methanol/oil molar ratio of 6:1, a reaction time of 120 min, a temperature of 87.5 °C, and a catalyst load of 0.5% by weight. The biodiesel produced met international fuel standards, such as Standard Test Method for Determining the Oxidation Stability of Diesel Fuels by Pressure Differential Scanning Calorimetry (PDSC) (ASTM D-6571) [180] and European Standard for Automotive Fuels–Fatty Acid Methyl Ester (FAME) for Diesel Engines–Requirements and Test Methods (EN 14214) [181], confirming its quality and viability as an alternative to petrodiesel production [182].

4.2. Cosmetic Industry

C. aurantium is also a valuable raw material in perfumery due to the aromatic properties of its essential oils. The essential oil from its flowers, called neroli, is used to add a floral, sweet, and complex scent to perfumes. The essential oil from the peel is used for its citrus aroma. *C. aurantium* is nowadays popular in perfumery, because the essential oils can provide fresh, sweet, and citrus notes to fragrances [183,184].

Burnett et al. [185], has conducted extensive evaluations on the safety of *C. aurantium* components for cosmetic applications, providing valuable insights on the potential uses. In one study, 30 ingredients derived from citrus plants and seeds were investigated. Among these, *C. aurantium* oil has emerged as a widely used cosmetic ingredient, particularly in leave-on skin care products, with 295 reported applications (hair care products without coloring, skin care products that do not require rinsing, skin care products that are rinsed off, and diluted skin care products for use in the bath, to name a few). However, the research concluded that more data are needed to confirm the safety of bitter orange oil in cosmetics. In another study, *C. aurantium* var *amara* leaf/twig extract and leaf/twig oil were safe when properly formulated to avoid skin irritation or sensitization. These findings are supported by studies (one with 25 human subjects where occlusive patches of the oil were applied to the forearm or back for 48 h and a maximization test with 25 human subjects) that demonstrated petitgrain bigarade oil (another name for bitter orange leaf/twig oil) is non-irritating and non-sensitizing in humans. Additionally, it was found that the petitgrain bigarade oil has a dermal LD50 greater than 2 g/kg in rabbits and does not cause photosensitivity. The oil is even recognized as safe for use in food, demonstrating its versatility and safety under controlled conditions.

Another study [186] turned the attention to the safety of 14 essential oils derived from citrus peels, including *C. aurantium*. Using a cold-press extraction method, the oils from fresh fruit peels were obtained and their suitability was assessed for cosmetic use. Their findings revealed that these essential oils are safe in specific conditions. These conditions include ensuring non-irritating and non-sensitizing formulations and a strict control of impurities. A key safety condition is that finished products must not exceed 0.0015% (15 ppm) of 5-methoxypsoralen (5-MOP), a phototoxic compound that can trigger photosensitivity in the skin. This recommendation is based on guidelines from the International

Fragrance Association (IFRA) and the Cosmetic Ingredient Review (CIR) Expert Panel; both emphasize the importance of adhering to these limits to prevent phototoxic reactions.

These studies highlight not only the potential of *C. aurantium* as a sustainable and functional ingredient in cosmetics but also the critical importance of formulating products with rigorous attention to safety standards. The findings demonstrate how this citrus species can contribute to the cosmetic industry while ensuring consumer well-being.

4.3. Food Industry

Finally, in the food industry, the effectiveness and potential of essential oils have been highlighted thanks to their natural antibacterial properties. These properties can be used in food preservation, inhibiting the growth of bacteria that cause spoilage. Studies demonstrate the effectiveness of *C. aurantium* essential oils against bacteria such as *Pectobacterium carotovorum* and *Bacillus subtilis*. In addition, bitter orange essential oils have been investigated as a natural disinfectant in vegetable processing [187].

Authors such as He et al. [188] studied the effects of fermentation with lactic acid bacteria (LAB) on the peel and pomace of *C. aurantium* (Huyou), to determine if LAB fermentation could improve the bioactive composition, volatile compounds, and antioxidant activity of these by-products. For this, they mixed freeze-dried peel and pomace with a sucrose solution and sterilized them at 100 °C for 15 min. The mixtures were then inoculated with two LAB strains, *Lactobacillus plantarum* JYLP-375 and *Lactobacillus acidophilus* JYLA-126, and incubated at specific temperatures and times (33.9 °C for 48 h and 36.7 °C for 36 h, respectively). The results showed that LAB significantly improved the bioconversion of organic acids, increasing ascorbic acid content by 27.3 μg/g in the peel and 17.0 μg/g in the pomace. Furthermore, fermentation enhanced the levels of rhoifolin, quercitrin, and quercetin by over 100%; an increase in the volatile compounds was detected, particularly alcohols and terpenes, which improved the aroma. Antioxidant activity was notably enhanced, with fermented peel (FPL) showing the highest scavenging activity in all assays: 91.2% for the hydroxyl radical, 68.5% for the superoxide radical, 75.5% for the DPPH radical, and 57.4% for the ABTS+ radical. The authors suggest that LAB fermentation could be a promising strategy to valorize Huyou waste, reducing waste, and generating new functional food products with high nutritional value and health benefits.

Hashemi et al. [189] investigated the effect of *C. aurantium* (Bahar narang) flower extract in a traditional yogurt stew to determine if the extract could be used as a natural preservative to extend the shelf life of the product. The extract was tested with four different pathogenic bacteria: *Escherichia coli* O157:H7, *Pseudomonas aeruginosa*, *Staphylococcus aureus*, and *Bacillus cereus*. The results showed that Bahar narang extract was effective in inhibiting the growth of all four bacteria, with the greatest effectiveness observed in *S. aureus* at a concentration of 2000 ppm, after 28 days, with a count decreased by 4.78 logs. Similarly, the same concentration achieved an *E. coli* decreased by 4.11 logs, *P. aeruginosa* by 3.71 logs, and *B. cereus* by 3.91 logs. The extract also proved effectiveness in reducing lipid and protein oxidation in the stew. The peroxide value, which indicates the amount of hydroperoxides formed during the initial stages of lipid oxidation, was significantly lower in the samples treated with the extract. Similarly, the anisidine value, which specifically measures 2-alkenals among the secondary oxidation products of lipids, was also lower in the treated samples. The authors concluded that Bahar narang extract could be a promising natural additive for the food industry, not only to extend the shelf life of foods but also to provide health benefits to consumers.

Another study conducted by Kačániová et al. [190] focused on evaluating the antimicrobial activity of *C. aurantium* essential oil (CAEO) in two bacteria, *Stenotrophomonas maltophilia* and *Bacillus subtilis*, and three species of fungi: Penicillium species (*P. crustosum*,

P. expansum, and *P. citrinum*). To analyze antibacterial activity, they employed the disc diffusion method, which measures the diameter of the inhibition zone around a disc impregnated with CAEO, expressed in millimeters (mm). The highest antibacterial activity was observed against *Stenotrophomonas maltophilia*, with an inhibition zone of 17.67 ± 0.58 mm, less than chloramphenicol, which showed 33.33 ± 1.53 mm. Activity was also observed against *Bacillus subtilis* (15.67 ± 1.53 mm), *P. crustosum* (10.67 ± 1.15 mm), *P. expansum* (8.67 ± 0.58 mm), and *P. citrinum* (7.33 ± 1.53 mm). For the fungi, the vapor phase method was used, which evaluates the inhibition of fungal growth by the volatile compounds of CAEO. The authors suggest that CAEO could be applied in the food industry in the production and processing of plant-based and animal-based foods such as meat, fish, and cheese.

5. Conclusions

Citrus fruits are among the most widely harvested and consumed products globally, forming a staple food in the diets of millions due to their appealing flavor and aroma. *C. aurantium*, or bitter orange, offers tremendous potential across various industries such as foods, cosmetics, and pharmaceuticals. Its flowers, fruits, and leaves are a rich source of nutritional components such as vitamins, minerals, and proteins, alongside bioactive compounds like flavonoids, alkaloids, coumarins, essential oils, terpenoids, and tannins. These properties make *C. aurantium* highly valuable for use in cosmetics, pharmaceuticals, and food production. The bioactive profile of *C. aurantium* suggests promising applications in treating a range of health conditions. Its antioxidants, anti-inflammatory, antifungal, and antidiabetic properties, among others, present particular interest in developing natural therapeutic agents. Moreover, these compounds support the innovation of functional foods aimed to promote health and wellness. Modern extraction methods aligned with green chemistry principles, such as UAE, SFE, and DES, enable the efficient extraction of these valuable *C. aurantium* compounds. These methods not only maximize extraction yields but also prioritize consumer safety and environmental sustainability, offering eco-friendly and effective alternatives to traditional extraction processes.

6. Future Prospects

The potential for *C. aurantium* is promising, particularly in the biotechnology, medicine, and food industries. The use of its bioactive compounds, such as flavonoids, alkaloids, and essential oils, is expanding due to the growing demand for natural products with antioxidants, anti-inflammatory, and anticancer properties. Advanced extraction technologies, like NADESs in combination with UAE, are improving efficiency and selectivity in obtaining these compounds, opening new opportunities for their use in dietary supplements and functional treatments due to their combined effectiveness and the low risk they pose to humans. In medicine, research of its pharmacological properties, such as enhancing metabolism, relieving digestive and respiratory issues, and its potential role in preventing chronic diseases, is increasing and progressing. Additionally, the valorization of by-products like sour orange peels through sustainable processes could contribute to a circular economy, where resources from this plant are totally utilized, reducing waste and promoting sustainability. However, further research of its toxicity, biocompatibility, and the optimization of extraction methods is essential to ensure the safety and effectiveness of its applications on a larger scale.

Author Contributions: Conceptualization, I.M.R.-B. and J.F.-C.; methodology, J.F.-C. and I.M.R.-B.; software, J.F.-C.; validation, I.M.R.-B., M.O.R.-S. and J.V.C.-R.; formal analysis, I.M.R.-B. and K.A.A.-B.; investigation, J.F.-C.; resources, J.F.-C. and I.M.R.-B.; data curation, I.M.R.-B., M.O.R.-S. and K.A.A.-B.; writing—original draft preparation, J.F.-C. and K.A.A.-B.; writing—review and editing, J.F.-C.,

M.O.R.-S. and I.M.R.-B.; visualization, I.M.R.-B., M.O.R.-S., K.A.A.-B. and J.V.C.-R.; supervision, I.M.R.-B.; project administration, I.M.R.-B. and M.O.R.-S.; funding acquisition, I.M.R.-B. All authors have read and agreed to the published version of the manuscript.

Funding: Centro de Investigación y Asistencia en Tecnología y Diseño del Estado de Jalisco A.C. and the scholarship 1337626 for Joaquín Fernández-Cabal, financed by CONAHCYT.

Data Availability Statement: Data are contained within the article.

Conflicts of Interest: The authors declare no conflicts of interest.

References

1. Vargas-Canales, J.M.; Guido-López, D.L.; Rodríguez-Haros, B.; Bustamante-Lara, T.I.; Camacho-Vera, J.H.; Orozco-Cirilo, S. Evolución de la especialización y competitividad de la producción de limón en México. *Rev. Mex. de Cienc. Agrícolas* **2020**, *11*, 1043–1056. [CrossRef]
2. Liang, X.; Wang, Y.; Shen, W.; Liao, B.; Liu, X.; Yang, Z.; Cheng, J.; Zhao, C.; Liao, Z.; Cao, J.; et al. Genomic and metabolomic insights into the selection and differentiation of bioactive compounds in citrus. *Mol. Plant* **2024**, *17*, 1753–1772. [CrossRef] [PubMed]
3. Liu, X.; Wang, B.; Tang, S.; Yue, Y.; Xi, W.; Tan, X.; Li, G.; Bai, J.; Huang, L. Modification, biological activity, applications, and future trends of citrus fiber as a functional component: A comprehensive review. *Int. J. Biol. Macromol.* **2024**, *269*, 131798. [CrossRef] [PubMed]
4. Mora, J.J.; Tavares, H.M.; Curbelo, R.; Dellacassa, E.; Cassel, E.; Apel, M.A.; von Poser, G.L.; Vargas, R.M. Supercritical fluid extraction of coumarins and flavonoids from citrus peel. *J. Supercrit. Fluids* **2024**, *215*, 106396. [CrossRef]
5. Maqbool, Z.; Khalid, W.; Atiq, H.T.; Koraqi, H.; Javaid, Z.; Alhag, S.K.; Al-Shuraym, L.A.; Bader, D.M.D.; Almarzuq, M.; Afifi, M.; et al. Citrus Waste as Source of Bioactive Compounds: Extraction and Utilization in Health and Food Industry. *Molecules* **2023**, *28*, 1636. [CrossRef] [PubMed]
6. TROPICSAFE. Sector de Los Cítricos: Análisis de Mercado Y Aspectos Socioeconómicos. Marco General Del Sector de Los Cítricos en Cuba, Guadalupe Y España. 2022. Available online: https://www.tropicsafe.eu/wp-content/uploads/2022/02/Sector-de-los-c%C3%ADtricos-an%C3%A1lisis-de-mercado.pdf (accessed on 26 December 2024).
7. Gobierno de México. Planeación Agrícola Nacional (2016–2030). Available online: https://www.gob.mx/cms/uploads/attachment/file/257073/Potencial-C_tricos-parte_uno.pdf (accessed on 26 December 2024).
8. Rownaghi, M.; Niakousari, M. Sour orange (*Citrus aurantium*) seed, a rich source of protein isolate and hydrolysate—A thorough investigation. *Heliyon* **2024**, *10*, e32503. [CrossRef] [PubMed]
9. Maksoud, S.; Abdel-Massih, R.M.; Rajha, H.N.; Louka, N.; Chemat, F.; Barba, F.J.; Debs, E. *Citrus aurantium* L. Active Constituents, Biological Effects and Extraction Methods. An Updated Review. *Molecules* **2021**, *26*, 5832. [CrossRef] [PubMed]
10. Xie, D.; Zhou, Y.; Wang, Y.; Xu, L.; Yang, C.; Zhang, G.; Zhang, H.; Du, Q.; Jin, P. Insight into sensory characteristics and volatile differences of freeze-dried *Citrus aurantium* L. var. *amara* Engl. flower essential oil-casein phospholipid nanocomposites sustained-release scented tea by combining GC-E-nose and SBSE-GC-MS. *Food Control* **2024**, *166*, 110714.
11. de Jesus Oliveira, J.; Dos Passos, E.M.; Alves, S.M.; Sarmento, V.H.; Bjerk, T.R.; Cardoso, J.C.; Blanco-Llamero, C.; Soouto, E.B.; Severino, P.; da Costa Mendonça, M. Microemulsion of essential oil of *Citrus aurantium* var. dulcis for control of Aleurocanthus woglumi and evaluation of selectivity against *Aschersonia aleyrodis* and *Ceraeochrysa cornuta*. *Crop Prot.* **2024**, *178*, 106586.
12. Chakroun, I.; Bouraoui, Z.; Ayachi, T.; Hosni, K.; Guerbèj, H.; Snoussi, M.; Jebali, J.; Gharred, T. Phytochemical constituents and potential applications of Thomson navel orange (*Citrus*× *aurantium* var. *sinensis* L.) peel extracts: Antioxidant, antimicrobial and antiproliferative properties. *Ind. Crops Prod.* **2023**, *206*, 117597.
13. Edrisi, S.; Bakhshi, H. Separation of polyphenolic compounds from *Citrus aurantium* L. peel by deep eutectic solvents and their recovery using a new DES-based aqueous two-phase system. *J. Mol. Liq.* **2024**, *402*, 124790. [CrossRef]
14. Aboualsoltani, F.; Bastani, P.; Khodaie, L.; Fazljou, S.M.B. Therapeutic Effects of *Citrus aurantium* Components on Psychological States: A Systematic Review. *Crescent J. Med. Biol. Sci.* **2020**, *7*, 443–450.
15. Lin, S.X.; Yang, C.; Jiang, R.S.; Wu, C.; Lang, D.Q.; Wang, Y.L.; Li, X.Y.; Jiang, C.P.; Liu, Q.; Shen, C.Y. Flavonoid extracts of *Citrus aurantium* L. var. *amara* Engl. Promote browning of white adipose tissue in high-fat diet-induced mice. *J. Ethnopharmacol.* **2024**, *324*, 117749. [PubMed]
16. Zhou, P.; Li, X.; Zhou, J.; Yang, H.; Shen, L. A green approach for multivariate consecutive extraction of essential oils and flavonoids from *Citrus aurantium* L. var. *amara* Engl.: Process optimization and mechanistic insights based on machine learning methods. *Ind. Crops Prod.* **2023**, *206*, 117611.

17. Estrada-Sierra, N.A.; Rincon-Enriquez, G.; Urías-Silvas, J.E.; Bravo, S.D.; Villanueva-Rodríguez, S.J. Impact of ripening, harvest season, and the nature of solvents on antioxidant capacity, flavonoid, and p-synephrine concentrations in *Citrus aurantium* extracts from residue. *Future Foods* **2022**, *6*, 100153. [CrossRef]
18. Miao, W.; Liu, X.; Li, N.; Bian, X.; Zhao, Y.; He, J.; Zhou, T.; Wu, J.L. Polarity-extended composition profiling via LC-MS-based metabolomics approaches: A key to functional investigation of *Citrus aurantium* L. *Food Chem.* **2023**, *405*, 134988. [CrossRef]
19. Zhou, X.; Wang, J.; Wang, H.; Lu, P.; Huang, Q.; Huang, M.; Lv, R.; Liu, D.; Wang, W. Identification of extracted antioxidants from *Citrus aurantium* 'Changshanhuyou' residue against digestive enzyme activities and airway smooth muscle cells proliferation using UPLC-MS/MS. *Food Biosci.* **2024**, *61*, 104910. [CrossRef]
20. Cai, W.F.; Yan, M.M.; Wang, Z.; Jiang, M.P.; Yan, B.; Shen, C.Y. Optimization of the extract from flower of *Citrus aurantium* L. var. *amara* Engl. and its inhibition of lipid accumulation. *J. Food Biochem.* **2022**, *46*, e14332.
21. Elhawary, E.A.; Nilofar, N.; Zengin, G.; Eldahshan, O.A. Variation of the essential oil components of *Citrus aurantium* leaves upon using different distillation techniques and evaluation of their antioxidant, antidiabetic, and neuroprotective effect against Alzheimer's disease. *BMC Complement Med. Ther.* **2024**, *24*, 73. [CrossRef]
22. Ashmawy, N.S.; Nilofar, N.; Zengin, G.; Eldahshan, O.A. Metabolic profiling and enzyme inhibitory activity of the essential oil of *Citrus aurantium* fruit peel. *BMC Complement Med. Ther.* **2024**, *24*, 262. [CrossRef] [PubMed]
23. B'chir, F.; Maurice, J.A. Chemical profile and extraction yield of essential oils from peel of *Citrus limon*, *Citrus aurantium*, and *Citrus limetta*: A review. *Stud. Nat. Prod. Chem.* **2023**, *79*, 135–204.
24. Wen, L.; He, M.; Yin, C.; Jiang, Y.; Luo, D.; Yang, B. Phenolics in *Citrus aurantium* fruit identified by UHPLC-MS/MS and their bioactivities. *LWT* **2021**, *147*, 111671. [CrossRef]
25. Rahaman, S.T.; Praveen, V. An Overview on Chemical Constituents and Pharmacological Effects of *Citrus aurantium* Species. *Int. J. Pharm. Res.* **2018**, *10*, 15.
26. Suntar, I.; Khan, H.; Patel, S.; Celano, R.; Rastrelli, L. An overview on *Citrus aurantium* L.: Its functions as food ingredient and therapeutic agent. *Oxidative Med. Cell. Longev.* **2018**, *2018*, 7864269. [CrossRef] [PubMed]
27. Gao, L.; Zhang, H.; Yuan, C.H.; Zeng, L.H.; Xiang, Z.; Song, J.F.; Wang, H.G.; Jiang, J.P. *Citrus aurantium* 'Changshan-huyou'—An ethnopharmacological and phytochemical review. *Front. Pharmacol.* **2022**, *13*, 983470. [CrossRef] [PubMed]
28. Ishaq, A.N.; Sani, D.; Abdullhi, S.A.; Jatau, I.D. In vitro anticoccidial activity of ethanolic leaf extract of *Citrus aurantium* L. against *Eimeria tenella* oocysts. *Sokoto J. Vet. Sci.* **2022**, *20*, 37–43. [CrossRef]
29. Gunwantrao, B.B.; Bhausaheb, S.K.; Ramrao, B.S.; Subhash, K.S. Antimicrobial activity and phytochemical analysis of orange (*Citrus aurantium* L.) and pineapple (*Ananas comosus* (L.) Merr.) peel extract. *Ann. Phytomed.* **2016**, *5*, 156–160. [CrossRef]
30. Abdallah, H.B.; Abbassi, A.; Trabelsi, A.; Krichen, Y.; Chekir-Ghedira, L.; Ghedira, K. Optimization of ultrasound-assisted extraction of polyphenols and flavonoids from *Citrus aurantium* L. var. *amara* Engl. Fruit peel using response surface methodology. *Biomass Convers. Biorefinery* **2024**, *14*, 14139–14151. [CrossRef]
31. Mansour, R. Determination of nutritional composition in citrus fruits (*C. aurantium*) during maturity. *Nutr. Food Sci.* **2019**, *49*, 299–317. [CrossRef]
32. Badalamenti, N.; Bruno, M.; Schicchi, R.; Geraci, A.; Leporini, M.; Gervasi, L.; Tundis, R.; Loizzo, M.R. Chemical compositions and antioxidant activities of essential oils, and their combinations, obtained from flavedo by-product of seven cultivars of Sicilian *Citrus aurantium* L. *Molecules* **2022**, *27*, 1580. [CrossRef] [PubMed]
33. Pasandideh, S.; Amir, A. Evaluation of antioxidant and inhibitory properties of *Citrus aurantium* L. on the acetylcholinesterase activity and the production of amyloid nano–bio fibrils. *Int. J. Biol. Macromol.* **2021**, *182*, 366–372. [CrossRef]
34. Değirmenci, H.; Hatice, E. Relationship between volatile components, antimicrobial and antioxidant properties of the essential oil, hydrosol and extracts of *Citrus aurantium* L. flowers. *J. Infect. Public Health* **2020**, *13*, 58–67. [CrossRef] [PubMed]
35. Shen, C.Y.; Wang, T.X.; Jiang, J.G.; Huang, C.L.; Zhu, W. Bergaptol from blossoms of *Citrus aurantium* L. var. *amara* Engl inhibits LPS-induced inflammatory responses and ox-LDL-induced lipid deposition. *Food Funct.* **2020**, *11*, 4915–4926. [PubMed]
36. Lakache, Z.; Ounissi, M.; Aliboudhar, H.; Hacib, H.; Tounsi, H.; Kameli, A. Comparative study of the chemical composition and the antioxidant, antidiabetic, anti-inflammatory and cytotoxic activities of *Citrus aurantium* essential oils extracted by conventional hydrodistillation and microwave-assisted hydrodistillation methods from Algeria. *J. Essent. Oil Bear. Plants* **2023**, *26*, 1067–1087.
37. Benayad, O.; Bouhrim, M.; Tiji, S.; Kharchoufa, L.; Addi, M.; Drouet, S.; Hano, C.; Lorenzo, J.M.; Bendaha, H.; Bnouham, M.; et al. Phytochemical profile, α-glucosidase, and α-amylase inhibition potential and toxicity evaluation of extracts from *Citrus aurantium* (L) peel, a valuable by-product from Northeastern Morocco. *Biomolecules* **2021**, *11*, 1555. [CrossRef] [PubMed]
38. Ren, W.; Wang, S.; Zhang, J.; Liu, D. Ethnopharmacology, chemical composition and functions of *Citrus aurantium* L. *J. Food Meas. Charact.* **2024**, *18*, 8843–8864. [CrossRef]
39. de Oliveira, D.F.; Martins, J.A.; Oliveira, C.R. Pharmacological aspects of *Citrus aurantium* (RUTACEAE) in anxiety disorders. *Braz. J. Nat. Sci.* **2022**, *4*, E1532022-8. [CrossRef]
40. Rosa-Falero, C.; Peña-Jiménez, J.A.; Joubert-Miranda, D. Anticonvulsant interactions and anxiolytic effects of *Citrus aurantium* leaf infusion in the zebrafish. *Pharm. Pharmacol. Int. J.* **2023**, *11*, 87–91.

41. Neto, G.C.; de Andrade, H.H.N.; Braga, J.E.F.; Júnior, L.J.Q.; de Almeida, R.N.; Diniz, M.D.F.F.M. Non-clinical study of the anxiolytic effect of *Citrus aurantium* L. essential oil in an inclusion complex with 2-hydroxypropyl-β-cyclodextrin. *Rev. de Ciências Farm. Básica e Apl.* **2020**, *41*, 1–12.
42. Nagarajan, P.; Subramaniyan, V.; Elavarasan, V.; Mohandoss, N.; Subramaniyan, P.; Vijayakumar, S. Biofabricated aluminium oxide nanoparticles derived from *Citrus aurantium* L.: Antimicrobial, anti-proliferation, and photocatalytic efficiencies. *Sustainability* **2023**, *15*, 1743. [CrossRef]
43. Sevindik, E.; Aydın, S.; Sujka, M.; Apaydın, E.; Yıldırım, K.; Palas, G. GC-MS analysis and evaluation of antibacterial and antifungal activity of essential oils extracted from fruit Peel of *Citrus aurantium* L. (Rutaceae) Grown in the West Anatolian Area. *Erwerbs-Obstbau* **2021**, *63*, 135–142. [CrossRef]
44. Ellouze, I.; Akacha, B.B.; Mekinić, I.G.; Saad, R.B.; Kačániová, M.; Kluz, M.I.; Mnif, W.; Garzoli, S.; Hsouna, A.B. Enhancing Antibacterial Efficacy: Synergistic Effects of *Citrus aurantium* Essential Oil Mixtures against *Escherichia coli* for Food Preservation. *Foods* **2024**, *13*, 3093. [CrossRef] [PubMed]
45. Amalina, N.D.; Wahyuni, S. Cytotoxic effects of the synthesized *Citrus aurantium* peels extract nanoparticles against MDA-MB-231 breast cancer cells. *J. Phys. Conf. Ser.* **2021**, *1918*, 032006. [CrossRef]
46. Amala Dev, A.R.; Joseph, S.M. Citrus essential oils: A rational view on its chemical profiles, mode of action of anticancer effects/antiproliferative activity on various human cancer cell lines. *Cell Biochem. Biophys.* **2023**, *81*, 189–203. [CrossRef] [PubMed]
47. Kallel, I.; Tarhouni, N.; Elaguel, A.; Gargouri, B.; Hadrich, B.; Allouche, N.; Bayoudh, A. Exploration of the antiproliferative and antioxidant effects and the molecular docking study EGFR and VEGFR2 of essential oil from *Citrus aurantium* Peels. *J. Essent. Oil Bear. Plants* **2023**, *26*, 1130–1150. [CrossRef]
48. Uahomo, P.O.; Kpaduwa, S.; Daniel, C.; Ezerioha, C.E. Characterization and Comparative Assessment of the Essential Oil from Lime (*Citrus aurantifolia*) Exocarp Using Maceration and Soxhlet Extraction Methods. *Sch. Int. J. Chem. Mater. Sci.* **2023**, *6*, 126–134. [CrossRef]
49. Hamedi, A.; Zarshenas, M.M.; Jamshidzadeh, A.; Ahmadi, S.; Heidari, R.; Pasdaran, A. *Citrus aurantium* (bitter orange) seeds oil: Pharmacognostic, anti-inflammatory and anti-nociceptive properties. *Trends Pharm. Sci.* **2019**, *5*, 153–164.
50. Elachouri, M.; Chaachouay, N.; Zidane, L.; Ouasti, I.; Bussmann, R.W. *Citrus × aurantium* L. *Citrus × sinensis* (L.) OsbeckRutaceae. In *Ethnobotany of Northern Africa and Levant*; Springer International Publishing: Berlin/Heidelberg, Germany, 2023; pp. 1–13.
51. Yan, Y.; Zhou, H.; Wu, C.; Feng, X.; Han, C.; Chen, H.; Liu, Y.; Li, Y. Ultrasound-assisted aqueous two-phase extraction of synephrine, naringin, and neohesperidin from *Citrus aurantium* L. fruitlets. *Prep. Biochem. Biotechnol.* **2021**, *51*, 780–791. [CrossRef] [PubMed]
52. Rajabi, M.; Amiri, S.; Rezazadeh-Bari, M. Optimization of hesperidin extraction using hot methanol method assisted with ultrasound waves from the peel wastes of bitter orange (*Citrus aurantium*) and Persian orange (*Citrus reticulata*). *J. Food Meas. Charact.* **2023**, *17*, 5582–5593. [CrossRef]
53. García-Martín, J.F.; Feng, C.H.; Domínguez-Fernández, N.M.; Álvarez-Mateos, P. Microwave-Assisted Extraction of Polyphenols from Bitter Orange Industrial Waste and Identification of the Main Compounds. *Life* **2023**, *13*, 1864. [CrossRef]
54. Trabelsi, D.; Aydi, A.; Zibetti, A.W.; Della Porta, G.; Scognamiglio, M.; Cricchio, V.; Langa, E.; Abderrabba, M.; Mainar, A.M. Supercritical extraction from *Citrus aurantium amara* peels using CO_2 with ethanol as co-solvent. *J. Supercrit. Fluids* **2016**, *117*, 33–39. [CrossRef]
55. Jerković, I.; Družić, J.; Marijanović, Z.; Gugić, M.; Jokić, S.; Roje, M. GC-FID/MS Profiling of Supercritical CO_2 Extracts of Peels from *Citrus aurantium*, *C. sinensis* cv. Washington navel, *C. sinensis* cv. Tarocco and *C. sinensis* cv. Doppio Sanguigno from Dubrovnik Area (Croatia). *Nat. Prod. Commun.* **2015**, *10*, 1934578X1501000745. [CrossRef]
56. Flamerz, R.A.; Obid, S.S.; Jasim, W.M. Antibacterial activity of silver nanoparticles synthesized from *Citrus aurantium* and its synergistic effect of the combination of silver nanoparticles with ampicillin against Proteus mirabilis. *Med. J. Babylon* **2023**, *20* (Suppl. S1), S130–S135. [CrossRef]
57. Erdoğan, Ö.; Paşa, S.; Demirbolat, G.M.; Birtekocak, F.; Abbak, M.; Çevik, Ö. Synthesis, characterization, and anticarcinogenic potent of green-synthesized zinc oxide nanoparticles via *Citrus aurantium* aqueous peel extract. *Inorg. Nano-Met. Chem.* **2023**, 1–9. [CrossRef]
58. He, Q.; Tang, G.; Hu, Y.; Liu, H.; Tang, H.; Zhou, Y.; Deng, X.; Peng, D.; Qian, Y.; Guo, W.; et al. Green and highly effective extraction of bioactive flavonoids from *Fructus aurantii* employing deep eutectic solvents-based ultrasonic-assisted extraction protocol. *Ultrason. Sonochemistry* **2024**, *102*, 106761. [CrossRef] [PubMed]
59. Zhou, J.; Tang, J. Extraction Mechanism of Hesperidin from *Citrus aurantium* L. Using a Novel Deep Eutectic Solvent: Experimental and Theoretical Investigation. *J. Braz. Chem. Soc.* **2024**, *36*, e-20240175. [CrossRef]
60. Ramírez-Sucre, M.O.; Avilés-Betanzos, K.A.; López-Martínez, A.; Rodríguez-Buenfil, I.M. Evaluation of Polyphenol Profile from Citrus Peel Obtained by Natural Deep Eutectic Solvent/Ultrasound Extraction. *Processes* **2024**, *12*, 2072. [CrossRef]

61. Du, H.; Wu, W.; Wang, Y.; Tang, P.; Wu, Y.; Qiu, J.; Xu, C.; Li, L.; Yang, M. Natural deep eutectic solvent-ultrasound for the extraction of flavonoids from *Fructus aurantii*: Theoretical screening, experimental and mechanism. *Arab. J. Chem.* **2024**, *17*, 105886. [CrossRef]
62. Koina, I.M.; Sarigiannis, Y.; Hapeshi, E. Green extraction techniques for the determination of active ingredients in tea: Current state, challenges, and future perspectives. *Separations* **2023**, *10*, 121. [CrossRef]
63. Yeasmen, N.; Orsat, V. Green extraction and characterization of leaves phenolic compounds: A comprehensive review. *Crit. Rev. Food Sci. Nutr.* **2023**, *63*, 5155–5193. [CrossRef] [PubMed]
64. Tapia-Quirós, P.; Granados, M.; Sentellas, S.; Saurina, J. Microwave-assisted extraction with natural deep eutectic solvents for polyphenol recovery from agrifood waste: Mature for scaling-up? *Sci. Total Environ.* **2024**, *912*, 168716. [CrossRef] [PubMed]
65. Panahi, Z.; Khoshbakht, R.; Javadi, B.; Firoozi, E.; Shahbazi, N. The effect of sodium alginate coating containing citrus (*Citrus aurantium*) and Lemon (*Citrus lemon*) extracts on quality properties of chicken meat. *J. Food Qual.* **2022**, *2022*, 6036113. [CrossRef]
66. Aboualsoltani, F.; Bastani, P.; Khodaie, L.; Fazljou, S.M.B.; Salehi-Pourmehr, H. The effect of *Citrus aurantium* L. flower extract on the severity of primary dysmenorrhoea: A double-blind, randomised, controlled clinical trial. *J. Herb. Med.* **2024**, *45*, 100878. [CrossRef]
67. Burnett, C.L.; Bergfeld, W.F.; Belsito, D.V.; Hill, R.A.; Klaassen, C.D.; Liebler, D.C.; Marks, J.G.; Shank, R.C.; Slaga, T.J.; Synder, P.W.; et al. Safety assessment of citrus peel-derived ingredients as used in cosmetics. *Int. J. Toxicol.* **2021**, *40* (Suppl. S3), 77S–99S. [CrossRef]
68. Twaij, B.M.; Hasan, M.N. Bioactive secondary metabolites from plant sources: Types, synthesis, and their therapeutic uses. *Int. J. Plant Biol.* **2022**, *13*, 4–14. [CrossRef]
69. Divekar, P.A.; Narayana, S.; Divekar, B.A.; Kumar, R.; Gadratagi, B.G.; Ray, A.; Singh, A.K.; Rani, V.; Singh, V.; Singh, A.K.; et al. Plant secondary metabolites as defense tools against herbivores for sustainable crop protection. *Int. J. Mol. Sci.* **2022**, *23*, 2690. [CrossRef] [PubMed]
70. Pang, Z.; Chen, J.; Wang, T.; Gao, C.; Li, Z.; Guo, L.; Xu, J.; Cheng, Y. Linking plant secondary metabolites and plant microbiomes: A review. *Front. Plant Sci.* **2021**, *12*, 621276. [CrossRef] [PubMed]
71. Bhatla, S.C.; Lal, M.A. Secondary Metabolites. In *Plant Physiology, Development and Metabolism*; Springer Nature Singapore: Singapore, 2023; pp. 765–808.
72. Lin, M.; Xu, C.; Gao, X.; Zhang, W.; Yao, Z.; Wang, T.; Feng, X.; Wang, Y. Comparative study on secondary metabolites from different citrus varieties in the production area of Zhejiang. *Front. Nutr.* **2023**, *10*, 1159676. [CrossRef] [PubMed]
73. Reshi, Z.A.; Ahmad, W.; Lukatkin, A.S.; Javed, S.B. From Nature to lab: A review of secondary metabolite biosynthetic pathways, environmental influences, and in vitro approaches. *Metabolites* **2023**, *13*, 895. [CrossRef] [PubMed]
74. Corona-España, A.M.; Garcia-Ramirez, M.A.; Romo-Gonzalez, R.; Rodriguez-Buenfil, I.M.; Reynoso, O.G. Phytochemicals from Secondary Metabolism and Their Role as Antioxidant and Anti-Inflammatory Molecules. In *Recent Advances in Phytochemical Research*; Intech Open: Rijeka, Croatia, 2024.
75. Santos-Sánchez, N.F.; Salas-Coronado, R.; Hernández-Carlos, B.; Villanueva-Cañongo, C. Shikimic acid pathway in biosynthesis of phenolic compounds. In *Plant Physiological Aspects of Phenolic Compounds*; Intech Open: Rijeka, Croatia, 2019; Volume 1, pp. 1–15.
76. Letchuman, S.; Madhuranga, H.D.; Kaushalya MB, L.N.; Premarathna, A.D.; Saravanan, M. Alkaloids Unveiled: A Comprehensive Analysis of Novel Therapeutic Properties, Mechanisms, and Plant-Based Innovations. *Intell. Pharm.* 2024, *in press*. [CrossRef]
77. Faisal, S.; Badshah, S.L.; Kubra, B.; Emwas, A.H.; Jaremko, M. Alkaloids as potential antivirals. A comprehensive review. *Nat. Prod. Bioprospect.* **2023**, *13*, 4. [CrossRef] [PubMed]
78. Akinboye, A.J.; Kim, K.; Choi, S.; Yang, I.; Lee, J.G. Alkaloids in food: A review of toxicity, analytical methods, occurrence and risk assessments. *Food Sci. Biotechnol.* **2023**, *32*, 1133–1158. [CrossRef]
79. Rao, M.J.; Wu, S.; Duan, M.; Wang, L. Antioxidant metabolites in primitive, wild, and cultivated citrus and their role in stress tolerance. *Molecules* **2021**, *26*, 5801. [CrossRef] [PubMed]
80. Liao, L.; Tang, Y.; Li, B.; Tang, J.; Xu, H.; Zhao, K.; Zhang, X. Stachydrine, a potential drug for the treatment of cardiovascular system and central nervous system diseases. *Biomed. Pharmacother.* **2023**, *161*, 114489. [CrossRef] [PubMed]
81. Ruiz-Moreno, C.; Del Coso, J.; Giráldez-Costas, V.; González-García, J.; Gutiérrez-Hellín, J. Effects of p-synephrine during exercise: A brief narrative review. *Nutrients* **2021**, *13*, 233. [CrossRef] [PubMed]
82. Gutiérrez-Hellín, J.; Baltazar-Martins, G.; Rodríguez, I.; Lara, B.; Ruiz-Moreno, C.; Aguilar-Navarro, M.; Del Coso, J. *p*-Synephrine, the main protoalkaloid of *Citrus aurantium*, raises fat oxidation during exercise in elite cyclists. *Eur. J. Sport Sci.* **2020**, *21*, 1273–1282. [CrossRef]
83. Aluko, B.T.; Smith, Y.R.A.; Ogundare, M.A.B. Anti-Inflammatory and Antioxidant Activities of Alkaloid Extract of *Citrus aurantium* Leaves. *Int. J. Novel Res. Life Sci.* **2023**, *10*, 11–18.
84. Liu, Y.; Luo, J.; Peng, L.; Zhang, Q.; Rong, X.; Luo, Y.; Li, J. Flavonoids: Potential therapeutic agents for cardiovascular disease. *Heliyon* **2024**, *10*, e32563. [CrossRef] [PubMed]

85. Vicente, O.; Boscaiu, M. Flavonoids: Antioxidant Compounds for Plant Defence and for a Healthy Human Diet. *Not. Bot. Horti Agrobot. Cluj-Napoca* **2017**, *46*, 14–21. [CrossRef]
86. González-Sarrías, A.; Tomás-Barberán, F.A.; García-Villalba, R. Structural diversity of polyphenols and distribution in foods. In *Dietary Polyphenols: Their Metabolism and Health Effects*; Wiley: Hoboken, NJ, USA, 2020; pp. 1–29.
87. Hosseinzade, A.; Sadeghi, O.; Naghdipour Biregani, A.; Soukhtehzari, S.; Brandt, G.S.; Esmaillzadeh, A. Immunomodulatory effects of flavonoids: Possible induction of T CD_4^+ regulatory cells through suppression of mTOR pathway signaling activity. *Front. Immunol.* **2019**, *10*, 51. [CrossRef] [PubMed]
88. Sok Yen, F.; Shu Qin, C.; Tan Shi Xuan, S.; Jia Ying, P.; Yi Le, H.; Darmarajan, T.; Gunasekaran, B.; Salvamani, S. Hypoglycemic effects of plant flavonoids: A review. *Evid.-Based Complement Altern. Med.* **2021**, *2021*, 2057333. [CrossRef]
89. Shamsudin, N.F.; Ahmed, Q.U.; Mahmood, S.; Ali Shah, S.A.; Khatib, A.; Mukhtar, S.; Alsharif, M.A.; Parveen, H.; Zakaria, Z.A. Antibacterial effects of flavonoids and their structure-activity relationship study: A comparative interpretation. *Molecules* **2022**, *27*, 1149. [CrossRef] [PubMed]
90. de Luna, F.C.F.; Ferreira, W.A.S.; Casseb, S.M.M.; de Oliveira, E.H.C. Anticancer potential of flavonoids: An overview with an emphasis on tangeretin. *Pharmaceuticals* **2023**, *16*, 1229. [CrossRef] [PubMed]
91. Zhang, Q.; Yan, Y. The role of natural flavonoids on neuroinflammation as a therapeutic target for Alzheimer's disease: A narrative review. *Neural Regen. Res.* **2023**, *18*, 2582–2591. [CrossRef] [PubMed]
92. Lu, K.; Yip, Y.M. Therapeutic potential of bioactive flavonoids from citrus fruit peels toward obesity and diabetes mellitus. *Future Pharmacol.* **2023**, *3*, 14–37. [CrossRef]
93. Liga, S.; Paul, C.; Péter, F. Flavonoids: Overview of biosynthesis, biological activity, and current extraction techniques. *Plants* **2023**, *12*, 2732. [CrossRef]
94. Addi, M.; Elbouzidi, A.; Abid, M.; Tungmunnithum, D.; Elamrani, A.; Hano, C. An overview of bioactive flavonoids from citrus fruits. *Appl. Sci.* **2021**, *12*, 29. [CrossRef]
95. Maurya, A.; Prasad, J.; Das, S.; Dwivedy, A.K. Essential oils and their application in food safety. *Front. Sustain. Food Syst.* **2021**, *5*, 653420. [CrossRef]
96. Sharma, S.; Barkauskaite, S.; Jaiswal, A.K.; Jaiswal, S. Essential oils as additives in active food packaging. *Food Chem.* **2021**, *343*, 128403. [CrossRef] [PubMed]
97. Bolouri, P.; Salami, R.; Kouhi, S.; Kordi, M.; Asgari Lajayer, B.; Hadian, J.; Astatkie, T. Applications of essential oils and plant extracts in different industries. *Molecules* **2022**, *27*, 8999. [CrossRef] [PubMed]
98. Mohamed, A.A.; Alotaibi, B.M. Essential oils of some medicinal plants and their biological activities: A mini review. *J. Umm Al-Qura Univ. Appl. Sci.* **2023**, *9*, 40–49. [CrossRef]
99. Brah, A.S.; Armah, F.A.; Obuah, C.; Akwetey, S.A.; Adokoh, C.K. Toxicity and therapeutic applications of citrus essential oils (CEOs): A review. *Int. J. Food Prop.* **2023**, *26*, 301–326. [CrossRef]
100. Oulebsir, C.; Mefti-Korteby, H.; Djazouli, Z.E.; Zebib, B.; Merah, O. Essential oil of *Citrus aurantium* L. leaves: Composition, antioxidant activity, elastase and collagenase inhibition. *Agronomy* **2022**, *12*, 1466. [CrossRef]
101. Flores-Morales, V.; Villasana-Ruíz, A.P.; Garza-Veloz, I.; González-Delgado, S.; Martinez-Fierro, M.L. Therapeutic effects of coumarins with different substitution patterns. *Molecules* **2023**, *28*, 2413. [CrossRef] [PubMed]
102. Todorov, L.; Saso, L.; Kostova, I. Antioxidant activity of coumarins and their metal complexes. *Pharmaceuticals* **2023**, *16*, 651. [CrossRef] [PubMed]
103. Heghes, S.C.; Vostinaru, O.; Mogosan, C.; Miere, D.; Iuga, C.A.; Filip, L. Safety profile of nutraceuticals rich in coumarins: An update. *Front. Pharmacol.* **2022**, *13*, 803338. [CrossRef] [PubMed]
104. Razuvaeva, Y.G.; Toropova, A.A.; Salchak, S.M.; Olennikov, D.N. Coumarins of Ferulopsis hystrix: LC–MS profiling and gastroprotective and antioxidant activities of skimmin and peucenidin. *Appl. Sci.* **2023**, *13*, 9653. [CrossRef]
105. Jin, A.; Wang, Y.; Tong, L.; Liu, G.; Feng, J.; Li, Y.; Shen, C.; Wu, W. Coumarins and flavones from *Ficus erecta* and their anti-inflammatory activity. *J. Ethnopharmacol.* **2024**, *333*, 118472. [CrossRef] [PubMed]
106. Martin, A.L.A.; De Menezes, I.R.; Sousa, A.K.; Farias, P.A.; Dos Santos, F.A.; Freitas, T.S.; Figueredo, F.G.; Ribeiro-Filho, J.; Carvalho, D.T.; Coutinho, H.D.M.; et al. In vitro and in silico antibacterial evaluation of coumarin derivatives against MDR strains of *Staphylococcus aureus* and *Escherichia coli*. *Microb. Pathog.* **2023**, *177*, 106058. [CrossRef] [PubMed]
107. Zhang, Z.; Geng, D.; Yang, Z.; Pan, L.; Jin, L. Synthesis and Antifungal Activity of Coumarin Derivatives Containing Hydrazone Moiety. *Chem. Biodivers.* **2024**, *21*, e202400583. [CrossRef] [PubMed]
108. Orioli, R.; Belluti, F.; Gobbi, S.; Rampa, A.; Bisi, A. Naturally Inspired Coumarin Derivatives in Alzheimer's Disease Drug Discovery: Latest Advances and Current Challenges. *Molecules* **2024**, *29*, 3514. [CrossRef] [PubMed]
109. Husain, A.S.; Gawhale, N.D.; Khan, S.L.; Murlidhar, D.V.; Siddiqui, F.A.; Kazi, A.A. An Overview of Natural and Synthetic Coumarin Derivatives as Potential Antidiabetic Agents. *J. Pharm. Negat. Results* **2022**, *13*, 739–744. [CrossRef]

110. Ouabane, M.; Alaqarbeh, M.; Hajji, H.; Tabti, K.; Ajana, M.A.; Sbai, A.; Sekkate, C.; Lakhlifi, T.; Bouachrine, M. Quality Control of Coumarins, Furocoumarins and Polymethoxyflavones in Citrus Essential Oils: In Silico Analysis. *ChemistrySelect* **2024**, *9*, e202303037. [CrossRef]
111. Li, S.; Kelly, C.; Knob, R.; McConnell, K.; Barrett, J. Analysis of Coumarin-Based Phototoxins in Citrus-Derived Essential Oils Using Liquid Chromatography-Mass Spectrometry. *Chromatographia* **2023**, *87*, 59–69. [CrossRef]
112. Huang, Y.; Xie, F.-J.; Cao, X.; Li, M.-Y. Research progress in biosynthesis and regulation of plant terpenoids. *Biotechnol. Biotechnol. Equip.* **2021**, *35*, 1799–1808. [CrossRef]
113. Kamran, S.; Sinniah, A.; Abdulghani, M.A.M.; Alshawsh, M.A. Therapeutic Potential of Certain Terpenoids as Anticancer Agents: A Scoping Review. *Cancers* **2022**, *14*, 1100. [CrossRef] [PubMed]
114. Yang, W.; Chen, X.; Li, Y.; Guo, S.; Wang, Z.; Yu, X. Advances in Pharmacological Activities of Terpenoids. *Nat. Prod. Commun.* **2020**, *15*, 1934578X20903555. [CrossRef]
115. Avalos, M.; Garbeva, P.; Vader, L.; van Wezel, G.P.; Dickschat, J.S.; Ulanova, D. Biosynthesis, evolution and ecology of microbial terpenoids. *Nat. Prod. Rep.* **2021**, *39*, 249–272. [CrossRef] [PubMed]
116. Jahangeer, M.; Fatima, R.; Ashiq, M.; Basharat, A.; Qamar, S.A.; Bilal, M.; Iqbal, H.M. Therapeutic and Biomedical Potentialities of Terpenoids—A Review. *J. Pure Appl. Microbiol.* **2021**, *15*, 471–483. [CrossRef]
117. Luo, Q.; Tian, Z.; Zheng, T.; Xu, S.; Ma, Y.; Zou, S.; Zuo, Z. Terpenoid composition and antioxidant activity of extracts from four chemotypes of *Cinnamomum camphora* and their main antioxidant agents. *Biofuels Bioprod. Biorefining* **2021**, *16*, 510–522. [CrossRef]
118. Hu, Y.J.; Chen, M.L.; Liang, D. Lignans and terpenoids from *Gaultheria leucocarpa* var. *yunnanensis* and their anti-inflammatory and antioxidant activities. *Fitoterapia* **2022**, *162*, 105293. [PubMed]
119. Kobayashi, Y.; Sato, H.; Yorita, M.; Nakayama, H.; Miyazato, H.; Sugimoto, K.; Jippo, T. Inhibitory effects of geranium essential oil and its major component, citronellol, on degranulation and cytokine production by mast cells. *Biosci. Biotechnol. Biochem.* **2016**, *80*, 1172–1178. [CrossRef]
120. Mittu, B.; Chaubey, N.; Singh, M.; Begum, Z. Cosmeceutical applications of terpenes and terpenoids. In *Specialized Plant Metabolites as Cosmeceuticals*; Elsevier: Amsterdam, The Netherlands, 2024; pp. 25–41.
121. Han, M.; Zhao, L.; Cheng, H.; Qi, Z. Enhancing fractionation of terpenoids and terpenes in citrus essential oils by a biphasic extraction system. *Chem. Eng. Sci.* **2024**, *299*, 120476. [CrossRef]
122. Okla, M.K.; Alamri, S.A.; Salem, M.Z.; Ali, H.M.; Behiry, S.I.; Nasser, R.A.; Alaraidh, I.A.; Al-Ghtani, S.M.; Soufan, W. Yield, Phytochemical Constituents, and Antibacterial Activity of Essential Oils from the Leaves/Twigs, Branches, Branch Wood, and Branch Bark of Sour Orange (*Citrus aurantium* L.). *Processes* **2019**, *7*, 363. [CrossRef]
123. Nidhi, P.; Rolta, R.; Kumar, V.; Dev, K.; Sourirajan, A. Synergistic potential of *Citrus aurantium* L. essential oil with antibiotics against *Candida albicans*. *J. Ethnopharmacol.* **2020**, *262*, 113135. [CrossRef]
124. Segun, P.A.; Ismail, F.M.; Ogbole, O.O.; Nahar, L.; Evans, A.R.; Ajaiyeoba, E.O.; Sarker, S.D. Acridone alkaloids from the stem bark of *Citrus aurantium* display selective cytotoxicity against breast, liver, lung and prostate human carcinoma cells. *J. Ethnopharmacol.* **2018**, *227*, 131–138. [CrossRef] [PubMed]
125. Arbo, M.D.; Schmitt, G.C.; Limberger, M.F.; Charão, M.F.; Moro, Â.M.; Ribeiro, G.L.; Dallegrave, E.; Garcia, S.C.; Leal, M.B.; Limberger, R.P. Subchronic toxicity of *Citrus aurantium* L. (Rutaceae) extract and *p*-synephrine in mice. *Regul. Toxicol. Pharmacol.* **2009**, *54*, 114–117. [CrossRef] [PubMed]
126. Park, K.I.; Park, H.S.; Kim, M.K.; Hong, G.E.; Nagappan, A.; Lee, H.J.; Yumnam, S.; Lee, W.S.; Won, C.K.; Shin, S.C.; et al. Flavonoids identified from Korean *Citrus aurantium* L. inhibit Non-Small Cell Lung Cancer growth in vivo and in vitro. *J. Funct. Foods* **2014**, *7*, 287–297. [CrossRef]
127. Trabelsi, D.; Hamdane, A.M.; Said, M.B.; Abdrrabba, M. Chemical composition and antifungal activity of essential oils from flowers, leaves and peels of Tunisian *Citrus aurantium* against *Penicillium digitatum* and *Penicillium italicum*. *J. Essent. Oil Bear. Plants* **2016**, *19*, 1660–1674. [CrossRef]
128. Mencherini, T.; Campone, L.; Piccinelli, A.L.; García Mesa, M.; Sánchez, D.M.; Aquino, R.P.; Rastrelli, L. HPLC-PDA-MS and NMR characterization of a hydroalcoholic extract of *Citrus aurantium* L. var. *amara* peel with antiedematogenic activity. *J. Agric. Food Chem.* **2013**, *61*, 1686–1693. [CrossRef]
129. Jha, A.K.; Sit, N. Extraction of bioactive compounds from plant materials using combination of various novel methods: A review. *Trends Food Sci. Technol.* **2022**, *119*, 579–591. [CrossRef]
130. Waseem, M.; Majeed, Y.; Nadeem, T.; Naqvi, L.H.; Khalid, M.A.; Sajjad, M.M.; Sultan, M.; Khan, M.U.; Khayrullin, M.; Shariati, M.A.; et al. Conventional and advanced extraction methods of some bioactive compounds with health benefits of food and plant waste: A comprehensive review. *Food Front.* **2023**, *4*, 1681–1701. [CrossRef]
131. Usman, M.; Nakagawa, M.; Cheng, S. Emerging trends in green extraction techniques for bioactive natural products. *Processes* **2023**, *11*, 3444. [CrossRef]
132. More, P.R.; Jambrak, A.R.; Arya, S.S. Green, environment-friendly and sustainable techniques for extraction of food bioactive compounds and waste valorization. *Trends Food Sci. Technol.* **2022**, *128*, 296–315. [CrossRef]

133. Verep, D.; Ateş, S.; Karaoğul, E. A Review of Extraction Methods for Obtaining Bioactive Compounds in Plant-Based Raw Materials. *Bartın Orman Fakültesi Derg.* **2023**, *25*, 492–513. [CrossRef]
134. Sheldon, R.A.; Bode, M.L.; Akakios, S.G. Akakios. Metrics of green chemistry: Waste minimization. *Curr. Opin. Green Sustain. Chem.* **2022**, *33*, 100569. [CrossRef]
135. Martínez, J.; Cortés, J.F.; Miranda, R. Green chemistry metrics, a review. *Processes* **2022**, *10*, 1274. [CrossRef]
136. Yusoff, I.M.; Taher, Z.M.; Rahmat, Z.; Chua, L.S. A review of ultrasound-assisted extraction for plant bioactive compounds: Phenolics, flavonoids, thymols, saponins and proteins. *Food Res. Int.* **2022**, *157*, 111268. [CrossRef] [PubMed]
137. Nour, A.H.; Oluwaseun, A.R.; Nour, A.H.; Omer, M.S.; Ahmed, N. Microwave-assisted extraction of bioactive compounds. In *Microwave Heating. Electromagnetic Fields Causing Thermal and Non-Thermal Effects*; Intech Open: Rijeka, Croatia, 2021; pp. 1–31.
138. Uwineza, P.A.; Waśkiewicz, A. Recent advances in supercritical fluid extraction of natural bioactive compounds from natural plant materials. *Molecules* **2020**, *25*, 3847. [CrossRef] [PubMed]
139. Soltanmohammadi, F.; Jouyban, A.; Shayanfar, A. New aspects of deep eutectic solvents: Extraction, pharmaceutical applications, as catalyst and gas capture. *Chem. Pap.* **2021**, *75*, 439–453. [CrossRef]
140. Carreira-Casais, A.; Otero, P.; Garcia-Perez, P.; Garcia-Oliveira, P.; Pereira, A.G.; Carpena, M.; Soria-Lopez, A.; Simal-Gandara, J.; Prieto, M.A. Benefits and drawbacks of ultrasound-assisted extraction for the recovery of bioactive compounds from marine algae. *Int. J. Environ. Res. Public Health* **2021**, *18*, 9153. [CrossRef] [PubMed]
141. Wang, G.H.; Huang, C.T.; Huang, H.J.; Tang, C.H.; Chung, Y.C. Biological Activities of *Citrus aurantium* Leaf Extract by Optimized Ultrasound-Assisted Extraction. *Molecules* **2023**, *28*, 7251. [CrossRef] [PubMed]
142. Cui, Q.; Jiang, L.J.; Wen, L.L.; Tian, X.L.; Yuan, Q.; Liu, J.Z. Metabolomic profiles and differential metabolites of volatile components in *Citrus aurantium* Changshan-huyou pericarp during different growth and development stages. *Food Chem. X* **2024**, *23*, 101631. [CrossRef] [PubMed]
143. López-Salazar, H.; Camacho-Díaz, B.H.; Ocampo, M.A.; Jiménez-Aparicio, A.R. Microwave-assisted extraction of functional compounds from plants: A Review. *Bioresources* **2023**, *18*, 6614. [CrossRef]
144. Nonglait, D.L.; Gokhale, J.S. Review insights on the demand for natural pigments and their recovery by emerging microwave-assisted extraction (MAE). *Food Bioprocess Technol.* **2024**, *17*, 1681–1705. [CrossRef]
145. Alvi, T.; Asif, Z.; Khan, M.K.I. Clean label extraction of bioactive compounds from food waste through microwave-assisted extraction technique—A review. *Food Biosci.* **2022**, *46*, 101580. [CrossRef]
146. Agustin, M.N.P.; Negoro, A.P.; Putri, D.K.Y.; Rizkiana, M.F.; Rahmawati, I. Extraction of Essential Oils from Sweet Orange Leaves (*Citrus aurantium*) Using the Microwave Hydrodistillation (MHD) Method. *J. Biobased Chem.* **2023**, *3*, 76–95. [CrossRef]
147. Pavlova, P.L.; Minakov, A.V.; Platonov, D.V.; Zhigarev, V.A.; Guzei, D.V. Supercritical fluid application in the oil and gas industry: A comprehensive review. *Sustainability* **2022**, *14*, 698. [CrossRef]
148. Lopez-Hortas, L.; Rodriguez, P.; Diaz-Reinoso, B.; Gaspar, M.C.; de Sousa, H.C.; Braga, M.E.; Dominguez, H. Supercritical fluid extraction as a suitable technology to recover bioactive compounds from flowers. *J. Supercrit. Fluids* **2022**, *188*, 105652. [CrossRef]
149. Ray, A.; Dubey, K.K.; Marathe, S.J.; Singhal, R. Supercritical fluid extraction of bioactives from fruit waste and its therapeutic potential. *Food Biosci.* **2023**, *52*, 102418. [CrossRef]
150. Ahangari, H.; King, J.W.; Ehsani, A.; Yousefi, M. Supercritical fluid extraction of seed oils—A short review of current trends. *Trends Food Sci. Technol.* **2021**, *111*, 249–260. [CrossRef]
151. Jiang, X.; Liu, Q.; Fei, F.; Chen, Z.; Shu, C.; Jie, X.; Tao, Y.; Feng, P.; Yao, L.; Zhou, W.; et al. Supercritical Fluid-extracted *Citrus aurantium* L. var. *amara* Engl. Essential Oil Nanoemulsion: Preparation, Characterization, and Its Sleep-promoting Effect. *J. Oleo Sci.* **2024**, *73*, 773–786. [PubMed]
152. Ferreira, C.; Sarraguça, M. A Comprehensive Review on Deep Eutectic Solvents and Its Use to Extract Bioactive Compounds of Pharmaceutical Interest. *Pharmaceuticals* **2024**, *17*, 124. [CrossRef] [PubMed]
153. Omar, K.A.; Sadeghi, R. Physicochemical properties of deep eutectic solvents: A review. *J. Mol. Liq.* **2022**, *360*, 119524. [CrossRef]
154. Liu, Y.; Zhang, H.; Yu, H.; Guo, S.; Chen, D. Deep eutectic solvent as a green solvent for enhanced extraction of narirutin, naringin, hesperidin and neohesperidin from *Aurantii Fructus*. *Phytochem. Anal.* **2019**, *30*, 156–163. [CrossRef] [PubMed]
155. Cannavacciuolo, C.; Pagliari, S.; Frigerio, J.; Giustra, C.M.; Labra, M.; Campone, L. Natural deep eutectic solvents (NADESs) combined with sustainable extraction techniques: A review of the green chemistry approach in food analysis. *Foods* **2022**, *12*, 56. [CrossRef] [PubMed]
156. Hikmawanti, N.P.E.; Ramadon, D.; Jantan, I.; Mun'im, A. Natural deep eutectic solvents (NADES): Phytochemical extraction performance enhancer for pharmaceutical and nutraceutical product development. *Plants* **2021**, *10*, 2091. [CrossRef] [PubMed]
157. Rownaghi, M.; Niakousari, M.; Golmakani, M.T.; Hosseini, S.M.H. Sour orange fruit (*Citrus aurantium*) seeds: Humble seeds bursting with goodness. *J. Food Bioprocess Eng.* **2022**, *5*, 67–75.
158. Hao, K.X.; Shen, C.Y.; Zhong, R.F.; Jiang, J.G. Multiple bioactivities of choerospondin obtained from blossom of *Citrus aurantium* L. Var. *Amara* engl. And its anti-inflammation and anti-atherosclerosis action pathways. *J. Funct. Foods* **2024**, *112*, 105941.

159. Bozinou, E.; Athanasiadis, V.; Chatzimitakos, T.; Ganos, C.; Gortzi, O.; Diamantopoulou, P.; Papanikolaou, S.; Chinou, I.; Lalas, S.I. Essential Oil of Greek *Citrus sinensis* cv New Hall-*Citrus aurantium* Pericarp: Effect upon Cellular Lipid Composition and Growth of *Saccharomyces cerevisiae* and Antimicrobial Activity against Bacteria, Fungi, and Human Pathogenic Microorganisms. *Processes* **2023**, *11*, 394. [CrossRef]
160. Di Napoli, M.; Castagliuolo, G.; Badalamenti, N.; Maresca, V.; Basile, A.; Bruno, M.; Varcamonti, M.; Zanfardino, A. *Citrus aurantium* 'crispifolia' essential oil: A promise for nutraceutical applications. *Nutraceuticals* **2023**, *3*, 153–164. [CrossRef]
161. Nogueira, G.F.; Oliveira, R.A.D.; Velasco, J.I.; Fakhouri, F.M. Methods of incorporating plant-derived bioactive compounds into films made with agro-based polymers for application as food packaging: A brief review. *Polymers* **2020**, *12*, 2518. [CrossRef] [PubMed]
162. Jiang, Y.; Zhang, Y.; Deng, Y. Latest Advances in Active Materials for Food Packaging and Their Application. *Foods* **2023**, *12*, 4055. [CrossRef] [PubMed]
163. Barbinta-Patrascu, M.E.; Bita, B.; Negut, I. From nature to technology: Exploring the potential of plant-based materials and modified plants in biomimetics, bionics, and green innovations. *Biomimetics* **2024**, *9*, 390. [CrossRef] [PubMed]
164. Naturitas. Aceite Esencial de Naranja Agria 10 ML de Aceite Esencial Naranja Pranarom. Available online: https://www.naturitas.mx/p/suplementos/fitoterapia/aceites-esenciales/aceite-esencial-de-naranja-agria-10-ml-de-aceite-esencial-naranja-pranarom (accessed on 21 December 2024).
165. loom Perfumery. Lacura. Available online: https://bloomperfume.com/products/lacura (accessed on 21 December 2024).
166. L'Occitane. Eau de Toilette Néroli & Orchidée. Available online: https://mx.loccitane.com/products/eau-de-toilette-neroli-y-orquidea (accessed on 21 December 2024).
167. Chás do Mundo. Té de Naranja Amarga, Cascas (*Citrus aurantium* L.). Available online: https://chasdomundo.pt/es/tes-medicinales/te-de-naranja-amarga-citrus-aurantium (accessed on 21 December 2024).
168. Cremas Caseras. Aceite Esencial Neroli 1 mL. Available online: https://www.cremas-caseras.es/aceites-esenciales/176-aceite-esencial-neroli-1ml.html (accessed on 21 December 2024).
169. Naturitas Citrus aurantium 60 Cápsulas Biform. Available online: https://www.naturitas.mx/p/suplementos/fitoterapia/compuestos-herbarios/citrus-aurantium-60-capsulas-biform (accessed on 21 December 2024).
170. Cosmotienda. Exfoliante Naranja Amarga 1000 P. Available online: https://www.cosmotienda.com/tienda/exfoliante-naranja-amarga-1000-p-5271.html?osCsid=ruqjj7splkjtjlt78717t8b9a3 (accessed on 21 December 2024).
171. Naturmed Scientific. Extracto de Naranja Amarga (*Citrus aurantium*). Available online: https://naturmedscientific.com/es/PRODUCTO/Extracto-de-naranja-amarga-Citrus-aurantium (accessed on 21 December 2024).
172. Benditaluz. Crema de Manos y Cuerpo de Naranja Amarga. Available online: https://benditaluz.myshopify.com/products/crema-de-manos-y-cuerpo-de-naranja (accessed on 21 December 2024).
173. Han, M.; Qing, Y.; Fu, J.; He, W.; Huang, J.; Zhu, X.; Yang, L.; Yao, L.; Peng, T.; Wang, Z.; et al. Mechanism of Jizhi syrup's prevention and treatment of acute bronchitis based on LPS-iNOS inflammatory mediators' signalling. *J. Ethnopharmacol.* **2025**, *337*, 118708. [CrossRef] [PubMed]
174. Ubuy. Pure Original Ingredients Hesperidin Powder 4 oz—*Citrus aurantium* Fruit Peel Bioflavonoid. Available online: https://www.ubuy.com.mx/sp/product/FSTGQ1Y6M-pure-original-ingredients-hesperidin-powder-4-oz-citrus-aurantium-fruit-peel-bioflavonoid (accessed on 21 December 2024).
175. Cosme De. WELEDA Citrus Refreshing Oil. Available online: https://us.cosme-de.com/product-page?pdid=17268&src=googlefeed&srsltid=AfmBOopFnmnKnR6F2mREuM6lb9eb59h-zwfG3sjgVxizErWosG4y_g3Qltw (accessed on 21 December 2024).
176. El Nemr, A.; Aboughaly, R.M.; El Sikaily, A.; Ragab, S.; Masoud, M.S.; Ramadan, M.S. Utilization of *Citrus aurantium* peels for sustainable production of high surface area type I microporous nano activated carbons. *Biomass Convers. Biorefinery* **2021**, *13*, 1613–1631. [CrossRef]
177. Navaei Shoorvarzi, S.; Shahraki, F.; Shafaei, N.; Karimi, E.; Oskoueian, E. *Citrus aurantium* L. bloom essential oil nanoemulsion: Synthesis, characterization, cytotoxicity, and its potential health impacts on mice. *J. Food Biochem.* **2020**, *44*, e13181. [CrossRef]
178. Al-Hamdani, H.; Ahmed, S.; Hameed, R.; Getan, A. The Effect of Environmentally safe nanosynthesis with copper particles by using *Citrus aurantium* fruit extract against harmful mosquitoes. *Asian J. Water Environ. Pollut.* **2021**, *18*, 59–67. [CrossRef]
179. Hafez Ghoran, S.; Fadaei Dashti, M.; Maroofi, A.; Shafiee, M.; Zare-Hoseinabadi, A.; Behzad, F.; Mehrabi, M.; Jangjou, A.; Jamali, K. Biosynthesis of zinc ferrite nanoparticles using polyphenol-rich extract of *Citrus aurantium* flowers. *Nanomed. Res. J.* **2020**, *5*, 20–28.
180. Jain, S.; Sharma, M.P. Review of different test methods for the evaluation of stability of biodiesel. *Renew. Sustain. Energy Rev.* **2010**, *14*, 1937–1947. [CrossRef]
181. *EN 14214:2012*; Liquid Petroleum Products—Fatty Acid Methyl Esters (FAME) for Use in Diesel Engines and Heating Applications—Requirements and Test Methods. European Committee for Standardization (CEN): Brussels, Belgium, 2012.

182. Ahmad, M.; Zafar, M. Conversion of waste seed oil of *Citrus aurantium* into methyl ester via green and recyclable nanoparticles of zirconium oxide in the context of circular bioeconomy approach. *Waste Manag.* **2021**, *136*, 310–320.
183. Ferrer, V.; Costantino, G.; Paoli, M.; Paymal, N.; Quinton, C.; Ollitrault, P.; Tomi, F.; Luro, F. Intercultivar diversity of sour orange (*Citrus aurantium* L.) based on genetic markers, phenotypic characteristics, aromatic compounds and sensorial analysis. *Agronomy* **2021**, *11*, 1084. [CrossRef]
184. Mohagheghniapour, A.; Saharkhiz, M.J.; Haghighi, T.M. A Compositional Perspective of Sour Orange (*Citrus aurantium* L.) Flowers Essential Oil under Different Storage Conditions. *Food Sci. Eng.* **2022**, *3*, 154–169. [CrossRef]
185. Burnett, C.L.; Bergfeld, W.F.; Belsito, D.V.; Hill, R.A.; Klaassen, C.D.; Liebler, D.C.; Marks, J.G.; Shank, R.C.; Salga, T.J.; Synder, P.W.; et al. Safety Assessment of Citrus Plant-and Seed-Derived Ingredients as Used in Cosmetics. *Int. J. Toxicol.* **2021**, *40* (Suppl. S3), 39S–52S. [CrossRef] [PubMed]
186. Burnett, C.L.; Bergfeld, W.F.; Belsito, D.V.; Hill, R.A.; Klaassen, C.D.; Liebler, D.C.; Marks, J.G.; Shank, R.C.; Salga, T.J.; Synder, P.W.; et al. Safety Assessment of Citrus Fruit-Derived Ingredients as Used in Cosmetics. *Int. J. Toxicol.* **2021**, *40* (Suppl. S3), 5S–38S. [CrossRef] [PubMed]
187. Kačániová, M.; Verešová, A.; Čmiková, N. The Antimicrobial Activity of *Citrus aurantium* Amara. *Sci. Pap. Anim. Sci. Biotechnol.* **2024**, *57*, 32–37.
188. He, Y.; Zhu, Y.; Lv, J.; Gu, Y.; Wang, T.; Chen, J. Effects of lactic acid bacteria fermentation on the bioactive composition, volatile compounds and antioxidant activity of Huyou (*Citrus aurantium* 'Changshan-huyou') peel and pomace. *Food Qual. Saf.* **2023**, *7*, fyad003. [CrossRef]
189. Hashemi, S.M.B.; Amininezhad, R.; Shirzadinezhad, E.; Farahani, M.; Yousefabad, S.H.A. The antimicrobial and antioxidant effects of *Citrus aurantium* L. Flowers (Bahar Narang) extract in traditional yoghurt Stew during refrigerated storage. *J. Food Saf.* **2016**, *36*, 153–161. [CrossRef]
190. Kačániová, M.; Terentjeva, M.; Galovičová, L.; Ivanišová, E.; Štefániková, J.; Valková, V.; Borotová, P.; Kowalczewski, P.Ł.; Kunová, S.; Felšöciová, S.; et al. Biological activity and antibiofilm molecular profile of *Citrus aurantium* essential oil and its application in a food model. *Molecules* **2020**, *25*, 3956. [CrossRef] [PubMed]

processes MDPI

Review

Active Polymer Films with Olive Leaf Extract: Potential for Food Packaging, Biomedical, and Cosmetic Applications

Sylwia Grabska-Zielińska

Faculty of Chemical Technology and Engineering, Bydgoszcz University of Science and Technology, Seminaryjna 3, 85-326 Bydgoszcz, Poland; sylwia.grabska-zielinska@pbs.edu.pl

Abstract: This review paper highlights the latest advancements in polymer films modified with olive leaf extract (OLE) for various applications, particularly in food packaging. The overview is focused on the preparation, properties, and multifunctionality of OLE-enhanced polymer materials. Olive leaf extract, known for its antibacterial, antifungal, and antioxidant properties, is also shown to enhance the physicochemical characteristics of polymer films. This review consolidates current knowledge on using OLE as a bioactive additive in polymer-based packaging materials, improving their structural integrity and functionality. In addition to food packaging, the review explores other applications of OLE-modified polymer films in industries such as biomedicine, pharmaceuticals, and cosmetics. The paper also addresses future perspectives in polymer food packaging, suggesting that OLE-modified films offer significant potential for industrial and academic research due to their enhanced properties. Overall, this review provides a comprehensive overview of the role of olive leaf extract in improving the performance of polymer films in diverse fields.

Keywords: olive leaf extract; polymers; films; food packaging; biomedical applications

1. Introduction

Olea europaea L., known as olive, is an evergreen tree belonging to the *Oleaceae* family. It has great commercial and historical importance, especially in the Mediterranean [1,2]. It is an essential plant to extract olive oil, but practically not only olive fruits can be used in various applications. Olive leaves, olive tree pruning biomass, olive pomace, olive mill wastewater (OMWW), and olive stones could be used in different ways, technologies, and industries [3,4].

Olive fruits and olive oil are important elements of the healthy daily diet of a large part of the world's population, while the olive leaves are of interest to scientists and companies dealing with supplementation, packaging materials, and the food industry, thanks to their high amount of polyphenolic compounds [1]. In the past, olive leaves were commonly used as a remedy for fever and other infections, including malaria, in Mediterranean countries such as Spain, Italy, France, Greece, Morocco, Tunisia, Turkey, and Palestine. Nowadays, the use of fresh or dried olive leaves to make olive tea is popular, as well as consuming their dried, powdered, capsulated form as dietary supplements [5]. Olive leaves are becoming more popular as tea and dietary supplements because they are believed to have antimicrobial, antifungal, antioxidant, hypoglycemic, anti-hypertensive, hypocholesterolemia, and anti-inflammatory properties [5–8].

Olive leaf extract is rich in a large variety of phenolic derivatives, which consist of simple phenols (the most common and important low-molecular-weight phenolic compounds), flavonoids (flavones, flavanones, flavonols, and flavan-3-ols), secoiridoids, and other compounds [2]. Several studies focus on the extraction of olive leaves and the analysis of the phenolic compound composition in the resulting extracts. However, the analysis of these results indicates that the presence of different phenolic compounds in OLE depends on such factors as olive tree cultivation, harvest season and period, environmental and climatic conditions, plant diseases, maturity, moisture content, time and temperature of extraction, solvent, pH, solvent-to-substrate ratio, adsorbent [8–10]. In addition, it was determined that the extraction method significantly affects the amount and composition of extract obtained from olive leaves. Scientists have adopted numerous extraction methods (conventional and latest) to obtain OLE. Among these techniques maceration, cold solvent extraction, Soxhlet extraction, microwave-assisted extraction, ultrasound-assisted extraction, supercritical fluid extraction, and pressurized liquid methods are commonly used [10–13]. Taking into account the above, it is important to evaluate the composition of the particularly obtained and used OLE. An example list of compounds present in the chloroformic OLE is given in Table 1. The study was performed by LC-MS/MS using negative ionization according to the method described in detail elsewhere [14].

Table 1. Phenolic compounds identified by LC-MS/MS using the negative ionization mode of Chemlali olive leaf cultivar extract. Olive leaves have been collected from Sfax (Tunisia) in July 2019. Adapted from S. Grabska-Zielińska et al., "Polylactide Films with the Addition of Olive Leaf Extract—Physicochemical Characterization" [9].

N°	Molar Mass	Chemical Formula	Compound
1	170.0215	$C_7H_6O_5$	Gallic acid
2	154.0630	$C_8H_{10}O_3$	Hydroxytyrosol
3	180.0425	$C_9H_8O_4$	Caffeic acid
4	640.2003	$C_{29}H_{36}O_{16}$	β-hydroxyverbascoside III
5	610.1534	$C_{27}H_{30}O_{16}$	Quercetin 3-*O*-rutinoside (rutin)
6	624.2054	$C_{29}H_{36}O_{15}$	Verbascoside
7	448.1006	$C_{21}H_{20}O_{11}$	Luteolin 7-*O*-glucoside
8	702.2371	$C_{31}H_{42}O_{18}$	Oleuropein hexoside I
9	624.2054	$C_{29}H_{36}O_{15}$	Isoverbascoside
10	432.1056	$C_{21}H_{20}O_{10}$	Apigenin 7-*O*-glucoside
11	540.1843	$C_{25}H_{32}O_{13}$	Oleuropein I
12	540.1843	$C_{25}H_{32}O_{13}$	Oleuropein II
13	558.2309	$C_{26}H_{38}O_{13}$	6′-*O*-[(2*E*)-2,6-dimethyl-8-hydroxy-2-octenoyloxy]-secologanoside
14	926.3056	$C_{42}H_{54}O_{23}$	Jaspolyoside III

It has been repeatedly reported that oleuropein is the main compound found in olive leaf extract and its content is estimated to be in the range of 17–23% depending upon the harvesting time of the leaves [9,15–17]. Oleuropein is a phenolic secoiridoid glycoside that consists of hydroxytyrosol, elenolic acid, and a glucose molecule. It has been shown to possess anticancer properties, particularly against breast, pancreatic, and prostate cancers. Additionally, oleuropein exhibits various biological activities, including antimicrobial, hypoglycemic, antioxidant, antiviral, anti-hypertensive, hepatoprotective, and neuroprotective effects [17–20]. As the most abundant bioactive compound in olive leaves, oleuropein serves as a chemotaxonomic marker in the Oleaceae family. Numerous articles in the literature summarize its known biological activities and significance [17–22]. Olive leaf extract, particularly oleuropein, has been demonstrated to be non-toxic [23–26]. It showed no mutagenic or genotoxic effects in multiple tests, including the bacterial reverse mutation test, in vitro mammalian chromosomal aberration test, and in vivo mouse micronucleus test. The NOAEL was established at 1000 mg/kg bw/d in a 90-day study [26]. The safety of olive leaf extract (based on the Bonolive™—proprietary water-soluble extract of the leaves of the olive tree *Olea europaea* L.) is also corroborated by the lack of treatment-related adverse events in the long-term (12 months) clinical trial, the history of safe ingestion of olive leaves as animal feed, and consumption of olive-based products by humans that contain many of the same constituents, such as table olives, olive oils, and other olive-derived products with a lack of known toxicity [26].

Regardless of the extractant (water, chloroform, hexane, acetonitrile, diethyl ether, ethanol, 80% ethanol, 20% methanol), olive leaf extract exhibited antibacterial and antifungal activity [1]. Table 2 shows the microorganisms against which the extract was active. The antimicrobial activity studies were performed for the whole extracts, because extracts may be more beneficial than isolated constituents since a bioactive individual component can change its properties in the presence of other compounds present in the extracts [1,27]. Additionally, the additive and synergistic effects of phytochemicals in fruits and vegetables are responsible for their potent bioactive properties [1,28].

Table 2. The antimicrobial and antibacterial activity of olive leaf extract.

Type of Microorganism	Microorganism	References
Gram + bacteria	*Bacillus cereus*	[1,29]
	Bacillus subtilis	[1]
	Staphylococcus aureus	[1,29–34]
	Staphylococcus epidermidis	[34]
	Staphylococcus hominis	[6,35]
	Staphylococcus xylosus	[6,35]
	Staphylococcus capitis	[6,35]
	Listeria monocytogenes	[32,33,36]
	Listeria innocua	[37]
	Enterococcus faecalis	[30]
	Kocuria rhizophila	[6,35]
	Lactobacillus acidophilus	[6,35]
	Lactobacillus casei	[6,35]
	Micrococcus luteus	[6,35]
	Streptococcus pyogenes	[6,35]
Gram − bacteria	*Escherichia coli*	[1,29–33,36]
	Pseudomonas aeruginosa	[1,30]
	Klebsiella pneumoniae	[1,29]
	Klebsiella aerogenes	[38]
	Yersinia enterocolitica	[33,34]
	Vibrio parahaemolyticus	[39]
	Enterobacter cloacae	[38]
	Campylobacter jejuni	[37]
	Salmonella Typhimurium	[29,33]
	Salmonella enterica	[40]
	Salmonella Typhi	[39]
	Salmonella Enteritidis	[32,36]
	Acinetobacter calcoaceticus	[6,35]
	Helicobacter pylori	[6,35]
	Serratia marcescens	[6,35]

Table 2. *Cont.*

Type of Microorganism	Microorganism	References
Fungi	*Cryptococcus neoformans*	[1,8]
	Aspergillus flavus	[41]
	Aspergillus niger	[42]
	Aspergillus fumigatus	[42]
	Candida albicans	[1,8,30]
	Candida glabrata	[8,30]
	Candida krusei	[30]
	Candida parapsilosis	[8,30]
	Candida oleophila	[6,43]
	Saccharomyces uvarum	[6,43]
	Saccharomyces cerevisiae	[6,43]
	Metschnikowia fructicola	[6,43]
	Kloeckera apiculata	[6,43]
	Schizosaccharomyces pombe	[6,43]

What is more, utilizing olive leaf extract is a smart way to repurpose agricultural waste. Large amounts of leaves are left over after tree maintenance, pruning, and olive harvesting, as olives are primarily used for oil pressing and direct consumption. By extracting valuable compounds from these leaves, it does not only reduce waste but also creates sustainable solutions for food packaging and preservation. This approach supports a circular economy, where by-products are efficiently used, contributing to environmental sustainability while adding value to the food industry [15,44,45].

An important aspect of using olive leaf extract in packaging materials is its application in biodegradable polymers. This is particularly significant given the global challenge of waste accumulation and the difficulties associated with recycling. The development of environmentally friendly packaging materials is a priority for researchers as they seek solutions to mitigate the impact of plastic waste on the environment. In this work, various commercial packaging modified with olive leaf extract is also discussed, highlighting the potential of plant-derived substances as natural preservatives with antioxidant and antimicrobial properties. These modifications demonstrate the viability of using plant-based extracts to enhance both the sustainability and functionality of packaging materials [44,45]. All the reasons and motivations for using olive leaf extract are pointed out in Figure 1.

All in all, this review summarizes the current research on polymer films modified with olive leaf extract for food packaging applications. In addition, it discusses other uses of olive leaf extract in the food industry, including its role as a preservative and antimicrobial agent in direct contact with food, as well as encapsulated olive leaf extract for preservation purposes. The article also explores the extract's applications in the biomedical and pharmaceutical industries and briefly mentions its utilization in cosmetics and the synthesis of silver and gold nanoparticles for medical applications.

The motivation for preparing this review stemmed from the lack of comprehensive articles on the use of olive leaf extract as a modifier for polymer films. In this work, a thorough overview of this topic is provided, as no existing review has yet summarized the research in this area. Additionally, the goal of this manuscript was to compile the various applications and studies conducted on olive leaf extract-modified materials in one place, accompanied by relevant references, making it easier for researchers to access and build upon this knowledge.

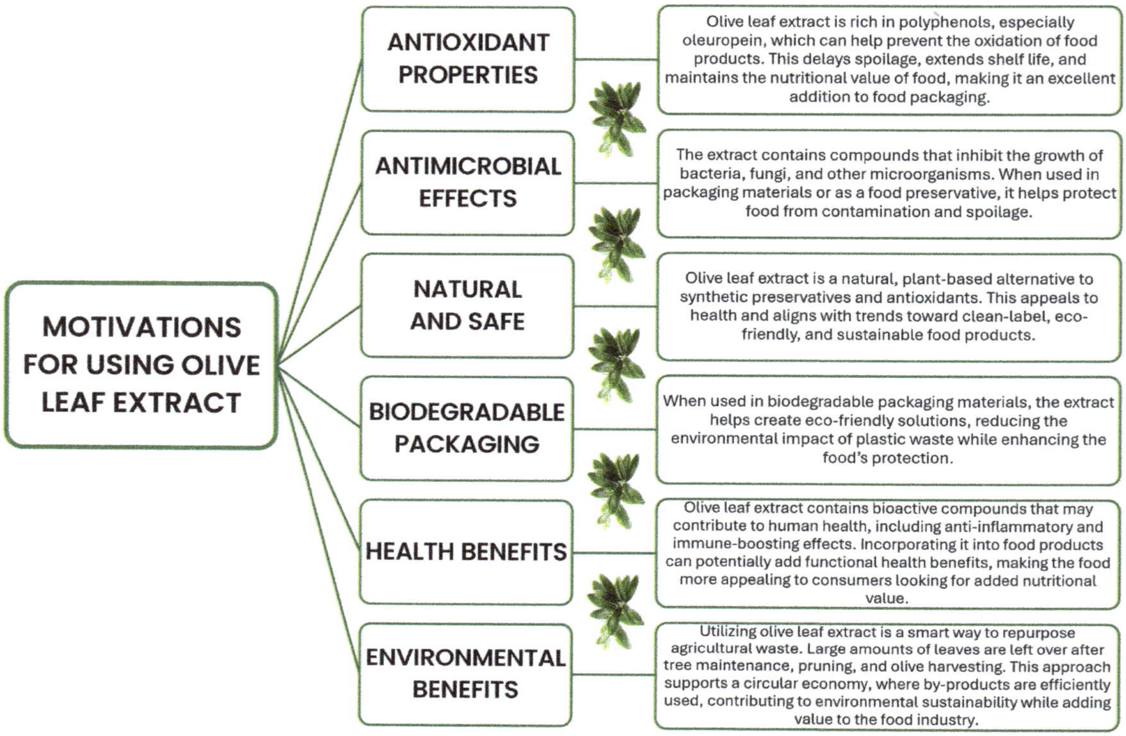

Figure 1. Motivations for using olive leaf extract. The diagram is the author's own work.

2. Incorporation of Olive Leaf Extract in Polymer Films for Food Packaging Applications

Nowadays, advances in food packaging play a primary role in keeping food fresh and safe, as well as in caring for the environment. Consumers want food that is healthy, fresh, high quality, and safe. Moreover, they increasingly pay attention to waste reduction and biodegradability of packaging.

Food packaging materials are very important to ensure the safety of food, to preserve the quality of food, and to extend their shelf life, as well as to prevent food deterioration, which happens as a result of microbial spoilage, metabolism, and oxidation. Food must be protected from the outside world, environmental pollution, and other factors such as light, temperature, humidity, physical damage, dust, odors, and microorganisms.

Epidemiological studies indicate that the number of food-borne illnesses caused by pathogens has increased significantly [46]. They are caused by various types of pathogens. Controlling pathogens could help significantly reduce food-borne epidemics and provide consumers with safe and healthy packaged food products. It can be found that active substances obtained from natural sources have antimicrobial and antibacterial activity. Such substances can serve as packaging additives to inhibit the action of food-borne microorganisms [47]. These naturally active substances can be found in essential oils, herbs, and plants. Thanks to such compounds, it is also possible to achieve longer food storage [46,47].

2.1. Active Packaging Films with OLE

Active packaging is one of the emerging areas where antimicrobial agents are embedded in the packaging materials so those agents can interact with the packaged food in a desirable way to control the growth of microbes. The area of research on food packaging with antimicrobial activity is currently a rapidly growing field. It is happening because it is regarded as an important tool in the combat against most potential food safety hazards [48–51]. Antimicrobial food packaging is one of the most innovative concepts of active packaging, where interactions with the product are envisioned to reduce, inhibit, or delay the growth of microbes that may be present in the packed food or on the surface of the packaging materials, and extend the shelf life of the preserved foodstuffs [48–50,52].

Taking into account the above, scientists are looking for plant extracts and active substances that can modify polymer materials, giving them antibacterial and antioxidant properties. Materials from various polymers, modified with olive leaf extract, are of great interest to science. Olive leaf extract was used to enrich the chitosan (CTS) [53–56], carrageenan (CN) [57–59], gelatin (GEL) [60], methylcellulose (MCEL) [61], sodium alginate (SA) [62], polylactide (PLA) [9,63], polyvinyl alcohol/cellulose (CEL) nanofibers [64], and cellulose nanocrystals/poly(butylene adipate-co-terephthalate) [65] films. Usually, each type of film enriched with OLE was characterized by antioxidant activity, thanks to the high polyphenols content in OLE.

CTS/OLE films exhibited reduced permeability against water vapor, higher solubility in food simulants, and antimicrobial activity against *L. monocytogenes* and *C. jejuni*, which was confirmed by in situ testing using chicken [54]. Nevertheless, the results of antibacterial tests suggest that CTS/OLE films may play an important role in inhibiting microorganisms capable of growing in food at refrigerated temperatures. Furthermore, OLE per se has an inhibitory effect but is not bactericidal at 41.5 °C [54]. When the effect of OLE-enriched CTS films on extending the shelf life of fish fillets was studied, better results were found for 1% than for 2% OLE addition, which can be caused by peroxidation property at high concentrations of the extract [53]. Moreover, the addition of OLE to CTS films improved tensile strength, elongation, and moisture retention capability [55].

G.S. da Rosa et al. [58] and T.R. Martiny et al. [57,59] prepared bio-based active packaging CN films with OLE, and the films were suggested as good materials for the preservation of food products, including lamb meat. The CN/OLE films exhibited good barrier and mechanical properties, with increased stretching capacity, reduced tensile strength, and higher water vapor permeability [58]. Additionally, the CN films enriched with OLE were characterized by changed thickness and color parameters and showed antibacterial activity against *E. coli*, and the packaging demonstrated inhibitory activity of the microbial population in the packed meat product [59]. On the other hand, antibacterial GEL films modified with OLE have been proposed to enhance the quality of cold-smoked salmon. The films significantly reduced the growth of *L. monocytogenes* on the fish over storage [60].

A study of Kasar cheese storage using MCEL films with the addition of OLE is also known [61]. The addition of OLE to MCEL films caused a decrease in water vapor permeation and elongation, and an increase in tensile strength. Antibacterial activity of the films against *S. aureus* and a decrease in the count of *S. aureus* during 7 and 14 days of Kasar cheese storage in the tested films were also observed [61].

If it is about SA films, the addition of OLE improved the barrier properties and promoted changes in resistance and flexibility [62]. Apart from that, the increase in the opacity values after the addition of OLE can help reduce the passage of light and thus be useful to avoid food oxidation. The films showed antibacterial activity against *S. aureus*, and the materials can be advisable for foods with lower water content and potential lipid degradation because of the reduced moisture content and water vapor permeability values after OLE incorporation [62].

Antimicrobial PLA/OLE films showed an antimicrobial effect on *S. aureus* [63]. Moreover, the incorporation of OLE into PLA films had no effect [9] or enhanced [63] the water vapor permeability, while the color of the polymeric films was strongly dependent on the presence and content of the extract and changed during contact with different solutions (HCl, water, NaOH) [9].

Evaluation of microbial and oxidative changes of 100% Iberian Spanish salchichón in contact with a polyvinyl alcohol/CEL nanofibers composite olive leaf extract food film and vacuum-packaged have been researched by M. Sánchez-Gutiérrez et al. [64]. Only a slight reduction of lipid oxidation and mesophilic aerobic bacteria was observed at 25 °C in the test sample, while at 5 °C no significant differences were found between control and test batches when the active films were tested in contact with sliced 100% Iberian salchichón [64]. CEL nanocrystals/poly(butylene adipate-co-terephthalate) packaging materials have been obtained by melt processing, and their physicochemical properties have been studied [65]. The results corroborate the incorporation of OLE in poly(butylene adipate-co-terephthalate) for food packaging with stable mechanical performance and antioxidant activity and suggest its use in active food packaging [65].

2.2. Edible Films with OLE

Edible films based on chitosan [66–68], sodium alginate [66,69], chitosan/hydroxyethyl cellulose (HCEL) [70], corn starch (CST) [71], corn zein (CZN) [72], and pectin (PEC) [73] enriched with olive leaf extract have been obtained, and their physicochemical and antibacterial properties were studied. Generally, each system was characterized by antioxidant properties, whereas CTS and CTS/HCEL-based films demonstrated antibacterial activity against Gram-negative (*S.* Enteritidis, *S. enterica*, *E. coli*, *P. aeruginosa*) and Gram-positive (*E. faecalis*, *S. aureus*, *B. cereus*, *L. monocytogenes*) bacteria, and antifungal activity against *A. flavus*, *A. niger* [68,70]. CST edible films with OLE showed homogeneity and high visual transparency with significantly improved thermo-oxidative resistance with temperature compared to the control film (non-modified CST film) [71]. The CTS coating with OLE could improve the oxidative status of the pork burgers, facilitating the production of healthier pork products, even though the lack of a noticeable inhibition of microbial growth (Aerobic Psychrotrophic Bacteria and Lactic Acid Bacteria). Nevertheless, incorporation of OLE into CTS coatings is beneficial to limit the changes in texture (hardness, gumminess, and chewiness) of meat that occur over storage, as well as oxidative reactions [67]. The edible CTS or SA-based films enriched with OLE were proven to retard the ripening process of sweet cherries [66], while edible SA coatings with OLE were found to be effective in reducing lipid oxidation in chicken nuggets stored in the deep freezer [69].

Another invention, where active PEC films with OLE were used as soluble sachets for chicken stock powder, was developed and performed by M. Sabbah et al. [73]. The properties of such films were compared to the properties of PEC films enriched with guava leaf extract (GLE). The results indicated that incorporating OLE or GLE into the pectin-based films significantly increased their opacity, greenness, and antioxidant activity, and the films had lower water vapor permeability than the control film. Additionally, sachets with chicken stock powder, after being introduced to boiling water, rapidly dissolved within seconds [73]. These studies suggested the potential of edible films incorporated with a natural olive leaf extract as sustainable systems to wrap hamburgers, chicken nuggets, fruits, vegetables, and sour cream packaging [66–71], as well as to use as soluble sachets for chicken stock powder, for example [73]. Additionally, edible films with olive leaf extract could be used as supplements to reinforce the human diet with polyphenols [68].

Apart from the above-described examples where scientists extracted olive leaves to use extract in their experiments, there are also examples of researchers using commercially available olive leaf extract. For example, the aim of the C. Fioretini et al. study [74], was to test the application on fresh pork burgers of PLA/polyhydroxybutyrate-based films coated with CTS/commercial OLE or MCEL/commercial OLE to evaluate their effect on meat quality preservation. The sample where CTS/commercial OLE coating was used showed a slight slowdown in *Enterobacteriaceae* growth, revealing promising results in maintaining the meat quality longer [74].

2.3. Conventional Materials Modified with OLE

In addition to the fact that scientists are working on creating new polymer films or edible coatings modified with olive leaf extract, they are also working on modifying existing films. An example of such a solution is supercritical solvent impregnation (SSI) of polyethylene terephthalate/polypropylene (PET/PP) commercial plastic films by olive leaf extract [75–77]. The authors carried out a study of OLE active substances migration into food simulants, and they obtained the highest migration rate into 95% ethanol (stimulant C, for food with high lipid content). The effectiveness of the optimum active packaging evaluated against lipid oxidation of sunflower seeds was confirmed. This effectiveness estimation could be possible thanks to the study of the peroxide value of sunflower seeds. The peroxide value of samples packed with impregnated PET/PP films was significantly lower than that obtained on samples packed with non-impregnated film and unpacked samples, which can lead to the opinion that foods with high lipid content are susceptible to the effect of this kind of active films [75].

In another study by F. Khanum et al., three types of commercial plastic films based on different mixtures of nylon, polyethylene, and polypropylene were coated by mixtures of olive leaf extract, methylcellulose, hydroxypropyl methylcellulose, and polyethylene glycol, and the performed studies found significant antioxidant and antibacterial (against *S. aureus*, *E. coli*, *S.* Typhimurium) potential of obtained packaging materials [78].

A study of active substances migration into food simulants (water, 10% ethanol, 50% ethanol) has also been performed for oriented polypropylene (OPP) films coated by shellac and cellulose nitrate lacquers with incorporated OLE [79]. This work provides information not about the possibility of using such a combination of substrates to prepare packaging but about the possibility of using such a research method for the comparison of the retention performances of various coating matrices [79].

Polyamide/polyethylene (20/70) based films with a thickness of 70 μm, coated by OLE, have been used for high-pressure packaging of sliced Iberian ham [80]. Active packaging was not effective because most of the sensory parameters (appearance, taste, flavor, texture) did not show significant differences due to packaging with or without OLE [80], and none of the antioxidant and antimicrobial effects on meat even applying a high dose of the lyophilized OLE [81]. Likely, oxidative characteristics of shoulders from pigs reared outdoors would be partly responsible for the high stability of this product during processing and storage [80].

In the next interesting study, olive leaf extract was incorporated into an aqueous dispersion formula, which was used to glue two different layers of polyethylene/polyethylene (PE/PE) [82,83] or polyethylene/paper (PE/P) [83] materials in such a way that components of OLE were sandwiched in the middle of the two layers. In this study, migration of the phenolic compounds from materials modified with OLE into two simulants was performed and demonstrated a non-migrating behavior. Phenolic compounds from OLE were anchored in the adhesive used to build the multilayer materials; PE/OLE/PE film efficiently enhanced the stability of fresh meat against oxidation processes [82], and it can be proposed as a packaging material for food products [83], especially for fresh minced pork meat.

2.4. Other OLE Applications for Extending Food Shelf Life

When summarizing the potential uses of olive leaf extract as an additive to various types of packaging films, it is also important to briefly mention other applications of this extract.

Olive leaf extract was used as an additive to food as a preservative, antibacterial, and antifungal agent [36,84–87]. For example, poultry meat slices from the chicken breast were dipped for 15 min in OLE at different concentrations (0.25%, 0.5%, 1%). The results of various tests revealed that the OLE addition reduced microbial growth successfully and maintained the chemical quality and sensory attributes of poultry meat [88]. The mixture of olive leaf extract and pomegranate peel extract was used as an antibacterial additive to poultry, yellow poultry, and rabbit minced meat. Various tests of studied meat were prepared, and the results showed the potential of this mixture in food preservation and shelf life [89]. In addition to meat [88–94], olive leaf extract has also been tested as a preservative of pasteurized milk [95,96], marinated anchovies [97], vacuum-packed salmon burgers [98], raw peeled undeveined shrimps [39], frankfurter type sausages [99], baked snacks [100], as well as gluten-free bread [101], vegetable pate [102], tomato [41], and pepper pastes [103]. Furthermore, OLE has been tested as an agent to improve the quality of refined oils and extra virgin olive oil [8], sunflower [104] and soybean [105] oils.

Other examples will be chitosan-coated nanoliposomes (chitosomes) containing olive leaf extract [106], as well as OLE encapsulated in chitosan [107], alginate [107], or gelatin/tragacanth gum [108] microbeads. Chitosomes were proposed as promising nanocarriers for retention of OLE optimized by response surface methodology (RSM) based on central composite design. The authors suggested encapsulation of OLE to use in different food systems because it is a reasonable solution to overcome some limitations of OLE, like bitter taste or the possibility of degradation of phenolic content due to the external and storage conditions [106,107].

In summary, olive leaf extract extends food shelf life, exhibits antibacterial and antifungal properties, and can be proposed as a food preservative. However, further studies on other types of food and additional microbiological testing are still needed. Table 3 provides a summary of the materials described above, along with their applications and a list of studies performed to characterize them.

Table 3. The summary of the food packaging based on polymers enriched with olive leaf extract.

Polymer Base	Application	Studies	Reference
Chitosan	Coatings for improving food shelf life of fish fillets	Studies of fish samples (fat, protein, thiobarbituric acid, volatile nitrogen, total microbial count on fish fillets)	[53]
	Bioactive films to ensure the microbiological safety of food and prevent earlier oxidative degradation	Thickness; measurements of solubility in aqueous solutions simulating foods; water vapor transmission rate (WVPR); mechanical properties; antimicrobial properties against *L. monocytogenes*, *E. coli* and *C. jejuni*	[54]
	Films for active food packaging	Thickness; light barrier properties; FTIR spectroscopy; X-ray diffraction analysis; mechanical properties; WVPR; moisture retention capability; antioxidant activity (AA); antibacterial properties against *S. aureus*	[55]
	Film coatings to preserve apple fruit	Analysis of coated apple measurements (anthocyanins content; carotenoids and chlorophylls content; total phenolic compounds (TPC) content; total flavonoids content); analysis of apple samples (AA of acetonic extracts; decay apple area determination; microstructure analysis)	[56]
	Edible film coatings to shelf life sweet cherries	Analysis of the sweet cherries (weight loss; total soluble solids and acidity; ascorbic acid content; AA; TPC; anthocyanins determination)	[66]
	Coatings for improving food shelf life of pork burgers	Headspace gas analysis, analysis of meat samples (microbial counts, color and texture measurements, lipid and protein oxidation measurements)	[67]

Table 3. *Cont.*

Polymer Base	Application	Studies	Reference
Chitosan	Edible films as bioplastics to wrap meat hamburgers; supplements to reinforce the human diet with polyphenols	AA; antimicrobial activity against *S.* Enteritidis, *S. enterica*, *E. faecalis*; in vitro oral digestion; mechanical properties	[68]
	Coating to PLA/polyhydroxybutyrate packaging film to test the application on fresh pork burgers	Analysis of meat samples (pH; color measurements; thiobarbituric acid reactive substances measurements, microbiological analysis, sensory analysis—appearance and odor)	[74]
	Chitosan-coated nanoliposomes (chitosomes) containing OLE to food shelf life	TPC; electrical conductivity; accelerated centrifugal stability of chitosomes; encapsulation efficiency; particle size; zeta potential; SEM; TEM; FTIR spectroscopy; stability of formulation during storage; release behavior of chitosomes	[106]
	Nanocarriers containing OLE for food shelf life	TPC; oxidation test; encapsulation efficiency	[107]
Chitosan/hydroxyethyl cellulose	Biodegradable, edible films for sour cream packaging	FTIR spectroscopy; contact angle; mechanical properties; barrier properties; antimicrobial properties against *S.* Typhimurium, *L. monocytogenes*, *S. aureus*, *E. coli*, *P. aeruginosa*, *B. cereus*, *A. niger*, *A. flavus*; AA; analysis of sour cream samples (oxidative stability parameters; pH; microbiological analysis)	[70]
Gelatin	Antimicrobial films for enhancing the quality of cold-smoked Salmon	Antimicrobial activity against *L. monocytogenes*; TPC; AA; color measurements; WVPR; storage studies: inhibition of *L. monocytogenes* in salmon samples	[60]
Gelatin/tragacanth gum	Nanocarriers containing OLE for food shelf life	AA; study of the pH effect (turbidimetry); encapsulation efficiency; SEM; FTIR spectroscopy; DSC	[108]
Pectin	Soluble sachets for chicken stock powder	Mechanical properties; thickness; AA; opacity; moisture content; color measurements; barrier properties	[73]
Corn starch	Edible films to prevent the oxidative deterioration of packaged foodstuff such as nuts, seeds and sausages	Thickness; color measurements; ATR-FTIR spectroscopy; thermal properties; mechanical properties; barrier properties; AA; antimicrobial activity against *E. coli* and *S. aureus*	[71]
Corn zein	Edible coatings to foodstuff	Mechanical properties; color measurements; thickness; moisture content; barrier properties	[72]
Methylcellulose	Active films to control *S. aureus* during storage of Kasar cheese	Mechanical properties; water vapor permeation; antibacterial activity against *S. aureus*; Kasar cheese storage	[61]
	Coating to PLA/polyhydroxybutyrate packaging film to test the application on fresh pork burgers	Analysis of meat samples (pH; color measurements; thiobarbituric acid reactive substances measurements; microbiological analysis; sensory analysis—appearance and odor)	[74]

Table 3. *Cont.*

Polymer Base	Application	Studies	Reference
Sodium alginate	Packaging for foods with lower water content and potential lipid degradation	Thickness; WVPR; color measurements; opacity; moisture content; water solubility; mechanical properties; SEM imaging; FTIR spectroscopy; antibacterial activity against *L. monocytogenes*, *S. aureus*, *E. faecalis*, *S.* Typhimurium, and *E. coli*; migration tests	[62]
	Edible film coatings to shelf life sweet cherries	Analysis of sweet cherries (weight loss; total soluble solids and acidity; ascorbic acid content; AA; TPC; anthocyanins determination)	[66]
	Packaging to improve the quality of chicken nuggets during refrigerated and deep-freeze storage	Analysis of chicken nugget samples (moisture content, pH, color measurements, lipid oxidation analysis, microbiological analysis)	[69]
	Nanocarriers containing OLE for food shelf life	TPC; oxidation test; encapsulation efficiency	[107]
Carrageenan	Active films for lamb meat preservation	Thickness; WVPR; mechanical properties; color measurements; FTIR spectroscopy; antimicrobial activity of films in the packaging of lamb meat	[57]
	Films for food packaging applications	Thickness; WVPR; mechanical properties; color measurements; TPC; AA; SEM imaging; FTIR spectroscopy	[58]
	Active films for lamb meat preservation	Thickness; WVPR; mechanical properties; color measurements; water solubility of films; inhibition of psychrophiles in meat samples (storage studies)	[59]
Polylactide	Active food packaging films	TPC; AA; FTIR-ATR spectroscopy; thermal properties; mechanical properties; thickness; WVPR; color measurements; SEM imaging	[9]
	Antimicrobial food packaging to reduce post-process growth of *S. aureus*	WVPR; mechanical properties; antimicrobial activity against *S. aureus*; thermal properties; water absorption; degradability	[63]
Polyethylene terephthalate/ polypropylene	Films to extend food shelf life (e.g., cherry tomatoes)	AA; antimicrobial activity against *P. aeruginosa*, *S. aureus*, *E. coli* and *S.* Enteritidis; application of active film to the preservation of cherry tomatoes	[75]
	Films to extend food shelf life	AA; SEM imaging	[76]
	Films to extend food shelf life	Color parameters; UV barrier properties; WVPR; mass transfer of OLE into food simulant	[77]
Polyethylene/ polyethylene layers	Packaging for fresh minced pork meat	Analysis of meat samples (color measurements; assay of thiobarbituric acid reactive substances)	[82]
	Food multilayer films to scavenge the free radicals from the package headspace	AA; migration tests	[83]
Methylcellulose/ hydroxypropyl methylcellulose/ polyethylene glycol	Coating to commercial plastic packaging materials	Thickness; AA, antibacterial activity against *S. aureus*, *E. coli*, *S.* Typhimurium; WVPR; oxygen transmission rate	[78]
Polyamide/ polyethylene	Preservation of sliced dry-cured shoulders from Iberian pigs	Analysis of ham samples (color, lipid oxidation, protein oxidation, flavor, taste, texture)	[80]
Polyethylene/ paper layers	Food multilayer films to scavenge the free radicals from the package headspace	AA; migration tests	[83]

Table 3. *Cont.*

Polymer Base	Application	Studies	Reference
Oriented polypropylene films coated by shellac and cellulose nitrate lacquers	Antioxidant packaging for food	AA; migration tests	[79]
Cellulose nanocrystals/poly(butylene adipate-co-terephthalate)	Active packaging for food	Thermal properties; mechanical properties; SEM imaging; release tests in simulant food; AA; oxidation induction temperature; study of π-π stacking interactions	[65]
-	Preservative, antibacterial, and antifungal agent to food shelf life	Tests on poultry meat slices from the chicken breast; poultry, yellow poultry, and rabbit minced meat; pasteurized milk; marinated anchovies; vacuum-packed salmon burgers; raw, peeled undeveined shrimps; frankfurter-type sausages; baked snacks; gluten-free bread; vegetable pate; tomato and pepper pastes	[39,41,88–103]
-	Agent to improve the quality of oils	Tests on refined oils, extra virgin olive oil, sunflower and soybean oils	[8,104,105]

3. Biomedical and Pharmaceutical Applications of Olive Leaf Extract

When reviewing the available literature on the use of olive leaf extract as an enrichment for polymer films, the most commonly mentioned application is in food packaging films. This chapter will focus on the other applications of olive leaf extract, which are not as widely considered but can also be found in the literature—biomedical and pharmaceutical applications.

3.1. Wound-Healing Materials Modified with OLE

CTS [109], SA/PEC [110], and SA/CTS [111] blends have been used to obtain films for potential wound healing. The materials loaded with plant-derived compounds ensure faster tissue repair thanks to the fact that they show better biocompatibility and biological properties. If it is about the use of polymeric material in wound healing, it is important to develop films with good mechanical properties, antibacterial properties, and good stability in water conditions with good permeation ability.

Chitosan membrane enriched with olive leaf extract presented an increase in fluid drainage capacity, promoting an improvement in wound healing [109]. Thanks to the adequate absorption of exudates and mechanical properties, the developed membranes showed visible characteristics for use as wound dressing. Additionally, CTS/OLE films exhibited antibacterial activity against *E. coli* and *S. aureus*, the most common bacteria in skin wounds [109]. An innovative approach to creating a suitable material for wound healing was the development of the 3D-printed patch using SA, PEC, and OLE as an active additive [110]. The obtained materials showed high antioxidant activity with an almost complete release of the polyphenols within 48 h. Further, the patches demonstrated the capacity to absorb exudates and maintain an optimal moisture balance. Most importantly, in this experiment, SA/PEC/OLE films showed the ability to promote cell motility and wound healing and to increase collagen I expression in BJ fibroblast cells [110]. Another promising strategy for potential in skin lesions involves using double-layer membranes composed of CTS and SA, enhanced with the addition of OLE [111]. Thanks to the antioxidant capacity, antibacterial activity (against *E. coli*), good flexibility, ease of manipulation, excellent adhesion between the layers, fluid drainage capacity, and water vapor transmission rate, bilayer CTS/SA films with OLE can be promising for utilization as wound dressings [111].

The objective of A. Casado-Diaz et al.'s study was to evaluate the antioxidant and wound-healing properties of EHO-85, a novel multifunctional amorphous hydrogel in which olive leaf extract was incorporated [112,113]. Apart from the fact that the materials showed antioxidant activity due to the presence of OLE polyphenols, these compounds act as reactive-oxygen-species (ROS) scavengers, thus protecting and promoting the viability of the cell types involved in the structure and regeneration of the skin, such as dermal fibroblasts (HDFs) and keratinocytes (HaCaT) [112]. Nowel EHO-85 hydrogel can modulate the wound microenvironment, mainly through its antioxidant effect and pH and moistening regulation [113].

3.2. Tissue Engineering Materials Modified with OLE

In addition to film-form materials, electrospun fibers can also be used for tissue engineering. An example of such a solution is the incorporation of olive leaf extract inside poly(hydroxybutyrate-co-hydroxyvalerate) (PHBHV) [114], zein (ZN) [115], carboxymethylcellulose/alginate [116], silk fibroin (SF)/hyaluronic acid (HA) [117], or poly(ω-pentadecalactone-co-δ-valerolactone)/gelatin (enzymatically synthesized lactone-based copolymer and gelatin, PDL-VL/GEL) [118] fibers *via* electrospinning to obtain bioactive materials that are useful in biomedical applications. The PHBHV/OLE materials showed adequate stability in the presence of metalloproteinase 9 (MMP-9, an enzyme produced in chronic wounds), together with adequate cytocompatibility with human Caucasian foreskin fetal fibroblasts (HFFF2) and human keratinocytes (HaCaT) [114], while ZN/OLE materials were biocompatible with mouse fibroblasts (NIH/3T3) which proliferated and spread on fiber surface [115]. Additionally, the release of the OLE polyphenols occurred within 4–8 days, and in SEM imaging, the fibrous meshes resulted in smooth fiber surfaces [114,115,117]. Antimicrobial tests have shown that carboxymethylcellulose/alginate/OLE electrospun mats and PDL-VL/GEL/OLE electrospun membranes achieve reduction toward the tested microorganisms (*S. aureus* [116,118], *E. coli* [116,118], *E. faecalis* [116], *P. aeruginosa* [116], *B. subtilis* [118]) while the high antioxidant activity of OLE was preserved during the electrospinning procedure. SF/HA/OLE nanofiber webs can be good candidates for wound dressing applications as well as tissue engineering, thanks to the antioxidant activity, controlled release of OLE, and nontoxicity against the human epidermal keratinocytes [117].

3.3. Drug Delivery Systems with OLE

Mouth-dissolving films (MDFs) are a drug delivery system that can offer some benefits, like the possibility of dispensing the drug with a film dispersed in the mouth, which eliminates problems of conventional drug delivery methods such as swallowing common among pediatric and geriatric patients [119]. Carboxymethyl chitosan/okra gum films with the addition of olive leaf extract as mucoadhesive oral films have been prepared by casting method, and their physicochemical properties have been studied. The authors prepared 17 different propositions of films and performed numerous tests, and the results showed that the addition and amount of OLE directly related to the surface properties of obtained films and their disintegration time. The formulation with the most suitable properties as flexibility and homogeneity, the lower disintegration time (200 s), and a high antioxidant capacity (3.21 mM TEAC/g MDF) have been suggested as a possibility to deliver polyphenols with the mouth dissolving films [120].

Olive leaf extract has been used to obtain silver (AgNP) [121–124] and gold (AuNP) [125] nanoparticles via the environmentally friendly method. AgNPs have the potential for use in medicine, including anticancer therapy [121,124], and as antibacterial agents against drug-resistant bacterial isolates [122], while AuNPs have great potential for use in biomedical applications and will play an important role in future optoelectronic and biomedical device applications [125].

3.4. OLE in Human and Animal Dietary Supplementation

Nowadays, olive leaf extract is proposed and marked as an antioxidant, anti-aging, cardioprotective, immune stimulator, antibiotic, anti-inflammatory, and blood sugar-regulating agent [8]. Several studies have been conducted on olive leaf extract supplementation involving both animals (e.g., chickens, rabbits, rats, fish) and humans. Dietary supplementation of common carp (*Cyprinus carpio*) increases hematological, serum biochemical parameters, and immune-related gene expression levels [126], as well as enhances growth performance, digestive enzyme activity, and growth-related gene expression [127]. If it is about hens supplementation, G.A. Papadopoulos et al. evaluated the effects of OLE, when supplemented at different levels in laying hens' diets, on egg yolk antioxidant properties, egg quality, fatty acid content, and liver pathology characteristics [128]. P. Xie et al. studied the effect of OLE addition to broiler diets on growth performance, breast meat quality, and antioxidant capacity, as well as caecal bacterial populations [129]. On the other hand, regarding dietary supplementation in rabbits, G.E. Younan et al. evaluated the effect of OLE on productive performance, blood parameters, and carcass traits in growing rabbits [130]. Generally, olive leaf extract presented a positive effect in many aspects and is recommended for use in supplementation.

The anticancer effect of olive leaf extract has been confirmed [8,131]. Both inflammatory and cancer cell models have shown that olive leaf polyphenols are anti-inflammatory and protect against DNA damage initiated by free radicals [131]. Nevertheless, it is important to take into account that cell models are very different from the complex human body, and applying these findings to cancer outcomes in humans is difficult. To prove that OLE improves cancer outcomes in humans, clinical trials would be required [131]. The supplementation with OLE yielded beneficial effects on lipid profile and blood pressure in adults, especially in patients with hypertension [132–134], and may represent an effective adjunct therapy that normalizes glucose homeostasis in individuals with diabetes [135]. Moreover, the impact of OLE on upper respiratory illness (URI) has been controlled in high school athletes, and the results showed that OLE supplementation over a season did not significantly reduce URI incidence but did decrease duration in athletes, potentially aiding return to play [136].

To sum up, OLE demonstrated positive effects in terms of antinociceptive, anticancer, cardiovascular, and neuroprotective activities, but human studies on this topic are still extremely uncommon, and the data are still inadequate [8]. Table 4 provides a summary of the materials described above, along with their applications and a list of studies performed to characterize them.

Table 4. The summary of the materials enriched with olive leaf extract used in biomedical and pharmaceutical fields.

Polymer Base	Application	Studies	Reference
Chitosan	Membranes to wound healing	Thickness; mechanical properties; SEM imaging; WVPR; solubility; swelling; fluid drainage capacity; FTIR-ATR spectroscopy; antibacterial activity against *E. coli* and *S. aureus*	[109]
Sodium alginate/ pectin	3D-printed patches for wound healing	Thermal properties; SEM imaging; film expansion profile; WVPR; TPC; AA; evaluation of the anti-inflammatory properties: nitric oxide scavenging activity; in vitro release study; in vitro diffusion study; cell viability assay with BJ fibroblasts; wound-healing scratch assay; hemolysis assay	[110]
Chitosan/sodium alginate	Double-layer membranes for potential in skin lesions	Thickness; mechanical properties; fluid drainage capacity; WVPR; thermal stability; ATR-FTIR spectroscopy; SEM imaging; antimicrobial activity against *E. coli* and *S. aureus*	[111]

Table 4. *Cont.*

Polymer Base	Application	Studies	Reference
EHO-85	Novel Multifunctional Amorphous Hydrogel to wound healing	AA; protection against oxidative damage of OLE on cell cultures (HDFs and HaCaT); in vivo wound-healing activity; moistening ability; study of the effect of EHO-85 on lipid peroxidation and pH in the wound bed; effect of EHO-85 on wound pH in an animal model	[112,113]
Poly(hydroxybutyrate-co-hydroxyvalerate)	Electrospun fibers to wound healing	SEM imaging; FTIR spectroscopy; degradation study; in vitro phenol release study; cytocompatibility with HFFF2 fibroblasts and HaCaT	[114]
Zein	Electrospun fibers to wound healing	SEM imaging; FTIR spectroscopy; thermogravimetric analysis; in vitro release studies; cytocompatibility with NIH/3T3 mouse fibroblasts	[115]
Carboxylmethylcellulose/alginate	Electrospun mats for wound healing	AA; in vitro release studies; SEM imaging; antimicrobial tests against *S. aureus*, *E. coli*, *E. faecalis* and *P. aeruginosa*	[116]
Silk fibroin/hyaluronic acid	Electrospun nanofiber webs	AA; in vitro drug release tests; in vitro cytotoxicity tests with the human epidermal keratinocytes	[117]
Poly(ω-pentadecalactone-co-δ-valerolactone)/gelatin	Electrospun membranes to wound healing	AA; SEM imaging wettability; thermal behavior; antibacterial activity against *E. coli*, *S. aureus*, *B. subtilis*; cytotoxicity tests with fibroblasts	[118]
Carboxymethyl chitosan/okra gum	Mouth-dissolving films for drug delivery	Thickness and weight; disintegration time; pH; SEM imaging; FTIR spectroscopy; AFM; AA	[120]
-	Agent to obtain silver (AgNP) and gold (AuNP) nanoparticles via the environmentally friendly method	SEM; TEM; TGA; XRD; FTIR spectroscopy; UV-Vis spectroscopy; zeta potential; dynamic light scattering (DLS); antimicrobial tests; cytotoxicity tests on healthy lines (HDF) and cancerous cell lines (U118, CaCo-2, Skov-3, MCF-7)	[121–124]
-	Supplementation agent to animals and humans	Tests on chickens, rabbits, rats and fishes; cell culture; human studies	[126–136]

EHO-85—amorphous hydrogel composed of purified water, *Olea Europea* leaf extract, Fucocert®, glycerin, Carbopol 980®, triethanolamine, disodic ethylenediaminetetraacetic acid (EDTA), and Geogard ultra® [113].

4. Cosmetic Applications of OLE

An interesting application of olive leaf extract is its use, either in its pure form [137–144] or combined with nanoparticles [145], as a cosmetic ingredient. M.L. Cádiz-Gurrea et al. studied olive leaf extract enriched with oleuropein [137]. A total of 49 compounds were detected in olive leaf extract, using HPLC coupled with QTOF/MS, while 10 compounds were detected for the first time in olive leaves. All detected compounds demonstrated high antioxidant and antiradical activities. Regarding enzyme inhibition, the promising potential of OLE against skin damage related to aging by elastase and hyaluronidase inhibition was reported. Additionally, OLE did not lead to a decrease in keratinocyte viability [137].

Usually, solvents such as acetone, methanol, ethanol, diethyl ether, and their mixtures, among others, are used for the extraction of active compounds from olive leaves. The novelty lies in the use of eco-friendly solvents such as polypropylene glycol (PPG), lactic acid (LA), and water [138]. In vitro and anti-aging bioactivities of prepared extracts were evaluated, and excellent antioxidant activity was observed, as well as inhibition of elastase, collagenase, tyrosinase, and lipoxygenase. The obtained results suggest that the extracts could be directly used in cosmetic products [138].

Day creams, anti-aging creams, and shampoo with the addition of olive leaf extract were prepared [139–141,143]. No significant changes were observed in the physicochemical parameters of cosmetics enriched with olive leaf extract [139,141], apart from a slight reduction in viscosity by improving the extracts' concentrations in day creams [140]. Furthermore, stability studies showed stable homogenous appearance and effectiveness during a year storage period at room temperature [139]. Apart from that, after two months of anti-aging cream applications by experiment participants, transepidermal water loss decreased, skin hydration increased, wrinkles improved, and skin imaging showed improvement in skin texture [141].

Another study presents the encapsulation of olive leaf extract in biodegradable poly(lactic acid) nanoparticles, characterization, and incorporation of them in a cosmetic formulation [145]. The stability studies of the prepared o/w cream containing loaded nanoparticles demonstrated enhanced stability, particularly in terms of viscosity, pH, organoleptic properties, emulsion phases, and texture [145].

5. Conclusions and Future Perspectives

The use of olive leaf extract (OLE) as a bioactive component in packaging materials offers a natural, sustainable alternative to synthetic preservatives, promoting food safety and shelf life extension. The utilization of olive leaf extract as a bioactive component in packaging materials presents a sustainable and innovative approach to extending product shelf life through its natural antioxidant and antimicrobial properties. Derived from olive tree by-products, including leaves obtained during pruning and harvests, OLE offers a natural, eco-friendly alternative to synthetic preservatives and contributes to waste reduction. However, optimizing extraction efficiency while maintaining environmental and economic sustainability remains a crucial challenge.

Future studies should focus on several key areas:

(1) Maximizing extraction efficiency and functionality: future research should prioritize developing extraction methods that maximize both yield and functionality while minimizing costs. Although advances have been made in extraction techniques such as supercritical fluid extraction or ultrasound-assisted extraction, further efforts are needed to refine these processes to achieve greater reproducibility, flexibility, and scalability [146,147].

(2) Mechanism of action: The specific biochemical pathways through which OLE exerts its antioxidant, anti-inflammatory and antimicrobial actions may not be fully explored [148,149]. A complete understanding of these mechanisms is critical to improving the formulation of OLE-based treatments, ensuring their efficacy in real-world food and medical applications.

(3) Analytical and cultivar variability: One of the challenges in utilizing OLE is the variability in phenolic biosynthesis and bioactivity across different olive cultivars. To address this, innovative analytical techniques like metabolic fingerprinting can be used to classify cultivars based on their molecular profiles, identifying the most suitable varieties for OLE extraction [150]. Understanding the specific composition and activity of OLE components from different cultivars is critical to ensuring the consistency and effectiveness of OLE in packaging applications.

(4) Replacement of synthetic preservatives: OLE's natural preservative qualities can effectively replace synthetic antioxidants and antimicrobials in packaging. Its ability to inhibit the growth of bacteria such as *E. coli*, *S. aureus*, and *S.* Typhimurium makes it a valuable component in active packaging systems. These systems can help reduce the need for artificial preservatives, contributing to safer, cleaner packaging solutions and potentially lowering health risks associated with chemical additives [88].

(5) Economic benefits and minimizing environmental impact: a key consideration for future developments is minimizing the use of toxic chemicals in the extraction and production process. Non-conventional extraction methods that reduce the use of solvents and chemicals are increasingly being explored to present an innovative

alternative or complement to traditional extraction methods (supercritical fluid extraction; cloud point extraction) [146,151]. Reducing the use of solvents and chemical residues is critical not only for improving extraction efficiency and environmental sustainability but also for consumer safety. This makes the process more cost-effective and aligns with environmental sustainability goals, particularly in industries looking to minimize waste (food shelf life and packaging, as well as the biomedical field). Additional in vivo toxicological studies on olive leaves and their phenolic compounds are necessary to fully substantiate the safety of OLE in industrial applications.

(6) Sustainability and circular economy practices: OLE extraction aligns with the principles of the circular economy by utilizing olive tree by-products, which are abundant due to regular pruning and olive harvesting [152,153]. Future perspectives should emphasize integrated approaches to ensure the maximum reuse of biomass, contributing to economic and environmental sustainability. Research into the flexible processing of olive leaves and biomass stabilization techniques will be essential to make OLE extraction more efficient and economically feasible. Furthermore, sequential extraction of various fractions from olive residues should be explored to increase the recovery of valuable components for packaging, food, pharmaceutical, and nutraceutical applications [152,154].

(7) Regulations and market acceptance of OLE-modified packaging materials: OLE not only provides antimicrobial and antioxidant protection for food products but also helps reduce food waste by extending shelf life. However, for such solutions to gain market acceptance, they must meet both consumer expectations as well as regional and global regulatory requirements. Future studies should evaluate the long-term stability, effectiveness, and safety of OLE in packaging under real-world conditions while also addressing any regulatory hurdles and certification processes for its commercial use. Additionally, research into consumer preferences will be important to the commercial success of OLE-infused packaging.

Additionally, future directions that extend beyond the field of packaging materials could be mentioned. One of the most promising applications of OLE is its incorporation into functional foods and nutraceuticals. The development of foods enriched with OLE to provide antioxidant, hypocholesterolemic, and anti-inflammatory benefits represents a promising direction in promoting public health. Despite the compelling evidence of OLE's health benefits, it has yet to be fully integrated into modern medical practice. With additional research, OLE could find a place in the treatment of bacterial infections, inflammatory diseases, and chronic conditions, potentially as a complement to existing medications. To truly unlock the therapeutic potential of OLE, more human studies are needed. Current research is predominantly preclinical, with limited human trials addressing specific areas like skin protection, platelet aggregation, and hypoglycemic effects. Broader human studies are required to confirm OLE's anticancer, cardiovascular, and anti-inflammatory effects.

OLE's potential to replace synthetic antioxidants and antimicrobials positions it as an innovative solution in the packaging sector, aligning with global trends toward sustainability and responsible production. By focusing on integrated approaches that optimize extraction, minimize environmental impact, and maximize the reuse of agricultural by-products, OLE can become a cornerstone of future sustainable packaging technologies, contributing to both environmental and economic goals outlined in the United Nations' 2030 Agenda for Sustainable Development [15].

Funding: This research received no external funding.

Data Availability Statement: No new data were created or analyzed in this study. Data sharing is not applicable to this article.

Conflicts of Interest: The author declares no conflicts of interest.

References

1. Pereira, A.P.; Ferreira, I.C.F.R.; Marcelino, F.; Valentão, P.; Andrade, P.B.; Seabra, R.; Estevinho, L.; Bento, A.; Pereira, J.A. Phenolic Compounds and Antimicrobial Activity of Olive (*Olea europaea* L. Cv. Cobrançosa) Leaves. *Molecules* **2007**, *12*, 1153–1162. [CrossRef]
2. Borjan, D.; Leitgeb, M.; Knez, Ž.; Hrnčič, M.K. Microbiological and Antioxidant Activity of Phenolic Compounds in Olive Leaf Extract. *Molecules* **2020**, *25*, 5946. [CrossRef] [PubMed]
3. Dauber, C.; Parente, E.; Zucca, M.P.; Gámbaro, A.; Vieitez, I. *Olea europea* and By-Products: Extraction Methods and Cosmetic Applications. *Cosmetics* **2023**, *10*, 112. [CrossRef]
4. Khwaldia, K.; Attour, N.; Matthes, J.; Beck, L.; Schmid, M. Olive Byproducts and Their Bioactive Compounds as a Valuable Source for Food Packaging Applications. *Compr. Rev. Food Sci. Food Saf.* **2022**, *21*, 1218–1253. [CrossRef]
5. Bilel, H.; Yahia, M.; Moustafa, S.M.N. An Overview of Chemical Composition and Fungicidal Activity of Olive (*Olea europea* L.) Leaf Extract. *Egypt. J. Chem.* **2023**, *66*, 303–311. [CrossRef]
6. Rahmanian, N.; Jafari, S.M.; Wani, T.A. Bioactive Profile, Dehydration, Extraction and Application of the Bioactive Components of Olive Leaves. *Trends Food Sci. Technol.* **2015**, *42*, 150–172. [CrossRef]
7. Talhaoui, N.; Vezza, T.; Gómez-Caravaca, A.M.; Fernández-Gutiérrez, A.; Gálvez, J.; Segura-Carretero, A. Phenolic Compounds and in Vitro Immunomodulatory Properties of Three Andalusian Olive Leaf Extracts. *J. Funct. Foods* **2016**, *22*, 270–277. [CrossRef]
8. Selim, S.; Albqmi, M.; Al-Sanea, M.M.; Alnusaire, T.S.; Almuhayawi, M.S.; AbdElgawad, H.; Al Jaouni, S.K.; Elkelish, A.; Hussein, S.; Warrad, M.; et al. Valorizing the Usage of Olive Leaves, Bioactive Compounds, Biological Activities, and Food Applications: A Comprehensive Review. *Front. Nutr.* **2022**, *9*, 1008349. [CrossRef]
9. Grabska-Zielińska, S.; Gierszewska, M.; Olewnik-Kruszkowska, E.; Bouaziz, M. Polylactide Films with the Addition of Olive Leaf Extract— Physico-Chemical Characterization. *Materials* **2021**, *14*, 7623. [CrossRef]
10. Topuz, S.; Bayram, M. Oleuropein Extraction from Leaves of Three Olive Varieties (*Olea europaea* L.): Antioxidant and Antimicrobial Properties of Purified Oleuropein and Oleuropein Extracts. *J. Food Process Preserv.* **2022**, *46*, e15697. [CrossRef]
11. Martín-García, B.; Pimentel-Moral, S.; Gómez-Caravaca, A.M.; Arráez-Román, D.; Segura-Carretero, A. Box-Behnken Experimental Design for a Green Extraction Method of Phenolic Compounds from Olive Leaves. *Ind. Crops Prod.* **2020**, *154*, 112741. [CrossRef]
12. da Rosa, G.S.; Vanga, S.K.; Gariepy, Y.; Raghavan, V. Comparison of Microwave, Ultrasonic and Conventional Techniques for Extraction of Bioactive Compounds from Olive Leaves (*Olea europaea* L.). *Innov. Food Sci. Emerg.* **2019**, *58*, 102234. [CrossRef]
13. Khalil, A.A.; Rahman, M.M.; Rauf, A.; Islam, M.R.; Manna, S.J.; Khan, A.A.; Ullah, S.; Akhtar, M.N.; Aljohani, A.S.M.; Abdulmonem, W.A.; et al. Oleuropein: Chemistry, Extraction Techniques and Nutraceutical Perspectives—An Update. *Crit. Rev. Food Sci. Nutr.* **2023**, 1–22. [CrossRef]
14. Frikha, N.; Bouguerra, S.; Kit, G.; Abdelhedi, R.; Bouaziz, M. Smectite Clay KSF as Effective Catalyst for Oxidation of M-Tyrosol with H_2O_2 to Hydroxytyrosol. *React. Kinet. Mech. Cat.* **2019**, *127*, 505–521. [CrossRef]
15. Markhali, F.S.; Teixeira, J.A.; Rocha, C.M.R. Olive Tree Leaves-A Source of Valuable Active Compounds. *Processes* **2020**, *8*, 1177. [CrossRef]
16. Medina, E.; Romero, C.; García, P.; Brenes, M. Characterization of Bioactive Compounds in Commercial Olive Leaf Extracts, and Olive Leaves and Their Infusions. *Food Funct.* **2019**, *10*, 4716–4724. [CrossRef]
17. Afaneh, I.; Yateem, H.; Al-Rimawi, F. Effect of Olive Leaves Drying on the Content of Oleuropein. *Am. J. Analyt. Chem.* **2015**, *6*, 246–252. [CrossRef]
18. Otero, D.M.; Oliveira, F.M.; Lorini, A.; da Fonseca Antunes, B.; Oliveira, R.M.; Zambiazi, R.C. Oleuropein: Methods for Extraction, Purifying and Applying. *Rev. Ceres* **2020**, *67*, 315–329. [CrossRef]
19. Omar, S.H. Oleuropein in Olive and Its Pharmacological Effects. *Sci. Pharm.* **2010**, *78*, 133–154. [CrossRef]
20. Przychodzen, P.; Wyszkowska, R.; Gorzynik-Debicka, M.; Kostrzewa, T.; Kuban-Jankowska, A.; Gorska-Ponikowska, M. Anticancer Potential of Oleuropein, the Polyphenol of Olive Oil, with 2-Methoxyestradiol, Separately or in Combination, in Human Osteosarcoma Cells. *Anticancer Res.* **2019**, *39*, 1243–1251. [CrossRef]
21. Rishmawi, S.; Haddad, F.; Dokmak, G.; Karaman, R. A Comprehensive Review on the Anti-Cancer Effects of Oleuropein. *Life* **2022**, *12*, 1140. [CrossRef] [PubMed]
22. Nediani, C.; Ruzzolini, J.; Romani, A.; Calorini, L. Oleuropein, a Bioactive Compound from *Olea europaea* L., as a Potential Preventive and Therapeutic Agent in Non-Communicable Diseases. *Antioxidants* **2019**, *8*, 578. [CrossRef] [PubMed]
23. Acar-Tek, N.; Ağagündüz, D. Olive Leaf (*Olea europaea* L. Folium): Potential Effects on Glycemia and Lipidemia. *Ann. Nutr. Metab.* **2020**, *76*, 10–15. [CrossRef]
24. Hamdi, H.K.; Castellon, R. Oleuropein, a Non-Toxic Olive Iridoid, Is an Anti-Tumor Agent and Cytoskeleton Disruptor. *Biochem. Biophys. Res. Commun.* **2005**, *334*, 769–778. [CrossRef]
25. Guex, C.G.; Reginato, F.Z.; Figueredo, K.C.; da da Silva, A.R.H.; Pires, F.B.; Jesus, R.d.S.; Lhamas, C.L.; Lopes, G.H.H.; Bauermann, L.d.F. Safety Assessment of Ethanolic Extract of *Olea europaea* L. Leaves after Acute and Subacute Administration to Wistar Rats. *Reg. Toxicol. Pharm.* **2018**, *95*, 395–399. [CrossRef]
26. Clewell, A.E.; Béres, E.; Vértesi, A.; Glávits, R.; Hirka, G.; Endres, J.R.; Murbach, T.S.; Szakonyiné, I.P. A Comprehensive Toxicological Safety Assessment of an Extract of *Olea europaea* L. Leaves (Bonolive™). *Int. J. Toxicol.* **2016**, *35*, 208–221. [CrossRef]

27. Borchers, A.T.; Keen, C.L.; Gershwini, M.E. Mushrooms, Tumors, and Immunity: An Update. *Exp. Biol. Med.* **2004**, *229*, 393–406 [CrossRef]
28. Liu, R.H. Health Benefits of Fruit and Vegetables Are from Additive and Synergistic Combinations of Phytochemicals. *Am. J. Clin. Nutr.* **2003**, *78*, 517S–520S. [CrossRef]
29. Aliabadi, M.A.; Darsanaki, R.K.; Rokhi, M.L.; Nourbakhsh, M.; Raeisi, G. Antimicrobial Activity of Olive Leaf Aqueous Extract *Ann. Biol. Res.* **2012**, *3*, 4189–4191.
30. Brahmi, F.; Flamini, G.; Issaoui, M.; Dhibi, M.; Dabbou, S.; Mastouri, M.; Hammami, M. Chemical Composition and Biological Activities of Volatile Fractions from Three Tunisian Cultivars of Olive Leaves. *Med. Chem. Res.* **2012**, *21*, 2863–2872. [CrossRef]
31. Qabaha, K.; Al-Rimawi, F.; Qasem, A.; Naser, S.A. Oleuropein Is Responsible for the Major Anti-Inflammatory Effects of Olive Leaf Extract. *J. Med. Food* **2018**, *21*, 302–305. [CrossRef] [PubMed]
32. Baysal, G.; Kasapbaşı, E.E.; Yavuz, N.; Hür, Z.; Genç, K.; Genç, M. Determination of Theoretical Calculations by DFT Method and Investigation of Antioxidant, Antimicrobial Properties of Olive Leaf Extracts from Different Regions. *J. Food Sci. Technol.* **2021**, *58*, 1909–1917. [CrossRef] [PubMed]
33. Sánchez-Gutiérrez, M.; Bascón-Villegas, I.; Rodríguez, A.; Pérez-Rodríguez, F.; Fernández-Prior, Á.; Rosal, A.; Carrasco, E. Article Valorisation of *Olea europaea* L. Olive Leaves through the Evaluation of Their Extracts: Antioxidant and Antimicrobial Activity. *Foods* **2021**, *10*, 966. [CrossRef]
34. Giacometti, J.; Milovanović, S.; Jurčić Momčilović, D.; Bubonja-Šonje, M. Evaluation of Antioxidant Activity of Olive Leaf Extract Obtained by Ultrasound-Assisted Extraction and Their Antimicrobial Activity against Bacterial Pathogens from Food. *Int. J. Food Sci. Technol.* **2021**, *56*, 4843–4850. [CrossRef]
35. Sudjana, A.N.; D'Orazio, C.; Ryan, V.; Rasool, N.; Ng, J.; Islam, N.; Riley, T.V.; Hammer, K.A. Antimicrobial Activity of Commercial *Olea europaea* (Olive) Leaf Extract. *Int. J. Antimicrob. Agents* **2009**, *33*, 461–463. [CrossRef]
36. Liu, Y.; McKeever, L.C.; Malik, N.S.A. Assessment of the Antimicrobial Activity of Olive Leaf Extract against Foodborne Bacterial Pathogens. *Front. Microbiol.* **2017**, *8*, 113. [CrossRef]
37. Šimat, V.; Skroza, D.; Tabanelli, G.; Čagalj, M.; Pasini, F.; Gómez-Caravaca, A.M.; Fernández-Fernández, C.; Sterniša, M.; Smole Možina, S.; Ozogul, Y.; et al. Antioxidant and Antimicrobial Activity of Hydroethanolic Leaf Extracts from Six Mediterranean Olive Cultivars. *Antioxidants* **2022**, *11*, 1656. [CrossRef]
38. Keskin, D.; Ceyhan, N.; Uğur, A.; Dbeys, A.D. Antimicrobial Activity and Chemical Constitutions of West Anatolian Olive (*Olea europaea* L.) Leaves. *J. Food Agric. Environ.* **2012**, *10*, 99–102.
39. Ahmed, A.M.; Rabii, N.S.; Garbaj, A.M.; Abolghait, S.K. Antibacterial Effect of Olive (*Olea europaea* L.) Leaves Extract in Raw Peeled Undeveined Shrimp (*Penaeus semisulcatus*). *Int. J. Vet. Sci. Med.* **2014**, *2*, 53–56. [CrossRef]
40. Debib, A.; Boukhatem, M.N. Phenolic Content, Antioxidant and Antimicrobial Activities of "Chemlali" Olive Leaf (*Olea europaea* L.) Extracts. *Int. J. Pharm. Phytochem. Ethnomed.* **2017**, *6*, 38–46. [CrossRef]
41. Ganje, M.; Jafari, S.M.; Dusti, A.; Dehnad, D.; Amanjani, M.; Ghanbari, V. Modeling Quality Changes in Tomato Paste Containing Microencapsulated Olive Leaf Extract by Accelerated Shelf Life Testing. *Food Bioprod. Process* **2016**, *97*, 12–19. [CrossRef]
42. Shialy, Z.; Zarrin, M.; Sadeghi-Nejad, B.; Yusef Naanaie, S. In Vitro Antifungal Properties of Pistacia Atlantica and Olive Extracts on Different Fungal Species. *Curr. Med. Mycol.* **2015**, *1*, 40–45. [CrossRef] [PubMed]
43. Korukluoglu, M.; Sahan, Y.; Yigit, A.; Karakas, R. Antifungal Activity of Olive Leaf (*Olea europaea* L.) Extracts from the Trilye Region of Turkey. *Ann. Microbiol.* **2006**, *56*, 359–362. [CrossRef]
44. Espeso, J.; Isaza, A.; Lee, J.Y.; Sörensen, P.M.; Jurado, P.; de Jesús Avena-Bustillos, R.; Olaizola, M.; Arboleya, J.C. Olive Leaf Waste Management. *Front. Sustain. Food Syst.* **2021**, *5*, 660582. [CrossRef]
45. Abbattista, R.; Ventura, G.; Calvano, C.D.; Cataldi, T.R.I.; Losito, I. Bioactive Compounds in Waste By-Products from Olive Oil Production: Applications and Structural Characterization by Mass Spectrometry Techniques. *Foods* **2021**, *10*, 1236. [CrossRef]
46. Elbehiry, A.; Abalkhail, A.; Marzouk, E.; Elmanssury, A.E.; Almuzaini, A.M.; Alfheeaid, H.; Alshahrani, M.T.; Huraysh, N.; Ibrahem, M.; Alzaben, F.; et al. An Overview of the Public Health Challenges in Diagnosing and Controlling Human Foodborne Pathogens. *Vaccines* **2023**, *11*, 725. [CrossRef]
47. Hou, T.; Sana, S.S.; Li, H.; Xing, Y.; Nanda, A.; Netala, V.R.; Zhang, Z. Essential Oils and Its Antibacterial, Antifungal and Anti-Oxidant Activity Applications: A Review. *Food Biosci.* **2022**, *47*, 101716. [CrossRef]
48. Ahmed, J.; Mulla, M.; Arfat, Y.A. Application of High-Pressure Processing and Polylactide/Cinnamon Oil Packaging on Chicken Sample for Inactivation and Inhibition of *Listeria monocytogenes* and *Salmonella* Typhimurium, and Post-Processing Film Properties. *Food Control* **2017**, *78*, 160–168. [CrossRef]
49. Olewnik-Kruszkowska, E.; Burkowska-But, A.; Tarach, I.; Walczak, M.; Jakubowska, E. Biodegradation of Polylactide-Based Composites with an Addition of a Compatibilizing Agent in Different Environments. *Int. Biodeterior. Biodegrad.* **2020**, *147*, 104840. [CrossRef]
50. Ahmed, J.; Mulla, M.Z.; Arfat, Y.A. Thermo-Mechanical, Structural Characterization and Antibacterial Performance of Solvent Casted Polylactide/Cinnamon Oil Composite Films. *Food Control* **2016**, *69*, 196–204. [CrossRef]
51. Ahmed, J.; Hiremath, N.; Jacob, H. Antimicrobial, Rheological, and Thermal Properties of Plasticized Polylactide Films Incorporated with Essential Oils to Inhibit *Staphylococcus aureus* and *Campylobacter jejuni*. *J. Food Sci.* **2016**, *81*, E419–E429. [CrossRef] [PubMed]

52. Ahmed, J.; Hiremath, N.; Jacob, H. Efficacy of Antimicrobial Properties of Polylactide/Cinnamon Oil Film with and without High-Pressure Treatment against *Listeria monocytogenes* and *Salmonella* Typhimurium Inoculated in Chicken Sample. *Food Packag. Shelf Life* **2016**, *10*, 72–78. [CrossRef]
53. Meherpour, A.; Ghafouri-O, H.; Nasiri, N.; Kioumarsi, H. Effect of Chitosan Coating with Olive Leaf Extract on Shelf Life of Bighead Fish (*Hypophthalmichthys nobilis*) Chilled Fillet. *Res. J. Med. Plants* **2020**, *14*, 111–118. [CrossRef]
54. Musella, E.; El Ouazzani, I.C.; Mendes, A.R.; Rovera, C.; Farris, S.; Mena, C.; Teixeira, P.; Poças, F. Preparation and Characterization of Bioactive Chitosan-Based Films Incorporated with Olive Leaves Extract for Food Packaging Applications. *Coatings* **2021**, *11*, 1339. [CrossRef]
55. Kazan, A.; Demirci, F. Olive Leaf Extract Incorporated Chitosan Films for Active Food Packaging. *Konya J. Eng. Sci.* **2023**, *11*, 1061–1072. [CrossRef]
56. Khalifa, I.; Barakat, H.; El-Mansy, H.A.; Soliman, S.A. Preserving Apple (*Malus domestica* Var. Anna) Fruit Bioactive Substances Using Olive Wastes Extract-Chitosan Film Coating. *Inf. Proc. Agric.* **2017**, *4*, 90–99. [CrossRef]
57. Martiny, T.R.; Pacheco, B.S.; Pereira, C.M.P.; Mansilla, A.; Astorga–España, M.S.; Dotto, G.L.; Moraes, C.C.; Rosa, G.S. A Novel Biodegradable Film Based on κ-Carrageenan Activated with Olive Leaves Extract. *Food Sci. Nutr.* **2020**, *8*, 3147–3156. [CrossRef]
58. da Rosa, G.S.; Vanga, S.K.; Gariepy, Y.; Raghavan, V. Development of Biodegradable Films with Improved Antioxidant Properties Based on the Addition of Carrageenan Containing Olive Leaf Extract for Food Packaging Applications. *J. Polym. Environ.* **2020**, *28*, 123–130. [CrossRef]
59. Martiny, T.R.; Raghavan, V.; de Moraes, C.C.; da Rosa, G.S.; Dotto, G.L. Bio-Based Active Packaging: Carrageenan Film with Olive Leaf Extract for Lamb Meat Preservation. *Foods* **2020**, *9*, 1759. [CrossRef]
60. Albertos, I.; Avena-Bustillos, R.J.; Martín-Diana, A.B.; Du, W.X.; Rico, D.; McHugh, T.H. Antimicrobial Olive Leaf Gelatin Films for Enhancing the Quality of Cold-Smoked Salmon. *Food Packag. Shelf Life* **2017**, *13*, 49–55. [CrossRef]
61. Ayana, B.; Turhan, K.N. Use of Antimicrobial Methylcellulose Films to Control Staphylococcus Aureus during Storage of Kasar Cheese. *Packag. Technol. Sci.* **2009**, *22*, 461–469. [CrossRef]
62. Moura-Alves, M.; Souza, V.G.L.; Silva, J.A.; Esteves, A.; Pastrana, L.M.; Saraiva, C.; Cerqueira, M.A. Characterization of Sodium Alginate-Based Films Blended with Olive Leaf and Laurel Leaf Extracts Obtained by Ultrasound-Assisted Technology. *Foods* **2023**, *12*, 4076. [CrossRef] [PubMed]
63. Erdohan, Z.Ö.; Çam, B.; Turhan, K.N. Characterization of Antimicrobial Polylactic Acid Based Films. *J. Food Eng.* **2013**, *119*, 308–315. [CrossRef]
64. Sánchez-Gutiérrez, M.; Rivera-Ruiz, A.; Sánchez-Fernández, L.; Rodríguez, A.; Carrasco, E. Evaluation of Microbial and Oxidative Changes of 100% Iberian Spanish Salchichón in Contact with a Composite Olive-Leaf-Extract Food Film and Vacuum-Packaged. *Future Foods* **2024**, *10*, 100417. [CrossRef]
65. De Cristofaro, G.A.; Paolucci, M.; Pappalardo, D.; Pagliarulo, C.; Sessini, V.; Lo Re, G. Interface Interactions Driven Antioxidant Properties in Olive Leaf Extract/Cellulose Nanocrystals/Poly(Butylene Adipate-Co-Terephthalate) Biomaterials. *Int. J. Biol. Macromol.* **2024**, *272*, 132509. [CrossRef]
66. Zam, W. Effect of Alginate and Chitosan Edible Coating Enriched with Olive Leaves Extract on the Shelf Life of Sweet Cherries (*Prunus avium* L.). *J. Food Qual.* **2019**, *2019*, 8192964. [CrossRef]
67. Carrapiso, A.I.; Pimienta, M.; Martín, L.; Cardenia, V.; Andrés, A.I. Effect of a Chitosan Coating Enriched with an Olive Leaf Extract on the Characteristics of Pork Burgers. *Foods* **2023**, *12*, 3757. [CrossRef]
68. Famiglietti, M.; Savastano, A.; Gaglione, R.; Arciello, A.; Naviglio, D.; Mariniello, L. Edible Films Made of Dried Olive Leaf Extract and Chitosan: Characterization and Applications. *Foods* **2022**, *11*, 2078. [CrossRef]
69. Ozvural, E.B. Fabrication of Olive Leaf Extract and Hazelnut Skin Incorporated Films to Improve the Quality of Nuggets during Refrigerated and Deep Freeze Storage. *Br. Poult. Sci.* **2019**, *60*, 708–715. [CrossRef]
70. El-Sayed, S.M.; El-Sayed, H.S.; Hashim, A.F.; Youssef, A.M. Valorization of Edible Films Based on Chitosan/Hydroxyethyl Cellulose/Olive Leaf Extract and TiO_2-NPs for Preserving Sour Cream. *Int. J. Biol. Macromol.* **2024**, *268*, 131727. [CrossRef]
71. García, A.V.; Álvarez-Pérez, O.B.; Rojas, R.; Aguilar, C.N.; Garrigós, M.C. Impact of Olive Extract Addition on Corn Starch-Based Active Edible Films Properties for Food Packaging Applications. *Foods* **2020**, *9*, 1339. [CrossRef] [PubMed]
72. Mousavi, S.B.; Mirzaei, H.; Kashiri, M.; Ziaiifar, A.M. The Effect of Olive Leaf Extract on Physical and Mechanical Properties of Zein Corn Edible Film. *J. Food Sci. Technol.* **2019**, *16*, 313–324.
73. Sabbah, M.; Al-Asmar, A.; Younis, D.; Al-Rimawi, F.; Famiglietti, M.; Mariniello, L. Production and Characterization of Active Pectin Films with Olive or Guava Leaf Extract Used as Soluble Sachets for Chicken Stock Powder. *Coatings* **2023**, *13*, 1253. [CrossRef]
74. Fiorentini, C.; Bassani, A.; Zaccone, M.; Luana, M.; Díaz, E.; Apodaca, D.; Spigno, G. Active Coated PLA-PHB Film with Formulations Containing a Commercial Olive Leaf Extract to Improve Quality Preservation of Fresh Pork Burgers. *Chem. Eng. Trans.* **2023**, *102*, 85–90. [CrossRef]
75. Cejudo Bastante, C.; Casas Cardoso, L.; Fernández Ponce, M.T.; Mantell Serrano, C.; Martínez de la Ossa-Fernández, E.J. Characterization of Olive Leaf Extract Polyphenols Loaded by Supercritical Solvent Impregnation into PET/PP Food Packaging Films. *J. Supercrit. Fluids* **2018**, *140*, 196–206. [CrossRef]

76. Cejudo Bastante, C.; Casas Cardoso, L.; Fernández-Ponce, M.T.; Mantell Serrano, C.; Martínez de la Ossa, E.J. Supercritical Impregnation of Olive Leaf Extract to Obtain Bioactive Films Effective in Cherry Tomato Preservation. *Food Packag. Shelf Life* **2019**, *21*, 100338. [CrossRef]
77. Bastante, C.C.; Cran, M.J.; Cardoso, L.C.; Serrano, C.M.; Bigger, S.W. Mass Transfer and Optical Properties of Active PET/PP Food-Grade Films Impregnated with Olive Leaf Extract. *Polymers* **2022**, *14*, 84. [CrossRef]
78. Khanum, F.; Zahoor, T.; Khan, M.I.; Asghar, M.; Sablani, S.S. Antioxidant, Antibacterial and Functional-Food-Packaging Potential of Leaf Extract from Pakistani Olive Cultivars. *Pack. J. Agric. Sci.* **2020**, *57*, 735–742. [CrossRef]
79. Licciardello, F.; Wittenauer, J.; Saengerlaub, S.; Reinelt, M.; Stramm, C. Rapid Assessment of the Effectiveness of Antioxidant Active Packaging-Study with Grape Pomace and Olive Leaf Extracts. *Food Packag. Shelf Life* **2015**, *6*, 1–6. [CrossRef]
80. Amaro-Blanco, G.; Delgado-Adámez, J.; Martín, M.J.; Ramírez, R. Active Packaging Using an Olive Leaf Extract and High Pressure Processing for the Preservation of Sliced Dry-Cured Shoulders from Iberian Pigs. *Innov. Food Sci. Emerg. Technol.* **2018**, *45*, 1–9. [CrossRef]
81. Delgado-Adámez, J.; Bote, E.; Parra-Testal, V.; Martín, M.J.; Ramírez, R. Effect of the Olive Leaf Extracts In Vitro and in Active Packaging of Sliced Iberian Pork Loin. *Packag. Technol. Sci.* **2016**, *29*, 649–660. [CrossRef]
82. Moudache, M.; Nerín, C.; Colon, M.; Zaidi, F. Antioxidant Effect of an Innovative Active Plastic Film Containing Olive Leaves Extract on Fresh Pork Meat and Its Evaluation by Raman Spectroscopy. *Food Chem.* **2017**, *229*, 98–103. [CrossRef]
83. Moudache, M.; Colon, M.; Nerín, C.; Zaidi, F. Phenolic Content and Antioxidant Activity of Olive By-Products and Antioxidant Film Containing Olive Leaf Extract. *Food Chem.* **2016**, *212*, 521–527. [CrossRef]
84. Korukluoglu, M.; Sahan, Y.; Yigit, A. Antifungal Properties of Olive Leaf Extracts and Their Phenolic Compounds. *J. Food Saf.* **2008**, *28*, 76–87. [CrossRef]
85. Korukluoglu, M.; Sahan, Y.; Yigit, A.; Ozer, E.T.; GÜCer, S. Antibacterial Activity and Chemical Constitutions of *Olea europaea* L. Leaf Extracts. *J. Food Process Preserv.* **2010**, *34*, 383–396. [CrossRef]
86. Difonzo, G.; Squeo, G.; Pasqualone, A.; Summo, C.; Paradiso, V.M.; Caponio, F. The Challenge of Exploiting Polyphenols from Olive Leaves: Addition to Foods to Improve Their Shelf-Life and Nutritional Value. *J. Sci. Food Agric.* **2021**, *101*, 3099–3116. [CrossRef]
87. Thielmann, J.; Kohnen, S.; Hauser, C. Antimicrobial Activity of *Olea europaea* Linné Extracts and Their Applicability as Natural Food Preservative Agents. *Int. J. Food Microbiol.* **2017**, *251*, 48–66. [CrossRef]
88. Saleh, E.; Morshdy, A.E.; El-Manakhly, E.; Al-Rashed, S.; Hetta, H.F.; Jeandet, P.; Yahia, R.; El-Saber Batiha, G.; Ali, E. Effects of Olive Leaf Extracts as Natural Preservative on Retailed Poultry Meat Quality. *Foods* **2020**, *9*, 1017. [CrossRef]
89. Forgione, G.; De Cristofaro, G.A.; Sateriale, D.; Pagliuca, C.; Colicchio, R.; Salvatore, P.; Paolucci, M.; Pagliarulo, C. Pomegranate Peel and Olive Leaf Extracts to Optimize the Preservation of Fresh Meat: Natural Food Additives to Extend Shelf-Life. *Microorganisms* **2024**, *12*, 1303. [CrossRef]
90. Rubel, S.A.; Yu, Z.N.; Murshed, H.M.; Islam, S.M.A.; Sultana, D.; Rahman, S.M.E.; Wang, J. Addition of Olive (*Olea europaea*) Leaf Extract as a Source of Natural Antioxidant in Mutton Meatball Stored at Refrigeration Temperature. *J. Food Sci. Technol.* **2021**, *58*, 4002–4010. [CrossRef]
91. Elama, C.; Tarawa, M.; Al-Rimawi, F. Oleuropein from Olive Leaf Extract as Natural Antioxidant of Frozen Hamburger. *J. Food Sci. Eng.* **2017**, *7*, 406–412. [CrossRef]
92. El-Sawah, K.T.; El-Shahawy, R.M.; Nageeb, A.I.; Atalla, K.M. Antimicrobial Activity of Olive Leaves Extracts and Application of Leaves Powder in Meat Preservation. *Fayoum J. Agric. Res. Dev.* **2024**, *38*, 45–55. [CrossRef]
93. Altemimi, A. A Study of the Protective Properties of Iraqi Olive Leaves against Oxidation and Pathogenic Bacteria in Food Applications. *Antioxidants* **2017**, *6*, 34. [CrossRef] [PubMed]
94. Shalaby, A.R.; Anwar, M.M.; Sallam, E.M. Improving Quality and Shelf-Life of Minced Beef Using Irradiated Olive Leaf Extract. *J. Food Process Preserv.* **2018**, *42*, e13789. [CrossRef]
95. El Dessouky Abdel-Aziz, M.; Darwish, M.S.; Mohamed, A.H.; El-Khateeb, A.Y.; Hamed, S.E. Potential Activity of Aqueous Fig Leaves Extract, Olive Leaves Extract and Their Mixture as Natural Preservatives to Extend the Shelf Life of Pasteurized Buffalo Milk. *Foods* **2020**, *9*, 615. [CrossRef]
96. Palmeri, R.; Parafati, L.; Trippa, D.; Siracusa, L.; Arena, E.; Restuccia, C.; Fallico, B. Addition of Olive Leaf Extract (OLE) for Producing Fortified Fresh Pasteurized Milk with an Extended Shelf Life. *Antioxidants* **2019**, *8*, 255. [CrossRef]
97. Testa, B.; Lombardi, S.J.; Macciola, E.; Succi, M.; Tremonte, P.; Iorizzo, M. Efficacy of Olive Leaf Extract (*Olea europaea* L. Cv Gentile Di Larino) in Marinated Anchovies (*Engraulis encrasicolus*, L.) Process. *Heliyon* **2019**, *5*, e01727. [CrossRef]
98. Khemakhem, I.; Fuentes, A.; Lerma-García, M.J.; Ayadi, M.A.; Bouaziz, M.; Barat, J.M. Olive Leaf Extracts for Shelf Life Extension of Salmon Burgers. *Food Sci. Technol. Int.* **2019**, *25*, 91–100. [CrossRef]
99. Alirezalu, K.; Hesari, J.; Eskandari, M.H.; Valizadeh, H.; Sirousazar, M. Effect of Green Tea, Stinging Nettle and Olive Leaves Extracts on the Quality and Shelf Life Stability of Frankfurter Type Sausage. *J. Food Process Preserv.* **2017**, *41*, e13100. [CrossRef]
100. Difonzo, G.; Pasqualone, A.; Silletti, R.; Cosmai, L.; Summo, C.; Paradiso, V.M.; Caponio, F. Use of Olive Leaf Extract to Reduce Lipid Oxidation of Baked Snacks. *Food Res. Int.* **2018**, *108*, 48–56. [CrossRef]
101. Moghaddam, M.F.T.; Jalali, H.; Nafchi, A.M.; Nouri, L. Evaluating the Effects of Lactic Acid Bacteria and Olive Leaf Extract on the Quality of Gluten-Free Bread. *Gene Rep.* **2020**, *21*, 100771. [CrossRef]

102. Difonzo, G.; Squeo, G.; Calasso, M.; Pasqualone, A.; Caponio, F. Physico-Chemical, Microbiological and Sensory Evaluation of Ready-to-Use Vegetable Pâté Added with Olive Leaf Extract. *Foods* **2019**, *8*, 138. [CrossRef] [PubMed]
103. Gamli, Ö.F.; Eraslan, Z.; Akben, S.B. Determination of the Protective Effects of Olive Leaf Extracts on Microbiological and Physicochemical Properties of Pepper Paste Using the Image Processing Methods. *J. Food Process Eng.* **2018**, *41*, e12861. [CrossRef]
104. Rafiee, Z.; Jafari, S.M.; Alami, M.; Khomeiri, M. Antioxidant Effect of Microwave-Assisted Extracts of Olive Leaves on Sunflower Oil. *J. Agric. Sci. Technol.* **2012**, *14*, 1497–1509.
105. Mohammadi, A.; Jafari, S.M.; Esfanjani, A.F.; Akhavan, S. Application of Nano-Encapsulated Olive Leaf Extract in Controlling the Oxidative Stability of Soybean Oil. *Food Chem.* **2016**, *190*, 513–519. [CrossRef]
106. Katouzian, I.; Taheri, R.A. Preparation, Characterization and Release Behavior of Chitosan-Coated Nanoliposomes (Chitosomes) Containing Olive Leaf Extract Optimized by Response Surface Methodology. *J. Food Sci. Technol.* **2021**, *58*, 3430–3443. [CrossRef]
107. Toprakçı, İ.; Şahin, S. Encapsulation of Olive Leaf Antioxidants in Microbeads: Application of Alginate and Chitosan as Wall Materials. *Sustain. Chem. Pharm.* **2022**, *27*, 100707. [CrossRef]
108. Oliveira, F.M.; Oliveira, R.M.; Gehrmann Buchweitz, L.T.; Pereira, J.R.; Cristina dos Santos Hackbart, H.; Nalério, É.S.; Borges, C.D.; Zambiazi, R.C. Encapsulation of Olive Leaf Extract (*Olea europaea* L.) in Gelatin/Tragacanth Gum by Complex Coacervation for Application in Sheep Meat Hamburger. *Food Control* **2022**, *131*, 108426. [CrossRef]
109. Alves, R.C.; Contessa, C.R.; Moraes, C.C.; da Rosa, G.S. Biopolymeric Membranes with Active Principle of Olive Leaves (*Olea europaea* L.) for Potential Topical Application. *Macromol* **2023**, *3*, 314–325. [CrossRef]
110. Patitucci, F.; Motta, M.F.; Dattilo, M.; Malivindi, R.; Leonetti, A.E.; Pezzi, G.; Prete, S.; Mileti, O.; Gabriele, D.; Parisi, O.I.; et al. 3D-Printed Alginate/Pectin-Based Patches Loaded with Olive Leaf Extracts for Wound Healing Applications: Development, Characterization and In Vitro Evaluation of Biological Properties. *Pharmaceutics* **2024**, *16*, 99. [CrossRef]
111. Paulino, L.G.S.; Avila, L.B.; Moraes, C.C.; Khan, M.R.; Manoharadas, S.; dos Reis, G.S.; Dotto, G.L.; da Rosa, G.S. Double-Layer Membranes of Chitosan and Sodium Alginate Added to Natural Olive Leaf Extract for Potential Use in Skin Lesions. *Resources* **2023**, *12*, 97. [CrossRef]
112. Casado-Diaz, A.; Moreno-Rojas, J.M.; Verdú-Soriano, J.; Lázaro-Martínez, J.L.; Rodríguez-Mañas, L.; Tunez, I.; La Torre, M.; Pérez, M.B.; Priego-Capote, F.; Pereira-Caro, G. Evaluation of Antioxidant and Wound-Healing Properties of EHO-85, a Novel Multifunctional Amorphous Hydrogel Containing *Olea europaea* Leaf Extract. *Pharmaceutics* **2022**, *14*, 349. [CrossRef]
113. Casado-Díaz, A.; La Torre, M.; Priego-Capote, F.; Verdú-Soriano, J.; Lázaro-Martínez, J.L.; Rodríguez-Mañas, L.; Berenguer Pérez, M.; Tunez, I. EHO-85: A Multifunctional Amorphous Hydrogel for Wound Healing Containing *Olea europaea* Leaf Extract: Effects on Wound Microenvironment and Preclinical Evaluation. *J. Clin. Med.* **2022**, *11*, 1229. [CrossRef] [PubMed]
114. De la Ossa, J.G.; Danti, S.; Esposito Salsano, J.; Azimi, B.; Tempesti, V.; Barbani, N.; Digiacomo, M.; Macchia, M.; Uddin, M.J.; Cristallini, C.; et al. Electrospun Poly(3-Hydroxybutyrate-Co-3-Hydroxyvalerate)/Olive Leaf Extract Fiber Mesh as Prospective Bio-Based Scaffold for Wound Healing. *Molecules* **2022**, *27*, 6208. [CrossRef] [PubMed]
115. Erdogan, I.; Demir, M.; Bayraktar, O. Olive Leaf Extract as a Crosslinking Agent for the Preparation of Electrospun Zein Fibers. *J. Appl. Polym. Sci.* **2015**, *132*, 41338. [CrossRef]
116. Peršin, Z.; Ravber, M.; Stana Kleinschek, K.; Knez, Ž.; Škerget, M.; Kurečič, M. Bio-Nanofibrous Mats as Potential Delivering Systems of Natural Substances. *Text. Res. J.* **2017**, *87*, 444–459. [CrossRef]
117. Basal, G.; Tetik, G.D.; Kurkcu, G.; Bayraktar, O.; Gurhan, I.D.; Atabey, A. Olive Leaf Extract Loaded Silk Fibroin/Hyaluronic Acid Nanofiber Webs for Wound Dressing Applications. *Dig. J. Nanomater. Biostruct.* **2016**, *11*, 1113–1123.
118. Ulker Turan, C.; Derviscemaloglu, M.; Guvenilir, Y. Enzymatically Synthesized Lactone-Based Copolymer and Gelatin Nanofibrous Blends Loaded with an Olive Leaf Phenolic Compound. *Mater. Today Commun.* **2024**, *38*, 108215. [CrossRef]
119. Sayed, A.M.; Kulkarni, A.D.; Pardeshi, P.U.; Kapile, C.R.; Nehe, A.D. Oral Fast Disintegrating Films of Phytochemicals: A Novel Drug Delivery System. *J. Drug Deliv. Ther.* **2022**, *12*, 226–232. [CrossRef]
120. Köse, M.D.; Gümüş Işık, Ş.; Bayraktar, O. Olive Leaf Polyphenols Loaded Mucoadhesive Oral Films. *J. Eng. Sci. Des.* **2021**, *9*, 366–380. [CrossRef]
121. Atalar, M.N.; Baran, A.; Baran, M.F.; Keskin, C.; Aktepe, N.; Yavuz, Ö.; İrtegun Kandemir, S. Economic Fast Synthesis of Olive Leaf Extract and Silver Nanoparticles and Biomedical Applications. *Part. Sci. Technol.* **2022**, *40*, 589–597. [CrossRef]
122. Khalil, M.M.H.; Ismail, E.H.; El-Baghdady, K.Z.; Mohamed, D. Green Synthesis of Silver Nanoparticles Using Olive Leaf Extract and Its Antibacterial Activity. *Arab. J. Chem.* **2014**, *7*, 1131–1139. [CrossRef]
123. Ramazanli, V.N.; Ahmadov, I.S. Synthesis of Silver Nanoparticles by Using Extract of Olive Leaves. *Adv. Biol. Earth Sci.* **2022**, *7*, 238–244.
124. Rashidipour, M.; Heydari, R. Biosynthesis of Silver Nanoparticles Using Extract of Olive Leaf: Synthesis and in Vitro Cytotoxic Effect on MCF-7 Cells. *J. Nanostruct. Chem.* **2014**, *4*, 112. [CrossRef]
125. Khalil, M.M.H.; Ismail, E.H.; El-Magdoub, F. Biosynthesis of Au Nanoparticles Using Olive Leaf Extract. 1st Nano Updates. *Arab. J. Chem.* **2012**, *5*, 431–437. [CrossRef]
126. Zemheri-Navruz, F.; Acar, Ü.; Yılmaz, S. Dietary Supplementation of Olive Leaf Extract Increases Haematological, Serum Biochemical Parameters and Immune Related Genes Expression Level in Common Carp (*Cyprinus carpio*) Juveniles. *Fish Shellfish Immunol.* **2019**, *89*, 672–676. [CrossRef]

127. Zemheri-Navruz, F.; Acar, Ü.; Yılmaz, S. Dietary Supplementation of Olive Leaf Extract Enhances Growth Performance, Digestive Enzyme Activity and Growth Related Genes Expression in Common Carp *Cyprinus carpio*. *Gen. Comp. Endocrinol.* **2020**, *296*, 113541. [CrossRef]
128. Papadopoulos, G.A.; Lioliopoulou, S.; Nenadis, N.; Panitsidis, I.; Pyrka, I.; Kalogeropoulou, A.G.; Symeon, G.K.; Skaltsounis, A.L.; Stathopoulos, P.; Stylianaki, I.; et al. Effects of Enriched-in-Oleuropein Olive Leaf Extract Dietary Supplementation on Egg Quality and Antioxidant Parameters in Laying Hens. *Foods* **2023**, *12*, 4119. [CrossRef]
129. Xie, P.; Deng, Y.; Huang, L.; Zhang, C. Effect of Olive Leaf (*Olea europaea* L.) Extract Addition to Broiler Diets on the Growth Performance, Breast Meat Quality, Antioxidant Capacity and Caecal Bacterial Populations. *Ital. J. Anim. Sci.* **2022**, *21*, 1246–1258 [CrossRef]
130. Younan, G.; Mohamed, M.; Morsy, W. Effect of Dietary Supplementation of Olive Leaf Extract on Productive Performance, Blood Parameters and Carcass Traits of Growing Rabbits. *Egypt. J. Nutr. Feeds* **2019**, *22*, 173–182. [CrossRef]
131. Boss, A.; Bishop, K.S.; Marlow, G.; Barnett, M.P.G.; Ferguson, L.R. Evidence to Support the Anti-Cancer Effect of Olive Leaf Extract and Future Directions. *Nutrients* **2016**, *8*, 513. [CrossRef]
132. Razmpoosh, E.; Abdollahi, S.; Mousavirad, M.; Clark, C.C.T.; Soltani, S. The Effects of Olive Leaf Extract on Cardiovascular Risk Factors in the General Adult Population: A Systematic Review and Meta-Analysis of Randomized Controlled Trials. *Diabetol. Metab. Syndr.* **2022**, *14*, 151. [CrossRef] [PubMed]
133. Perrinjaquet-Moccetti, T.; Busjahn, A.; Schmidlin, C.; Schmidt, A.; Bradl, B.; Aydogan, C. Food Supplementation with an Olive (*Olea europaea* L.) Leaf Extract Reduces Blood Pressure in Borderline Hypertensive Monozygotic Twins. *Phytother. Res.* **2008**, *22*, 1239–1242. [CrossRef]
134. Lockyer, S.; Rowland, I.; Spencer, J.P.E.; Yaqoob, P.; Stonehouse, W. Impact of Phenolic-Rich Olive Leaf Extract on Blood Pressure, Plasma Lipids and Inflammatory Markers: A Randomised Controlled Trial. *Eur. J. Nutr.* **2017**, *56*, 1421–1432. [CrossRef] [PubMed]
135. Wainstein, J.; Ganz, T.; Boaz, M.; Bar Dayan, Y.; Dolev, E.; Kerem, Z.; Madar, Z. Olive Leaf Extract as a Hypoglycemic Agent in Both Human Diabetic Subjects and in Rats. *J. Med. Food* **2012**, *15*, 605–610. [CrossRef]
136. Somerville, V.; Moore, R.; Braakhuis, A. The Effect of Olive Leaf Extract on Upper Respiratory Illness in High School Athletes: A Randomised Control Trial. *Nutrients* **2019**, *11*, 358. [CrossRef]
137. Cádiz-Gurrea, M.d.l.L.; Pinto, D.; Delerue-Matos, C.; Rodrigues, F. Olive Fruit and Leaf Wastes as Bioactive Ingredients for Cosmetics—A Preliminary Study. *Antioxidants* **2021**, *10*, 245. [CrossRef]
138. Marijan, M.; Mitar, A.; Jakupović, L.; Kardum, J.P.; Končić, M.Z. Optimization of Bioactive Phenolics Extraction and Cosmeceutical Activity of Eco-Friendly Polypropylene-Glycol–Lactic-Acid-Based Extracts of Olive Leaf. *Molecules* **2022**, *27*, 529. [CrossRef]
139. Al-rimawi, F.; Odeh, I.; Bisher, A.; Yateem, H.; Taraweh, M. Natural Antioxidants, Antibacterials from Olive Leaf Extracts Used in Cosmetics, Pharmaceutical, and Food Industries. In Proceedings of the Qatar Foundation Annual Research Conference Proceedings, Doha, Qatar, 19–20 March 2018; Hamad bin Khalifa University Press (HBKU Press): Ar Rayyan, Qatar, 2019; Volume 2014, p. HBPP0116.
140. d'Avanzo, N.; Mancuso, A.; Mare, R.; Silletta, A.; Maurotti, S.; Parisi, O.I.; Cristiano, M.C.; Paolino, D. Olive Leaves and Citrus Peels: From Waste to Potential Resource for Cosmetic Products. *Cosmetics* **2024**, *11*, 41. [CrossRef]
141. Wanitphakdeedecha, R.; Ng, J.N.C.; Junsuwan, N.; Phaitoonwattanakij, S.; Phothong, W.; Eimpunth, S.; Manuskiatti, W. Efficacy of Olive Leaf Extract–Containing Cream for Facial Rejuvenation: A Pilot Study. *J. Cosmet. Dermatol.* **2020**, *19*, 1662–1666. [CrossRef]
142. Oliveira, A.L.S.; Gondim, S.; Gómez-García, R.; Ribeiro, T.; Pintado, M. Olive Leaf Phenolic Extract from Two Portuguese Cultivars –Bioactivities for Potential Food and Cosmetic Application. *J. Environ. Chem. Eng.* **2021**, *9*, 106175. [CrossRef]
143. Sekar, M.; Noordin, H.A.M.N.N.M. Formulation and Evaluation of Herbal Shampoo Containing Rambutan Leaves Extract. *Int. J. Pharma Bio Sci.* **2016**, *7*, P146–P151. [CrossRef]
144. Rodrigues, F.; Pimentel, F.B.; Oliveira, M.B.P.P. Olive By-Products: Challenge Application in Cosmetic Industry. *Ind. Crops Prod.* **2015**, *70*, 116–124. [CrossRef]
145. Kesente, M.; Kavetsou, E.; Roussaki, M.; Blidi, S.; Loupassaki, S.; Chanioti, S.; Siamandoura, P.; Stamatogianni, C.; Philippou, E.; Papaspyrides, C.; et al. Encapsulation of Olive Leaves Extracts in Biodegradable PLA Nanoparticles for Use in Cosmetic Formulation. *Bioengineering* **2017**, *4*, 75. [CrossRef] [PubMed]
146. Le Floch, F.; Tena, M.T.; Ríos, A.; Valcárcel, M. Supercritical Fluid Extraction of Phenol Compounds from Olive Leaves. *Talanta* **1998**, *46*, 1123–1130. [CrossRef]
147. Ilgaz, C.; Kelebek, H.; Kadiroglu, P. Ultrasound-Assisted Extraction of Hydroxytyrosol from Lactiplantibacillus Plantarum Fermented Olive Leaves: Process Optimization and Bioactivity Assessment. *Fermentation* **2023**, *9*, 514. [CrossRef]
148. Al-Rimawi, F.; Sbeih, M.; Amayreh, M.; Rahhal, B.; Mudalal, S. Evaluation of the Antibacterial and Antifungal Properties of Oleuropein, *Olea europea* Leaf Extract, and *Thymus vulgaris* Oil. *BMC Complement. Med. Ther.* **2024**, *24*, 297. [CrossRef]
149. Owen, R.W.; Giacosa, A.; Hull, W.E.; Haubner, R.; Würtele, G.; Spiegelhalder, B.; Bartsch, H. Olive-Oil Consumption and Health: The Possible Role of Antioxidants. *Lancet Oncol.* **2000**, *1*, 107–112. [CrossRef]
150. Di Donna, L.; Mazzotti, F.; Naccarato, A.; Salerno, R.; Tagarelli, A.; Taverna, D.; Sindona, G. Secondary Metabolites of *Olea europaea* Leaves as Markers for the Discrimination of Cultivars and Cultivation Zones by Multivariate Analysis. *Food Chem.* **2010**, *121*, 492–496. [CrossRef]
151. Arya, S.S.; Kaimal, A.M.; Chib, M.; Sonawane, S.K.; Show, P.L. Novel, Energy Efficient and Green Cloud Point Extraction: Technology and Applications in Food Processing. *J. Food Sci. Technol.* **2019**, *56*, 524–534. [CrossRef]

152. Martínez, M.M.M.; Buitrago, E.M.E.; Yñiguez, R.; Puig-Cabrera, M. Circular Economy and Agriculture: Mapping Circular Practices, Drivers, and Barriers for Traditional Table-Olive Groves. *Sustain. Prod. Consum.* **2024**, *46*, 430–441. [CrossRef]
153. Lo Giudice, V.; Faraone, I.; Bruno, M.R.; Ponticelli, M.; Labanca, F.; Bisaccia, D.; Massarelli, C.; Milella, L.; Todaro, L. Olive Trees By-Products as Sources of Bioactive and Other Industrially Useful Compounds: A Systematic Review. *Molecules* **2021**, *26*, 5081. [CrossRef] [PubMed]
154. Berbel, J.; Posadillo, A. Review and Analysis of Alternatives for the Valorisation of Agro-Industrial Olive Oil By-Products. *Sustainability* **2018**, *10*, 237. [CrossRef]

processes

MDPI

Review

Chemical Composition and Biological Activities of the *Cnidoscolus quercifolis*: A Review

Joice Barbosa do Nascimento [1,2], Maria Inácio da Silva [1,2], Johnatan Wellisson da Silva Mendes [1,2], Alexandro Rodrigues Dantas [1,2], Fabíola Fernandes Galvão Rodrigues [1,2], Domenico Montesano [3,*], Monica Gallo [4], Paolo Trucillo [5,*], Gokhan Zengin [6] and José Galberto Martins da Costa [1,2]

1 Postgraduate Program in Biological Chemistry, Department of Biological Chemistry, Regional University of Cariri, Crato 63105-000, CE, Brazil; joicebarbosa.bio@gmail.com (J.B.d.N.); maria.i.silva@urca.br (M.I.d.S.); johnatansmendes@outlook.com (J.W.d.S.M.); alex.dantas@urca.br (A.R.D.); fabiolafer@gmail.com (F.F.G.R.); galberto.martins@gmail.com (J.G.M.d.C.)

2 Research Laboratory of Natural Products, Department of Biological Chemistry, Regional University of Cariri, Crato 63105-000, CE, Brazil

3 Department of Research & Development, Erbagil s.r.l., Via L. Settembrini 13, 82034 Telese Terme, Italy

4 Department of Molecular Medicine and Medical Biotechnology, University of Naples Federico II, Via Pansini 5, 80131 Naples, Italy; mongallo@unina.it

5 Department of Chemical, Materials and Industrial Production Engineering, University of Naples Federico II, Piazzale V. Tecchio 80, 80125 Naples, Italy

6 Physiology and Biochemistry Research Laboratory, Department of Biology, Science Faculty, Selcuk University, Campus, Konya 42130, Turkey; gokhanzengin@selcuk.edu.tr

* Correspondence: d.montesano@erbagil.com (D.M.); paolo.trucillo@unina.it (P.T.)

Abstract: *Cnidoscolus quercifolius*, commonly known as "favela", "faveleira", "urtiga-branca", and "cansanção", is a plant that is native to the Caatinga biome. The species is extremely tolerant to adverse weather conditions and is of great importance for the population of the semi-arid region, as it has uses in afforestation, the recovery of degraded areas, sawmills, fuels, animal feed, and food production. In addition, the species is popularly known for its medicinal uses, and it has been scientifically tested for such purposes. The objective of the research was to compile updated information about the chemical composition, biological activities, and botanical characteristics of the species, in addition to information about its use in folk medicine. It was observed that *C. quercifolius* has a strong usage among people in the Brazilian Caatinga for ophthalmic and other medical conditions, including inflammation in general, scarring, and infections. Studies involving the species have shown its effectiveness as antinociceptive, cytotoxic agent, antioxidant, and insecticide, as also thanks to its anti-inflammatory, hypoglycemic, and repellent characteristics. Other tests have indicated that the vegetable oil from the seed is promising for food consumption. This work demonstrates that further investigations are still necessary to determine the chemical composition and the toxicological characteristics of the species in order to support subsidies for the possible development of new drugs. Such future investigations may include the isolation of its substances, an analysis of its pharmacological activities, and a deepening of the understanding of the mechanisms of action of its various plant products.

Keywords: *Cnidoscolus quercifolius*; faveleira; popular use; phytochemistry; ethnopharmacology; bioactivity

1. Introduction

Empirical knowledge about the properties of medicinal plants helps in the development of research around the world. Ethnopharmacological studies help in the scientific search for new bioactive compounds and pharmacological properties that have not yet been identified. In Brazil, the vast biodiversity of flora and cultural diversity drive the pharmacological exploration of the country's different biomes [1,2].

The Caatinga, also known as the white forest, is an exclusively Brazilian biome [3] that is considered to be the main ecosystem in the northeastern region, occupying more than 50% of that region's area [4]. Despite being the least explored and the least scientifically known Brazilian biome [5,6], the Caatinga is home to a variety of landscapes, with considerable biodiversity and a biological heritage that includes a significant number of rare and endemic plant species with various uses and strong potential for the development of scientific research [7].

The genus *Cnidoscolus* includes several species, known above all for their health properties, due to the myriad of chemical constituents. The genus includes *C. aconitifolius*, *C. chayamansa* Mc Vaugh, *C. multilobus* (Pax) Johnst., *C. quercifolius* Pohl (synonym *C. phyllacanthus* (Mull. Arg) Pax and Hoffm), *C. urens* (L.) Arthur, *C. infestus* Pax and K. Hoffman, *C. pubescens* Pohl, and *C. souzae* Mc Vaugh. Among the native species of the Caatinga biome, *Cnidoscolus quercifolius* Pohl is one of the most important from the scientific and commercial points of view [6]. It is popularly known as "favela", "faveleira", "urtiga-branca", "cansanção", "mandioca-brava", "favela-de-cachorro", or "favela-de-galinha" [7–11]. *C. quercifolius* is a species that is extremely tolerant to adverse climatic conditions and has great socioeconomic importance for the semi-arid region, as it has several uses in afforestation, the recovery of degraded areas, sawmills, fuels, and human and animal food (goats, cattle, and sheep) [12]. Traditionally, plants belonging to the *Cnidoscolus* genus are often used as fodder for the animals of the Caatinga biome, mainly in periods of low rainfall. In addition, faveleira is commonly used in traditional medicine in the treatment of hemorrhoids, kidney problems, ophthalmic infections, urinary infections, inflammation in general, genitourinary problems (uterus, ovaries, and prostate), and other diseases [13].

Although this species represents good possibilities for the development of the semi-arid region and it is widely used in traditional medicine [6], most of the studies related to faveleira are focused on its use in agriculture and livestock [12]. Accordingly, this species is still largely unexplored, with little literature investigating bioactive compounds and their potential use in biotechnological areas. The enormous diversity and the high quantity of bioactive compounds of this plant, and the challenges in identifying them, probably contribute to this absence of literature, so that even today there is only obscure knowledge of this genus. Among the main activities that have been studied are its usage as an antioxidant [14,15], an antimicrobial agent [8,16], a cytotoxic agent [17], and its hypoglycemic [18], anti-inflammatory [19], and antinociceptive properties [20–22].

This manuscript seeks to critically review the literature regarding the pharmacological properties, ethnobotanical uses, and phytochemical analyses of *C. quercifolius*, with the aim of establishing a basis for further research into the use of the products of this species.

2. Methodology

The research was carried out in the Science Direct, Pubmed, Scielo, Web of Science, PMC, DOAJ, Google Scholar, and Lilacs databases. The consulted platforms covered the literature published between the 2000 and 2023 and considered articles published in English and Portuguese. The data collection period for this research was between June 2022 and June 2023 and the searches were carried out using the following terms. in Portuguese and English: *Cnidoscolus quercifolius*, *Cnidoscolus phyllacanthus*, faveleira, biological activity, and chemical composition. The terms were targeted for titles, abstracts, and keywords, alone or in combination.

The primary search identified 2391 results, with 51 from Science Direct, 13 from PUBMED, 14 from SCIELO, 62 from Web of Science, 38 from PMC, 46 from DOAJ, and 2,159 from Google Scholar. However, among these, some articles were indexed in two or more databases; therefore, they were considered only once.

The selected articles were manually reviewed to identify and exclude works that did not meet the inclusion criteria of the study, which were as follows: the period (the last 23 years); the languages (English and Portuguese); originality; and research that reported biological assays, chemical composition, or isolation of compounds from the species *C. quercifolius*. The exclusion criteria were as follows: works in which the content was not within the scope of the study, such as research that investigated the combination of *C. quercifolius* with other species; articles that did not report biological assays, composition chemistry, or compound isolation; and unreliable "publications" such as drafts, preprints of submitted papers, duplicate papers, and conference papers. After the initial screening of titles, abstracts, texts, and times of publication, 41 articles were selected; the rest did not meet the inclusion criteria (n = 2110).

3. Botanical Aspects of *Cnidoscolus quercifolius* Pohl

The species *Cnidoscolus quercifolius* Pohl stands out in the Caatinga biome because of its adaptation to adverse climatic conditions and its wide distribution in urban and rural areas of the Brazilian semi-arid region. It can be found in northeastern Brazil in the states of Ceará, Rio Grande do Sul, North, Paraíba, Pernambuco, Sergipe, Alagoas, Piauí, and Bahia, and in the state of Minas Gerais in the southeastern region of Brazil [23].

C. quercifolius (Figures 1 and 2) is a deciduous, heliophytic, fast-growing plant with an elongated, sparse, and irregularly branched canopy. Its flowering and fruiting can occur between the months of January and March [24], and the species can reach 2 to 5 m in height. It is morphologically classified as a tree, although it presents itself as a shrub or a tree, depending on the location and environmental conditions [9,25].

Figure 1. Tree of *C. quercifolius*. Source: Núcleo de Ecologia e Monitoramento Ambiental (Nema) da Universidade Federal do Vale do São Francisco (Univasf). Espécie do mês: Faveleira, 2019.

Figure 2. Inflorescence of *C. quercifolius*. Source: Núcleo de Ecologia e Monitoramento Ambiental (Nema) da Universidade Federal do Vale do São Francisco (Univasf). Espécie do mês: Faveleira, 2019.

One of its most striking characteristics of the species is the presence of aciculiform stinging trichomes that are present in structures such as the branches, petioles, leaf blades, perianths, and fruits. These thorns act as a highly efficient defense mechanism—when in contact with the skin, they cause allergic reactions such as intense and localized pain that can last for many days [9,26–28]. However, there are less common species, which are the result of natural mutation and which do not have spines in their structure [29,30].

The leaves of *C. quercifolius* have a size ranging from 8 to 16 cm, having a long, thick, lanceolate structure with stems on the blades and spiny endings with small pointed structures on their margins [31]. As a strategy to withstand water restriction, at the end of the rainy season, the leaves mature and fall, appearing again only in the next rainy season, right after the flowers and fruits [31,32]. The inflorescences of the *C. quercifolius* are composed of small white flowers organized in bunches, 4 cm in diameter, being dioecious, with pentamerous petals [31]. Its fruits are dry capsules, dehiscent with urticating hairs, and the dehiscence of the fruit occurs between 56 and 57 days after fertilization, with the explosive opening of the fruit throwing the seeds a distance of more than 30 m away from the mother plant, which facilitates the dissemination of this species in the region [31,32].

The oily seed of the *C. quercifolius* has recognized relevance for the people in the areas where this species is found, as it is the part that is most used for human consumption. It has a grayish-brown color with a streaky appearance; it is ovoid, rigid, and smooth, and has a considerable amount of lipids and proteins [33]. Its length varies from 1.5 to 2.0 cm and its kernel, which corresponds to 56% of its total weight, has a low-intensity yellow color [24,34]. Its roots are tuberous and have an internal viscous liquid; they can reach the deepest layers of the soil, which facilitates the use of rainwater [35]. The food reserves produced in the rainy season are stored in the roots, to be used for subsistence in the dry season [30]. In this way, due to its xerophilic character, the plant is able to grow and reproduce even in periods of prolonged drought [31].

4. Popular Uses of *C. quercifolius*

C. quercifolius is considered to be an ecologically and economically sustainable species [30] that has several forms of use. Due to its adaptation to water stress and high temperatures [23], this species is widely used in the afforestation and reforestation of degraded areas of different soil habitats in the Caatinga [12,25,36,37].

The species has a high economic value as a food source for local populations [23] because, in addition to its oilseeds being consumed in natura or used as ingredients in the preparation of food products (flour, cakes, and breads) [33], the oil extracted from the seeds is edible, having a pleasant taste and odor [29]. In addition, the seeds do not contain toxic proteins [31], and they have high quality and energy value in the production of cooking oil [38].

According to Medeiros et al. [39], the addition of oil from *C. quercifolius* to the diet of goats can increase the nutritional value of milk and of the cheese produced with this milk; therefore, it can be used as a dietary supplement [39]. Faveleira oil is also considered to be a sustainable alternative for the production of biodiesel and biomass [6,40]; however, the technological potential and industrial applications of this seed have not yet been fully explored and understood [15].

In addition to these potential uses, the use of faveleira in popular therapeutic procedures in northeastern Brazil has been reported. The species is commonly used for antimicrobial, expectorant, homeostatic [41], antiseptic, and antitumor purposes, and for the treatment of stomach infections, rheumatism [42,43], hemorrhoids, ophthalmic infections, injuries, fractures, urinary infections, inflammation in general, and genitourinary disorders (uterus, ovaries, and prostate) [13].

Its use is also indicated for toothache [44], ear pain, back pain, dysentery, appendicitis, flu, and cough [45]. The bark poultice is commonly used for healing [46] and the latex produced throughout the entire length of the plant is widely used against dermatoses and warts [45] and for cauterization and wound coagulation [45,47].

It is important to mention that most of the popularly explored therapeutic applications are passed down from generation to generation in different communities, especially in the interior of Brazil, without any scientific validation regarding their pharmacological and toxicological properties to guarantee safety in the use of these preparations. In this sense, ethnopharmacological studies and literature reviews can help to direct the analyses of these products and, consequently, to make a scientific and social contribution by not only seeking knowledge from within the communities, but also directing the research results to them and to ensure the application appropriate and safe use of these products for the health and wellbeing of the population.

5. Chemical Study of *C. quercifolius*

Analyses of chemical constituents have been shown to be of great importance in identifying the biotechnological potential of several plant species. *C. quercifolius* is still little explored with regard to the characterization, identification, and isolation of its bioactive compounds, as well as their use in biotechnological activities. There have been few reports on these aspects, as shown in Table 1. In the paragraphs that follow, the numbers in parentheses refer to the apex numbers reported for substances in Table 1.

In the extracts of the seeds and the pressed cake, several phenolic compounds were observed, including syringic acid (1), ellagic acid (2), catechin (3), quercetin (4), vanillin (5), eugenol (6), vanillic acid (7), and, especially, gallic acid (8) [46]. Alves et al. [48] investigated the chemical composition of essential oils from various parts of *C. quercifolius* (leaves, flowers and bark) obtained by hydrodistillation and, using GC-MS, identified 31 compounds in the leaves (9–11, 39–66), 30 compounds in the flowers (10, 12, 13, 42, 43, 44, 46, 49, 52, 54, 55, 58, 60, 67–83), and 18 compounds in the stem bark (15, 41, 64, 84–98). The main constituents identified in the essential oil of the leaves were phytol (9) (42.1%), α-terpineol (10) (10.9%), and 11,12-dihydroxyvalencene (11) (7.8%). In the flowers, γ-terpinene (12) (20.5%) and β-pinene (13) (9.6%) stood out. The diterpenes dehydroabietal (14) (29.9%) and abietadiene (15) (21.4%) were already in the peels. Oliveira-Júnior et al. [49] also identified several compounds in leaves (16, 17, 99–126) and stem bark (18–23, 99–103, 105, 106, 108, 112, 115, 116, 118, 119–121, 122, 126–151), including the oxidative stress indicators β-ionone (16) and dihydroactinidiolide (17) and the triterpenes lupeol (18) and diploptene (19). In addition, the authors identified, for the first time in a species of the genus *Cnidoscolus*, the diterpenes sandaracopimaradiene (20), 13-methyl-17-norcaur-15-ene (21), kaur-16-ene (22), and dehydroabiethane (23).

Literature data report the isolation of some chemical constituents for the species. Paredes et al. [50] isolated favelin (24), *O*-methyl favelin (25), deoxofavelin (26), and neofavelanone (27) from the root bark, but they also extracted linamarin (28), *trans*-cinnamic acid (29), a mixture of the steroids β-sitosterol (30) and stigmasterol (31). A mixture of the triterpenes lupeol-3β-*O*-cinnamate (32) and lupeol-3β-*O*-dihydrocinnamate (33) were isolated from the leaves [51] and phylacantone (34) from the stem bark [49].

Paula et al. [52] carried out a phytochemical investigation of the hexanic extract of faveleira stem bark, resulting in the isolation of the compounds 3-β-*O*-nanoyl lupeol (35), a mixture of the triterpenes of 3-β-*O*-cinnamoyl lupeol (37) and 3-β-*O*-dihydrocinnamoyl lupeol (38), lupeol (18), and a mixture of β-sitosterol (30) and stigmasterol (31). In the ethanolic extract, bis-nor-diterpenes deoxofaveline (26) and methyl faveline (36) were isolated.

Other authors also analyzed the chemical composition of various plant parts of this species [18,25,53–55] and identified the presence of several metabolites, including xanthones, steroids, triterpenoids, tannins, coumarins, lignans, monoterpenes/diterpenes, naphthoquinones, flavonoids (flavonols, flavones, catechins, and anthocyanins), and phenolic acids (gallic acid). The presence of anthracene derivatives, anthraquinones, xanthines, and saponins was also observed [11]. In this sense, it can be emphasized that the main representatives of the classes of secondary metabolites found in *C. quercifolius* are flavonoids and terpenes (triterpenes and diterpenes) [17,51]. A compilation of all chemical compounds reported for the species is provided in Table 1, and their respective structural formulas are shown in the Supplementary Materials, Figure S1.

The physicochemical characteristics of the oil from the seeds of *C. quercifolius* were also analyzed in some studies [17,27,33,36,39,54–56]. In a study carried out with 100 Caatinga species belonging to the Euphorbiaceae family, Silva et al. [57] reported that faveleira was the species with the highest oil content (22.6 and 33.5% *w*/*w*). The oil obtained from this species has low acidity, low peroxide index [15,58,59], low moisture [15], good thermo-oxidative stability, and low content of carotenoids, β-carotene and chlorophyll [19]. In addition, it has a yellow color [29] or a green color [58], with a density of 0.9122 ± 0.00 g/cm^3 or 912.20 ± 0.00 kg/m^3, viscosity of 0.0525 ± 0.0001 Pa/s [19], and a high saponification index, with a pleasant flavor and aroma [38].

However, the composition of the faveleira seed can undergo some variations. According to Medeiros et al. [35], during the dry season, there is a reduction in humidity that causes a concentration of nutrients, except for lipids. In addition, the composition of the faveleira seed can also be affected by the presence or absence of thorns. Cavalcanti et al. [29] compared the physicochemical characteristics of the oil from the seeds of *C. quercifolius* and observed some differences in its composition with and without thorns. The seed oil of the thorny variation showed higher values than the seed oil without thorns, in terms of density and viscosity, and lower values in terms of refraction, acidity, and saponification. The iodine and peroxide indices did not show significant differences.

A variety of techniques have been applied for the extraction of faveleira seed oil, including mechanical pressing [35,60], solvent extraction [59], extraction assisted by ultrasound [61], and supercritical carbon dioxide extraction [62]. Santos et al. [30] investigated the existence of differences in the composition of faveleira oil extracted by the conventional Soxhlet method with alternative solvents (ethyl acetate, isopropanol, and ethanol) and by the non-conventional pressurized liquid-extraction method. The results of that study showed that extraction with pressurized ethanol at 10 MPa (40 °C, 60 °C, and 80 °C) showed high oil yields (up to 92% efficiency) with lower solvent consumption (58 g versus ~160 g) and shorter extraction time (19 min versus 360 min); however, there was no significant difference in oil quality compared to the quality obtained by the conventional method.

Table 1. Chemical compounds identified and/or isolated in the species *C. quercifolius*. The numbers reported at the apex of following substances are related to molecule numbers reported in Figure S1.

Substance	Part of the Plant	Identification/Isolation Technique	Type of Extract/Fraction	Ref.
Syringic acid [1]	Seed and press cake	UHPLC-DAD	Acetone/water extract	[51]
Ellagic acid [2]	Seed and press cake	UHPLC-DAD	Acetone/water extract	[51]
Catechin [3]	Seed and press cake	UHPLC-DAD	Acetone/water extract	[51]
Quercetin [4]	Seed and press cake	UHPLC-DAD	Acetone/water extract	[51]
Vanillin [5]	Seed and press cake	UHPLC-DAD	Acetone/water extract	[51]
Eugenol [6]	Seed and press cake	UHPLC-DAD	Acetone/water extract	[51]
Vanillic acid [7]	Seed and press cake	UHPLC-DAD	Acetone/water extract	[51]
Gallic acid [8]	Seed and press cake	UHPLC-DAD	Acetone/water extract	[51]
Phytol [9]	Leaves	GC-MS	Essential oil	[48]
α-Terpineol [10]	Leaves, flowers	GC-MS	Essential oil	[48]
11,12-dihydroxyvalencene [11]	Leaves	GC-MS	Essential oil	[48]
γ-Terpinene [12]	Flowers	GC-MS	Essential oil	[48]
β-Pinene [13]	Flowers	GC-MS	Essential oil	[48]
Dehydroabietal [14]	Stem bark	GC-MS	Essential oil	[48]
Abietadiene [15]	Stem bark	GC-MS	Essential oil	[48]
β-Ionone [16]	Leaves	GC-MS	Hexane fraction	[49]
Dihydroactinidiolide [17]	Leaves	GC-MS	Hexane fraction	[49]

Table 1. *Cont.*

Substance	Part of the Plant	Identification/Isolation Technique	Type of Extract/Fraction	Ref.
Lupeol [18]	Stem bark	GC-MS; CLAE-UV-Vis	Hexane fraction; Hexane:AcOEt fraction	[49,52]
Diploptene [19]	Stem bark	GC-MS	Hexane fraction	[49]
Sandaracopimaradiene [20]	Stem bark	GC-MS	Hexane fraction	[49]
13-methyl-17-norkaur-15-ene [21]	Stem bark	GC-MS	Hexane fraction	[49]
Kaur-16-ene [22]	Stem bark	GC-MS	Hexane fraction	[49]
Dehydroabietano [23]	Stem bark	GC-MS	Hexane fraction	[49]
Faveline [24]	Root bark	HPLC–DAD, 1H- and 13C-NMR	Chloroform fraction	[50]
O-Methyl faveline [25]	Root bark	HPLC–DAD, 1H- and 13C-NMR	Chloroform fraction	[50]
Deoxofaveline [26]	Stem bark, root bark	CLAE- UV-Vis; HPLC–DAD, 1H- and 13C-NMR	Hexane:AcOEt fraction; Chloroform fraction	[50,52]
Neofavelanone [27]	Root bark	HPLC–DAD, 1H- and 13C-NMR	Chloroform fraction	[50]
Linamarin [28]	Leaves	IV, MS and NMR	Ethanolic extract	[63]
trans-cinnamic acid [29]	Leaves	IV, MS and NMR	Ethanolic extract	[63]
β-Sitosterol [30]	Leaves, stem bark	IV, MS and NMR; CLAE- UV-Vis	Hexane:AcOEt fraction	[52,63]
Stigmasterol [31]	Leaves, stem bark	IV, MS and NMR; CLAE- UV-Vis	Ethanolic extract; Hexane:AcOEt fraction	[52,63]
Lupeol-3β-*O*-cinnamate [32]	Leaves, stem bark	IV, MS and NMR; GC-MS	Ethanolic extract; Hexano:AcOEt fraction	[49,63]
Lupeol-3β-*O*-dihydrocinnamate [33]	Leaves, stem bark	IV, MS and NMR; GC-MS	Ethanolic extract; Hexane fraction	[49,63]
Phyllcanthon [34]	Stem bark	GC-MS	Hexano:AcOEt fraction	[49]
3-β-*O*-nanoyl-lupeol [35]	Stem bark	CLAE-UV-Vis	Hexane:CH2Cl2 fraction	[52]
Methyl favelin [36]	Stem bark	CLAE-UV-Vis	Hexane:AcOEt fraction	[52]
3-β-*O*-cinnamoyl-lupeol [37]	Stem bark	CLAE-UV-Vis	Hexane:CH2Cl2 fraction	[52]
3-β-*O*-dihydrocinnamoyl-lupeol [38]	Stem bark	CLAE-UV-Vis	Hexane:CH2Cl2 fraction	[52]
1,4-cineole [39]	Leaves	GC-MS	Essential oil	[48]
cis-arbusculone [40]	Leaves	GC-MS	Essential oil	[48]
trans-pinene hydrate [41]	Leaves, flowers, stem bark	GC-MS	Essential oil	[48]
Dihydrolinalool [42]	Leaves, flowers	GC-MS	Essential oil	[48]
Menthone [43]	Leaves, flowers	GC-MS	Essential oil	[48]
Tetrahydrolavandulol [44]	Leaves, flowers	GC-MS	Essential oil	[48]
Neo-dihydrocarveol [45]	Leaves	GC-MS	Essential oil	[48]
Shisofuran [46]	Leaves, flowers	GC-MS	Essential oil	[48]
cis-4-caranone [47]	Leaves	GC-MS	Essential oil	[48]
Citronellol [48]	Leaves	GC-MS	Essential oil	[48]
Thymol [49]	Leaves, flowers	GC-MS	Essential oil	[48]
trans-verbenyl acetate [50]	Leaves	GC-MS	Essential oil	[48]
Silfiperfol-4,7(14)-diene [51]	Leaves	GC-MS	Essential oil	[48]
(*Z*)-β-Damascone [52]	Leaves, flowers	GC-MS	Essential oil	[48]
α-Funebrene [53]	Leaves	GC-MS	Essential oil	[48]
Sesquithujene [54]	Leaves, flowers	GC-MS	Essential oil	[48]

Table 1. *Cont.*

Substance	Part of the Plant	Identification/Isolation Technique	Type of Extract/Fraction	Ref.
(*E*)-*β*-farnesene [55]	Leaves, flowers	GC-MS	Essential oil	[48]
Methyl-*β*-(*E*)-ionol [56]	Leaves	GC-MS	Essential oil	[48]
(*E*)-*β*-Ionene [57]	Leaves	GC-MS	Essential oil	[48]
Biciclogermacrene [58]	Leaves, flowers	GC-MS	Essential oil	[48]
β-Bisabolene [59]	Leaves	GC-MS	Essential oil	[48]
α-Cadinene [60]	Leaves, flowers	GC-MS	Essential oil	[48]
(*E*)-nerolidol [61]	Leaves	GC-MS	Essential oil	[48]
Curcumenol [62]	Leaves	GC-MS	Essential oil	[48]
Vetivenic acid [63]	Leaves	GC-MS	Essential oil	[48]
Cryptomeridiol [64]	Leaves, stem bark	GC-MS	Essential oil	[48]
β-Vetivone [65]	Leaves	GC-MS	Essential oil	[48]
Isophytol [66]	Leaves	GC-MS	Essential oil	[48]
α-Pinene [67]	Flowers	GC-MS	Essential oil	[48]
Sabinene [68]	Flowers	GC-MS	Essential oil	[48]
Myrcene [69]	Flowers	GC-MS	Essential oil	[48]
Dehydroxy-trans-Linalool oxide [70]	Flowers	GC-MS	Essential oil	[48]
Meta-mentha 1-(7),8-diene [71]	Flowers	GC-MS	Essential oil	[48]
δ-2-Carene [72]	Flowers	GC-MS	Essential oil	[48]
α-Phellandrene [73]	Flowers	GC-MS	Essential oil	[48]
o-Cresol methyl ether [74]	Flowers	GC-MS	Essential oil	[48]
iso-sylvestrene [75]	Flowers	GC-MS	Essential oil	[48]
δ-3-Carene [76]	Flowers	GC-MS	Essential oil	[48]
α-Terpinene [77]	Flowers,	GC-MS	Essential oil	[48]
Limonene [78]	Flowers	GC-MS	Essential oil	[48]
Dihydrolinalool [79]	Flowers	GC-MS	Essential oil	[48]
cis-verbenol [80]	Flowers	GC-MS	Essential oil	[48]
cis-*β*-Terpineol [81]	Flowers	GC-MS	Essential oil	[48]
neo-3-Thujanol [82]	Flowers	GC-MS	Essential oil	[48]
cis-Di-hidro-*β*-terpineol [83]	Flowers	GC-MS	Essential oil	[48]
(8*S*),14-cedranediol [84]	Stem bark	GC-MS	Essential oil	[48]
Isopimara-9(11),15-diene [85]	Stem bark	GC-MS	Essential oil	[48]
Beyerene [86]	Stem bark	GC-MS	Essential oil	[48]
Isohibaene [87]	Stem bark	GC-MS	Essential oil	[48]
Palmitic acid [88]	Stem bark	GC-MS	Essential oil	[48]
Cembrene [89]	Stem bark	GC-MS	Essential oil	[48]
13-*epi*-Dolabadiene [90]	Stem bark	GC-MS	Essential oil	[48]
Polygodial [91]	Stem bark	GC-MS	Essential oil	[48]
Phyllocladene [92]	Stem bark	GC-MS	Essential oil	[48]
Manool [93]	Stem bark	GC-MS	Essential oil	[48]
Linoleic acid [94]	Stem bark	GC-MS	Essential oil	[48]
Oleic acid [95]	Stem bark	GC-MS	Essential oil	[48]
Sandaracopimanal [96]	Stem bark	GC-MS	Essential oil	[48]
Phyllocladanol [97]	Stem bark	GC-MS	Essential oil	[48]
Dehydro-abietal [98]	Stem bark	GC-MS	Essential oil	[48]

Table 1. *Cont.*

Substance	Part of the Plant	Identification/Isolation Technique	Type of Extract/Fraction	Ref.
Tetradecane [99]	Leaves, stem bark	GC-MS	Hexane fraction	[49]
Pentadecane [100]	Leaves, stem bark	GC-MS	Hexane fraction	[49]
3-methylpentadecane [101]	Leaves, stem bark	GC-MS	Hexane fraction	[49]
Hexadecane [102]	Leaves, stem bark	GC-MS	Hexane fraction	[49]
2,6,10-trimethyl-pentadecane [103]	Leaves, stem bark	GC-MS	Hexane fraction	[49]
1-cyclohexyl decane [104]	Leaves	GC-MS	Hexane fraction	[49]
Heptadecane [105]	Leaves, stem bark	GC-MS	Hexane fraction	[49]
2,6,10,14-tetramethylpentadecane [106]	Leaves, stem bark	GC-MS	Hexane fraction	[49]
4-ethylheptadecane [107]	Leaves	GC-MS	Hexane fraction	[49]
3-methylheptadecane [108]	Leaves, stem bark	GC-MS	Hexane fraction	[49]
2,6,10,14-tetramethylhexadecane [109]	Leaves	GC-MS	Hexane fraction	[49]
4-cyclohexyl-tridecane [110]	Leaves	GC-MS	Hexane fraction	[49]
Hexadecane-1-ol [111]	Leaves	GC-MS	Hexane fraction	[49]
Nonadecan [112]	Leaves, stem bark	GC-MS	Hexane fraction	[49]
Octasane [113]	Leaves	GC-MS	Hexane fraction	[49]
3-methyl-nonadecane [114]	Leaves	GC-MS	Hexane fraction	[49]
1-tricosene [115]	Leaves, stem bark	GC-MS	Hexane fraction	[49]
Eicosane [116]	Leaves, stem bark	GC-MS	Hexane fraction	[49]
Octadecan-1-ol [117]	Leaves	GC-MS	Hexane fraction	[49]
Heneicosan [118]	Leaves, stem bark	GC-MS	Hexane fraction	[49]
Docosan [119]	Leaves, stem bark	GC-MS	Hexane fraction	[49]
Pentacosan [120]	Leaves, stem bark	GC-MS	Hexane fraction	[49]
Nonacosane [121]	Leaves	GC-MS	Hexane fraction	[49]
Hexacosane [122]	Leaves, stem bark	GC-MS	Hexane fraction	[49]
Heptacosan-1-ol [123]	Leaves	GC-MS	Hexane fraction	[49]
Tetracontane [124]	Leaves	GC-MS	Hexane fraction	[49]
Tetratetracontane [125]	Leaves	GC-MS	Hexane fraction	[49]
Squalene [126]	Leaves, stem bark	GC-MS	Hexane fraction	[49]
1-tridecene [127]	Stem bark	GC-MS	Hexane fraction	[49]
2,3,7-trimethyl-decane [128]	Stem bark	GC-MS	Hexane fraction	[49]
2-methyl-tetradecane [129]	Stem bark	GC-MS	Hexane fraction	[49]
3-methyl-tetradecane [130]	Stem bark	GC-MS	Hexane fraction	[49]
2-methylhexadecane-1-ol [131]	Stem bark	GC-MS	Hexane fraction	[49]
5-methylpentadecane [132]	Stem bark	GC-MS	Hexane fraction	[49]
2-methylpentadecane [133]	Stem bark	GC-MS	Hexane fraction	[49]
3-methylpentadecane [134]	Stem bark	GC-MS	Hexane fraction	[49]
3- hexadecene [135]	Stem bark	GC-MS	Hexane fraction	[49]
1-hexadecene [136]	Stem bark	GC-MS	Hexane fraction	[49]
3-hexyl-1,1,2-trimethyl-cyclobutane [137]	Stem bark	GC-MS	Hexane fraction	[49]
1-decyl-cyclopentane [138]	Stem bark	GC-MS	Hexane fraction	[49]
2-methylhexadecane [139]	Stem bark	GC-MS	Hexane fraction	[49]

Table 1. *Cont.*

Substance	Part of the Plant	Identification/Isolation Technique	Type of Extract/Fraction	Ref.
2-methylheptadecane [140]	Stem bark	GC-MS	Hexane fraction	[49]
2-phenyldodecane [141]	Stem bark	GC-MS	Hexane fraction	[49]
7-methylhexadecane [142]	Stem bark	GC-MS	Hexane fraction	[49]
1-octadecene [143]	Stem bark	GC-MS	Hexane fraction	[49]
Octadecane [144]	Stem bark	GC-MS	Hexane fraction	[49]
2-methyl-eicosane [145]	Stem bark	GC-MS	Hexane fraction	[49]
1-nonadecene [146]	Stem bark	GC-MS	Hexane fraction	[49]
Heneicosan-1-ol [147]	Stem bark	GC-MS	Hexane fraction	[49]
7-hexil-eicosano [148]	Stem bark	GC-MS	Hexane fraction	[49]
7-hexyl-docosane [149]	Stem bark	GC-MS	Hexane fraction	[49]
Heptacosan-1-ol [150]	Stem bark	GC-MS	Hexane fraction	[49]
Tetracontane [151]	Stem bark	GC-MS	Hexane fraction	[49]

UHPLC—ultra-high performance liquid chromatography; GC—gas chromatography; MS—mass spectrometry; UV-Vis—ultraviolet-visible; HPLC—high performance liquid chromatography; DAD—diode array detection; 1H- and 13C-NMR—1H and 13C nuclear magnetic resonance spectroscopy; IV—infra-red.

Regarding the lipid profile of this seed oil, some authors have claimed that *C. quercifolius* has a large amount of unsaturated fatty acids, especially polyunsaturated ones, and that the predominant fatty acid is linoleic acid, followed by oleic acid [15,34,57]. Although it is necessary to carry out a more detailed bioprospecting of the compounds present in the chemical constitution, it is possible to state that the seed oil of *C. quercifolius* has several physicochemical characteristics that make it suitable for use as an edible oil [40].

6. Biological Activities of *C. quercifolius*

6.1. Antinociceptive and Anti-Inflammatory Activities

The study by Ribeiro et al. [19] investigated the anti-inflammatory capacity of *C. quercifolius* seed oil. At a concentration of 500 mg/kg, the oil inhibited paw edema better than indomethacin (20 mg/kg); moreover, the oral administration of the oil significantly reduced the sensation of pain in mice, demonstrating its antinociceptive potential. It was also observed that the oil significantly reduced the pro-inflammatory mediator TNF-α in the plasma of mice.

The authors indicated that quercetin, found in the oil, may be related to anti-inflammatory properties, in view of its already established action in inhibiting the metabolism of arachidonic acid and in the production of eucosanoids, which are important inflammatory mediators.

Some other in vivo methods using models of peritonitis and myeloperoxidase, can help to verify the extent of the anti-inflammatory potential of the oil, In addition, the determination and quantification of cytokines and mediators can help to direct the inhibition pathways and elucidate the oil's mechanisms of action.

It was established by Gomes et al. [20] that the ethanolic extract of the bark and leaves of *C. quercifolius* has an antinociceptive capacity, using established models such as writhing induced by acetic acid, hot plate testing, and formalin testing. In all trials, an effect was observed when using doses that ranged from 100 to 400 mg/kg.

According to Ribeiro et al. [19], the extract could inhibit the production of prostaglandins, which is an important mediator of nociceptive processes. The tests also demonstrated that the extract possibly acts at a central level of pain control, possibly by interacting with opioid receptors. In this sense, the verification of these hypotheses and the elucidation of the antinociceptive mechanisms of the extract can be carried out using tests with opioid receptor antagonists and through the quantification of inflammatory mediators.

6.2. Cytostatic Activity

An investigation into the antiproliferative activity of four species of *Cnidoscolus* referenced in ethnopharmacological surveys in the Caatinga was carried out by Peixoto-Sobrinho et al. [64]. *C. quercifolius* was among the four species that wee studied. The authors found that faveleira bark extract was active against HT-29 (human colon adenocarcinoma) and Hep-2 (human laryngeal squamous cell carcinoma) strains. Paredes et al. [50] evaluated the cytotoxic activity of five extract fractions of the root bark of *C. quercifolius* (hexane, chloroform, ethyl acetate, methanol, and water) against several cell lines. Among the tested extracts, only the chloroform fraction showed good cytotoxic activity against different cancer cell lines—colon carcinoma (HCT-116), ovarian carcinoma (OVCAR-8), and human glioblastoma (SF-295). Better results were achieved for the cell lines HCT-116 and OVCAR-8. These findings were in agreement with the study by Alves et al. [45], who also found that *C. quercifolius* has cytostatic activity against tumor cells.

Furthermore, Oliveira-Júnior et al. [17] reported that the ethyl acetate fraction obtained from the leaves of *C. quercifolius* demonstrated strong cytotoxicity against prostate cancer cell lines (PC3 and PC3-M) and breast cancer (MCF-7), presenting IC_{50} values between 15.75 and 46.97 µg/mL. It was also observed that the ethyl acetate fraction had the highest flavonoid content; according to the authors, these compounds may be related to the cytotoxic activity of the species.

It was found in other studies that compounds isolated from parts of this plant also have this biological activity. The anti-melanoma potential of a terpenoid filacantone (34) obtained from *C. quercifolius* was investigated by Oliveira-Júnior et al. [65] and showed promising cytotoxic activity (IC_{50} 40.9 µM) against chemoresistant human melanoma cells. The antiproliferative activity of linamarin (28) and lupeol derivatives (lupeol-3β-*O*-cinnamate (32) and lupeol-3β-*O*-dihydrocinnamate (33)) were evaluated in human tumor cells; however, the compounds did not show significant inhibition of cell growth when compared to the antitumor used as a control (doxorubicin) [63,66].

Other compounds isolated from faveleira stem bark, such as deoxofaveline (26), 3-β-*O*-nanoyl-lupeol (35), 3-β-*O*-cinnamoyl-lupeol (37), the β-sitosterol mixture (30), and 3-β-*O*-dihydrocinnamoyl-lupeol (38) were also evaluated for their cytotoxic potential against tumor cell lines HL60 (promyelocytic leukemia), MCF-7 (breast carcinoma), and NCIH292 (lung cancer). It was observed in that study that the compound deoxofaveline was considered the most active, with IC_{50} values ranging from 2.7 to 8.9 µg/mL. Faveline methyl was selective for leukemia cells (IC_{50} 1.6µg/mL), showing weak activity in other cells [63].

Some authors indicated that the antiproliferative activity of *C. quercifolius* can be attributed to the presence of cytotoxic substances identified in the species, such as faveline methyl ether, deoxofaveline, favelanone, and neofavelanone, as well as the terpenoids filacantone, 3β-*O*-cinamoyl -lupeol, and 3β-*O*-dihydrocinnamoyl-lupeol. The results of these studies showed that the species has very promising activity for the treatment of neoplastic conditions and may guide the discovery of bioactive compounds for the development of new anti-cancer drugs. However, more specific investigations are needed to determine the mechanism of action of the extracts and their isolated substances.

6.3. Hypoglycemic Effect

C. quercifolius also seems to be promising for the development of new drugs in the treatment of diabetes. It was demonstrated by Lira et al. [54] that the aqueous extract of the leaves of *C. quercifolius* at a dose of 200 mg/kg of body weight has a hypoglycemic effect in diabetic mice, without the presence of oral toxicity. Brito et al. [67], when analyzing the hypoglycemic effect of the aqueous and methanolic extracts of the leaves of *C. quercifolius* in the treatment of diabetic mice, observed that after 30 days all the mice that were treated with the extracts presented a higher percentage of decrease in glycemia than the mice who were treated with metformin (28.67%) at a dose of 200 mg/kg. The percentage of reduction of the aqueous extract was 42.77% and the percentage of reduction of the methanolic extract was 40.25%. This activity was attributed to the presence of gallic acid in both extracts, identified by HPLC. This phenolic compound had already established hypoglycemic activity, by reducing insulin resistance [67].

Study by Moura et al. [8], showed that a protein isolated from the seed of *C. quercifolius* called Cq-IMP (*Cnidoscolus quercifolius*—insulin mimetic protein) has cross-reactivity of the anti-insulin antibody with hypoglycemic potential. This is considered to be an insulin mimetic protein with potent hypoglycemic activity. Moreover, it was able to reduce glycemia in vitro, using the glucose-responsive cell line RIN-5f. Protein toxicity was also evaluated and, according to the authors, Cq-IMP did not show cytotoxicity even at the highest concentration tested (1 mg/mL). The authors also showed that the protein has primary structural characteristics similar to those of sucrose-binding proteins, requiring further investigation to assess the mechanisms involved in the activity.

6.4. Antimicrobial Activity

Through the minimum inhibitory concentration (MIC) and the minimum bactericidal concentration (MBC), Oliveira-Junior et al. [6] verified that three terpenoids (lupeol-3β-*O*-cinnamate (32), lupeol-3β-*O*-dihydrocinnamate (33), and filacantone (34)) isolated from the stem bark of *C. quercifolius* showed significant antibacterial action—especially filacantone (0.25 mg/mL^{-1}), which was considered to be more effective than the standard drug (gentamicin) (0.40 mg/mL^{-1}) against *Enterococcus faecalis* and *E.coli*. The hexanic and methanolic extracts, on the other hand, showed low antibacterial capacity compared to the isolated compounds, both in MIC (5.0 mg/mL^{-1}, both extracts) and in MBC (>10 mg/mL^{-1}, both extracts). Carvalho et al. [68] investigated the hydroalcoholic extract of leaves against 18 bacteria isolated from the teats of goat; however, among the strains analyzed, the extract was effective only on *Staphylococcus* sp. negative coagulase.

The methanolic extract of the bark and the leaves was evaluated by Sobrinho et al. [9], and only the bark extract showed antibacterial activity. It was active against strains of *Staphylococcus aureus*, with MIC between 250 and 500 μg/mL. Its dichloromethane fraction was the most active among the tested samples, with MIC values ranging from 62.5 to 250 μg/mL. Fernandes et al. [51] carried out antimicrobial activity tests with the ethanolic extract of the leaves and with compounds obtained from the extract (linamarin, trans-cinnamic acid, a mixture of two steroids, and a mixture of triterpenes) and observed that linamarin acted as a moderate inhibitor of strains of *E.coli* but was a weak inhibitor of *S. aureus* and *P. aeruginosa*. Thus, the authors stated that the antimicrobial potential of *C. quercifolius* may be partially related to these compounds.

Methanolic extracts of the leaves, roots, and root bark of *C. quercifolius* were evaluated using Gram-positive and Gram-negative bacteria and fungi. Extracts from the leaves and root bark showed similar activity, inhibiting the growth of *Enterococcus faecium*, *E. faecalis*, *Staphylococcus epidermidis*, and *Pseudomonas aeruginosa*, while the extract from the roots inhibited only the *S. epidermidis* strain. According to the authors, the methanolic extracts of *C. quercifolius* showed better inhibition against Gram-positive bacterial strains, as none of the extracts tested managed to inhibit the growth of *E. coli* and *Klebsiella pneumoniae*. Regarding the inhibition of fungal growth, it was observed that the leaf extract showed inhibition against *Lasiodiplodia theobromae* LF11, *L. theobromae* LF124, and *Colletotrichum gloeosporioides* LF50, while root and root bark extracts showed inhibition only against *Colletotrichum gloeosporioides* LF50.

Other authors who investigated the antimicrobial activity of extracts or oils from *C. quercifolius* did not obtain the same results. According to de Souza Eller et al. [69], the crude hydroalcoholic extract of faveleira bark did not show antimicrobial activity, nor did the ethanolic extract of the leaves [43]. Furthermore, Ribeiro et al. [19] found that the seed of *C. quercifolius* does not have antimicrobial activity, as it is not able to inhibit the growth of *S. aureus*, *Listeria monocytogenes*, *Bacillus cereus*, *E. faecalis*, *P. aeroginosa*, *Enterobacter cloacae*, *E. coli*, *Salmonella typhimurium*, or *Enterobacter aerogenescultures*.

These studies suggested that *C. quercifolius* can be considered a source of new antibacterial agents; however, further investigations against different microorganisms and drugs are necessary, as well as other more specific methods of analysis aimed at determining the effectiveness of this species and identifying its mechanism of action in bacterial inhibition.

6.5. Antioxidant Potential

The antioxidant potential of *C. quercifolius* seeds was investigated in some studies [19,28,32,58,59]. The methanolic extract of the leaves, roots, and root bark was evaluated. The root bark extract showed the best antioxidant results, with an IC_{50} of 21.56 ± 0.71 g·mL^{-1}, followed by the extract from the leaf (IC_{50} 133.30 ± 0.73 g·mL^{-1}) and the extract from the root (IC_{50} 171.82 ± 0.69 g·mL^{-1}) [10].

Ribeiro et al. [19] stated that the seed oil, its fractions, and the *C. quercifolius* press cake have antioxidant activity, with the press cake presenting higher potential in the test of ferric reducing antioxidant power (FRAP) (p = 0.0000008), oxygen radical absorbance capacity (ORAC) (p = 0.0015), free radical scavenging DPPH• (p = 0.0011), and total antioxidant activity (p = 0.0023). The antioxidant activity was evaluated using the phosphomolybdenum method. When analyzing the antioxidant activity of seed and pressed cake extracts, Ribeiro et al. [46] also observed similar behavior. The percentage of DPPH• inhibition in the seed and press cake extract was $81.53 \pm 1.80\%$ and $96.63 \pm 1.62\%$, respectively.

According to the authors, the greater inhibition found in the pressed cake may have been related to the lower lipid content of the press cake, which was capable of interfering with the analysis of antioxidant activity. Although interferences may occur, chemical investigation using GC/MS or LC/MS is necessary in order to elucidate which compounds are present in these extracts. In general, hexane extracts are rich in fatty materials and are generally cited as important interferences in the most diverse types of analyses. However, depending on the method applied and the type of compound present in the extract, it is possible that some of these have activity.

This condition was observed in studies by Morais et al. [55], who investigated the antioxidant action of leaves, branches, and roots extracted in hexane, ethanol, and water using the free radical scavenging method, DPPH•. The extracts showed relatively low antioxidant capacity, with a significant potential identified only in the hexane extract of the leaves (IC_{50} of 58.3 ppm).

In most studies, phenolic compounds such asflavonoids were identified as responsible for antioxidant action of *C. quercifolius*, [39,46]. Torres et al. **[60]** suggested that higher antioxidant activity occurs when conditions that favor the extraction of the flavonoid rutin are employed, indicating that this may be partly related to the antioxidant potential of the species.

Naturally, this factor is not an absolute rule. For example, the ethanolic extract of stem bark has been investigated for its antioxidant potential by Nunes et al. [70], using the free radical scavenging methods DPPH• (radical 1,1-diphenyl-2-picrylhydrazil) and ABTS$^+$ (radical 2,2′azinobis-(3-ethylbenzthiazoline-6-sulfonic acid)). This extract showed no detectable antioxidant activity, such as the ethanolic extract of the leaves [19] and the methanolic extract of the bark and leaves [26]. It is important to emphasize that, generally, a higher content of phenolic compounds and flavonoids is found in these types of extracts.

Despite these results, several classes of compounds can exert antioxidant activity, and it is essential to verify different methods that evaluate other types of mechanisms. It can be seen that most of the mentioned studies analyzed their products through free radical scavenging assays; however, other methods, such as deoxyribose protection assays and β-carotene/linoleic acid co-oxidation, can verify the real extent of the antioxidant potential of these products and all their mechanisms of action.

6.6. Insecticidal and Repellent Potential

The insecticidal potential of the seed oil and crude extract of the bark of *C. quercifolius* on the pupae and larvae (late L3 and early L4 stages) of *Aedes aegypti* was investigated by Candido et al. [20]. The authors observed that the studied concentrations were not toxic for the L3 larvae; however, they were lethal for the pupae of this vector, demonstrating that the insecticide potential of *C. quercifolius* may vary according to the period of development of this insect's life cycle. In addition, it was verified that there was similarity between the pupal mortality of the extract and the oil after 24 h. The authors also verified that the seed oil has a repellent action against the oviposition of *A. aegypti* [71] and that the bark extract has an insecticide action, showing efficacy in the control of *Neoleucinodes elegantalis* and *Bemisia* sp. in tomato cultivation [64].

Despite promising results in controlling pests and vectors, the application of these products as natural insecticides requires further evaluation, considering two main factors. The first factor concerns the efficiency of these products, because establishing their active components and their concentration is essential for the proper use of a product that requires constant application. The second factor is related to the inhibition mechanism of the product, especially when it comes to a complex mixture of substances that act synergistically and can cause unwanted effects. Some mechanisms are already established and can be explored in terms of investigations in this direction, either by inhibiting the process of nutrition, development, or reproduction, or by acting directly on specific targets [72]. It is essential to establish its mode of action in order to guarantee effectiveness and safety in the use of these products.

6.7. Hepatoprotective Activity

Hepatoprotective and antioxidant activity in D-galactosamine-induced hepatitis in rats was investigated by Kolli et al. [73]. Using doses of 200 and 400 mg/kg of ethanolic extract of *C. phyllacanthus* (syn. *C. quercifolius* Pohl) leaves, the authors found that the treatment at both doses had a hepatoprotective effect against liver damage, reducing the amount of enzymes and serum markers of liver damage. Such treatments also demonstrated antioxidant activity, with the dose of 400 mg/kg/body weight proving to be more promising and presenting results similar to those with the standard medication (Vitamin-E).

A similar study was carried out by Sharma et al. [74], in which it was observed that the ethanolic extract of the leaves at dosages of 200 mg/kg and 400 mg/kg attenuated the increase in the serum levels of liver enzymes. The reduction of alanine aminotransferase (ALT), aspartate aminotransferase (AST), and alkaline phosphatase (ALP) to normal values indicated that the repair of damaged tissues had occurred. In addition, the histopathological study showed liver repair, demonstrating that the extract has a hepatoprotective effect.

In these studies, the authors suggested that the ethanolic extract of the leaves of *C. quercifolius* acts by preserving the structural integrity of the hepatocyte membrane and, possibly, in regenerating the cell parenchyma; however, histological analyses and a deeper biochemical evaluation are necessary in order to determine how this process occurs. It was also demonstrated that the ability to eliminate free radicals is directly related to the hepatoprotection presented by the extract, which proved to be capable of helping to maintain the function of enzymes such as superoxide dismutase, catalase, and glutathione peroxidase, which are fundamental in the antioxidant defense system.

6.8. Toxicity

In some regions, during periodic droughts, farmers use *C. quercifolius* to feed their animals. However, in a study carried out by Oliveira et al. [24], intoxication in goats due to the ingestion of leaves of *C. phyllacanthus* was reported. Toxicity was caused in proportion to the amount of leaves consumed and their degree of maturation. High doses of the fresh plant caused more intense signs of intoxication, even leading to the death of some animals. According to the authors, this intoxication was caused by the presence of compounds that contain hydrocyanic acid (HCN) in their structures.

The toxicity of the methanolic extract of leaves, roots, and root bark was evaluated by testing against *Artemia* sp. [10]. The results revealed that the leaf extracts (LC_{50} = 1079.78 g·mL^{-1}) and the root bark extracts (LC_{50} = 341.45 g·mL^{-1}) showed low toxicity, while the root extracts showed significant toxicity with LC_{50} = 84.76 g·mL^{-1} [10]. The hydroalcoholic extract of the barks was also evaluated in rodents and presented relatively low acute oral toxicity, with an LD_{50} greater than 2 g/kg [69].

The first study on the toxicity of faveleira seed oil in vitro and in vivo was carried out by Ribeiro et al. [59]. According to the authors, the oil does not have cytotoxic potential against non-tumor cells at a maximum concentration of 5000 µg/mL. Furthermore, in the acute oral toxicity test, no deaths or behavioral changes that would indicate toxicity were observed in any of the mice treated with doses ranging from 10 to 5000 mg/kg. However, further research is needed to analyze the toxicity of the species.

Most of the studies presented in this section bring important toxicological analyses to ensure the adequate evolution of research, through effective and safe concentrations and doses, especially in in vivo studies. It is important to mention that different methods can be applied to investigate the toxicity of natural products. Recently, our research group has been working with models using zebrafish [75], as an alternative with high reproducibility and good efficacy in the analyses.

7. Conclusions

C. quercifoliusis is a species of high value for the populations of the semi-arid region, mainly because of its therapeutic uses in the treatment of diseases and its potential use as a source of food. However, despite being traditionally used in several communities in northeastern Brazil, few of its properties have been scientifically evaluated and determined by pharmacological studies. In addition, there are few studies that address the potential of faveleira for human consumption. Therefore, it is necessary to deepen the scientific investigations involving the chemical composition, biological properties, and nutritional effects of the species, and to isolate and characterize the substances responsible for the claimed effects of the species. Such studies are extremely important, not only in proving the efficacy and promoting the safe use of this species, but also to elucidate its various bioactivities and to support subsidies for the development of new drugs with therapeutic properties.

Supplementary Materials: The following supporting information can be downloaded at: https://www.mdpi.com/article/10.3390/pr11072203/s1, Figure S1: Structural formulas of chemical compounds identified and/or isolated in the species *C. quercifolius*.

Author Contributions: J.B.d.N.: conceptualization, investigation, and writing—original draft preparation; M.I.d.S.: writing—review and editing; J.W.d.S.M.: writing—review and editing; A.R.D.: writing—review and editing; F.F.G.R.: writing—review and editing; D.M.: investigation and writing—original draft preparation; M.G.: investigation and writing—original draft preparation; P.T.: investigation and writing—original draft preparation; G.Z.: investigation and writing—original draft preparation; J.G.M.d.C.: supervision, conceptualization, investigation and writing—original draft preparation. All authors have read and agreed to the published version of the manuscript.

Funding: This research received no external funding.

Institutional Review Board Statement: Not applicable.

Informed Consent Statement: Not applicable.

Data Availability Statement: Not applicable.

Conflicts of Interest: The authors declare no conflict of interest.

References

1. Robert, A.P.; Sieben, P.G.; de Lima, A. Etnobotânica da planta Tabernaemontana sananho Ruiz & Pavon (Apocynaceae)-Revisão Integrativa. *Revista Fitos* **2023**.
2. Sganzerla, C.M.; Predebom, A.J.; Veloso, J.; da Silva Corralo, V.; Junior, W.A.R. Revisão integrativa aplicada a levantamentos etnobotânicos de plantas medicinais no Brasil. *Rev. Acta Ambient. Catarin.* **2022**, *19*, 1–16. [CrossRef]
3. Leal, I.R.; Da Silva, J.M.C.; Tabarelli, M.; Lacher, T.E., Jr. Changing the course of biodiversity conservation in the Caatinga of northeastern Brazil. *Conserv. Biol.* **2005**, *19*, 701–706. [CrossRef]
4. IBGE, Instituto Brasileiro de Geografia e Estatística. Mapa de Biomas e de Vegetação. 2014. Available online: http://www.ibge.gov.br/home/presidencia/noticias/21052004biomashtml.shtm (accessed on 5 July 2022).

5. Angelotti, F.; Sá, I.B.; de Meio, R.F. Mudanças Climáticas e Desertificação no Semi-Árido Brasileiro. 2009. Available online: https://ainfo.cnptia.embrapa.br/digital/bitstream/item/142624/1/ID-41687.pdf (accessed on 1 July 2023).
6. Bezerra, P.D.F. *Viabilidade da Cultura Cnidoscolus Quercifolius Pohl Para Produção de Biodiesel No Semiárido Nordestino*; Universidade Federal do Rio Grande do Norte: Natal, Brazil, 2011.
7. Maciel, B. *Uso Sustentável e Conservação Dos Recursos Florestais da Caatinga*; Serviço Florestal Brasileiro, Misistério do Meio Ambiente: Santarém, Brazil, 2010; pp. 76–81.
8. de Oliveira-Júnior, R.G.; Ferraz, C.A.A.; Pontes, M.C.; Cavalcante, N.B.; da Cruz Araujo, E.C.; de Oliveira, A.P.; Picot, L.; Rolim, L.A.; da Silva Almeida, J.R.G. Antibacterial activity of terpenoids isolated from Cnidoscolus quercifolius Pohl (Euphorbiaceae), a Brazilian medicinal plant from Caatinga biome. *Eur. J. Integr. Med.* **2018**, *24*, 30–34. [CrossRef]
9. Melo, A.L.d.; Sales, M.F.d. O gênero Cnidoscolus Pohl (Crotonoideae-Euphorbiaceae) no Estado de Pernambuco, Brasil. *Acta Bot. Bras.* **2008**, *22*, 806–827. [CrossRef]
10. Moura, L.F.W.G.; da Silva Costa, H.P.; da Silva Neto, J.X.; Dias, L.P.; Magalhães, F.E.A.; van Tilburg, M.F.; Florean, E.O.P.T.; de Oliveira, J.T.A.; de Sousa, D.d.O.B.; Guedes, M.I.F. Orally hypoglycemic activity of an insulin mimetic glycoprotein isolated from Cnidoscolus quercifolius Pohl.(Euphorbiaceae) seeds, Cq-IMP. *Int. J. Biol. Macromol.* **2020**, *159*, 886–895. [CrossRef]
11. Sobrinho, T.; Castro, V.; Saraiva, A.; Almeida, D.; Tavares, E.; Pisciottano, M.; Amorim, E. Phytochemical screening and antibacterial activity of four *Cnidoscolus* species (Euphorbiaceae) against standard strains and clinical isolates. *J. Med. Plant Res.* **2012**, *6*, 3742–3748.
12. Paredes, P.F.M.; Vasconcelos, F.R.; Paim, R.T.T.; Marques, M.M.M.; De Morais, S.M.; Lira, S.M.; Braquehais, I.D.; Vieira, Í.G.P.; Mendes, F.N.P.; Guedes, M.I.F. Screening of bioactivities and toxicity of *Cnidoscolus quercifolius* Pohl. *Evid.-Based Complement. Altern. Med.* **2016**, *2016*, 7930563. [CrossRef]
13. De Albuquerque, U.P.; De Medeiros, P.M.; De Almeida, A.L.S.; Monteiro, J.M.; Neto, E.M.d.F.L.; de Melo, J.G.; Dos Santos, J.P. Medicinal plants of the caatinga (semi-arid) vegetation of NE Brazil: A quantitative approach. *J. Ethnopharmacol.* **2007**, *114*, 325–354. [CrossRef]
14. Sousa, F.d.C.A.; dos Santos, I.G.; de Sousa, M.W.; da Silva, E.G.; dos Santos, B.N.G.; de Medeiros, M.d.G.F.; de Brito, M.d.R.M.; da Silva, W.C.; Siqueira, H.D.a.S.; Siqueira, F.F.F.S. Atividade antioxidante in vitro de Lippia origanoides HBK. *Res. Soc. Dev.* **2021**, *10*, e2810816716. [CrossRef]
15. Ribeiro, P.P.C.; Silva, D.M.d.L.e.; Assis, C.F.d.; Correia, R.T.P.; Damasceno, K.S.F.d.S.C. Bioactive properties of faveleira (*Cnidoscolus quercifolius*) seeds, oil and press cake obtained during oilseed processing. *PloS ONE* **2017**, *12*, e0183935. [CrossRef]
16. Agra, M.d.F. *Plantas Da Medicina Popular Dos Cariris Velhos*; João Pessoa, Editora União: Paraíba, Brasil, 1996.
17. Oliveira, J.; Ferraz, C.; Pereira, E.; Sampaio, P.; Silva, M.; Pessoa, C.; Rolim, L.; Almeida, J.d.S. Phytochemical analysis and cytotoxic activity of *Cnidoscolus quercifolius* Pohl (Euphorbiaceae) against prostate (PC3 and PC3-M) and breast (MCF-7) cancer cells. *Pharmacogn. Mag.* **2019**, *15*, 24–28.
18. Brito, F.C.R.; Rodrigues, P.A.S.; Lira, S.M.; Paredes, P.F.M.; do Vale Canabrava, N.; da Silva, J.Y.G. Effects of Cnidoscolus Quercifolius Pohl leaves extracts on glucemia reduction in diabetic mice Efeitos das folhas de Cnidoscolus Quercifolius Pohl sobre a redução da glucemia em ratos diabéticos. *Braz. J. Dev.* **2022**, *8*, 16159–16174. [CrossRef]
19. Ribeiro, P.P.C.; da Silva Chaves, K.S.F.; de Veras, B.O.; de Oliveira, J.R.S.; de Menezes Lima, V.L.; de Assis, C.R.D.; da Silva, M.V.; de Sousa Júnior, F.C.; de Assis, C.F.; de Araújo Padilha, C.E. Chemical and biological activities of faveleira (Cnidoscolus quercifolius Pohl) seed oil for potential health applications. *Food Chem.* **2021**, *337*, 127771. [CrossRef] [PubMed]
20. Candido, L.P.; Cavalcanti, M.T.; Beserra, E.B. Bioactivity of plant extracts on the larval and pupal stages of *Aedes aegypti* (Diptera, Culicidea). *Rev. Da Soc. Bras. De Med. Trop.* **2013**, *46*, 420–425. [CrossRef]
21. de Araújo Gomes, L.M.; de Lima-Saraiva, S.R.G.; de Andrade, T.M.D.; Silva, J.C.; Diniz, T.C.; Barreto, V.N.S.; Mendes, R.L.; Quintans-Júnior, L.J.; de Sousa Quintans, J.S.; de Lima, J.T. Antinociceptive activity of the ethanolic extract from barks and leaves of *Cnidoscolus quercifolius* (Euphorbiaceae) in mice. *J. Young Pharm.* **2014**, *6*, 64. [CrossRef]
22. de Lima, E.R.; da Silva, B.N.; Oliveira, R.A.S.; van der Linden, L.A.; da Silva, V.C.L.; de Azevedo Rêgo, M.S.; Marinho, M.L. Avaliação do efeito antimicrobiano in vitro dos extratos etanólicos de Plectranthus neochilus e Cnidoscolus quercifolius. *Med. Veterinária (UFRPE)* **2020**, *14*, 248–253. [CrossRef]
23. Ribeiro, P.P.C.; de Medeiros, J.M.S.; da Silva Chaves, K.S.F. *Cnidoscolus quercifolius*: Nutritional value, bioactive activity and potential application of seed and its derivatives in human nutrition. *Trends Food Sci. Technol.* **2020**, *105*, 70–75. [CrossRef]
24. Drumond, M.A.; Salviano, L.M.; Cavalcanti, N.d.B.; Pereira, L.G. Produção, distribuição da biomassa e composição bromatológica da parte aérea da faveleira. *Rev. Bras. De Ciências Agrárias* **2007**, *2*, 308–310. [CrossRef]
25. Fernandes, A.F.C. Caracterização Físico-Química Da Droga Vegetal, Estudo Fitoquímico E Farmacológico de Cnidoscolus Quercifolius Pohl (Euphorbiaceae). 2019. Available online: http://tede.bc.uepb.edu.br/jspui/handle/tede/3914 (accessed on 1 July 2023).
26. Oliveira, D.M.; Pimentel, L.A.; Araújo, J.A.; Rosane, M.; Dantas, A.F.; Riet-Correa, F. Intoxicação por Cnidoscolus phyllacanthus (Euphorbiaceae) em caprinos. *Pesqui. Veterinária Bras.* **2008**, *28*, 36–42. [CrossRef]
27. Oliveira Júnior, R. Estudo Fitoquímico E Avaliação Do Efeito Citotóxico de Cnidoscolus Quercifolius Pohl (Euphorbiaceae) Em Células de Melanoma Humano. Master's Thesis, Universidade Federal do Vale do São Francisco, Juazeiro, Brazil, 2017.

28. Peixoto-Sobrinho, T. Estudo Químico E Biológico de Espécies Do Gênero Cnidoscolus Presentes No Ecossistema Caatinga Com Potencial Atividade Terapêutica. Master's Thesis, Universidade Federal de Pernambuco, Recife, Brazil, 2011.
29. Cavalcanti, M.T.; da Silveira, D.C.; Florêncio, I.M.; Florentino, E.R.; Maracajá, P.B. Caracteristicas físico-químicas do óleo das sementes da faveleira, Cnidoscolus phyllacanthus, Mart., Pax et K. Hoffm., com e sem espinho. *Rev. Verde De Agroecol. E Desenvolv. Sustentável* **2011**, *6*, 49.
30. Santos, K.A.; de Aguiar, C.M.; da Silva, E.A.; da Silva, C. Evaluation of favela seed oil extraction with alternative solvents and pressurized-liquid ethanol. *J. Supercrit. Fluids* **2021**, *169*, 105125. [CrossRef]
31. Lisboa, M.C.; Wiltshire, F.M.S.; Fricks, A.T.; Dariva, C.; Carrière, F.; Lima, Á.S.; Soares, C.M.F. Oleochemistry potential from Brazil northeastern exotic plants. *Biochimie* **2020**, *178*, 96–104. [CrossRef] [PubMed]
32. Ribeiro, P.P.C. Caracterização Físico-Químicas Do óleo Da Semente de Faveleira (Cnidoscolus Quercifolius) E Avaliação Das Propriedades Bioativas Da Semente, Do Óleo E Da Torta. Brasil. 2016. Available online: https://repositorio.ufrn.br/handle/123456789/22188 (accessed on 1 July 2023).
33. Santos, A.; Araújo, M.; Sousa, R.; Lemos, J. Plantas medicinais conhecidas na zona urbana de Cajueiro da Praia, Piauí, Nordeste do Brasil. *Rev. Bras. De Plantas Med.* **2016**, *18*, 442–450. [CrossRef]
34. Santos, K.A.; Aragão Filho, O.P.; Aguiar, C.M.; Milinsk, M.C.; Sampaio, S.C.; Palú, F.; da Silva, E.A. Chemical composition, antioxidant activity and thermal analysis of oil extracted from favela (*Cnidoscolus quercifolius*) seeds. *Ind. Crops Prod.* **2017**, *97*, 368–373. [CrossRef]
35. Medeiros, J.M.S.d.; Ribeiro, P.P.C.; Freitas, E.P.S.; Santos, J.A.B.d.; Damasceno, K.S.F.d.S.C. Chemical composition of faveleira (*Cnidoscolus phyllacanthus*) seeds collected in different seasons. *Rev. Ceres* **2018**, *65*, 286–290. [CrossRef]
36. Coelho, M.; Belmiro, M.; Santos, J.; Fernandes, G. Herbivory among habitats on the Neotropical tree Cnidoscolus quercifolius Pohl. in a seasonally deciduous forest. *Braz. J. Biol.* **2012**, *72*, 453–457. [CrossRef]
37. Pereira, V.L.A.; Alves, F.A.L.; da Silva, V.M.; de Oliveira, J.C.V. Valor nutritivo e consumo voluntário do feno de faveleira fornecido a ovinos no semiárido pernambucano. *Rev. Caatinga* **2012**, *25*, 96–101.
38. Santos, J.; Dantas, J.; Medeiros, C.; Athaíde-Filho, P.; Conceição, M.M.; Santos, J., Jr.; Souza, A. Thermal analysis in sustainable development: Thermoanalytical study of faveleira seeds (*Cnidoscolus quercifolius*). *J. Therm. Anal. Calorim.* **2005**, *79*, 271–275. [CrossRef]
39. Medeiros, E.; Queiroga, R.; Oliveira, M.; Medeiros, A.; Sabedot, M.; Bomfim, M.; Madruga, M. Fatty acid profile of cheese from dairy goats fed a diet enriched with castor, sesame and faveleira vegetable oils. *Molecules* **2014**, *19*, 992–1003. [CrossRef]
40. Freitas, M.d.L. Estudos Das Espécies Cnidoscolus Quercifolius Pax Et K. Hoffm E *Annona muricata* L. Para Geração de Energia. 2013. Available online: https://www.repositorio.ufal.br/handle/riufal/1188 (accessed on 1 July 2023).
41. Matos, F. *Farmácias Vivas: Sistema de Utilização de Plantas Medicinais Projetado Para Pequenas Comunidades*; Editora UFC, Universidade Federal do Ceará: Fortaleza, Brazil, 2002.
42. Agra, M.d.F.; Baracho, G.S.; Nurit, K.; Basílio, I.J.L.D.; Coelho, V. Medicinal and poisonous diversity of the flora of "Cariri Paraibano", Brazil. *J. Ethnopharmacol.* **2007**, *111*, 383–395. [CrossRef]
43. Agra, M.d.F.; Silva, K.N.; Basílio, I.J.L.D.; Freitas, P.F.d.; Barbosa-Filho, J.M. Survey of medicinal plants used in the region Northeast of Brazil. *Rev. Bras. De Farmacogn.* **2008**, *18*, 472–508. [CrossRef]
44. Regis, M.; Pinto, T.; Almeida, G.; Lima, E. Perfil Do Uso de Plantas Medicinais Em Saúde Bucal No Município de Caicó/Rn. 2021. Available online: https://downloads.editoracientifica.org/articles/201102328.pdf (accessed on 1 July 2023).
45. Alves, E.P.; de F, L.R.; de Almeida, C.M.; Freires, I.A.; Rosalen, P.L.; Ruiz, A.L.; Granville-Garcia, A.F.; Godoy, G.P.; Pereira, J.V.; de Brito Costa, E.M. Antimicrobial and Antiproliferative Activity of Bauhinia forficata Link and Cnidoscolus quercifolius Extracts commonly Used in Folk Medicine. *J. Contemp. Dent. Pract.* **2017**, *18*, 635–640.
46. Ribeiro Filho, N.; Caldeira, V.; Florêncio, I.; Azevedo, D.; Dantas, J. Avaliação comparada dos índices químicos nitrogênio e fósforo nas porções morfológicas das espécimes de faveleira com espinhos e sem espinhos. *Rev. Bras. De Prod. Agroind.* **2007**, *9*, 149–160. [CrossRef]
47. Rêgo, M.; Franco, E.; Oliveira, R.; Linden, L.; Silva, V.; Maia, C.; Teixeira, M.; Marinho, M.; Lima, E. Evaluation of tissue repair using phytotherapeutic gel from *Plectranthus neochilus, Schlechter* (boldo-gambá) and *Cnidoscolus quercifolius* Pohl (favela) in Wistar rats. *Arq. Bras. De Med. Veterinária E Zootec.* **2021**, *73*, 395–405. [CrossRef]
48. Alves, A.; de Moraes, M.; da Camara, C.; Lucena, M. Chemical Composition of the Essential Oil of Cnidoscolus quercifolius from Brazil. *Chem. Nat. Compd.* **2020**, *56*, 933–936. [CrossRef]
49. Oliveira-Junior, R.; Ferraz, C.; Oliveira, A.; Alencar-Filho, J.; Almeida, J.; Brasileiro, C. Chemical Constituents of Non-Polar Fractions Obtained from Cnidoscolus quercifolius Pohl (Euphorbiaceae), a Medicinal Plant Native from the Brazilian Caatinga Biome. *Rev. Virtual De Química* **2019**, *11*, 498–516. [CrossRef]
50. Paredes, P.F.M.; de Morais, S.M.; Brito, F.C.R.; Moura, L.F.W.G.; de Araújo Rodrigues, P.; Benjamin, S.R.; Magalhães, F.E.A.; Florean, E.O.P.T.; Guedes, M.I.F. Characterization of *Cnidoscolus quercifolius* Pohl bark root extract and evaluation of cytotoxic effect on human tumor cell lines. *Asian Pac. J. Trop. Biomed.* **2018**, *8*, 345.
51. Ribeiro, P.P.C.; Sousa Júnior, F.C.d.; Assis, C.F.d.; Veras, B.O.d.; Padilha, C.E.d.A.; Stamford, T.C.M.; Damasceno, K.S.F.d.S.C. Phenolic profiles of faveleira (*Cnidoscolus quercifolius* Pohl) seed and press cake extracts: Potential for a new trend in functional food. *Braz. J. Food Technol.* **2020**, *23*, 1–8. [CrossRef]

52. Paula, A.C.; Melo, K.M.; da Silva, A.M.; Ferreira, D.A.; Monte, F.J.; Santiago, G.M.; Lemos, T.L.; Braz-Filho, R.; Militao, G.C.; da Silva, P.B. Chemical constituents and cytotoxic activity of Cnidoscolus phyllacanthus. *Rev. Virtual De Quim.* **2016**, *8*, 231–241. [CrossRef]
53. de Araújo Gomes, L.M.; de Andrade, T.M.; Silva, J.C.; de Lima, J.T.; Quintans-Junior, L.J.; da Silva Almeida, J.R.G. Phytochemical screening and anti-inflammatory activity of Cnidoscolus quercifolius (Euphorbiaceae) in mice. *Pharmacogn. Res.* **2014**, *6*, 345.
54. Lira, S.; Canabrava, N.; Benjamin, S.; Silva, J.; Viana, D.; Lima, C.; Paredes, P.; Marques, M.; Pereira, E.; Queiroz, E. Evaluation of the toxicity and hypoglycemic effect of the aqueous extracts of *Cnidoscolus quercifolius* Pohl. *Braz. J. Med. Biol. Res.* **2017**, *50*. [CrossRef]
55. Morais, N.; Oliveira Neto, F.; Melo, A.; Bertini, L.; Silva, F.; Alves, L. Prospecção fitoquímica e avaliação do potencial antioxidante de *Cnidoscolus phyllacanthus* (müll. Arg.) Pax & k.hoffm. Oriundo de apodi—RN. *Rev. Bras. De Plantas Med.* **2016**, *18*, 180–185.
56. Sobrinho, T.; Castro, V.; Saraiva, A.M.; Almeida, D.M.; Tavares, E.A.; Amorim, E.L. Phenolic content and antioxidant capacity of our *Cnidoscolus* species (Euphorbiaceae) used as ethnopharmacologicals in Caatinga. *Afr. J. Pharm. Pharmacol.* **2011**, *5*, 2310–2316.
57. Silva, S.I.; Oliveira, A.F.M.; Negri, G.; Salatino, A. Seed oils of Euphorbiaceae from the Caatinga, a Brazilian tropical dry forest. *Biomass Bioenergy* **2014**, *69*, 124–134. [CrossRef]
58. Ribeiro, P.P.C. Caracterização E Utilização Sustentável Do Óleo E Do Resíduo Da Semente de Faveleira (Cnidoscolus quercifolius): Agregação de Valor a Uma Forrageira Do Semiárido. 2021. Available online: https://repositorio.ufpe.br/handle/123456789/40770 (accessed on 1 July 2023).
59. Ribeiro, P.P.C.; Silva, D.M.d.L.; Dantas, M.M.; Ribeiro, K.D.d.S.; Dimenstein, R.; Damasceno, K.S.F.d.S.C. Determination of tocopherols and physicochemical properties of faveleira (*Cnidoscolus quercifolius*) seed oil extracted using different methods. *Food Sci. Technol.* **2019**, *39*, 280–285. [CrossRef]
60. Torres, D.d.S.; Pereira, E.C.; Sampaio, P.A.; de Souza, N.A.; Ferraz, C.A.; Oliveira, A.P.d.; Moura, C.A.; Almeida, J.R.; Rolim-Neto, P.J.; de Oliveira-Júnior, R.G. Influência do método extrativo no teor de flavonoides de Cnidoscolus quercifolius POHL (Euphorbiaceae) e atividade antioxidante. *Química Nova* **2018**, *41*, 743–747. [CrossRef]
61. Santos, K.A.; da Silva, E.A.; da Silva, C. Ultrasound-assisted extraction of favela (*Cnidoscolus quercifolius*) seed oil using ethanol as a solvent. *J. Food Process. Preserv.* **2021**, *45*, e15497. [CrossRef]
62. Santos, K.A.; da Silva, E.A.; da Silva, C. Supercritical CO_2 extraction of favela (*Cnidoscolus quercifolius*) seed oil: Yield, composition, antioxidant activity, and mathematical modeling. *J. Supercrit. Fluids* **2020**, *165*, 104981. [CrossRef]
63. Fernandes, A.F.C.; Silvestre, G.F.G.; Rocha Júnior, A.C.S.; de Souza Lima, T.K.; Ruiz, A.L.T.G.; Alves, H.d.S. Antiproliferative, antileishmanial and antimicrobial studies on *Cnidoscolus quercifolius* Pohl (Euphorbiaceae). *Nat. Prod. Res.* **2021**, *35*, 5339–5343. [CrossRef]
64. Peixoto, M.d.S.R.M.; de Lima, V.L.A.; Dantas, J.P.; Souza, S.S. Eficiência de extratos vegetais e urina de vaca no controle de Neoleucinodes elegantalis, Guenée, 1854, Lepidoptera, Pyralidae, e Bemisia sp, Hemiptera, Aleurodidae, em tomateiro orgânico. *Rev. Verde De Agroecol. E Desenvolv. Sustentável* **2013**, *8*, 16.
65. Oliveira-Júnior, R.G.; Ferraz, C.A.A.; de Oliveira, A.P.; da Cruz Araújo, E.C.; Prunier, G.; Beaugeard, L.; Groult, H.; Picot, L.; de Alencar Filho, E.B.; El Aouad, N. Bis-nor-diterpene from *Cnidoscolus quercifolius* (Euphorbiaceae) induces tubulin depolymerization-mediated apoptosis in BRAF-mutated melanoma cells. *Chem.-Biol. Interact.* **2022**, *355*, 109849. [CrossRef] [PubMed]
66. Corrêa, J.C.R.; Salgado, H.D.N. Atividade inseticida das plantas e aplicações: Revisão. *Rev. Bras. De Plantas Med.* **2011**, *13*, 500–506. [CrossRef]
67. Brito, I.L.; Yilmaz, S.; Huang, K.; Xu, L.; Jupiter, S.D.; Jenkins, A.P.; Naisilisili, W.; Tamminen, M.; Smillie, C.S.; Wortman, J.R.; et al. Mobile genes in the human microbiome are structured from global to individual scales. *Nature* **2016**, *535*, 435–439. [CrossRef]
68. Carvalho, C.B.; Santos, C.S.; Viana, F.A.; Diniz, J.C.; Rocha, S.A.S.; Amóra, S.S.A.; Alves, N.D.; Feijó, F.M.C. Ação dos extratos de favela (*Cnidoscolus phyllacanthus*), catingueira (*Caesalpinia pyramidalis*) e nim (*Azadiracta indica*) sobre bactérias isoladas de cabras de aptidão leiteira: Action of slum (*Cnidoscolus phyllacanthus*), catingueira (*Caesalpinia pyramidalis*) and neem (*Azadiracta indica*) extracts on bacteria isolated from dairy goats. *Lat. Am. J. Dev.* **2021**, *3*, 2297–2306.
69. de Souza Eller, S.C.W.; Feitosa, V.A.; Arruda, T.A.; Antunes, R.M.P.; Catão, R.M.R. Avaliação antimicrobiana de extratos vegetais e possível interação farmacológica in vitro. *Rev. De Ciências Farm. Básica E Apl.* **2015**, *36*, 131–136.
70. Nunes, F.; Dias, H.; Cavalcante, G. Investigação das atividades antioxidante e antimicrobiana de duas espécies arbóreas ocorrentes no bioma caatinga. *Estação Científica (Unifap)* **2016**, *6*, 81–90. [CrossRef]
71. Candido, L.P.; Beserra, E.B. Repellent activity of *Cnidoscolus phyllacanthus* Mart. and *Ricinus communis* L. extracts against *Aedes aegypti* L. oviposition behavior. *Biotemas* **2015**, *28*, 105–112. [CrossRef]
72. e Castro, A.; Filho, J.V.; Militão, G.C.G.; da Silva, T.G.; de Amorim, E.L.C. Antiproliferative activity of species of the genus Cnidoscolus against HT-29, Hep-2 and NCI-H292 cells. *Mol. Clin. Pharm.* **2012**, *3*, 55–61.
73. Kolli, P.; Kancharla, S.; Gopaiah, K.V. *C. phyllacanthus* antioxidant & protective activity in albino wister rats. *Int. J. Pharmacogn. Chem.* **2021**, *2*, 14–24.

74. Sharma, R.S.; Tyagi, B.; Chouhan, P. Hepatoprotective activity of Cnidoscolus Phyllacanthus leaves against D-galactosamine induced hepatotoxicity in Rats. *J. Appl. Pharm. Sci. Res.* **2021**, *4*, 21–25. [CrossRef]
75. Nonato, C.d.F.A.; de Melo, E.V.S.; Camilo, C.J.; Ferreira, M.K.A.; de Meneses, J.E.A.; da Silva, A.W.; dos Santos, H.S.; Ribeiro-Filho, J.; Silva, J.P.R.; Tavares, J.F.; et al. Antibacterial Activity and Anxiolytic Effect in Adult Zebrafish of Genus *Lippia* L. Species. *Plants* **2023**, *12*, 1675. [CrossRef] [PubMed]

MDPI AG
Grosspeteranlage 5
4052 Basel
Switzerland
Tel.: +41 61 683 77 34

Processes Editorial Office
E-mail: processes@mdpi.com
www.mdpi.com/journal/processes

www.ingramcontent.com/pod-product-compliance
Lightning Source LLC
Chambersburg PA
CBHW061820210926
51850CB00006B/758
9783725861453